T0178474

Lecture Notes in Computer Science 12743

More information about this subseries at http://www.springer.com/series/7407

Maciej Paszynski · Dieter Kranzlmüller ·
Valeria V. Krzhizhanovskaya ·
Jack J. Dongarra · Peter M. A. Sloot (Eds.)

Computational Science – ICCS 2021

21st International Conference
Krakow, Poland, June 16–18, 2021
Proceedings, Part II

 Springer

Editors
Maciej Paszynski (iD)
AGH University of Science and Technology
Krakow, Poland

Valeria V. Krzhizhanovskaya (iD)
University of Amsterdam
Amsterdam, The Netherlands

Peter M. A. Sloot (iD)
University of Amsterdam
Amsterdam, The Netherlands

ITMO University
St. Petersburg, Russia

Nanyang Technological University
Singapore, Singapore

Dieter Kranzlmüller (iD)
Ludwig-Maximilians-Universität München
Munich, Germany

Leibniz Supercomputing Center (LRZ)
Garching bei München, Germany

Jack J. Dongarra (iD)
University of Tennessee at Knoxville
Knoxville, TN, USA

ISSN 0302-9743 ISSN 1611-3349 (electronic)
Lecture Notes in Computer Science
ISBN 978-3-030-77963-4 ISBN 978-3-030-77964-1 (eBook)
https://doi.org/10.1007/978-3-030-77964-1

LNCS Sublibrary: SL1 – Theoretical Computer Science and General Issues

This Springer imprint is published by the registered company Springer Nature Switzerland AG
The registered company address is: Gewerbestrasse 11, 6330 Cham, Switzerland

Preface

Welcome to the proceedings of the 21st annual International Conference on Computational Science (ICCS 2021 - https://www.iccs-meeting.org/iccs2021/).

In preparing this edition, we had high hopes that the ongoing COVID-19 pandemic would fade away and allow us to meet this June in the beautiful city of Kraków, Poland. Unfortunately, this is not yet the case, as the world struggles to adapt to the many profound changes brought about by this crisis. ICCS 2021 has had to adapt too and is thus being held entirely online, for the first time in its history.

These challenges notwithstanding, we have tried our best to keep the ICCS community as dynamic and productive as always. We are proud to present the proceedings you are reading as a result of that.

ICCS 2021 was jointly organized by the AGH University of Science and Technology, the University of Amsterdam, NTU Singapore, and the University of Tennessee.

The International Conference on Computational Science is an annual conference that brings together researchers and scientists from mathematics and computer science as basic computing disciplines, as well as researchers from various application areas who are pioneering computational methods in sciences such as physics, chemistry, life sciences, engineering, arts, and humanitarian fields, to discuss problems and solutions in the area, identify new issues, and shape future directions for research.

Since its inception in 2001, ICCS has attracted an increasing number of attendees and higher quality papers, and this year is not an exception, with over 350 registered participants. The proceedings have become a primary intellectual resource for computational science researchers, defining and advancing the state of the art in this field.

The theme for 2021, "**Computational Science for a Better Future**," highlights the role of computational science in tackling the current challenges of our fast-changing world. This conference was a unique event focusing on recent developments in scalable scientific algorithms, advanced software tools, computational grids, advanced numerical methods, and novel application areas. These innovative models, algorithms, and tools drive new science through efficient application in physical systems, computational and systems biology, environmental systems, finance, and other areas.

ICCS is well known for its excellent lineup of keynote speakers. The keynotes for 2021 were given by

- **Maciej Besta**, ETH Zürich, Switzerland
- **Marian Bubak**, AGH University of Science and Technology, Poland | Sano Centre for Computational Medicine, Poland
- **Anne Gelb**, Dartmouth College, USA
- **Georgiy Stenchikov**, King Abdullah University of Science and Technology, Saudi Arabia
- **Marco Viceconti**, University of Bologna, Italy

- **Krzysztof Walczak**, Poznan University of Economics and Business, Poland
- **Jessica Zhang**, Carnegie Mellon University, USA

This year we had 635 submissions (156 submissions to the main track and 479 to the thematic tracks). In the main track, 48 full papers were accepted (31%); in the thematic tracks, 212 full papers were accepted (44%). A high acceptance rate in the thematic tracks is explained by the nature of these tracks, where organisers personally invite many experts in a particular field to participate in their sessions.

ICCS relies strongly on our thematic track organizers' vital contributions to attract high-quality papers in many subject areas. We would like to thank all committee members from the main and thematic tracks for their contribution to ensure a high standard for the accepted papers. We would also like to thank *Springer, Elsevier,* and *Intellegibilis* for their support. Finally, we appreciate all the local organizing committee members for their hard work to prepare for this conference.

We are proud to note that ICCS is an A-rank conference in the CORE classification.

We wish you good health in these troubled times and look forward to meeting you at the conference.

June 2021

Maciej Paszynski
Dieter Kranzlmüller
Valeria V. Krzhizhanovskaya
Jack J. Dongarra
Peter M. A. Sloot

Organization

Local Organizing Committee at AGH University of Science and Technology

Chairs

Maciej Paszynski
Aleksander Byrski

Members

Marcin Łos
Maciej Woźniak
Leszek Siwik
Magdalena Suchoń

Thematic Tracks and Organizers

Advances in High-Performance Computational Earth Sciences: Applications and Frameworks – IHPCES

Takashi Shimokawabe
Kohei Fujita
Dominik Bartuschat

Applications of Computational Methods in Artificial Intelligence and Machine Learning – ACMAIML

Kourosh Modarresi
Paul Hofmann
Raja Velu
Peter Woehrmann

Artificial Intelligence and High-Performance Computing for Advanced Simulations – AIHPC4AS

Maciej Paszynski
Robert Schaefer
David Pardo
Victor Calo

Biomedical and Bioinformatics Challenges for Computer Science – BBC

Mario Cannataro
Giuseppe Agapito

Mauro Castelli
Riccardo Dondi
Italo Zoppis

Classifier Learning from Difficult Data – CLD^2

Michał Woźniak
Bartosz Krawczyk

Computational Analysis of Complex Social Systems – CSOC

Debraj Roy

Computational Collective Intelligence – CCI

Marcin Maleszka
Ngoc Thanh Nguyen
Marcin Hernes
Sinh Van Nguyen

Computational Health – CompHealth

Sergey Kovalchuk
Georgiy Bobashev
Stefan Thurner

Computational Methods for Emerging Problems in (dis-)Information Analysis – DisA

Michal Choras
Robert Burduk
Konstantinos Demestichas

Computational Methods in Smart Agriculture – CMSA

Andrew Lewis

Computational Optimization, Modelling, and Simulation – COMS

Xin-She Yang
Leifur Leifsson
Slawomir Koziel

Computational Science in IoT and Smart Systems – IoTSS

Vaidy Sunderam
Dariusz Mrozek

Computer Graphics, Image Processing and Artificial Intelligence – CGIPAI

Andres Iglesias
Lihua You
Alexander Malyshev
Hassan Ugail

Data-Driven Computational Sciences – DDCS

Craig Douglas

Machine Learning and Data Assimilation for Dynamical Systems – MLDADS

Rossella Arcucci

MeshFree Methods and Radial Basis Functions in Computational Sciences – MESHFREE

Vaclav Skala
Marco-Evangelos Biancolini
Samsul Ariffin Abdul Karim
Rongjiang Pan
Fernando-César Meira-Menandro

Multiscale Modelling and Simulation – MMS

Derek Groen
Diana Suleimenova
Stefano Casarin
Bartosz Bosak
Wouter Edeling

Quantum Computing Workshop – QCW

Katarzyna Rycerz
Marian Bubak

Simulations of Flow and Transport: Modeling, Algorithms and Computation – SOFTMAC

Shuyu Sun
Jingfa Li
James Liu

Smart Systems: Bringing Together Computer Vision, Sensor Networks and Machine Learning – SmartSys

Pedro Cardoso
Roberto Lam

João Rodrigues
Jânio Monteiro

Software Engineering for Computational Science – SE4Science

Jeffrey Carver
Neil Chue Hong
Anna-Lena Lamprecht

Solving Problems with Uncertainty – SPU

Vassil Alexandrov
Aneta Karaivanova

Teaching Computational Science – WTCS

Angela Shiflet
Nia Alexandrov
Alfredo Tirado-Ramos

Uncertainty Quantification for Computational Models – UNEQUIvOCAL

Wouter Edeling
Anna Nikishova

Reviewers

Ahmad Abdelfattah
Samsul Ariffin Abdul
   Karim
Tesfamariam Mulugeta
   Abuhay
Giuseppe Agapito
Elisabete Alberdi
Luis Alexandre
Vassil Alexandrov
Nia Alexandrov
Julen Alvarez-Aramberri
Sergey Alyaev
Tomasz Andrysiak
Samuel Aning
Michael Antolovich
Hideo Aochi
Hamid Arabnejad
Rossella Arcucci
Costin Badica
Marina Balakhontceva

Bartosz Balis
Krzysztof Banas
Dariusz Barbucha
Valeria Bartsch
Dominik Bartuschat
Pouria Behnodfaur
Joern Behrens
Adrian Bekasiewicz
Gebrail Bekdas
Mehmet Belen
Stefano Beretta
Benjamin Berkels
Daniel Berrar
Sanjukta Bhowmick
Georgiy Bobashev
Bartosz Bosak
Isabel Sofia Brito
Marc Brittain
Jérémy Buisson
Robert Burduk

Michael Burkhart
Allah Bux
Krisztian Buza
Aleksander Byrski
Cristiano Cabrita
Xing Cai
Barbara Calabrese
Jose Camata
Almudena Campuzano
Mario Cannataro
Alberto Cano
Pedro Cardoso
Alberto Carrassi
Alfonso Carriazo
Jeffrey Carver
Manuel Castañón-Puga
Mauro Castelli
Eduardo Cesar
Nicholas Chancellor
Patrikakis Charalampos

Bartosz Krawczyk
Dariusz Krol
Valeria Krzhizhanovskaya
Adam Krzyzak
Pawel Ksieniewicz
Marek Kubalcík
Sebastian Kuckuk
Eileen Kuehn
Michael Kuhn
Michal Kulczewski
Julian Martin Kunkel
Krzysztof Kurowski
Marcin Kuta
Bogdan Kwolek
Panagiotis Kyziropoulos
Massimo La Rosa
Roberto Lam
Anna-Lena Lamprecht
Rubin Landau
Johannes Langguth
Shin-Jye Lee
Mike Lees
Leifur Leifsson
Kenneth Leiter
Florin Leon
Vasiliy Leonenko
Roy Lettieri
Jake Lever
Andrew Lewis
Jingfa Li
Hui Liang
James Liu
Yen-Chen Liu
Zhao Liu
Hui Liu
Pengcheng Liu
Hong Liu
Marcelo Lobosco
Robert Lodder
Chu Kiong Loo
Marcin Los
Stephane Louise
Frederic Loulergue
Hatem Ltaief
Paul Lu
Stefan Luding

Laura Lyman
Scott MacLachlan
Lukasz Madej
Lech Madeyski
Luca Magri
Imran Mahmood
Peyman Mahouti
Marcin Maleszka
Alexander Malyshev
Livia Marcellino
Tomas Margalef
Tiziana Margaria
Osni Marques
M. Carmen Márquez
    García
Paula Martins
Jaime Afonso Martins
Pawel Matuszyk
Valerie Maxville
Pedro Medeiros
Fernando-César
    Meira-Menandro
Roderick Melnik
Valentin Melnikov
Ivan Merelli
Marianna Milano
Leandro Minku
Jaroslaw Miszczak
Kourosh Modarresi
Jânio Monteiro
Fernando Monteiro
James Montgomery
Dariusz Mrozek
Peter Mueller
Ignacio Muga
Judit Munoz-Matute
Philip Nadler
Hiromichi Nagao
Jethro Nagawkar
Kengo Nakajima
Grzegorz J. Nalepa
I. Michael Navon
Philipp Neumann
Du Nguyen
Ngoc Thanh Nguyen
Quang-Vu Nguyen

Sinh Van Nguyen
Nancy Nichols
Anna Nikishova
Hitoshi Nishizawa
Algirdas Noreika
Manuel Núñez
Krzysztof Okarma
Pablo Oliveira
Javier Omella
Kenji Ono
Eneko Osaba
Aziz Ouaarab
Raymond Padmos
Marek Palicki
Junjun Pan
Rongjiang Pan
Nikela Papadopoulou
Marcin Paprzycki
David Pardo
Anna Paszynska
Maciej Paszynski
Abani Patra
Dana Petcu
Serge Petiton
Bernhard Pfahringer
Toby Phillips
Frank Phillipson
Juan C. Pichel
Anna
    Pietrenko-Dabrowska
Laércio L. Pilla
Yuri Pirola
Nadia Pisanti
Sabri Pllana
Mihail Popov
Simon Portegies Zwart
Roland Potthast
Malgorzata
    Przybyla-Kasperek
Ela Pustulka-Hunt
Alexander Pyayt
Kun Qian
Yipeng Qin
Rick Quax
Cesar Quilodran Casas
Enrique S. Quintana-Orti

Ewaryst Rafajlowicz
Ajaykumar Rajasekharan
Raul Ramirez
Célia Ramos
Marcus Randall
Lukasz Rauch
Vishal Raul
Robin Richardson
Sophie Robert
João Rodrigues
Daniel Rodriguez
Albert Romkes
Debraj Roy
Jerzy Rozenblit
Konstantin Ryabinin
Katarzyna Rycerz
Khalid Saeed
Ozlem Salehi
Alberto Sanchez
Aysin Sanci
Gabriele Santin
Rodrigo Santos
Robert Schaefer
Karin Schiller
Ulf D. Schiller
Bertil Schmidt
Martin Schreiber
Gabriela Schütz
Christoph Schweimer
Marinella Sciortino
Diego Sevilla
Mostafa Shahriari
Abolfazi
   Shahzadeh-Fazeli
Vivek Sheraton
Angela Shiflet
Takashi Shimokawabe
Alexander Shukhman
Marcin Sieniek
Nazareen
   Sikkandar Basha
Anna Sikora
Diana Sima
Robert Sinkovits
Haozhen Situ
Leszek Siwik

Vaclav Skala
Ewa
   Skubalska-Rafajlowicz
Peter Sloot
Renata Slota
Oskar Slowik
Grazyna Slusarczyk
Sucha Smanchat
Maciej Smolka
Thiago Sobral
Robert Speck
Katarzyna Stapor
Robert Staszewski
Steve Stevenson
Tomasz Stopa
Achim Streit
Barbara Strug
Patricia Suarez Valero
Vishwas Hebbur Venkata
Subba Rao
Bongwon Suh
Diana Suleimenova
Shuyu Sun
Ray Sun
Vaidy Sunderam
Martin Swain
Jerzy Swiatek
Piotr Szczepaniak
Tadeusz Szuba
Ryszard Tadeusiewicz
Daisuke Takahashi
Zaid Tashman
Osamu Tatebe
Carlos Tavares Calafate
Andrei Tchernykh
Kasim Tersic
Jannis Teunissen
Nestor Tiglao
Alfredo Tirado-Ramos
Zainab Titus
Pawel Topa
Mariusz Topolski
Pawel Trajdos
Bogdan Trawinski
Jan Treur
Leonardo Trujillo

Paolo Trunfio
Ka-Wai Tsang
Hassan Ugail
Eirik Valseth
Ben van Werkhoven
Vítor Vasconcelos
Alexandra Vatyan
Raja Velu
Colin Venters
Milana Vuckovic
Jianwu Wang
Meili Wang
Peng Wang
Jaroslaw Watróbski
Holger Wendland
Lars Wienbrandt
Izabela Wierzbowska
Peter Woehrmann
Szymon Wojciechowski
Michal Wozniak
Maciej Wozniak
Dunhui Xiao
Huilin Xing
Wei Xue
Abuzer Yakaryilmaz
Yoshifumi Yamamoto
Xin-She Yang
Dongwei Ye
Hujun Yin
Lihua You
Han Yu
Drago Žagar
Michal Zak
Gabor Závodszky
Yao Zhang
Wenshu Zhang
Wenbin Zhang
Jian-Jun Zhang
Jinghui Zhong
Sotirios Ziavras
Zoltan Zimboras
Italo Zoppis
Chiara Zucco
Pavel Zun
Pawel Zyblewski
Karol Zyczkowski

Contents – Part II

**Artificial Intelligence and High-Performance Computing
for Advanced Simulations**

Biomedical and Bioinformatics Challenges for Computer Science

Advances in High-Performance Computational Earth Sciences: Applications and Frameworks

Large-Scale Stabilized Multi-physics Earthquake Simulation for Digital Twin

Ryota Kusakabe[1]([⊠])(iD), Tsuyoshi Ichimura[1], Kohei Fujita[1], Muneo Hori[2], and Lalith Wijerathne[1]

[1] The University of Tokyo, Bunkyo-ku, Tokyo, Japan
{ryota-k,ichimura,fujita,lalith}@eri.u-tokyo.ac.jp
[2] Japan Agency for Marine-Earth Science and Technology, Yokosuka-shi, Kanagawa, Japan
horimune@jamstec.go.jp

Abstract. The development of computing environments and computational techniques, together with data observation technology, big data and extreme-scale computing (BDEC) has gained immense attention. An example of BDEC is the digital twin concept of a city, a high-fidelity model of the city developed based on a computing system for the BDEC. The virtual experiments using numerical simulations are performed there, whose results are used in decision making. The earthquake simulation, which entails the highest computational cost among numerical simulations in the digital twin, was targeted in this study. In the multi-physics earthquake simulation considering soil liquefaction, the computation could become unstable when a high resolution is used for spatial discretization. In the digital twin, high-resolution large-scale simulation is performed repeatedly, and thus it is important to avoid such instability due to the discretization setting. In this study, an earthquake simulation method was developed to stably perform high-resolution large-scale simulations by averaging the constitutive law spatially based on a non-local approach. The developed method enables us to stably perform simulations with high-resolution of the order of 0.1 m and obtain a converged solution.

Keywords: Digital twin · Finite element method · Non-local approach

1 Introduction

With the development of computing environments and computational techniques, large-scale physics-based simulations have become possible. In addition, as the huge quantity of data has been and will be collected, the big data and extreme-scale computing (BDEC) [2] has gained attention in a wide range of research fields. In the BDEC, it is expected that a huge amount of observed data and machine learning results from these observed data are integrated into physics-based simulations, which enables us to perform more reliable simulations.

© Springer Nature Switzerland AG 2021
M. Paszynski et al. (Eds.): ICCS 2021, LNCS 12743, pp. 3–15, 2021.
https://doi.org/10.1007/978-3-030-77964-1_1

One example of the BDEC is the digital twin concept of a city, where high-fidelity model of the city is developed on a computing system for the BDEC. The internal state of the digital twin, such as river level and groundwater level, is updated based on the data observed in real time. Virtual experiments on the digital twin are performed using numerical simulations, whose results are used for decision making.

In this study, the earthquake simulation, which requires the highest computational cost among virtual experiments of the digital twin, was targeted. To overcome huge computational cost of the earthquake simulation, various methods have been developed using different discretization scheme such as the finite element method (FEM) [7,9] and finite different method [4,5,12,13]. We have developed fast soil liquefaction simulation methods for CPUs [15] and GPUs [16] based on FEM-based scalable earthquake simulation methods [12,13]. In this study, based on the GPU-based soil liquefaction simulation method [16], we developed a method for computing the risk from earthquakes considering soil liquefaction, which fluctuates as the groundwater level changes over a year. It is assumed that the groundwater level is one of the internal state variables of the digital twin and is updated based on groundwater level measured at several observation points. As frequency of localized torrential rain is increasing due to the climate change, it is concerned that the water level exceeds the assumption during the seismic design for a certain period of time. In other words, the risk from earthquakes is not static, but it changes dynamically. The earthquake simulations considering complex multi-physics, such as the one presented in this study, are expected to play an important role in assessing such dynamic risk. In previous studies, the groundwater level was often set arbitrarily such as "2 m below the ground surface". In this study, the groundwater level is computed using the seepage analysis to perform more accurate earthquake simulations.

Various kinds of constitutive laws have been developed to model a complex nonlinear behavior of materials in numerical simulations. However, some of them could cause instability and divergence of the computation in high-resolution simulations. The constitutive law [10,11] used in this study is for multi-physics soil liquefaction analysis considering highly nonlinear soil behavior and the influence of excess pore water pressure, and can cause simulation instability. In the digital twin, it is required to stably perform large-scale simulations many times with high resolution using different sets of parameters, and therefore it is crucial to avoid such instability of simulation caused by the discretization setting.

We developed a stabilization method to overcome this problem. In the developed method, the constitutive law is spatially averaged based on a non-local approach, which has been developed in the field of continuum damage mechanics [8,14]. By this averaging, the constitutive law is computed in a certain volume, not in a point, and thus, the developed method is expected to avoid the instability of simulation. Besides, the generalized alpha method [6], which can prescribe numerical dissipation for high-frequency modes, is adopted for time integration to further stabilize the analysis.

We performed earthquake simulation considering soil liquefaction using soil models with a different spatial resolution to demonstrate the capability of the developed method. The simulations with the conventional method got unstable for high-resolution simulations and the results did not converge. On the other hand, the simulations using the developed method were stable even for the highest resolution of 0.125 m and the results converged as the resolution increased. For example, we performed a large-scale earthquake simulation using a soil-structure model that mimics an actual site near an estuary with 702 million degrees of freedom (DOFs). Here, 480 Nvidia Tesla V100 GPUs on AI Bridging Cloud Infrastructure [1] were used for computation.

2 Methodology

In the digital twin, the internal state is updated using observed data. In order to accurately assess the risk from an earthquake, the earthquake simulation needs to be performed repeatedly to reassess the risk as the internal state of the digital twin is updated; thus, fast and stable methods for earthquake simulation are required. In this study, the multi-physics earthquake simulation considering soil liquefaction was targeted as one of the problem settings in which the simulation instability could occur. We have developed fast soil liquefaction simulation methods [15,16] based on large-scale simulation methods [12,13]. In this study, we developed a stabilization method for multi-physics earthquake simulation of the digital twin based on the GPU-based fast earthquake simulation method [16] using a non-local approach [8,14], which has been developed in the field of continuum damage mechanics.

2.1 Problem Settings of the Earthquake Simulation

The earthquake simulation in this study uses the finite element method (FEM) with second-order unstructured tetrahedral elements. The FEM allows the use of models with complex 3D geometry and easy handling of nonlinear materials and a traction-free boundary condition. The generalized alpha method [6] adopted as the time integration method prescribes numerical dissipation for high-frequency modes and suppresses the instability caused by the high-frequency modes due to numerical errors. The target equation, which is the motion equation discretized by the FEM and the generalized alpha method, becomes

$$A\delta u = b, \tag{1}$$

where,

$$A = \frac{1 - \alpha_m}{\beta \, dt^2} M + \frac{(1 - \alpha_f)\gamma}{\beta \, dt} C^{(n)} + (1 - \alpha_f) K^{(n)}, \tag{2}$$

$$b = (1 - \alpha_f) f^{(n+1)} + \alpha_f f^{(n)} - q^{(n)}$$

$$+ M \left[\left(\frac{1 - \alpha_m}{2\beta} - 1 \right) a^{(n)} + \frac{1 - \alpha_m}{\beta \, dt} v^{(n)} \right]$$

$$+ C \left[(1 - \alpha_f) \left(\frac{\gamma}{2\beta} - 1 \right) dt a^{(n)} + \left(\frac{(1 - \alpha_f)\gamma}{\beta} - 1 \right) v^{(n)} \right]. \tag{3}$$

Here, M, C, and K are the matrices of mass, Rayleigh damping, and stiffness, respectively. δu, f, q, v, and a are the vectors of the displacement increment, external force, inner force, velocity, and acceleration, respectively, and dt is the time increment. $*^{(n)}$ is the variable $*$ in the n-th time step. $\alpha_m, \alpha_f, \beta$, and γ are parameters for the generalized alpha method, where $\alpha_m = (2\rho_\infty - 1)/(\rho_\infty + 1)$, $\alpha_f = \rho_\infty/(\rho_\infty + 1)$, $\beta = (1 - \alpha_m + \alpha_f)^2/4$, $\gamma = 1/2 - \alpha_m + \alpha_f$. ρ_∞ is the spectral radius in the high-frequency limit, which was set 0.8 in this study. (For $\alpha_m = \alpha_f = 0$, the Eq. (1) matches the target equation that uses the Newmark beta method for the time integration.)

The target Eq. (1) is solved every time step to obtain δu, and the variables are updated as follows:

$$u^{(n+1)} = u^{(n)} + \delta u, \tag{4}$$

$$v^{(n+1)} = \left(1 - \frac{\gamma}{2\beta} \right) dt a^{(n)} + \left(1 - \frac{\gamma}{\beta} \right) v^{(n)} + \frac{\gamma}{\beta \, dt} \delta u, \tag{5}$$

$$a^{(n+1)} = \left(1 - \frac{1}{2\beta} \right) a^{(n)} - \frac{1}{\beta \, dt} v^{(n)} + \frac{1}{\beta \, dt^2} \delta u, \tag{6}$$

$$K^{(n+1)} = \sum_e \int_{V_e} B^T D^{(n+1)} B \, dV, \tag{7}$$

$$q^{(n+1)} = \sum_e \int_{V_e} B^T \sigma^{(n+1)} \, dV. \tag{8}$$

Here B is the matrix used to convert the displacement into the strain. D is the elasto-plastic matrix and σ is the total stress, calculated from displacement using the constitutive law. The elasto-plastic matrix and total stress for soil above the groundwater level are calculated using the stiffness parameters that depend on the confining stress. Those for soil below the groundwater level are calculated using the stiffness parameter that are calculated with the excess pore water pressure model [11], in which the accumulated elastic shear work is the index of the progress of liquefaction. $\int_{V_e} * \, dV$ represents volume integration in the e-th element.

2.2   Seepage Analysis to Compute the Groundwater Distribution

In the digital twin, the internal state of the model is updated based on the data observed in real time. In this study, the groundwater level, which fluctuates over a year, is targeted as an internal state variable. In conventional earthquake simulations, the groundwater level was often set arbitrarily such as "2 m below the ground surface". In this study, the groundwater level was computed through the seepage analysis to perform a more reliable simulation.

In the seepage analysis, water is assumed to be incompressible, and a steady flow is computed. By neglecting the vertical flow and integrating the equation of conservation of mass with respect to vertical direction, the target equation becomes

$$\nabla \cdot (\boldsymbol{T}(h)\nabla h) - Q(h) = 0, \tag{9}$$

where,

$$\boldsymbol{T}(h) = \int_0^h \boldsymbol{K}\,dz, \tag{10}$$

$$Q(h) = \int_0^h q\,dz. \tag{11}$$

Here, h is the total hydraulic head, $\boldsymbol{K}$ is the tensor of permeability coefficient, and q is the outflow of water to the outside of the system. The two-dimensional distribution of the total hydraulic head is computed by discretizing Eq. (9) using the FEM with a structured grid, and is used as the groundwater level distribution. The groundwater level distribution is set constant over time during an earthquake simulation, assuming that an earthquake occurs in a relatively short time period compared to the groundwater level fluctuation.

2.3 Stabilization Method for the Earthquake Simulation by Averaging the Constitutive Law

In the earthquake simulations where the soil behaves highly nonlinear, the computation could get unstable and diverge, especially when the high resolution is used for spatial discretization. One cause of instability of such a simulation is the assumption that the constitutive law holds at all points in the domain, even though the soil is an inhomogeneous material and the constitutive law for soil describes the "average" relationship between the strain and stress in a certain volume (This means there is a range of resolution in which the constitutive law holds validly). In simulations with a relatively low resolution of the order of 1 to 10 m, the above assumption does not cause serious problems because each point represents a certain volume. However, in simulations with a high resolution of the order of 0.1 m, this assumption could lead to instability of simulation as the resolution for simulation is outside the valid range of resolutions, for which the constitutive law holds. Even though the simulation does not always get unstable when the resolution is outside the valid range, large-scale simulations are more likely to get unstable compared to small-scale simulations. In the digital twin, it is required to stably perform large-scale simulations many times with high resolution; thus it is important to avoid such instability of simulation caused by the discretization setting.

In this study, the multi-physics earthquake simulation considering soil liquefaction was targeted as one of the problem settings in which the simulation instability can occur. The constitutive law [10,11] that considers the high nonlinearity of soil behavior and the influence of excess pore water pressure was used. The constitutive law consists of the excess pore water pressure model and multi-spring model, and describes soil liquefaction under undrained conditions. Based on the excess pore water pressure model, the soil stiffness parameters θ (bulk modulus, shear modulus, and shear strength) are calculated from the accumulated value of plastic shear work, which is used as the index of the progress of soil liquefaction. With the multi-spring model f, the stress σ and elasto-plastic matrix D are calculated using the stiffness parameters θ and strain ε as follows:

$$\sigma = f(\varepsilon; \theta) \tag{12}$$

$$D = \frac{\partial f(\varepsilon; \theta)}{\partial \varepsilon} \tag{13}$$

When the conventional method is used for simulation with high resolution of the order of 0.1 m, the instability could occur due to extremely large strain in several elements caused by the following cycle: (1) the soil liquefies; (2) the soil stiffness is reduced; (3) the strain gets large; (4) the plastic shear work (which is used as the index of the progress of liquefaction) is accumulated; (5) the soil liquefies further. This problem happens due to the assumption that the constitutive law holds at all points in the domain, which is not suitable in high-resolution simulations.

In the developed method, the constitutive law is averaged using a non-local approach [8, 14], which has been developed in the field of continuum damage mechanics, according to the following steps:

1. The stiffness parameters $\theta(x)$ at each position x are calculated based on the excess pore water pressure model in the same way as the conventional constitutive law.
2. The stiffness parameters θ are averaged and the non-local stiffness parameters $\bar{\theta}$ are calculated from Eq. (14).

$$\bar{\theta}(x) = \int_V \alpha(\xi; x)\theta(\xi) \, d\xi, \tag{14}$$

Here $\alpha(\xi; x)$ is a non-negative function that has the maximum value at $\xi = x$, monotonically decreases with respect to $\|\xi - x\|$ and satisfies $\int_V \alpha(\xi; x) \, d\xi = 1$. The function below was used in this study.

$$\alpha(\xi; x) = \frac{\alpha_0(\|\xi - x\|)}{\int_V \alpha_0(\|\zeta - x\|) \, d\zeta} \tag{15}$$

$$\alpha_0(r) = \begin{cases} \left(1 - \dfrac{r^2}{R^2}\right)^2 & \text{if } 0 \leq r \leq R \\ 0 & \text{otherwise} \end{cases} \tag{16}$$

Here R is the parameter that determines the range of averaging. $R = 1$ m was used in this study.

3. The stress σ and elasto-plastic matrix D are calculated using the non-local stiffness parameters $\bar{\theta}$ instead of the stiffness parameters θ.

$$\sigma = f(\varepsilon; \bar{\theta}) \tag{17}$$

$$D = \frac{\partial f(\varepsilon; \bar{\theta})}{\partial \varepsilon} \tag{18}$$

By this averaging, the constitutive law is calculated in a certain volume even in high-resolution simulations, which is expected to avoid the instability of simulation. Specifically, even if the soil liquefies locally, the soil stiffness reduction is suppressed by the averaging. Thus, in the cycle that causes extreme strain (i.e., (1) the soil liquefies; (2) the soil stiffness is reduced; (3) the strain gets large; (4) the plastic shear work is accumulated; (5) the soil liquefies further), the step from (1) to (2) is less likely to happen and the extreme strain is expected to be prevented. The computational complexity of this algorithm is $O((R/\Delta x)^3)$ and the amount of MPI communication is $O(R/\Delta x^3)$, where Δx is spatial resolution of the soil-structure model.

3 Numerical Experiment

Numerical experiments were carried out to demonstrate the capability of the developed method. It is shown that the developed stabilization method was able to stably perform the simulation; however the conventional method failed to perform when the spatial resolution was high. A large-scale high-resolution earthquake simulation with a soil-structure model mimicking an actual site near an estuary was performed. Here, the distribution of groundwater level computed by the seepage analysis was used. The seismic wave observed during the 1995 Hyogo-ken Nambu earthquake [3] was used as an input wave for all simulations.

3.1 Comparison of the Conventional Method and Developed Method

Earthquake simulations considering soil liquefaction were performed using the ground model shown in Fig. 1. Table 1 shows the material properties. The groundwater level was set 2 m below the ground surface. The time increment was 0.001 s. Simulations with the conventional method and developed method were performed with four different spatial resolutions of 1 m, 0.5 m, 0.25 m, and

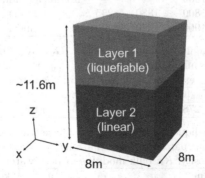

Fig. 1. The model used in the comparison of the developed method and conventional method.

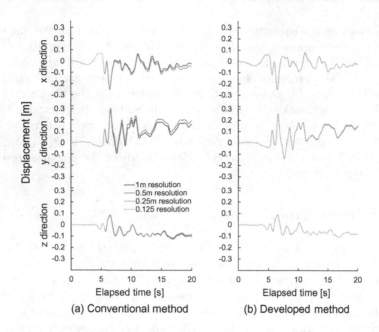

Fig. 2. Comparison of the time history of displacement at the center of the ground surface.

0.125 m. The difference between the conventional and developed methods is (a) the computation of stress $\boldsymbol{\sigma}$ and elasto-plastic matrix $\boldsymbol{D}$ with the constitutive law and (b) the time integration method. In the conventional method, the stress

Table 1. Model properties

(a) Soil profile properties. ρ: density, V_p, V_s: velocity of primary and secondary waves.

	$\rho[\mathrm{kg/m^3}]$	$V_p[\mathrm{m/s}]$	$V_s[\mathrm{m/s}]$	constitutive law
Layer1	1500	—	—	nonlinear (liquefiable)
Layer2	1800	1380	255	linear
Bedrock	1900	1770	490	linear

(b) Parameters for the nonlinear constitutive law. G_{ma}, K_{ma}: elastic shear modulus and bulk modulus at a confining pressure of σ'_{ma}, σ'_{ma}: reference confining pressure, m_K, m_G: parameters for nonlinearity, ϕ_f: shear resistance angle.

$G_{\mathrm{ma}}[\mathrm{GPa}]$	$K_{\mathrm{ma}}[\mathrm{GPa}]$	$\sigma'_{\mathrm{ma}}[\mathrm{kPa}]$	m_G	m_K	ϕ_f
106.6	278.0	−37	0.5	0.5	40°

(c) Parameters for liquefiable propety. ϕ_p: phase transformation angle, S_{min}, p_1, p_2, c_1 and w_1: parameters for dilatancy, ρ_f: density of pore water, n: porosity, K_f: bulk modulus of pore water.

ϕ_p	S_{min}	p_1	p_2	c_1	w_1	$\rho_f[\mathrm{kg/m^3}]$	n	$K_f[\mathrm{GPa}]$
28°	0.01	0.5	0.65	3.97	7.0	1000	0.45	2200

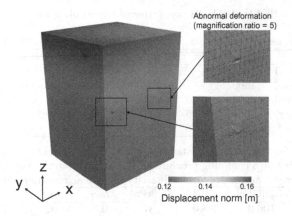

Fig. 3. Displacement norm and deformation when the computation was abnormally terminated in the simulation with 0.125-m resolution using the conventional method.

σ and elasto-plastic matrix D are computed using the stiffness parameters θ, which are not averaged, using Eqs. (12, 13), and the Newmark beta method was used for time integration. In the developed method, the stress σ and elasto-plastic matrix D are computed using the non-local stiffness parameters $\bar{\theta}$ using Eqs. (17, 18), and the generalized alpha method is used for time integration.

Figure 2 shows the time history of the displacement at the center of the ground surface. In the conventional method, the iteration of the conjugate gradient method was not converged in solving the target equation and the computation was abnormally terminated around $t = 5.6$ s in simulations with a resolution of 0.25 m and 0.125 m. Also, the simulation results did not converge as the resolution increased (Fig. 2(a)). Figure 3 depicts the displacement when the computation was abnormally terminated in the 0.125-m-resolution simulation. Several elements deformed to an extreme extent due to the instability. On the other hand, with the developed method, the abnormal termination did not occur and the simulation results converged, as shown in Fig. 2(b). Hence, the developed method enables to stably perform high-resolution simulation and obtain converged results.

3.2 Application: Multi-physics Earthquake Simulation with a Large-Scale High-Resolution Soil-Structure Model

Earthquake simulation using a large-scale high-resolution soil-structure model which mimicked an actual site near an estuary was performed. The spatial resolution was 0.5 m, with which the simulation using the conventional method could be unstable. The soil structure model shown in Fig. 4 was used; the model has 702,192,969 DOFs and 172,929,616 elements. Table 2 shows the material properties and the same parameters for liquefiable property as the previous section (shown in Table 1(c)) were used. The seepage analysis of the groundwater was

Table 2. Model properties of the application

(a) Soil profile properties. ρ: density, V_p, V_s: velocity of primary and secondary waves.

	$\rho[\mathrm{kg/m^3}]$	$V_p[\mathrm{m/s}]$	$V_s[\mathrm{m/s}]$	constitutive law
Layer 1	1500	—	—	nonlinear(liquefiable)
Layer 2	1500	—	—	nonlinear(non-liquefiable)
Bedrock	1900	1770	490	linear
water	1000	1500	100	linear
wharf	2100	3378	2130	linear

(b) Parameters for the nonlinear constitutive law. $G_{\mathrm{ma}}, K_{\mathrm{ma}}$: elastic shear modulus and bulk modulus at a confining pressure of σ'_{ma}, σ'_{ma}: reference confining pressure, m_K, m_G: parameters for nonlinearity, ϕ_f: shear resistance angle.

	$G_{\mathrm{ma}}[\mathrm{GPa}]$	$K_{\mathrm{ma}}[\mathrm{GPa}]$	$\sigma'_{\mathrm{ma}}[\mathrm{kPa}]$	m_G	m_K	ϕ_f
layer 1	106.6	278.0	−37	0.5	0.5	35°
layer 2	117.0	327.2	−419	0.5	0.5	40°

conducted assuming that layer 1 was an unconfined aquifer to obtain the groundwater level distribution, as shown in Fig. 5. Using the developed method, the seismic response considering soil liquefaction for 20 s (20,000 time steps) was simulated with the time increment of 0.001 s. Figure 6 shows the displacement on the ground surface at $t = 20$ s.

The computation was parallelized with 480 MPI processes × 1 GPU per 1 MPI process = 480 GPUs, and was conducted on 120 computing nodes of a GPU-based supercomputer AI Bridging Cloud Infrastructure (ABCI)[1]. (Each computing node of ABCI consists of four NVIDIA Tesla V100 GPUs and two Intel Xeon Gold 6148 CPUs.) The computation time was 4 h 20 min. The computation time for the non-local approach was 18 min (6 min for the initial setting and 0.023 s/time step × 30,000 time steps = 12 min), which is 7% of the computation time for the whole analysis. It is indicated that the developed method

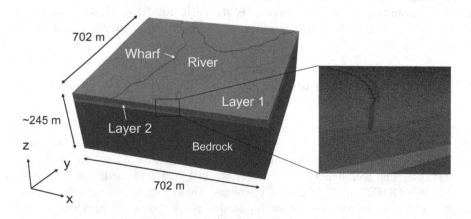

Fig. 4. The model used in the application.

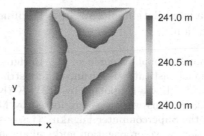

Fig. 5. Groundwater level distribution computed by the seepage analysis.

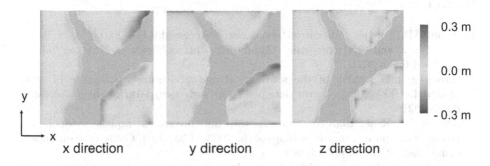

x direction   y direction   z direction

Fig. 6. Displacement on the ground surface at $t = 20$ s in the application.

enables us to stably perform large-scale high-resolution earthquake simulation only with a slight increase in computation time compared to the previous fast analysis method [16].

4   Conclusion

In this study, we developed a stabilization method for multi-physics earthquake simulation of the digital twin. In the multi-physics simulation considering soil liquefaction, the computation instability can occur when a high spatial resolution is used. In the developed method, the constitutive law is averaged using a non-local approach. With this method, the simulation was stably performed even with a high spatial resolution of 0.125 m, and the simulation results were converged. The multi-physics earthquake simulation with a 702 million DOF soil-structure model with 0.5 m resolution was performed to show that the developed method enables us to stably perform the large-scale high-resolution analysis.

Even though the implementation of the constitutive law [10,11] is presented in this paper, the developed method is a versatile method, and it can be applied to other constitutive laws that could cause the simulation instability. Future work involves the validation of simulation results of the developed method, the use of 3D seepage analysis for the distribution of groundwater level, and soil liquefaction analysis under drained conditions. Furthermore, integration with observed data using data assimilation and machine learning is expected to result in a more reliable simulation.

Acknowledgments. Computational resource of AI Bridging Cloud Infrastructure (ABCI) provided by National Institute of Advanced Industrial Science and Technology (AIST) was used. This work was supported by JSPS KAKENHI Grant Numbers JP18H05239 and JP20J22348. This work was supported by MEXT as "Program for Promoting Researches on the Supercomputer Fugaku" (Large-scale numerical simulation of earthquake generation, wave propagation and soil amplification: hp200126).

References

1. About ABCI. https://abci.ai/en/about_abci/. Accessed 6 Feb 2021
2. Big Data and Extreme-scale Computing. https://www.exascale.org/bdec/. Accessed 6 Feb 2021
3. Strong ground motion of the southern Hyogo prefecture earthquake in 1995 observed at Kobe. http://www.data.jma.go.jp/svd/eqev/data/kyoshin/. Accessed 6 Feb 2021
4. Akcelik, V., et al.: High resolution forward and inverse earthquake modeling on terascale computers. In: Proceedings of the 2003 ACM/IEEE Conference on Supercomputing, SC 2003, p. 52 (2003)
5. Bao, H., et al.: Large-scale simulation of elastic wave propagation in heterogeneous media on parallel computers. Comput. Methods Appl. Mech. Eng. **152**(1), 85–102 (1998)
6. Chung, J., Hulbert, G.M.: A time integration algorithm for structural dynamics with improved numerical dissipation: the generalized-α method. J. Appl. Mech. **60**(2) (1993)
7. Cui, Y., et al.: Physics-based seismic hazard analysis on petascale heterogeneous supercomputers. In: Proceedings of the International Conference on High Performance Computing, Networking, Storage and Analysis, SC 2013 (2013)
8. de Vree, J., Brekelmans, W., van Gils, M.: Comparison of nonlocal approaches in continuum damage mechanics. Comput. Struct. **55**(4), 581–588 (1995)
9. Frankel, A., Vidale, J.: A three-dimensional simulation of seismic waves in the Santa Clara Valley, California, from a Loma Prieta aftershock. Bull. Seismol. Soc. Am. **82**(5), 2045–2074 (1992)
10. Iai, S.: Three dimensional formulation and objectivity of a strain space multiple mechanism model for sand. Soils Found. **33**(1), 192–199 (1993)
11. Iai, S., Matsunaga, Y., Kameoka, T.: Strain space plasticity model for cyclic mobility. Soils Found. **32**(2), 1–15 (1992)
12. Ichimura, T., et al.: Physics-based urban earthquake simulation enhanced by 10.7 BlnDOF × 30 K time-step unstructured FE non-linear seismic wave simulation. In: Proceedings of the International Conference for High Performance Computing, Networking, Storage and Analysis, SC 2014, pp. 15–26 (2014)
13. Ichimura, T., et al.: Implicit nonlinear wave simulation with 1.08 T DOF and 0.270 T unstructured finite elements to enhance comprehensive earthquake simulation. In: Proceedings of the International Conference for High Performance Computing, Networking, Storage and Analysis, SC 2015, pp. 1–12 (2015)
14. Jirásek, M., Marfia, S.: Non-local damage model based on displacement averaging. Int. J. Numer. Method Eng. **63**(1), 77–102 (2005)

15. Kusakabe, R., Fujita, K., Ichimura, T., Hori, M., Wijerathne, L.: A fast 3D finite-element solver for large-scale seismic soil liquefaction analysis. In: Rodrigues, J.M.F., et al. (eds.) ICCS 2019. LNCS, vol. 11537, pp. 349–362. Springer, Cham (2019). https://doi.org/10.1007/978-3-030-22741-8_25
16. Kusakabe, R., Fujita, K., Ichimura, T., Yamaguchi, T., Hori, M., Wijerathne, L.: Development of regional simulation of seismic ground-motion and induced lique-faction enhanced by GPU computing. Earthquake Eng. Struct. Dyn. **50**, 197–213 (2021)

On the Design of Monte-Carlo Particle Coagulation Solver Interface: A CPU/GPU Super-Droplet Method Case Study with PySDM

Piotr Bartman$^{(\boxtimes)}$ ⓘ and Sylwester Arabas ⓘ

Jagiellonian University, Kraków, Poland
piotr.bartman@doctoral.uj.edu.pl, sylwester.arabas@uj.edu.pl

Abstract. Super-Droplet Method (SDM) is a probabilistic Monte-Carlo-type model of particle coagulation process, an alternative to the mean-field formulation of Smoluchowski. SDM as an algorithm has linear computational complexity with respect to the state vector length, the state vector length is constant throughout simulation, and most of the algorithm steps are readily parallelizable. This paper discusses the design and implementation of two number-crunching backends for SDM implemented in PySDM, a new free and open-source Python package for simulating the dynamics of atmospheric aerosol, cloud and rain particles. The two backends share their application programming interface (API) but leverage distinct parallelism paradigms, target different hardware, and are built on top of different lower-level routine sets. First offers multi-threaded CPU computations and is based on Numba (using Numpy arrays). Second offers GPU computations and is built on top of ThrustRTC and CURandRTC (and does not use Numpy arrays). In the paper, the API is discussed focusing on: data dependencies across steps, parallelisation opportunities, CPU and GPU implementation nuances, and algorithm workflow. Example simulations suitable for validating implementations of the API are presented.

Keywords: Monte-Carlo · Coagulation · Super-Droplet Method · GPU

1 Introduction

The Super-Droplet Method (SDM) introduced in [20] is a computationally efficient Monte-Carlo type algorithm for modelling the process of collisional growth (coagulation) of particles. SDM was introduced in the context of atmospheric modelling, in particular for simulating the formation of cloud and rain through particle-based simulations. Such simulations couple a grid-based computational fluid-dynamics (CFD) core with a probabilistic, so-called super-particle (hence the algorithm name), representation of the particulate phase, and constitute

Funding: Foundation for Polish Science (POIR.04.04.00-00-5E1C/18-00).

M. Paszynski et al. (Eds.): ICCS 2021, LNCS 12743, pp. 16–30, 2021.
https://doi.org/10.1007/978-3-030-77964-1_2

a tool for comprehensive modelling of aerosol-cloud-precipitation interactions (e.g., [1,3,10,15]; see [13] for a review). The probabilistic character of SDM is embodied, among other aspects, in the assumption of each super-particle representing a multiple number of modelled droplets with the same attributes (including particle physicochemical properties and position in space). The super-droplets are thus a coarse-grained view of droplets both in physical and attribute space.

The probabilistic description of collisional growth has a wide range of applications across different domains of computational sciences (e.g., astrophysics, aerosol/hydrosol technology including combustion). While focused on SDM and depicted with atmospheric phenomena examples, the material presented herein is generally applicable in development of software implementing other Monte-Carlo type schemes for coagulation (e.g., Weighted Flow Algorithms [8] and other, see Sect. 1 in [21]), particularly when sharing the concept of super-particles.

The original algorithm description [20], the relevant patent applications (e.g., [22]) and several subsequent works scrutinising SDM (e.g., [2,9,18,25] expounded upon the algorithm characteristics from users' (physicists') point of view. There are several CFD packages implementing SDM including: SCALE-SDM [19] and UWLCM [10]; however the implementation aspects were not within the scope of the works describing these developments. The aim of this work is to discuss the algorithm from software developer's perspective. To this end, the scope of the discussion covers: data dependencies, parallelisation opportunities, state vector structure and helper variable requirements, minimal set of computational kernels needed to be implemented and the overall algorithm workflow. These concepts are depicted herein with pseudo-code-mimicking Python snippets (syntactically correct and runnable), but the solutions introduced are not bound to a particular language. In contrast, the gist of the paper is the language-agnostic API (i.e., application programming interface) proposal put forward with the very aim of capturing the implementation-relevant nuances of the algorithm which are, arguably, tricky to discern from existing literature on SDM, yet which have an impact on the simulation performance. The API provides an indirection layer separating higher-level physically-relevant concepts from lower-level computational kernels.

Validity of the proposed API design has been demonstrated with two distinct backend implementations included in a newly developed simulation package PySDM [5]. The two backends share the programming interface, while differing substantially in the underlying software components and the targeted hardware (CPU vs. GPU). PySDM is free/libre and open source software, presented results are readily reproducible with examples shipped with PySDM (https://github.com/atmos-cloud-sim-uj/PySDM).

The remainder of this paper is structured as follows. Section 2 briefly introduces the SDM algorithm through a juxtaposition against the alternative classic Smoluchowski's coagulation equation (SCE). Section 3 covers the backend API. Section 4 presents simulations performed with PySDM based on benchmark setups from literature and documenting CPU and GPU performance of the implementation. Section 5 concludes the work enumerating the key points brought out in the paper.

2   SDM as Compared to SCE

2.1   Mean-Field Approach: Smoluchowski's Coagulation Equation

Population balance equation which describes collisional growth is historically known as the Smoluchowski's coagulation equation (SCE) and was introduced in [23,24] (for a classic and a recent overview, see e.g. [7] and [14], respectively). It is formulated under the mean-field assumptions of sufficiently large well-mixed system and of neglected correlations between numbers of droplets of different sizes (for discussion in the context of SDM, see also [9]).

Let function $c(x,t) : \mathbb{R}^+ \times \mathbb{R}^+ \to \mathbb{R}^+$ correspond to particle size spectrum and describe the average concentration of particles with size defined by x at time t in a volume V. Smoluchowski's coagulation equation describes evolution of the spectrum in time due to collisions. For convenience, t is skipped in notation: $c(x) = c(x,t)$, while $\dot{c}$ denotes partial derivative with respect to time.

$$\dot{c}(x) = \frac{1}{2} \int_0^x a(y, x-y)c(y)c(x-y)dy - \int_0^\infty a(y,x)c(y)c(x)dy \qquad (1)$$

where $a(x_1, x_2)$ is the so-called kernel which defines the rate of collisions (and coagulation) between particles of sizes x_1 and x_2 and $a(x_1, x_2) = a(x_2, x_1)$. The first term on the right-hand side is the production of particles of size x by coalescence of two smaller particles and the factor $1/2$ is for avoiding double counting. The second term represents the reduction in number of colliding particles due to coagulation.

The Smoluchowski's equation has an alternative form that is discrete in size space. Let x_0 be the smallest considered difference of size between particles, $x_i = ix_0$, $i \in \mathbb{N}$ and $c_i = c(x_i)$ then:

$$\dot{c}_i = \frac{1}{2} \sum_{k=1}^{i-1} a(x_k, x_{i-k})c_k c_{i-k} - \sum_{k=1}^{\infty} a(x_k, x_i)c_k c_i \qquad (2)$$

Analytic solutions to the equation are known only for simple kernels [14], such as: constant $a(x_1, x_2) = 1$, additive $a(x_1, x_2) = x_1 + x_2$ (Golovin's kernel [12]) or multiplicative $a(x_1, x_2) = x_1 x_2$. Taking atmospheric physics as an example, collisions of droplets within cloud occur by differentiated movements of particles caused by combination of gravitational, electrical, or aerodynamic forces, where gravitational effects dominate. As such, sophisticated kernels are needed to describe these phenomena, and hence numerical methods are required for solving the coagulation problem. However, when multiple properties of particles (volume, chemical composition, etc.) need to be taken into account, the numerical methods for SCE suffer from the curse of dimensionality due to the need to distinguish particles of same size x but different properties.

Additionally, it is worth to highlight that, in practice, the assumptions of the Smoluchowski equation may be difficult to meet. First, the particle size changes at the same time due to processes other than coalescence (e.g., condensation/evaporation). Second, it is assumed that the system is large enough and

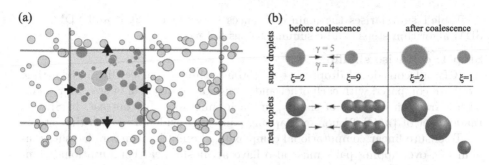

Fig. 1. (a): Schematic of how a CFD grid is populated with immersed super-particles; (b) schematic of super-particle collision (based on Fig. 1 in [20]), parameters γ and $\tilde{\gamma}$ are pertinent to representation of multiple collisions within a single coalescence event.

the droplets inside are uniformly distributed, which in turn is only true for a small volume in the atmosphere. Moreover, using the Smoluchowski's equation that describes evolution of the mean state of the system, leads to deterministic simulations. The alternative to Smoluchowski equation are Monte-Carlo methods based on stochastic model of the collision-coalescence process. Stochastic approach enables simulating ensembles of realisations of the process aiding studies of rare (far-from-the-mean) phenomena like rapid precipitation onset (see [13]).

2.2 Monte-Carlo Approach: Super-Droplet Method (SDM)

The stochastic coalescence model for cloud droplet growth was already analysed by Gillespie in [11]. There remained however still challenges (elaborated on below) related to computational and memory complexity which are addressed by SDM. The original formulation of SDM given in [20] was presented in the context of liquid-phase atmospheric clouds, and extended to cover clouds featuring ice phase of water in [21]).

Tracking all particles in large simulations has immeasurable cost, which is why the notion of super-particles is introduced. For convenience, let Ω be an indexed set of super-droplets for which collisions are considered (e.g., particles within one CFD grid cell, see Fig. 1a with the blue droplets in a shaded cell depicting set Ω). Each element of this set has unique index $i \in [0, 1, ..., n(\Omega) - 1]$, where $n(\Omega)$ is the number of considered super-droplets. Attributes related to a specific super-droplet i are denoted by $attr_{[i]}$. Note that only extensive attributes (volume, mass, number of moles) are considered within the coalescence scheme.

In SDM, each super-droplet represents a set of droplets with the same attributes (including position) and is assigned with a *multiplicity* denoted by $\xi \in \mathbb{N}$ and which can be different in each super-droplet. The multiplicity $\xi_{[i]}$ of super-droplet of size $x_{[i]}$ is conceptually related to $c_i = c(x_i)$ from Eq. (2). The number of super-droplets must be sufficient to cover the phase space of particle properties. The higher the number, the smaller the multiplicities and the higher fidelity of discretisation (see [20,21] for discussion).

Table 1 summarises the main differences between the SCE and SDM. The SDM algorithm steps can be summarised as follows.

Step 1: cell-wise shuffling

In SCE, all considered droplets (e.g., those within a grid cell in a given time step) are compared with each other and a fraction of droplets of size $x_{[i]}$ collides with a fraction droplets of size $x_{[j]}$ in each time step. In contrast, in SDM a random set of $\lfloor n(\Omega)/2 \rfloor$ non-overlapping pairs is considered in each time step.

To ensure linear computational complexity, the mechanism for choosing random non-overlapping pairs must also have at most linear computational complexity. It can be obtained by permuting an indexed array of super-droplets where each pair consist of super-droplets with indices $(2j)$ and $(2j + 1)$ where $j \in [0, 1, 2, \ldots, \lfloor n(\Omega)/2 \rfloor)$, see Appendix A in [20].

Step 2: collision probability evaluation

If a fraction of droplets is allowed to collide, a collision of two super-droplets can produce third super-droplet which has different size. To represent such collision event, a new super-droplet would need to be added to the system what leads to: increasing the number of super-droplets in the system, smaller and smaller multiplicities and increasing memory demands during simulation. Moreover, it is hard to predict at the beginning of a simulation how much memory is needed (see discussion in [16]). Adding new super-droplets can be avoided by mapping colliding super-droplets to a multi-dimensional grid (e.g., 2D for particle volume and its dry size), simulating coalescence on this grid using SCE-based approach and discretising the results back into Lagrangian particles as proposed in [1]. Such approach however, not only entails additional computational cost, but most importantly is characterised by the very curse of dimensionality and numerical diffusion that were meant to be avoided by using particle-based technique.

SDM avoids the issue of unpredictable memory complexity stemming from the need of allocating space for freshly collided super-particle of size that differs from the size of the colliding ones. To this end, in SDM only collisions of all of $\min\{\xi_{[i]}, \xi_{[j]}\}$ droplets are considered, and thus a collision of two super-droplets always produces one or two super-droplets (see Fig. 1b). This means there is no need for adding extra super-droplets to the system during simulation.

For each pair, probability of collision p is up-scaled by the number of real droplets represented by the super-droplet with higher multiplicity $(\max\{\xi_{[i]}, \xi_{[j]}\})$. As a result, computational complexity is $\mathcal{O}(n(\Omega))$ instead of $\mathcal{O}(n(\Omega)^2)$.

The evaluated probability of collision of super-droplets requires further up-scaling due to the reduced amount of sampled candidate pairs as outlined in Step 1 above. To this end, the probability is multiplied by the ratio of the number of all non-overlapping pairs $(n(\Omega)^2 - n(\Omega))/2$ to the number of considered candidate pairs $\lfloor n(\Omega)/2 \rfloor$.

Table 1. Comparison of SCE and SDM approaches to modelling collisional growth.

	SCE (mean-field)	SDM (probabilistic)
Considered pairs	All (i, j) pairs	Random set of $n(\Omega)/2$ non-overlapping pairs, probability up-scaled by $(n(\Omega)^2 - n(\Omega))/2$ to $n(\Omega)/2$ ratio
Comp. complexity	$\mathcal{O}(n(\Omega)^2)$	$\mathcal{O}(n(\Omega))$
Collisions	Colliding a fraction of $\xi_{[i]}, \xi_{[j]}$	Collide all of $\min\{\xi_{[i]}, \xi_{[j]}\}$ (all or nothing)
Collisions triggered	Every time step	By comparing probability with a random number

For each selected pair of j-th and k-th super-droplets, the probability p of collision of the super-droplets within time interval Δt is thus evaluated as:

$$p = a(v_{[j]}, v_{[k]}) \max\{\xi_{[j]}, \xi_{[k]}\} \frac{(n(\Omega)^2 - n(\Omega))/2}{\lfloor n(\Omega)/2 \rfloor} \frac{\Delta t}{\Delta V} \tag{3}$$

where a is a coalescence kernel and ΔV is the considered volume assumed to be mixed well enough to neglect the positions of particles when evaluating the probability of collision.

Step 3: collision triggering and attribute updates
In the spirit of Monte-Carlo methods, the collision events are triggered with a random number $\phi_\gamma \sim Uniform[0, 1)$, and the rate of collisions (per timestep) $\gamma = \lceil p - \phi_\gamma \rceil$ with $\gamma \in \mathbb{N}$. Noting that the rate γ can be greater than 1, further adjustment is introduced to represent multiple collision of super-particles:

$$\tilde{\gamma} = \min\{\gamma, \lfloor \xi_{[j]}/\xi_{[k]} \rfloor\} \tag{4}$$

where $\xi_{[j]} \geq \xi_{[k]}$ was assumed (without losing generality). The conceptual view of collision of two super-droplets is intuitively depicted in Fig. 1b, presented example corresponds to the case of $\gamma = 5$ and $\tilde{\gamma} = \min\{5, \lfloor 9/2 \rfloor\} = 4$.

As pointed out in Step 2 above, SDM is formulated in a way assuring that each collision event produces one or two super-droplets. A collision event results in an update of super-droplet attributes denoted with $A \in \mathbb{R}^{n_{attr}}$ where n_{attr} is the number of considered extensive particle properties. During coalescence, particle positions remain unchanged, values of extensive attributes of collided droplets add up, while the multiplicities are either reduced or remain unchanged. This corresponds to the two considered scenarios defined in points (5) (a) and (b) in Sect. 5.1.3 in [20] and when expressed using the symbols used herein, with attribute values after collision denoted by hat, gives:

1. $\xi_{[j]} - \tilde{\gamma}\xi_{[k]} > 0$

$$\hat{\xi}_{[j]} = \xi_{[j]} - \tilde{\gamma}\xi_{[k]} \qquad \hat{\xi}_{[k]} = \xi_{[k]}$$
$$\hat{A}_{[j]} = A_{[j]} \qquad \hat{A}_{[k]} = A_{[k]} + \tilde{\gamma}A_{[j]} \tag{5}$$

2. $\xi_{[j]} - \tilde{\gamma}\xi_{[k]} = 0$

$$\hat{\xi}_{[j]} = \lfloor \xi_{[k]}/2 \rfloor \qquad \hat{\xi}_{[k]} = \xi_{[k]} - \lfloor \xi_{[k]}/2 \rfloor$$
$$\hat{A}_{[j]} = \hat{A}_{[k]} \qquad \hat{A}_{[k]} = A_{[k]} + \tilde{\gamma}A_{[j]}$$
(6)

Case 1 corresponds to the scenario depicted in Fig. 1b. In case 2, all droplets following a coalescence event have the same values of extensive attributes (e.g., volume), thus in principle could be represented with a single super droplet. However, in order not to reduce the number of super-droplets in the system, the resultant super droplet is split in two. Since integer values are used to represent multiplicities, in the case of $\xi_{[k]} = 1$ in Eq. (6), splitting is not possible and the j-th super-droplet is removed from the system.

3   Backend API

The proposed API is composed of four data structures (classes) and a set of library routines (computational kernels). Description below outlines both the general, implementation-independent, structure of the API, as well as selected aspects pertaining to the experience from implementing CPU and GPU backends in the PySDM package. These were built on top of the Numba [17] and ThrustRTC Python packages, respectively. Numba is an just-in-time (JIT) compiler for Python code, it features extensive support for Numpy and features multi-threading constructs akin to the OpenMP infrastructure. ThrustRTC uses the NVIDIA CUDA real-time compilation infrastructure offering high-level Python interface for execution of both built-in and custom computational kernels on GPU.

In multi-dimensional simulations coupled with a grid-based CFD fluid flow solver, the positions of droplets within the physical space are used to split the super-particle population among grid cells (CFD solver mesh). Since positions of droplets change based on the fluid flow and droplet mass/shape characteristics, the particle-cell mapping changes throughout the simulation. In PySDM, as it is common in cloud physics applications, collisions are considered only among particles belonging to the same grid cell, and the varying number of particles within a cell is needed to be tracked at each timestep to evaluate the super-particle collision rates. Figure 1a outlines the setting.

3.1   Data Structures and Simulation State

The **Storage** class is a base container which is intended to adapt the interfaces of the underlying implementation-dependent array containers (in PySDM: Numpy or ThrustRTC containers for CPU and GPU backends, respectively). This make the API independent of the underlying storage layer.

The **Storage** class has 3 attributes: data (in PySDM: an instance of Numpy **ndarray** or an instance of ThrustRTC **DVVector**), shape (which specifies size and dimension) and dtype (data type: **float**, **int** or **bool**). The proposed API employs one- and two-dimensional arrays, implementations of the **Storage** class

```
 1 idx = Index(N_SD, int)
 2 multiplicities = IndexedStorage(idx, N_SD, int)
 3 attributes = IndexedStorage(idx, (N_ATTR, N_SD), float)
 4 volume_view = attributes[0:1, :]
 5
 6 cell_id = IndexedStorage(idx, N_SD, int)
 7 cell_idx = Index(N_CELL, int)
 8 cell_start = Storage(N_CELL + 1, int)
 9
10 pair_prob = Storage(N_SD//2, float)
11 pair_flag = PairIndicator(N_SD, bool)
12
13 u01 = Storage(N_SD, float)
```

Fig. 2. Simulation state example with N_SD super-particles, N_CELL grid cells and N_ATTR attributes.

feature an indirection layer handling the multi-dimensionality in case the underlying library supports one-dimensional indexing only (as in ThrustRTC). The two-dimensional arrays are used for representing multiple extensive attributes (with row-major memory layout). In general, structure-of-arrays layout is used within PySDM.

Storage handles memory allocation and optionally the host-device (CPU-accessible and GPU-accessible memory) data transfers. Equipping Storage with an override of the [ ] operator as done in PySDM can be helpful for debugging and unit testing (in Python, Storage instances may then be directly used with max(), min(), etc.), care needs to be taken to ensure memory-view semantics for non-debug usage, though. Explicit allocation is used only (once per simulation),

The IndexedStorage subclass of Storage is intended as container for super-particle attributes. In SDM, at each step of simulation a different order of particles needs to be considered. To avoid repeated permutations of the attribute values, the Index subclass of Storage is introduced. One instance of Index is shared between IndexedStorage instances and is referenced by the idx field.

The Index class features permutation and array-shrinking logic (allowing for removal of super-droplets from the system). To permute particle order, it is enough to shuffle Index. To support simulations in multiple physical dimensions, Index features sort-by-key logic where a cell id attribute is used as the key.

The PairIndicator class provides an abstraction layer facilitating pairwise operations inherent to several steps of the SDM workflow. In principle, it represents a Boolean flag per each super-particle indicating weather in the current state (affected by random shuffling and physical displacement of particles), a given particle is considered as first-in-a-pair. Updating PairIndicator, it must be ensured that the next particle according to a given Index is the second one in a pair – i.e., resides in the same grid cell. The PairIndicator is also used to handle odd and even counts of super-particles within a cell (see also Fig. 4). The rationale to store the pair-indicator flags, besides potential speedup, is to separate the cell segmentation logic pertinent to the internals of SDM implementation from potentially user-supplied kernel code.

Figure 2 lists a minimal set of instances of the data structures constituting an SDM-based simulation state.

```
 1 # step 0: removal of super-droplets with zero multiplicity
 2 remove_if_equal_0(idx,                                      # in/out
 3                     multiplicities)                         # in
 4
 5 # step 1: cell-wise shuffling, pair flagging
 6 urand(u01)
 7 shuffle_per_cell(cell_start,                                # out
 8                   idx,                                      # in/out
 9                   cell_idx, cell_id, u01)                   # in
10
11 flag_pairs(pair_flag,                                       # out
12             cell_id, cell_idx, cell_start)                 # in
13
14 # step 2: collision probability evaluation
15 coalescence_kernel(pair_prob,                               # out
16                     pair_flag, volume_view)                # in
17
18 times_max(pair_prob,                                        # in/out
19            multiplicities, pair_flag)                       # in
20
21 normalize(pair_prob,                                        # in/out
22            dt, dv, cell_id, cell_idx, cell_start)           # in
23
24 # step 3: collision triggering and attribute updates
25 urand(u01[:N_SD//2])
26 compute_gamma(pair_prob,                                    # in/out
27                u01[:N_SD//2])                               # in
28
29 update_attributes(multiplicities, attributes,              # in/out
30                    pair_prob)                               # in
```

Fig. 3. Algorithm workflow within a timestep, in/out comments mark argument intent.

3.2   Algorithm Workflow and API Routines

The algorithm workflow, coded in Python using the proposed data structures, and divided into the same algorithm steps as outlined in Sect. 2 is given in Fig. 3. An additional Step 0 is introduced to account for handling the removal of zero-multiplicity super-particles at the beginning of each timestep.

The `cell_id` attribute represents particle-cell mapping. Furthermore, the `cell_idx` helper Index instance can be used to specify the order in which grid cells are traversed – for parallel execution scheduling purposes. Both `cell_id` and `cell_idx` are updated before entering into the presented block of code.

Step 1 logic begins with generation of random numbers used for random permutation of the particles (in PySDM, the CURandRTC package is used). It serves as a mechanism for random selection of super-particle pairs in each grid cell. First, a shuffle-per-cell step is done in which the `cell_start` array is updated to indicate the location within cell-sorted `idx` where sets of super-particles belonging to a given grid cell start. Second, the `pair_flag` indicator is updated using the particle-cell mapping embodied in `cell_start`.

In PySDM, the shuffle-per-cell operation is implemented with two alternative strategies depending on the choice of the backend. A parallel sort with a random key is used on GPU, while the CPU backend uses a serial $\mathcal{O}(n)$ permutation algorithm (as in Appendix A in [20]). This choice was found to offer shortest

```
 1  _kernel = ThrustRTC.For(
 2      ['perm_cell_start', 'perm_cell_id', 'pair_flag', 'length'], "i", '''
 3      pair_flag[i] = (
 4          i < length - 1 &&
 5          perm_cell_id[i] == perm_cell_id[i+1] &&
 6          (i - perm_cell_start[perm_cell_id[i]]) % 2 == 0
 7      );
 8      ''')
 9
10  def flag_pairs(pair_flag, cell_start, cell_id, cell_idx):
11      perm_cell_id = ThrustRTC.DVPermutation(cell_id.data, cell_id.idx.data)
12      perm_cell_start = ThrustRTC.DVPermutation(cell_start.data, cell_idx.data)
13      d_length = ThrustRTC.DVInt64(len(cell_id))
14      _kernel.launch_n(len(cell_id),
15          [perm_cell_start, perm_cell_id, cell_idx.data,
16          pair_flag.indicator.data, d_length])
```

Fig. 4. GPU backend implementation of the pair-flagging routine.

```
 1  @numba.njit(parallel=True, error_model='numpy')
 2  def _update_attributes(length, n, attributes, idx, gamma):
 3      for i in prange(length//2):
 4          j = idx[2*i]
 5          k = idx[2*i + 1]
 6          if n[j] < n[k]:
 7              j, k = k, j
 8          g = min(int(gamma[i]), int(n[j] / n[k]))
 9          if g == 0:
10              continue
11          new_n = n[j] - g * n[k]
12          if new_n > 0:
13              n[j] = new_n
14              for attr in range(0, len(attributes)):
15                  attributes[attr, k] += g * attributes[attr, j]
16          else:  # new_n == 0
17              n[j] = n[k] // 2
18              n[k] = n[k] - n[j]
19              for attr in range(0, len(attributes)):
20                  attributes[attr, j] = attributes[attr, j] * g \
21                                      + attributes[attr, k]
22                  attributes[attr, k] = attributes[attr, j]
23
24  def update_attributes(n, intensive, attributes, gamma):
25      _update_attributes(len(n.idx),
26          n.data, intensive.data, attributes.data,    # in/out
27          n.idx.data, gamma.data)                      # in
```

Fig. 5. CPU backend implementation of the attribute-update routine. a

respective execution times for the considered setup. In general, the number of available threads and the number of droplets and cells considered will determine optimal choice. Furthermore, solutions such as MergeShuffle [4] can be considered for enabling parallelism within the permutation step.

Instructions in Step 2 and Step 3 blocks correspond to the evaluation of subsequent terms of Eq. (3) and the collision event handling. To exemplify the way the GPU and CPU backends are engineered in PySDM, the implementation of the pair flag routine with ThrustRTC, and the update-attributes routine with Numba are given in Fig. 4 and Fig. 5, respectively.

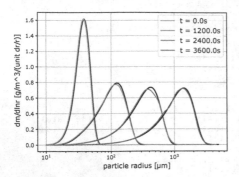

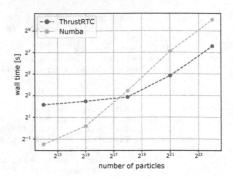

Fig. 6. Particle mass spectrum: SDM results (colour) vs. analytic solution (black). (Color figure online)

Fig. 7. Simulation wall time as a function of the number of super particles used.

4 Example Simulations

This section outlines a set of test cases useful in validating implementation of the proposed API (and included in the set of examples shipped with PySDM). First, a simulation constructed analogously as that reported in Fig. 2 in the original SDM paper [20] is proposed. The considered volume of $V = 10^6$ m^3 is populated with $N_0 \cdot V = 2^{23} \cdot 10^6$ particles. Sizes of the particles follow an exponential particle volume distribution $p(v_{[i]}) = \bar{v}^{-1} \exp(-v_{[i]}/\bar{v})$ where $\bar{v} = (30.531\mu m)^3 4\pi/3$. The simulation uses 2^{17} super-droplets initialised with equal multiplicities so that at $t = 0$ s each one represents a quantile of the total number of modeled particles (within an arbitrarily chosen percentile range from 0.001% to 99.999%). The timestep is 1 s. The Golovin [12] additive $a(v_1, v_2) = b(v_1 + v_2)$ coalescence kernel with $b = 1500\text{s}^{-1}$ is used for which an analytical solution to the Smoluchowski's equation is known (cf. Equation 36 in [12]):

$$\phi(v,t) = \frac{1 - \tau(t)}{v\sqrt{\tau(t)}} I_{ve}^1 \left(\frac{2v\sqrt{\tau(t)}}{\bar{v}} \right) \exp\left(\frac{-v(1 + \tau(t) - 2\sqrt{\tau(t)})}{\bar{v}} \right) \qquad (7)$$

where $\tau = 1 - \exp(-N_0 b \bar{v} t)$ and I_{ve}^1 stands for exponentially scaled modified Bessel function of the first kind (`scipy.special.ive`).

Plot 6 shows the relationship between the mass of droplets per unit of ln$r$ and the droplet radius r (assuming particle density of 1000 kg/m^3 as for liquid water). Results obtained with PySDM are plotted by aggregating super-particle multiplicities onto a grid of ca. 100 bins, and smoothing the result twice with a running average with a centered window spanning five bins.

Figure 7 documents simulation wall times for the above test case measured as a function of number of super-particles employed. Wall times for CPU (Numba) and GPU (ThrustRTC) backends are compared depicting an over four-fold GPU-to-CPU speedup for large state vectors (tested on commodity hardware: Intel Core i7-10750H CPU and NVIDIA GeForce RTX 2070 Super Max-Q GPU).

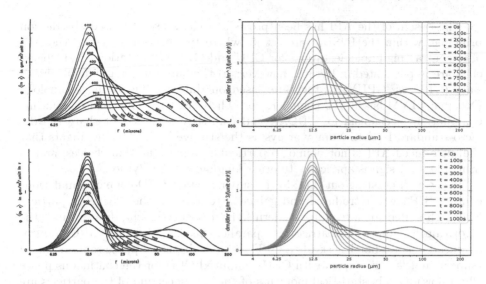

Fig. 8. Left panels show Figs. 5 and 8 from [6] (copyright: American Meteorological Society, used with permission). Right panels depict solutions obtained with PySDM. Top: gravitational kernel; bottom: kernel modelling electric field effect.

Figure 8 presents results visualised in analogous plots but from simulations with more physically-relevant coalescence kernels. Simulation setup follows the work of [6] and features so-called gravitational kernel involving a parametric form of particle terminal velocity dependence on its size, and a parametric kernel modeling the electric field effects on collision probability. Since the analytic solution is not known in such cases, the results are juxtaposed with figures reproduced from [6].

5   Summary and Discussion

This paper has discussed a set of data structures and computational kernels constituting a number-crunching backend API for the Super-Droplet Method Monte-Carlo algorithm for representing collisional growth of particles. Employment of the API assures separation of concerns, in particular separation of parallelisation logic embedded in the backend-level routines from domain logic pertinent to the algorithm workflow. This improves code readability, paves the way for modular design and testability, which all contribute to code maintainability.

The presented SDM algorithm and API descriptions discern data dependencies across the steps of the algorithm (including in/out parameter "intent") and highlight parallelisation opportunities in different steps. Most of the discerned steps of the SDM algorithm are characterised by some degree of freedom in terms of their implementation. Embracing the API shall help in introducing a dependency injection mechanism allowing unit testing of selected steps and profiling performance with different variants of implementation.

The design of the API has been proved, in the sense of reused abstraction principle, within the PySDM project [5] where two backends sharing the API offer contrasting implementations for CPU and GPU computations. Both backends are implemented in Python, however: (i) they are targeting different hardware (CPU vs. GPU), (ii) they are based on different underlying technology (Numba: LLVM-based JIT compiler, and ThrustRTC: NVRTC-based runtime compilation mechanism for CUDA), and (iii) they even do not share the de-facto-standard Python Numpy arrays as the storage layer. This highlights that the introduced API is not bound to particular implementation choices, and in principle its design is applicable to other languages than Python.

It is worth pointing out that, in the super-particle CFD-coupled simulations context SDM was introduced and gained attention in, the scheme is particularly well suited for leveraging modern hybrid CPU-GPU hardware. First, the algorithm is (almost) embarrassingly parallel. Second, the CPU-GPU transfer overhead is not a bottleneck when GPU-resident dispersed phase representation (super-particles) is coupled with CPU-computed CFD for the continuous phase (fluid flow) as only statistical moments of the size spectrum of the particles are needed for the CFD coupling. Third, the CPU and GPU resources on hybrid hardware can be leveraged effectively if fluid advection (CPU) and particle collisions (GPU) are computed simultaneously (see [10]).

Overall, the discussed API prioritises simplicity and was intentionally presented in a paradigm-agnostic pseudo-code-mimicking way, leaving such aspects as object orientation up to the implementation. Moreover, while the presented API includes data structures and algorithm steps pertinent to multi-dimensional CFD grid coupling, presented examples featured zero-dimensional setups, for brevity. In PySDM [5], the backend API is extended to handle general form of coalescence kernels, representation of particle displacement, condensational growth of particles, aqueous chemical reactions, and the examples shipped with the package include simulations in multiple physical dimensions.

Acknowledgements. We thank Shin-ichiro Shima and Michael Olesik for helpful discussions as well as an anonymous reviewer for insightful comments. Thanks are due to Numba and ThrustRTC developers for support in using the packages.

References

1. Andrejczuk, M., Grabowski, W.W., Reisner, J., Gadian, A.: Cloud-aerosol interactions for boundary layer stratocumulus in the Lagrangian Cloud Model. J. Geophys. Res. Atmos. **115** (2010). https://doi.org/10.1029/2010JD014248
2. Arabas, S., Jaruga, A., Pawlowska, H., Grabowski, W.W.: libcloudph++ 1.0: a single-moment bulk, double-moment bulk, and particle-based warm-rain microphysics library in C++. Geosci. Model Dev. **8** (2015). https://doi.org/10.5194/gmd-8-1677-2015
3. Arabas, S., Shima, S.: Large-eddy simulations of trade wind cumuli using particle-based microphysics with Monte Carlo coalescence. J. Atmos. Sci. **70** (2013). https://doi.org/10.1175/JAS-D-12-0295.1

4. Bacher, A., Bodini, O., Hollender, A., Lumbroso, J.: MergeShuffle: a very fast, parallel random permutation algorithm (2005). https://arxiv.org/abs/1508.03167
5. Bartman, P., et al.: PySDM v1: particle-based cloud modelling package for warm-rain microphysics and aqueous chemistry (2021). https://arxiv.org/abs/2103.17238
6. Berry, E.: Cloud droplet growth by collection. J. Atmos. Sci. **24** (1966). https://doi.org/10.1175/1520-0469(1967)024%3C0688:CDGBC%3E2.0.CO;2
7. Chandrasekhar, S.: Stochastic problems in physics and astronomy. Rev. Mod. Phys. **15** (1943). https://doi.org/10.1103/RevModPhys.15.1
8. DeVille, R.E., Riemer, N., West, M.: Weighted Flow Algorithms (WFA) for stochastic particle coagulation. J. Comput. Phys. **230** (2011). https://doi.org/10.1016/j.jcp.2011.07.027
9. Dziekan, P., Pawlowska, H.: Stochastic coalescence in Lagrangian cloud microphysics. Atmos. Chem. Phys. **17** (2017). https://doi.org/10.5194/acp-17-13509-2017
10. Dziekan, P., Waruszewski, M., Pawlowska, H.: University of Warsaw Lagrangian Cloud Model (UWLCM) 1.0: a modern large-eddy simulation tool for warm cloud modeling with Lagrangian microphysics. Geosc. Model Dev. **12** (2019). https://doi.org/10.5194/gmd-12-2587-2019
11. Gillespie, D.: The stochastic coalescence model for cloud droplet growth. J. Atmos. Sci. **29** (1972). https://doi.org/10.1175/1520-0469(1972)029%3C1496:TSCMFC%3E2.0.CO;2
12. Golovin, A.: The solution of the coagulation equation for raindrops. Taking condensation into account. Bull. Acad. Sci. SSSR Geophys. Ser. **148** (1963). http://mi.mathnet.ru/dan27630. (in Russian)
13. Grabowski, W., Morrison, H., Shima, S., Abade, G., Dziekan, P., Pawlowska, H.: Modeling of cloud microphysics: can we do better? Bull. Am. Meteorol. Soc. **100** (2019). https://doi.org/10.1175/BAMS-D-18-0005.1
14. Hansen, K.: Abundance distributions; large scale features. In: Hansen, K. (ed.) Statistical Physics of Nanoparticles in the Gas Phase. SSAOPP, vol. 73, pp. 205–251. Springer, Cham (2018). https://doi.org/10.1007/978-3-319-90062-9_8
15. Hoffmann, F., Noh, Y., Raasch, S.: The route to raindrop formation in a shallow cumulus cloud simulated by a Lagrangian cloud model. J. Atmos. Sci. **74** (2017). https://doi.org/10.1175/JAS-D-16-0220.1
16. Jensen, J.B., Lee, S.: Giant sea-salt aerosols and warm rain formation in marine stratocumulus. J. Atmos. Sci. **65** (2008). https://doi.org/10.1175/2008JAS2617.1
17. Lam, S., Pitrou, A., Seibert, S.: Numba: a LLVM-based Python JIT compiler. In: Proceedings of the Second Workshop on the LLVM Compiler Infrastructure in HPC, LLVM 2015. ACM (2015). https://doi.org/10.1145/2833157.2833162
18. Li, X.Y., Brandenburg, A., Haugen, N.E., Svensson, G.: Eulerian and Lagrangian approaches to multidimensional condensation and collection. J. Adv. Model. Earth Syst. **9** (2017). https://doi.org/10.1002/2017MS000930
19. Sato, Y., Shima, S., Tomita, H.: Numerical convergence of shallow convection cloud field simulations: comparison between double-moment Eulerian and particle-based Lagrangian microphysics coupled to the same dynamical core. J. Adv. Model. Earth Syst. **10** (2018). https://doi.org/10.1029/2018MS001285
20. Shima, S., Kusano, K., Kawano, A., Sugiyama, T., Kawahara, S.: The super-droplet method for the numerical simulation of clouds and precipitation: a particle-based and probabilistic microphysics model coupled with a non-hydrostatic model. Q. J. Royal Meteorol. Soc. **135** (2009). https://doi.org/10.1002/qj.441

21. Shima, S., Sato, Y., Hashimoto, A., Misumi, R.: Predicting the morphology of ice particles in deep convection using the super-droplet method: development and evaluation of SCALE-SDM 0.2.5-2.2.0, -2.2.1, and -2.2.2. Geosci. Model Dev. **13** (2020). https://doi.org/10.5194/gmd-13-4107-2020
22. Shima, S., Sugiyama, T., Kusano, K., Kawano, A., Hirose, S.: Simulation method, simulation program, and simulator (2007). https://data.epo.org/gpi/EP1847939A3
23. Smoluchowski, M.: Drei Vorträge über Diffusion, Brownsche Molekularbewegung und Koagulation von Kolloidteilchen I. Phys. Z. **22** (1916). https://fbc.pionier.net.pl/id/oai:jbc.bj.uj.edu.pl:387533. (in German)
24. Smoluchowski, M.: Drei Vorträge über Diffusion, Brownsche Molekularbewegung und Koagulation von Kolloidteilchen II. Phys. Z. **23** (1916). https://fbc.pionier.net.pl/id/oai:jbc.bj.uj.edu.pl:387534. (in German)
25. Unterstrasser, S., Hoffmann, F., Lerch, M.: Collisional growth in a particle-based cloud microphysical model: insights from column model simulations using LCM1D (v1.0). Geosci. Model Devel. **13** (2020). https://doi.org/10.5194/gmd-13-5119-2020

Applications of Computational Methods in Artificial Intelligence and Machine Learning

A Deep Neural Network Based on Stacked Auto-encoder and Dataset Stratification in Indoor Location

Jing Zhang and Ying Su[✉]

Department of Communication, Shanghai Normal
University, Shanghai 200234, People's Republic of China
{jannety,yingsu}@shnu.edu.cn

Abstract. Indoor location has become the core part in the large-scale location-aware services, especially in the extendable/scalable applications. Fingerprint location by using the signal strength indicator (RSSI) of the received WiFi signal has the advantages of full coverage and strong expansibility. It also has the disadvantages of requiring data calibration and lacking samples under the dynamic environment. This paper describes a deep neural network method used for indoor positioning (DNNIP) based on stacked auto-encoder and data stratification. The experimental results show that this DNNIP has better classification accuracy than the machine learning algorithms that are based on UJIIndoorLoc dataset.

Keywords: Indoor location · Deep Neural Network · Machine Learning algorithm

1 Introduction

The application of indoor location involves the integration of multiple interdisciplinary works. This location information meets the users' needs with·convenient navigations in large indoor situations. In the e-commerce application, specific recommendations with accurate location information can be presented to users. In the emergency rescue, fast effective actions can be taken with the obtained exact location of the rescued target. Also in hospitals, doctors can manage patients and medical supplies through location tracking. And in social activities, groups can be easily built based on the people's indoor locations, thus enriching the interactions among peoples [1].

In an indoor environment, the traditional GPS positioning is generally not suitable due to its large signal attenuation. Thus the indoor positioning introduces a variety of sensors and equipments. In the coverage of one wireless heterogeneous network, the fingerprint location is implemented with the received signal strength indicator (RSSI) values from at least three wireless transmitters. The use of WiFi signal is a popular solution, which complies with IEEE 802.11 standard and has high bandwidth and transmit rate. The advantage of using WiFi for positioning is its full WiFi signal coverage in most cities, providing the readiness of the infrastructure. In order to improve the positioning

© Springer Nature Switzerland AG 2021
M. Paszynski et al. (Eds.): ICCS 2021, LNCS 12743, pp. 33–46, 2021.
https://doi.org/10.1007/978-3-030-77964-1_3

accuracy, the acoustic ranging and WiFi are combined to test the relative distance information between reference nodes. Accelerometers and compasses are also used to help obtain accurate maps with large indoor areas [2].

The choice of the location of an access point is the key to improving the accuracy of positioning when applying the fingerprint matching for the location. Some literatures have analyzed the impact of the number of reference nodes on the accuracy, and proposed appropriate selection strategies. Some select reference nodes via gradient descent search during the training phase, RSSI surface fitting, and hierarchical clustering [3, 4]. Even some publishes propose a goal-driven model by combining multi-target reference node deployment with genetic algorithm [5]. After the selection of RSSI signals, the process algorithm is another factor that affects the accuracy of positioning. There are two types of intelligent location algorithms: neighbor selection and machine learning. K-Nearest Neighbor (KNN) is a common fingerprint matching algorithm. The accuracy can be improved through the minimum circle clustering and the adaptive weighted KNN matching [6]. Artificial neural network (ANN) is also introduced to support indoor and outdoor positioning with its particle swarm optimization to be used to optimize the neurons [7].

In large indoor stereo positioning scenarios, the problem becomes more complex due to the existence of multiple floors and multiple buildings. How to deal with the random fluctuation of the signal, the noise of the multipath effect, and the dependence on the equipment to be used are the main challenges. This paper [8] proposes a fingerprint based pipeline processing method, which firstly identifies the user's building, and then establishes the association among access points. Through the voting of these various networks, the buildings and their floors are correctly calculated [8]. The KNN algorithm mentioned above would make proper adjustment to the building specifically selected via grid search [9]. In addition, with the popularity of mobile devices, big data has become accessible. Due to the improvement of GPU performance and rich algorithm libraries, the deep learning can address these challenges effectively. Deep neural network (DNN) simply needs to be for the comparisons of location precisions within larger positioning areas. It needs fewer parameter adjustments but provides greater scalability [10–12].

In this research, the UJIIndoorLoc dataset was selected as the original records [13, 14]. Taken into the consideration of the cases like the possible classification errors of the buildings or the floors, the loss happening in the training stage will be the same as in the validation stage. Thus the dataset was segmented as layers based on the RSSI values from 520 wireless access points (WAP) and the three attributes as buildingID, floorID, and SpaceID. This paper proposes the use of deep neural network for indoor positioning (DNNIP) after the comparisons with the support vector machine (SVM), random forest and gradient boosting decision tree. The DNNIP approach ensures the accuracy and reduces the adjustment of parameters.

The rest of this paper is organized as follows: Sect. 2 introduces the hierarchical method for the dataset. Section 3 describes the DNNIP network structure and training algorithm. Then the experimental results are represented. The last section has the conclusion.

2 Fingerprint Localization and Datasets Stratification

Indoor fingerprint positioning consists of two stages as shown in Fig. 1. During the offline stage, the wireless map is created by measuring the RSSI values of different access points or reference nodes. The map includes not only the RSSI values, but also the details of the reference nodes placed: the building number, the specific floor, and the space details. While at the online stage, the location of the user is obtained by matching the RSSI value of the user with the wireless map through a location algorithm.

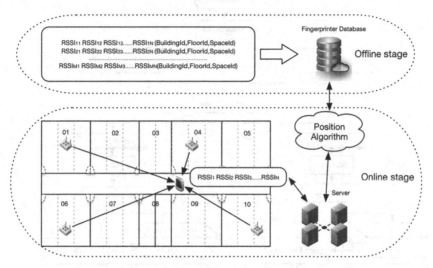

Fig. 1. Fingerprint positioning stages

2.1 Open Dataset of UJIIndoorLoc

The dataset of UJIIndoorLoc was collected from three buildings of Jaume I University, with four, four and five floors respectively. Data from 933 different reference points was measured against 20+ users who are using 25 different types of mobile terminals. The whole data set is divided into training subset and test subset with 19,937 records used for training and 1,111 records for testing. Each record contains 529 dimensions, including 520 RSSI values and 9 additional attributes, like BuildingID, FloorID, and SpaceID.

2.2 Dataset Stratification

The data preparation is based on the location stratification. Each sub-dataset is constructed by the building number, floor number and space number. The format of each sample is BuildingID, FloorID, SpaceID, WAP001, ..., WAP520. 90% of the whole training dataset of UJIIndoorLoc are classified as training dataset and test dataset, represented as DNNIP_Train and DNNIP_Test. The test dataset is used for testing, while the data of the original test set are transformed as validation set. Each dataset is then divided into three sub-datasets based on the BuildingID. And each sub-dataset is further divided according to the FloorID. Finally, the spaceID is added. The final bottom sub-dataset reflects the minimum positioning range. The hierarchical sub-datasets are illustrated in Fig. 2. From these sub-datasets, the building information is firstly discovered and predicted, and then the floors are classified. then each sample will be further classified by SpaceID.

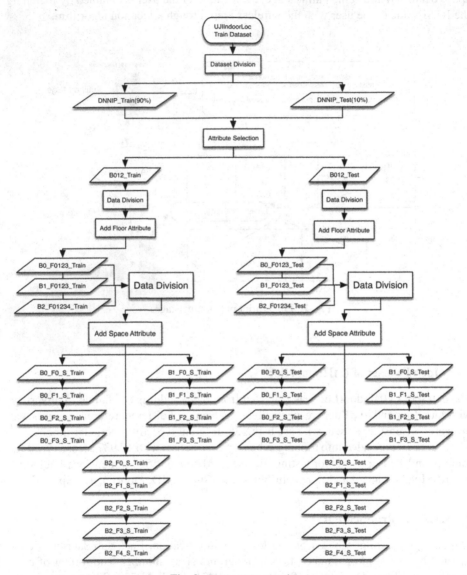

Fig. 2. Dataset preparation

3 DNNIP with Auto-encoders

3.1 Network Structure of DNNIP

DNNIP has multiple auto-encoders (AE) to learn features from large samples. The Stacked AE (SAE) makes the output vector value as close as possible to the input vector value. The output of the hidden-layer neurons is of the most research interest as these post-encoding parameters carry all of the input information in a compact feature form. When the number of AE increases, the feature representation becomes more abstract. Therefore, DNNIP is more suitable for the complex classification tasks.

The reason for applying SAE into DNNIP is that if a DNN is directly trained to perform the classification task through random initialization, the underlying error is almost zero and the error gradient can easily disappear. When using the AE structure for the unsupervised learning from the training data, the pre-trained network can make the training data to a certain extent such that the initial value of the whole network is within an appropriate state. Therefore, through the supervised learning the classification of the second stage can be made easier and the convergence speed can be accelerated. Also the activation function of a rectified linear unit (ReLu) is applied to the SAE neurons, i.e., $f(z) = \max(0, z)$. It is well known that in order to use the traditional backward-propagation learning algorithm, the activation function of neurons normally uses a sigmoid function. In DNN, however, the sigmoid function would appear soft saturated, and the error gradient would disappear easily.

As shown in Fig. 3, at the top of the DNNIP network, the classifier consists of a dropout layer. For strengthening the learning of redundancy, the dropout layer randomly deletes the connections between layers during the training process to achieve better generalization and avoid over-fitting. The final output layer is the Softmax layer that outputs the probability of classifications.

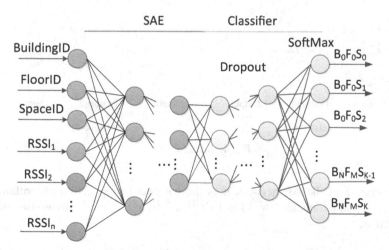

Fig. 3. The network structure of DNNIP

3.2   Auto-encoder and Training Algorithm

Every AE is a fully connected multi-layer forward network, and it maps the input to the hidden layer and then maps to the output. Basically the AE reconstructs the input with which the hidden layer maps the input to become low dimensional vectors. When the number of AEs increases, the number of hidden-layer neurons would decrease. The hidden layer can learn the features of $h = f_\theta(x)$, where θ represents the connection parameters, in which the weight w, the threshold b and the activation functions are included. Moreover, the input of the next AE is the output of the previous AE from its hidden layer.

Figure 4 shows the first two AEs of a DNNIP. The input layer of the first AE has 523 neurons to accept 523 dimensional RSSI values plus the three attributes - BuildingID, FloorID and SpaceID. The output layer also has 523 neurons, representing the reconstructed input to the next AE.

The SAE part is reconstructed as follows: Given the required input data $x = \{x^{(i)}|i = 1, \cdots, m\}$. Let the number of AEs be N_k. By feeding x to train the first AE, the network parameters $\left(w^{(1,1)}, b^{(1,1)}, w^{(1,2)}, b^{(1,2)}\right)$ along with the output of the hidden layer $a^{(1,2)}$ are obtained, as shown in Fig. 5(a). Afterwards, this $a^{(1,2)}$ is then used as the input of the second AE, as shown in Fig. 5(b). Another group of parameters $\left(w^{(2,1)}, b^{(2,1)}, w^{(2,2)}, b^{(2,2)}\right)$ and $a^{(2,2)}$ can be obtained. Repeat this procedure until the N_k number of AE trainings are reached. At this point, multiple groups of network parameters $\{(w^{(k,1)}, b^{(k,1)}, a^{(k,2)})|k = 1, \cdots, N_k\}$ are available.

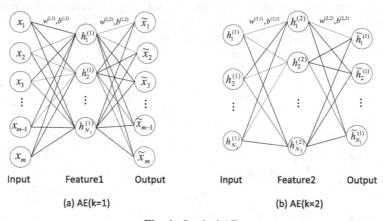

(a) AE(k=1)   (b) AE(k=2)

Fig. 4.   Stacked AEs

The training of the AE network uses the backward propagation algorithm. This training is to find parameters $(w, b) = \left(w^{(k,1)}, b^{(k,1)}, w^{(k,2)}, b^{(k,2)}\right)$ by minimizing the loss function,

$$\min_{\arg w,b} L(w, b) = \left[\frac{1}{m}\sum_{i=1}^{m} J\left(w, b; x^{(i)}, y^{(i)}\right)\right] + \frac{\lambda}{2}\sum_{l=1}^{n_l-1}\sum_{i=1}^{s_l}\sum_{j=1}^{s_l+1}\left(w_{ji}^{(k,l)}\right)^2 + \beta\sum_{j=1}^{s_l} KL(\rho||\overline{\rho}_j).$$

(1)

where $J(\cdot)$ is the mean squared error (MSE) between the input and the output, which is further formulated as in (2). In (1), $x^{(i)} \in R^{s_l \times 1}$ and $y^{(i)}$ denote the input and output of the ith sample, respectively; m, n_l and s_l represent the number of training samples, the number of network layers and the number of neurons in layer l, respectively; $w_{ji}^{(k,l)}$ denotes the connection weight between the ith neuron on the layer l and the jth neuron on the $l+1$ layer of the kth AE; $KL(\cdot)$ represents the sparseness constraint formulated in (3); λ, β and ρ represent the weight coefficients, respectively.

$$J\left(w, b; x^{(i)}, y^{(i)}\right) = \left\| \frac{1}{2} h_{w,b}\left(x^{(i)}\right) - y^{(i)} \right\|^2 \tag{2}$$

where $h_{w,b}\left(x^{(i)}\right)$ is the SAE output vector,

$$h_{w,b}\left(x^{(i)}\right) = a^{(k,3)} = f\left(w^{(k,2)} a^{(k,2)} + b^{(k,2)}\right), \tag{3}$$

where $b^{(k,2)}$ and $a^{(k,2)}$ represent the offset vector and output vector of the kth AE, respectively,

$$a^{(k,2)} = f\left(w^{(k,1)} x^{(i)} + b^{(k,1)}\right), \tag{4}$$

and

$$KL(\rho \| \bar{\rho}_j) = \rho \lg \frac{\rho}{\bar{\rho}_j} + (1-\rho) \lg \frac{1-\rho}{1-\bar{\rho}_j}. \tag{5}$$

where $\bar{\rho}_j$ is the average output of the jth neuron, $\bar{\rho}_j = \frac{1}{m} \sum_{i=1}^{m} [a_j^{(k,2)} x^{(i)}]$; wherein $a_j^{(k,2)}$ is the output of the jth hidden neuron.

For the sample $\left(x^{(i)}, y^{(i)}\right)$, the error gradient on the neuron n of the output layer is

$$\delta_n^{(k,3)} = \frac{\partial}{\partial Z_n^{(k,3)}} \left\{ \frac{1}{2} \left\| h_{w,b}\left(x^{(i)}\right) - y^{(i)} \right\|^2 \right\} = -\left(y_n^{(i)} - a_n^{(k,l)}\right) f'\left(z_n^{(k,3)}\right), \tag{6}$$

The gradient residual of the neuron n on the hidden layer k is

$$\delta_n^{(k,2)} = f'\left(Z_n^{(k,2)}\right) \cdot \left[\sum_{m=1}^{s_l} \left(w_{mn}^{(k,2)} \delta_n^{(k,3)}\right) + \beta \cdot \left(-\frac{\rho}{\bar{\rho}_m} + \frac{1-\rho}{1-\bar{\rho}_m}\right) \right], \tag{7}$$

In (6) and (7), the vector $z^{(k,l+1)} = w^{(k,l)} a^{(k,l)} + b^{(k,l)}$, $l = 1, 2$; $z_n^{(k,l)}$ is the nth element of $Z^{(k,l)}$.

To find the partial derivatives of the cost function yields.

$$\nabla_{w^{(k,l)}} J\left(w, b; RSSI^{(i)}, y^{(i)}\right) = \frac{\partial J\left(w, b; RSSI^{(i)}, y^{(i)}\right)}{\partial w^{(k,l)}} = \delta^{(k,l+1)} \left(a^{(k,l)}\right)^T, \tag{8}$$

$$\nabla_{b^{(k,l)}} J\left(w, b; RSSI^{(i)}, y^{(i)}\right) = \frac{\partial J\left(w, b; RSSI^{(i)}, y^{(i)}\right)}{\partial b^{(k,l)}} = \delta^{(k,l+1)}. \tag{9}$$

After obtaining the loss function and the partial derivative of the parameters, the gradient descent algorithm is to get the optimal parameters of the AE network. The training process is as follows.

1) For $l = 1$ to N_k, let the matrix $\Delta w^{(k,l)} = 0$, the vector $\Delta b^{(k,l)} = 0$, initialize the learning step α, $0 < \alpha < 1$.
2) For $i = 1$ to m, calculate

$$\Delta w^{(k,l)} := \Delta w^{(k,l)} + \nabla_{w^{(k,l)}} J\left(w, b; RSSI^{(i)}, y^{(i)}\right),$$

$$\Delta b^{(k,l)} := \Delta b^{(k,l)} + \nabla_{b^{(k,l)}} J\left(w, b; RSSI^{(i)}, y^{(i)}\right)$$

3) Update the parameters

$$w^{(k,l)} = w^{(k,l)} - \alpha\left[\left(\frac{1}{m}\Delta w^{(k,l)}\right) + \lambda w^{(k,l)}\right],$$

$$b^{(k,l)} = b^{(k,l)} - \alpha\left[\frac{1}{m}\Delta b^{(k,l)}\right].$$

4) Repeat 2) until the algorithm converges or reaches the maximum number of iterations and output $\left(w^{(k,l)}, b^{(k,l)}, w^{(k,l+1)}, b^{(k,l+1)}\right)$.

In the training process, the value of the cost function is determined by all the training samples. It has irrelevance with the training sequence. Inside the SAE, the output of the hidden layer $a^{(k,2)}$ is the feature of the whole training dataset.

3.3 Classification

After the SAE's unsupervised training process is completed, the decoder layer of each layer, i.e., the output layer is disconnected. The trained SAE is then connected to the classifier, as can be seen in Fig. 3. Generally the number of buildings is denoted by N, the number of floors in building i by M_i, and the number of spaces in floor j by K_j. Then the outputs of the classifiers are separated into $\sum_{i=1}^{N} \sum_{j=1}^{M_i} K_j$ classes.

4 Experimental Results Comparison

During the experiments, the Pandas library is selected for data processing, Keras library for DNN, TensorFlow for numerical computation, and Scikit-learn library for the typical machine learning algorithm computations. The relevant parameters of simulation are shown in Table 1.

To evaluate the performance of the DNNIP against other indoor positioning algorithms that include SVM, random forest algorithm and gradient-promotion decision tree, we calculate the accuracy according to (10) based on the definition of accuracy for classification problems used in statistical learning, and then select the true positive (TP), false positive (FP), false negative (FN) as well as true negative (TN) to ultimately measure the correct rate (CR).

$$CR = \frac{TP + TN}{TP + TN + FP + FN}. \tag{10}$$

Table 1. The DNNIP related parameters.

DNNIP parameters	Values
SAE hidden layers	64,128,256
SAE activation function	ReLu
SAE optimizer	ADAM
SAE loss	MSE
Classifier hidden layers	128–128
Classifier optimizer	ADAM
Classifier loss	Categorical Cross Entropy
Classifier dropout rate	0.18
Ratio of training data to overall data	0.90
Number of epochs	20
Batch size	10

In fact, TP and TN represent the correct classification numbers, while FP and FN indicate the wrong classification numbers.

Several distinct DNN structures are shown in Fig. 5. These networks are represented by having the numbers in parentheses, which indicate the number of neurons used in the hidden layer. The first DNN is a fully connected network with no SAE components, and it uses a dropout layer to prevent over-fitting. For each chosen structure, a variety of optimization strategies are used by starting with constant tuning and testing the dropout value to be within the range of 5% to 20%, The final settling is set on 18% as shown in Table 1. Meanwhile, the different values of the learning rate of the ADAM optimizer are tried and compared with the best value achieved through repeated adjustments.

4.1 Effect of the SAE Structure

Figure 5 displays the accuracy performances of different DNN networks. It clearly shows that the higher the number of hidden-layer neurons, the higher the accuracy. The accuracy degree of the classifications of the buildings and floors coming out of the test set can achieve 94.2%, while the accuracy against the validation dataset can only be close to 83.8%. The reason is because of the characteristics of the validation dataset - only part of the samples contains valid RSSI values, while other samples lack RSSI values, and the default value is 0. The results demonstrate that the SAE's network structure can effectively reduce the dimensioning of the input vector from 523 to 256, 128, and 64. The simplified results can then be linked to classifiers.

Figure 5 indicates that the SAE's identification accuracy with 256 and 128 hidden neurons superimposed on the validation data set can reach up to 98%, and the accuracy on the test set can be improved up to 89%. This proves that the SAE can learn even from a simplified representation of the input information and get better results than the DNN that does not have an AE network. The comparison of SAE (256-128) vs. SAE (128-64) would lead to a conclusion that the more neurons are put on the hidden-layer, the higher accuracy can be obtained. The SAE (256-128-64), however, achieves the similar performance as the SAE (256-128) does. This is mainly due to the fact that the more number of AEs the more complex the network would be, and thus consuming more time to adjust parameters, and causing the performance to improve slowly.

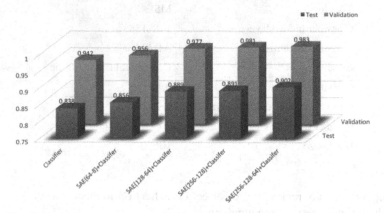

Fig. 5. Impact on the accuracy of Classifications of different DNN structures

Figure 6 shows the changes of the accuracy of the classifications of the buildings and the floors on both the test-set and the validation-set along with a training duration by SAE (256-128). The degree of the correct recognitions by the SAE achieves up 95% or higher after the 11th iteration.

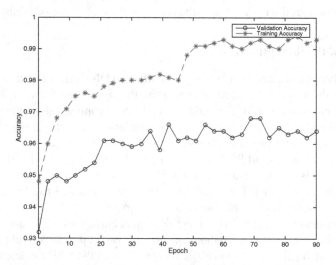

Fig. 6. Accuracy of test and validation dataset using DNNIP

4.2 Accuracy Comparison of Different Algorithms

The SAE that does not partition the dataset is labeled as DNLIP. And the algorithm of the experiment used for the building level and the floor level in the training and simulation in [11] is labeled as SVM, i.e., the random forest and decision tree algorithms. As shown in Fig. 7, DNNIP presents the highest accuracy of 88.9%, while SVM has the lowest accuracy of 82.5%. Compared with the DNLIP algorithm, the DNNIP algorithm has some improvement. This is because that when the DNLIP algorithm classifies the buildings and floors, it gets the same loss value in the training phase and cannot obtain the correct high recognition rate.

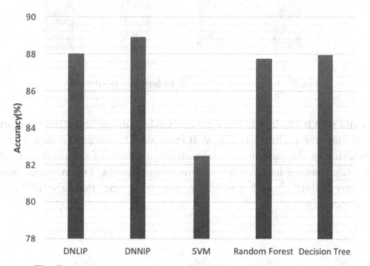

Fig. 7. Accuracy comparison of building and floor positioning

The first step of using the DNNIP method for the indoor positioning is the classification of the buildings, and the accuracy of the algorithm is shown in Fig. 8. It can be seen that the algorithms above can achieve very good accuracies. And the accuracy of the decision tree algorithm can be up to 99.2%, which is better than all other machine learning algorithms.

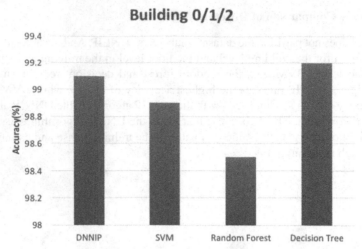

Fig. 8. Accuracy comparison of building positioning

The second step is to classify the floors of each building using the floor attributes. The resultant outcome is shown in Fig. 9. It indicates that when the data set is divided by the numbering of the buildings, the obtained accuracy of the proposed DNNIP is better than other machine learning algorithms in most cases. The averaged positioning accuracy of this DNNIP algorithm is 93.8%, which suggests that DNNIP has a certain capability in accuracy.

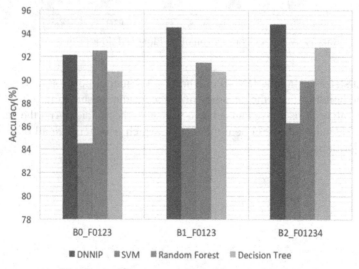

Fig. 9. Accuracy comparison of floor positioning

Finally, the accuracies of the classifications against the space among these algorithms are compared with results shown in Fig. 10. It shows that the DNNIP has the highest average positioning accuracy in some classifications. It can also be found that all the algorithms would get worse results when classifying the ground and top floors than classifying the middle floors for the same building.

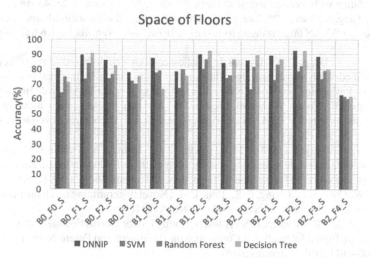

Fig. 10. Accuracy comparison of space positioning

5 Conclusion

In this paper, the DNNIP method is proposed to locate the user's position in large indoor buildings, and further used to segment and layer the original UJIIndoorLoc dataset. The DNNIP takes the network structure of the stacked auto-encoders and ReLu activation function so as to avoid the gradient disappearance in the training processes. The obtained classifying accuracy of the proposed DNNIP algorithm is the highest among all other machine learning algorithms. And after the training, this algorithm does not need to find the best match of samples within the database, which saves the time of manually adjusting the parameters. One disadvantage of the proposed DNNIP algorithm is that when being used against the hierarchical datasets for the training, the computation is much higher than that of the traditional machine learning algorithms, and more computing resources are needed when the training set is updated and adjusted.

References

1. Gu, Y., Lo, A., Niemegeers, I.: A survey of indoor positioning systems for wireless personal networks. IEEE Commun. Surv. Tutor. **11**(1), 13–32 (2009)
2. Machaj, J., Brida, P., Majer, N.: Challenges introduced by heterogeneous devices for Wi-Fi-based indoor localization. Concurr. Comput.: Pract. Exp. **32**, 1–10 (2019)

3. Miao, H., Wang, Z., Wang, J., Zhang, L., Zhengfeng, L.: A novel access point selection strategy for indoor location with Wi-Fi. In: China Control and Decision Conference (CCDC), pp. 5260–5265. IEEE, Changsha (2014)
4. Wang, B., Zhou, S., Yang, L.T., Mo, Y.: Indoor positioning via subarea fingerprinting and surface fitting with received signal strength. Pervasive Mob. Comput. **23**, 43–58 (2015)
5. Lin, T., Fang, S., Tseng, W., Lee, C., Hsieh, J.: A group-discrimination-based access point selection for WLAN fingerprinting localization. IEEE Trans. Veh. Technol. **63**(8), 3967–3976 (2014)
6. Liu, W., Fu, X., Deng, Z., Xu, L., Jiao, J.: Smallest enclosing circle-based fingerprint clustering and modified-WKNN matching algorithm for indoor positioning. In: International Conference on Indoor Positioning and Indoor Navigation (IPIN), pp. 1–6. IEEE, Alcala de Henares (2016)
7. Gharghan, S.K., Nordin, R., Ismail, M., Ali, J.A.: Accurate wireless sensor localization technique based on hybrid PSO-ANN algorithm for indoor and outdoor track cycling. IEEE Sens. J. **16**(2), 529–541 (2016)
8. Turgut, Z., Ustebay, S., Aydin, M.A., Aydin, Z.G., Sertbaş, A.: Performance analysis of machine learning and deep learning classification methods for indoor localization in Internet of things environment. Trans. Emerg. Telecommun. Technol. **30**(9), 1–18 (2019)
9. Ma, Y.-W., Chen, J.-L., Chang, F.-S., Tang, C.-L.: Novel fingerprinting mechanisms for indoor positioning. Int. J. Commun. Syst. **29**(3), 638–656 (2016)
10. Félix, G., Siller, M., Álvarez, E.N.: A fingerprinting indoor localization algorithm based deep learning. In: Eighth International Conference on Ubiquitous and Future Networks (ICUFN), pp. 1006–1011. IEEE, Vienna (2016)
11. Nowicki, M., Wietrzykowski, J.: Low-effort place recognition with WiFi fingerprints using deep learning. In: Szewczyk, R., Zieliński, C., Kaliczyńska, M. (eds.) ICA 2017. AISC, vol. 550, pp. 575–584. Springer, Cham (2017). https://doi.org/10.1007/978-3-319-54042-9_57
12. Zhang, W., Liu, K., Zhang, W.D., Zhang, Y., Gu, J.: Deep neural networks for wireless localization in indoor and outdoor environments. Neurocomputing **194**, 279–287 (2016)
13. Moreira, A., Nicolau, M.J., Meneses, F., Costa, A.: Wi-Fi fingerprinting in the real world-RTLS@UM at the EvAAL competition. In: International Conference on Indoor Positioning and Indoor Navigation (IPIN), pp. 1–10. IEEE, Banff (2015)
14. Torres-Sospedra, J., Montoliu, R., Martinez-Uso, A., et al.: UJIIndoorLoc: a new multi-building and multi-floor database for WLAN fingerprint-based indoor localization problems. In: International Conference on Indoor Positioning and Indoor Navigation (IPIN), pp. 261–270, IEEE, Busan (2014)

Recurrent Autoencoder
with Sequence-Aware Encoding

Robert Susik$^{(\boxtimes)}$ (iD)

Institute of Applied Computer Science, Łódź University of Technology,
Łódź, Poland
rsusik@kis.p.lodz.pl

Abstract. Recurrent Neural Networks (RNN) received a vast amount of
attention last decade. Recently, the architectures of Recurrent AutoEncoders (RAE) found many applications in practice. RAE can extract the
semantically valuable information, called context that represents a latent
space useful for further processing. Nevertheless, recurrent autoencoders
are hard to train, and the training process takes much time. This paper
proposes a new recurrent autoencoder architecture with sequence-aware
encoding (RAES), and its second variant which employs a 1D Convolutional layer (RAESC) to improve its performance and flexibility. We
discuss the advantages and disadvantages of the solution and prove that
the recurrent autoencoder with sequence-aware encoding outperforms a
standard RAE in terms of model training time in most cases. The extensive experiments performed on a dataset of generated sequences of signals
shows the advantages of RAES(C). The results show that the proposed
solution dominates over the standard RAE, and the training process is
the order of magnitude faster.

Keywords: Recurrent · AutoEncoder · RNN · Sequence

1 Introduction

Recurrent Neural Networks (RNN) [22,29] received a vast amount of attention last decade and found a wide range of applications such as language modelling [18,24], signal processing [8,23] and anomaly detection [19,25].

The RNN is (in short) a neural network adapted to sequential data that have
the ability to map sequences to sequences achieving excellent performance on
time series. The RNN can process the data of variable length or fixed length (in
this case, the computational graph can be unfolded and considered a feedforward
neural network). Multiple layers of RNN can be stacked to process efficiently
long input sequences [12,21]. The training process of deep recurrent neural network (DRNN) is difficult because the gradients (in backpropagation through
time [29]) either vanish or explode [3,9]. It means that despite the RNN can
learn long dependencies the training process may take a very long time or even
fail. The problem was resolved by the application of Long Short-Term Memory

© Springer Nature Switzerland AG 2021
M. Paszynski et al. (Eds.): ICCS 2021, LNCS 12743, pp. 47–57, 2021.
https://doi.org/10.1007/978-3-030-77964-1_4

(LSTM) [13] or much newer and simpler Gated Recurrent Units (GRU) [6]. Nevertheless, it is not easy to parallelize calculations in recurrent neural networks which impacts the training time.

A different and efficient approach was proposed by Aäron et al. [20] who proved that stacked 1D convolutional layers can process efficiently long sequences handling tens of thousands of time steps. The CNNs have also been widely applied to autoencoder architecture to solve problems such as outlier and anomaly detection [1,14,16], noise reduction [5], and more.

Autoencoders [4] are unsupervised algorithms capable of learning latent representations (called *context* or *code*) of the input data. The context is usually smaller than input data to extract only the semantically valuable information. Encoder-Decoder (Sequence-to-Sequence) [7,26] architecture looks very much like autoencoder and consists of two blocks: encoder, and decoder, both containing a couple of RNN layers. The encoder takes the input data and generates the code (a semantic summary) used to represent the input. Later, the decoder processes the code and generates the final output. The encoder-decoder approach allows having variable-length input and output sequences in contrast to classic RNN solutions. Several related attempts, including an interesting approach introduced by Graves [12] have been later successfully applied in practice in [2,17]. The authors proposed a novel differentiable attention mechanism that allows the decoder to focus on appropriate words at each time step. This technique improved the state of the art in neural machine translation (NMT) and was later applied even without any recurrent or convolutional layers [28]. Besides the machine translation, there are multiple variants and applications of the Recurrent AutoEncoders (RAE). In [10], the authors proposed the variational recurrent autoencoder (VRAE), which is a generative model that learns the latent vector representation of the input data used later to generate new data. Another variational autoencoder was introduced in [11,27] where authors apply convolutional layers and WaveNet for audio sequence. Interesting approach, the Feedback Recurrent AutoEncoder (FRAE) was presented in [30]. In short, the idea is to add a connection that provides feedback from decoder to encoder. This design allows efficiently compressing the sequences of speech spectrograms.

This paper presents an autoencoder architecture that applies a different context layout and employs a 1D convolutional layer to improve its flexibility and reduce the training time. We also propose a different interpretation of the context (the final hidden state of the encoder). We transform the context into the sequence that is passed to the decoder. This technical trick, even without changing other elements of architecture, improves the performance of recurrent autoencoder.

We demonstrate the power of the proposed architecture for time series reconstruction (the generated sequences of signal). We perform a wide range of experiments on a dataset of generated signals, and the results are promising.

Following contributions of this work can be enumerated: (i) We propose a recurrent autoencoder with sequence-aware encoding that trains much faster than standard RAE. (ii) We suggest an extension to proposed solution which employs the 1D convolutional layer to make the solution more flexible. (iii) We show that this architecture performs very well on univariate and multivariate time series reconstruction.

2   The Model

In this section, we describe our approach and its variants. We also discuss the advantages and disadvantages of the proposed architecture and suggest possible solutions to its limitation.

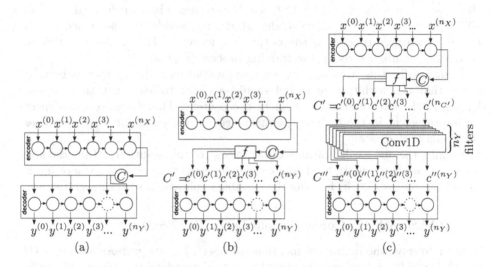

Fig. 1. Recurrent autoencoder architectures: (a) Recurrent AutoEncoder (RAE) [7, 26]; (b) Recurrent AutoEncoder with Sequence-aware encoding (RAES); (c) Recurrent AutoEncoder with Sequence-aware encoding and 1D Convolutional layer (RAESC).

2.1   Recurrent AutoEncoder (RAE)

The recurrent autoencoder generates an output sequence $Y = (y^{(0)}, y^{(1)}, \ldots, y^{(n_Y-1)})$ for a given an input sequence $X = (x^{(0)}, x^{(1)}, \ldots, x^{(n_X-1)})$, where n_Y and n_X are the sizes of output and input sequences respectively (both can be of the same or different size). Usually, $X = Y$ to force autoencoder learning the semantic meaning of data. First, the input sequence is encoded by the RNN encoder, and then the given fixed-size context variable C of size m_C (a single vector containing m_C features) is decoded by the decoder (usually also RNN), see Fig. 1a. If $m_C < n_X$, then the autoencoder is called undercomplete. On the other hand if $m_C > n_X$, then the autoencoder is called overcomplete. The first variant is much more popular as it allows the autoencoder to learn the semantically valuable information.

2.2   Recurrent AutoEncoder with Sequential Context (RAES)

We propose a recurrent autoencoder architecture (Fig. 1b) where the final hidden state of the encoder $C = (c_0, c_1, \ldots, c_{m_C-1})$ is interpreted as the sequence of time steps $C' = (c'^{(0)}, c'^{(1)}, \ldots, c'^{(n_C-1)})$, thus $c'^{(i)} = (c_0'^{(i)}, c_1'^{(i)}, \ldots, c_{m_{C'}-1}'^{(i)})$, where n_C is the number of time steps (equal to n_Y) and $m_{C'}$ is the number of features (called λ later). It is an operation performed on the context and may be defined as $f : C \mapsto C'$. The code $C = (c_i)_{i=0}^{n_C-1}$ is transformed to

$$C' = ((c_{i\lambda+j})_{j=0}^{\lambda-1})_{i=0}^{n_C/\lambda-1} \tag{1}$$

where $\lambda = n_C/n_X$ ($\lambda \in \mathbb{N}$). Once the context is transformed ($C' = (c'^{(0)}, c'^{(1)}, \ldots, c'^{(n_X-1)})$), the decoder starts to decode the sequence C' of $m_{C'} = \lambda$ features producing the output sequence Y. This technical trick in the data structure speeds up the training process (Sect. 3).

Additionally, this way, we put some sequential meaning to the context. It means that this architecture model should attempt to learn the time dependencies in the data and store them in the context. Therefore, we can expect the improvement in the training performance, in particular for long sequences (hundreds of elements).

Finally, the one easily solvable disadvantage of this solution is that the size of context must be multiple of input sequence length $n_C = \lambda n_X$, where n_C is the size of context C, which limits the possible applications of such architecture.

2.3   RAES with 1D Convolutional Layer (RAESC)

In order to solve the limitation mentioned in Sect. 2.2, we propose adding a 1D convolutional layer (and max-pooling layer) to the architecture right before the decoder (Fig. 1c). This approach gives the ability to control the number of output channels (also denoted as *feature detectors* or *filters*), defined as follows:

$$C''(i) = \sum_k \sum_l C'(i+k, l)w(k, l) \tag{2}$$

In this case, n_C does not have to be multiple of n_X, thus to have the desired output sequence of n_Y length, the number of filters should be equal to n_Y. Moreover, the output of the 1D convolution layer $C'' = \mathrm{conv1D}(C')$ should be transposed. Hence each channel becomes an element of the sequence as shown in Fig. 1c. Finally, the desired number of features on output Y can be configured with hidden state size of the decoder.

A different and simpler approach to solve the mentioned limitation is stretching the context C to the size of decoder input sequence and filling the gaps in with averages.

The described variant is very simplified and is only an outline of proposed recurrent autoencoder architecture (the middle part of it, to be more precise) which can be extended by adding pooling and recurrent layers or using different convolution parameters (such as stride, or dilation values). Furthermore, in our view, this approach could be easily applied to other RAE architectures (such as [11,30]).

The recurrent neural network gradually forgets some information at each time step, and may completely ignore the time dependencies between the beginning and end of the sequence. It is even worse in the recurrent autoencoder where the context size is usually limited. We believe that the proposed layout of the context enforces the model to store the time dependencies in the context, thus it works well on long sequences. Additionally, this characteristic may be very interesting in signal classification tasks where such memory may be crucial to identify an object.

3   Experiments

In order to evaluate the proposed approach, we run a few experiments, using a generated dataset of signals. We tested the following algorithms:

- Standard Recurrent AutoEncoder (RAE) [7,26].
- RAE with Sequence-aware encoding (RAES).
- RAES with Convolutional and max-pooling layer (RAESC).

The structure of decoder and encoder is the same in all algorithms. Both decoder and encoder are single GRU [6] layer, with additional time distributed fully connected layer in the output of the decoder. The algorithms were implemented in Python 3.7.4 with TensorFlow 2.3.0. The experiments were run on a GPU server with an AMD Epyc 7702P CPU (64 cores, 128 threads) clocked at 2.0 GHz with 4 MiB L1, 32 MiB L2 and 256 MiB L3 cache and 6x Quadro RTX 6000 graphic cards with 24220MiB VRAM each (only one was used). The test machine was equipped with 504 GB of RAM and running Ubuntu 18.04.4 64-bit OS. We trained the models with Adam optimizer [15] in batches of size 100 and Mean Squared Error (MSE) loss function. All the presented algorithms were implemented and the source codes (including all the datasets mentioned above) can be found at the following URL: https://github.com/rsusik/raesc. The dataset contains generated time series (sum of sine waves) that consists of 5000 sequences of length 200 with {1, 2, 4, 8} features and is published in the same repository along with source codes. The dataset was shuffled and split to training and validation sets in proportions of 80:20, respectively.

In the first set of analyses, we investigated the impact of context size and the number of features on performance. We noticed a considerable difference in training speed (number of epochs needed to achieve plateau) between the classic approach and ours. To prove whether our approach has an advantage over the RAE, we performed tests with different size of the context n_C and a different number of input features m_X. We set context size (n_C) proportionally to the size of the input and we denote it as:

$$\sigma = \frac{n_C}{m_X n_X} \tag{3}$$

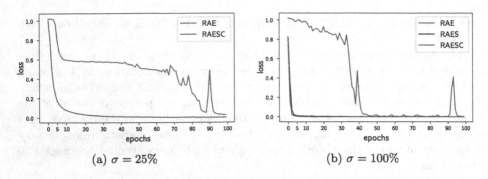

(a) $\sigma = 25\%$　　　　　　　　　　　(b) $\sigma = 100\%$

Fig. 2. Loss as function of epoch number for univariate data and $\sigma = \{25\%, 100\%\}$.

Figure 2 proves that the training process of the RAE needs much more epochs than RAESC to achieve a similar value of loss function. In chart a) the size of context is set to $\sigma = 25\%$ and in b) it is set to $\sigma = 100\%$ of the input size. For $\sigma = 25\%$ the RASEC achieves plateau after 20 epochs while the RAE does not at all (it starts decreasing after nearly 80 epochs, but behaves unstable). There is no RAES result presented in this plot because of the limitation mentioned in Sect. 2.2 (size of the code was too small to fit the output sequence length). For $\sigma = 100\%$ both RASEC and RAES achieve the plateau in less than five epochs (order of magnitude faster) while the RAE after about 35 epochs.

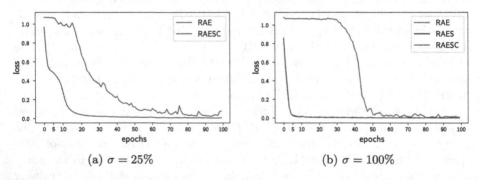

(a) $\sigma = 25\%$　　　　　　　　　　　(b) $\sigma = 100\%$

Fig. 3. Loss as function of epoch number for two features ($m_X = 2$) and $\sigma = \{25\%, 100\%\}$.

Figure 3 shows the loss in function of the number of epochs for two features in input data. This experiment confirms that both RAES and RAESC dominates in terms of training speed, but a slight difference can be noticed in comparison to univariate data (Fig. 2). It shows that the RAE achieves plateau in about 50 epochs for both cases while RAES and RAESC after 20 epochs for $\sigma = 25\%$ and in about five epochs for $\sigma = 100\%$.

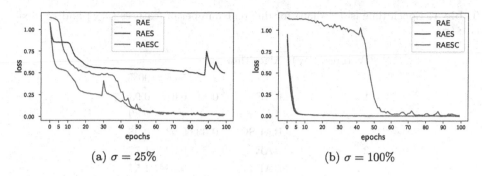

Fig. 4. Loss as function of epoch number for four features ($m_X = 4$) and $\sigma = \{25\%, 100\%\}$.

Figure 4 presents a loss in function of the number of epochs for four features. Comparing Fig. 4a to previous ones (Fig. 3a and Fig. 2a) we can clearly see a downward trend (for growing number of features) in all architectures' performance, but the slope for proposed ones look steeper than for RAE. On the other hand, it can not be observed for $\sigma = 100\%$ (Fig. 4b) because in this case, the difference is marginal.

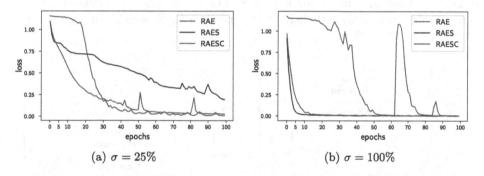

Fig. 5. Loss as function of epoch number for $m_X = 8$ and $\sigma = \{25\%, 100\%\}$.

Figure 5 shows a loss in function of the number of epochs for eight features. This figure is interesting in several ways comparing to the previous ones (Figs. 2, 3, 4). The chart a) shows that, for a much larger number of features and relatively small size of the context, the training of RAES variant takes much more epochs. The similar observation may be noticed for RAESC, where the loss drops much faster than the RAE at the beginning of the training but achieves the plateau at almost the same step. On the other hand, chart b) shows that for a larger size of context, the proposed solution dominates.

We measured each algorithm's training time to confirm that the proposed solution converges faster than RAE for the same size of context. Table 1 shows

Table 1. Epoch time [s] (median) for different number of features (m_X) and context size (σ).

Features (m_X)	Algorithm	σ		
		25%	50%	100%
1	RAE	0.61	0.61	0.93
	RAES	–	–	0.89
	RAESC	0.63	0.63	0.98
2	RAE	0.64	0.92	1.77
	RAES	–	0.88	1.57
	RAESC	0.65	0.96	1.85
4	RAE	0.91	1.74	4.65
	RAES	0.89	1.57	3.81
	RAESC	0.97	1.85	4.75
8	RAE	1.75	4.63	13.22
	RAES	1.56	3.80	10.07
	RAESC	1.85	4.74	13.47

the median of epoch time for a different number of features and context size. The table confirms that the RAES is faster than RAE by about 5% for univariate data and about 31% faster for $m_X = 8$. The training process (the epoch) of RAESC algorithm takes a slightly more time than RAE, which is marginal (less than 2%) for $m_X = 8$.

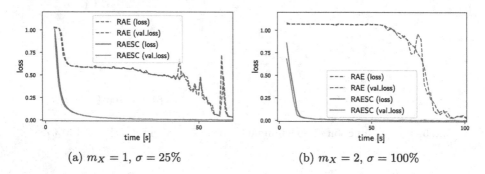

(a) $m_X = 1$, $\sigma = 25\%$ (b) $m_X = 2$, $\sigma = 100\%$

Fig. 6. Loss and validation loss as function of time [s] for $m_X = \{1, 2\}$ and $\sigma = \{25\%, 100\%\}$ respectively.

To confirm that the presented architectures do not tend to overfit, we compared a loss and validation loss functions. Figure 6 illustrates the loss and validation loss of RAE and RAESC (to make the chart more readable, RAES is excluded) in the function of time (in seconds). We can clearly see on both charts (Fig. 6a and b) that the validation loss goes approximately along with loss function (except for a few bounces). We observed similar behaviour for all the experiments performed.

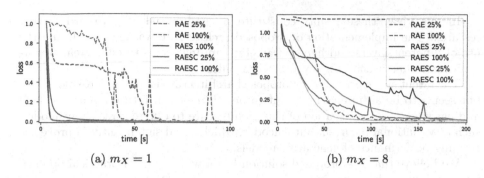

Fig. 7. Loss as function of time [s] for $m_X = \{1, 8\}$ and $\sigma = \{25\%, 100\%\}$.

To make it transparent and clear, we compared all the algorithms, including the training time for different σ. Figure 7 presents a loss in a function of time (in seconds) for $m_X = \{1, 8\}$ features. As expected, we can clearly see that proposed architecture dominates for univariate data disqualifying RAE (Fig. 7a). Interestingly, the RAE with larger context size ($\sigma = 100\%$) converges faster than the one with smaller context size ($\sigma = 25\%$). The fastest are RAES and RAESC ($\sigma = 100\%$ both), achieving almost the same results (the loss functions of both overlap on the chart). For the multivariate data (in Fig. 7b) on the contrary to univariate we can see that the RAE with smaller context size converges much faster than with a larger one. The most striking fact to emerge from these results is that the RAE 100% does not drop in the whole period. The training of RAE 25% is slower at the begining than proposed architecture but speeds up after 40 s achieving very similar result after 100 s.

In most of the charts presented it can be noticed that the training process of RAE fluctuates significantly on the contrary to the proposed solution where it is relatively stable. It is also worth mentioning that all the experiments were performed with a fixed filter size for both convolutional and max-pooling layers, and it is likely that we could achieve better results by tuning these hyperparameters.

4   Conclusions and future work

In this work, we proposed an autoencoder with sequence-aware encoding. We proved that this solution outperforms the RAE in terms of training speed in most cases.

The experiments confirmed that the training of proposed architecture takes less time than the standard RAE. It is a critical factor if the training time is limited, for example, in Automated Machine Learning (AutoML) tools or in a hyperparameter optimization. The context size and the number of features in the input sequence have a high impact on training performance. Only for a relatively large number of features and small size of the context the RAE achieves comparable results to the proposed solution. In other cases our solution dominates and the training time is an order of magnitude shorter.

In our view, these results constitute a good initial step toward further research. The implementation of proposed architecture was simplified, and the use of different layers and hyperparameter tunning seems to offer great opportunities to tune it achieving even better results.

The latent space produced by proposed architecture is still underexplored yet but seems to be an interesting point to be addressed in future research. These findings suggest an application of this architecture to different recurrent models such as variational recurrent autoencoder, which could significantly improve the training performance of generative models.

We believe that the proposed solution has a wide range of practical applications and is worth confirming.

References

1. An, J., Cho, S.: Variational autoencoder based anomaly detection using reconstruction probability. Spec. Lect. IE **2**(1), 1–18 (2015)
2. Bahdanau, D., Cho, K., Bengio, Y.: Neural machine translation by jointly learning to align and translate. arXiv preprint arXiv:1409.0473 (2014)
3. Bengio, Y., Simard, P., Frasconi, P.: Learning long-term dependencies with gradient descent is difficult. IEEE Trans. Neural Netw. **5**(2), 157–166 (1994)
4. Bourlard, H., Kamp, Y.: Auto-association by multilayer perceptrons and singular value decomposition. Biol. Cybern. **59**(4–5), 291–294 (1988). https://doi.org/10.1007/BF00332918
5. Chiang, H.T., Hsieh, Y.Y., Fu, S.W., Hung, K.H., Tsao, Y., Chien, S.Y.: Noise reduction in ECG signals using fully convolutional denoising autoencoders. IEEE Access **7**, 60806–60813 (2019)
6. Cho, K., van Merrienboer, B., Bahdanau, D., Bengio, Y.: On the properties of neural machine translation: encoder-decoder approaches. CoRR abs/1409.1259 (2014). http://arxiv.org/abs/1409.1259
7. Cho, K., et al.: Learning phrase representations using RNN encoder-decoder for statistical machine translation. arXiv preprint arXiv:1406.1078 (2014)
8. Ding, J., Wang, Y.: WiFi CSI-based human activity recognition using deep recurrent neural network. IEEE Access **7**, 174257–174269 (2019)
9. Doya, K.: Bifurcations of recurrent neural networks in gradient descent learning. IEEE Trans. Neural Netw. **1**(75), 218 (1993)
10. Fabius, O., van Amersfoort, J.R.: Variational recurrent auto-encoders. arXiv preprint arXiv:1412.6581 (2014)
11. Gârbacea, C., et al.: Low bit-rate speech coding with VQ-VAE and a wavenet decoder. In: ICASSP 2019–2019 IEEE International Conference on Acoustics, Speech and Signal Processing (ICASSP), pp. 735–739. IEEE (2019)
12. Graves, A.: Generating sequences with recurrent neural networks. arXiv preprint arXiv:1308.0850 (2013)
13. Hochreiter, S., Schmidhuber, J.: Long short-term memory. Neural Comput. **9**(8), 1735–1780 (1997)
14. Kieu, T., Yang, B., Guo, C., Jensen, C.S.: Outlier detection for time series with recurrent autoencoder ensembles. In: IJCAI, pp. 2725–2732 (2019)
15. Kingma, D.P., Ba, J.: Adam: a method for stochastic optimization. arXiv preprint arXiv:1412.6980 (2014)

16. Liao, W., Guo, Y., Chen, X., Li, P.: A unified unsupervised gaussian mixture variational autoencoder for high dimensional outlier detection. In: 2018 IEEE International Conference on Big Data (Big Data), pp. 1208–1217. IEEE (2018)
17. Luong, M.T., Pham, H., Manning, C.D.: Effective approaches to attention-based neural machine translation. arXiv preprint arXiv:1508.04025 (2015)
18. Mikolov, T., Kombrink, S., Burget, L., Černocký, J., Khudanpur, S.: Extensions of recurrent neural network language model. In: 2011 IEEE International Conference on Acoustics, Speech and Signal Processing (ICASSP), pp. 5528–5531. IEEE (2011)
19. Nanduri, A., Sherry, L.: Anomaly detection in aircraft data using recurrent neural networks (RNN). In: 2016 Integrated Communications Navigation and Surveillance (ICNS), pp. 5C2-1. IEEE (2016)
20. van den Oord, A., et al.: WaveNet: a generative model for raw audio (2016). https://arxiv.org/abs/1609.03499
21. Pascanu, R., Gulcehre, C., Cho, K., Bengio, Y.: How to construct deep recurrent neural networks. In: Proceedings of the Second International Conference on Learning Representations (ICLR 2014) (2014)
22. Rumelhart, D.E., Hinton, G.E., Williams, R.J.: Learning representations by back-propagating errors. Nature **323**(6088), 533–536 (1986)
23. Shahtalebi, S., Atashzar, S.F., Patel, R.V., Mohammadi, A.: Training of deep bidirectional RNNs for hand motion filtering via multimodal data fusion. In: GlobalSIP, pp. 1–5 (2019)
24. Shi, Y., Hwang, M.Y., Lei, X., Sheng, H.: Knowledge distillation for recurrent neural network language modeling with trust regularization. In: ICASSP 2019–2019 IEEE International Conference on Acoustics, Speech and Signal Processing (ICASSP), pp. 7230–7234. IEEE (2019)
25. Su, Y., Zhao, Y., Niu, C., Liu, R., Sun, W., Pei, D.: Robust anomaly detection for multivariate time series through stochastic recurrent neural network. In: Proceedings of the 25th ACM SIGKDD International Conference on Knowledge Discovery & Data Mining, pp. 2828–2837 (2019)
26. Sutskever, I., Vinyals, O., Le, Q.V.: Sequence to sequence learning with neural networks. In: Advances in Neural Information Processing Systems, pp. 3104–3112 (2014)
27. Van Den Oord, A., Vinyals, O., et al.: Neural discrete representation learning. In: Advances in Neural Information Processing Systems, pp. 6306–6315 (2017)
28. Vaswani, A., et al.: Attention is all you need. In: Advances in Neural Information Processing Systems, pp. 5998–6008 (2017)
29. Werbos, P.J.: Backpropagation through time: what it does and how to do it. Proc. IEEE **78**(10), 1550–1560 (1990)
30. Yang, Y., Sautière, G., Ryu, J.J., Cohen, T.S.: Feedback recurrent autoencoder. In: ICASSP 2020, 2020 IEEE International Conference on Acoustics, Speech and Signal Processing (ICASSP), pp. 3347–3351. IEEE (2020)

A Gist Information Guided Neural Network for Abstractive Summarization

Yawei Kong[1,2], Lu Zhang[1,2], and Can Ma[1(✉)]

[1] Institute of Information Engineering, Chinese Academy of Sciences, Beijing, China
{kongyawei,zhanglu0101,macan}@iie.ac.cn
[2] University of Chinese Academy of Sciences, Beijing, China

Abstract. Abstractive summarization aims to condense the given documents and generate fluent summaries with important information. It is challenging for selecting the salient information and maintaining the semantic consistency between documents and summaries. To tackle these problems, we propose a novel framework - Gist Information Guided Neural Network (GIGN), which is inspired by the process that people usually summarize a document around the gist information. First, we incorporate multi-head attention mechanism with the self-adjust query to extract the global gist of the input document, which is equivalent to a question vector questions the model "What is the document gist?". Through the interaction of the query and the input representations, the gist contains all salient semantics. Second, we propose the remaining gist guided module to dynamically guide the generation process, which can effectively reduce the redundancy by attending to different contents of gist. Finally, we introduce the gist consistency loss to improve the consistency between inputs and outputs. We conduct experiments on the benchmark dataset - CNN/Daily Mail to validate the effectiveness of our methods. The results indicate that our GIGN significantly outperforms all baseline models and achieves the state-of-the-art.

Keywords: Abstractive summarization · Multi-Head Attention · Global gist · Consistency loss

1 Introduction

Recently, document summarization has attracted growing research interest for its promising commercial values. It aims to produce fluent and coherent summaries with the original documents. Existing approaches for building a document summarization system can be categorized into two groups: extractive and abstractive methods. The extractive methods focus on extracting sentences from the original document, which can produce more fluent sentences and preserve the meaning of the original documents but tend to information redundancy and incoherence between sentences. In contrast, the abstractive methods effectively

---

Y. Kong and L. Zhang—Equal contribution.

© Springer Nature Switzerland AG 2021
M. Paszynski et al. (Eds.): ICCS 2021, LNCS 12743, pp. 58–71, 2021.
https://doi.org/10.1007/978-3-030-77964-1_5

avoid these problems by utilizing arbitrary words and expressions that are more consistent with the way of humans. However, the abstractive methods are much more challenging due to the sophisticated semantic organization.

Encouraged by the success of recurrent neural network (RNN) in NLP, most typical approaches [4, 14, 18, 20] employ the sequence-to-sequence (seq2seq) framework to model the document summarization system that consists of an encoder and a decoder. Given the input documents, the encoder first encodes them into semantic vectors and then the decoder utilizes these vectors to generate summaries in the decoding step. Although the popular methods are improved from various perspectives, such as introducing reinforcement learning [3, 16] or incorporating topic information [18, 23], they still fail to achieve convincing performance for ignoring the global gist information.

Intuitively, humans tend to generate a document summary around the gist information. Different from the topic that just focuses on the talking point of the original documents, the gist information represents the essence of text that contains more wealth of information. Thus, the gist is more suitable for the document summarization task and can guide the model how to generate relevant, diverse, and fluent summaries.

Towards filling this gap, we propose an effective Gist Information Guided Neural Network (GIGN) for abstractive summarization. To distill the gist information from the semantic representation of the original documents, we first introduce a Gist Summary Module (GSM) that consists of the multi-head attention mechanism with self-adjust query. The query effectively questions the model "What is the document gist?". Through the interaction between this query vector and the hidden state of each token in the document, we obtain a global representation of the gist, which contains several pieces of salient information. Obviously, the gist not only plays a global guidance role but also is required to have the capability of attending to different contents of gist during the process of decoding. Thus, we propose a Remaining Gist Information Guided Module (RGIGM). The remaining gist information is calculated by the global gist and the salient information of the generated summary, which dynamically guides the decoder to generate tokens at each step. And this mechanism can effectively reduce redundant information and makes the gist contents express completely. Furthermore, we also propose a gist consistency loss to guarantee that the main information of the generated summary is consistent with the input document. Finally, we introduce policy gradient reinforcement learning [16] to reduce the exposure bias problem.

We conduct experience on the benchmark CNN/Daily Mail dataset to validate the effectiveness of GIGN. The experimental results indicate that our GIGN significantly outperforms all baselines. In summary, this paper makes the following contributions:

- To the best of our knowledge, this is the first work to introduce the gist information for abstractive summarization.
- We introduce a gist summary module, which contains several pieces of salient information. And then, a remaining gist information guided mod-

ule is employed to attending to different contents of gist. They dynamically incorporate the gist information into the generation process, which effectively improves the model performance.
- We further propose a novel gist consistency loss to ensure the generated summary is coherent with the inputs. And the reinforced learning can further improve performance by reducing the exposure bias problem.
- The empirical results demonstrate that our approach outperforms all baselines in both automatic metrics and human judgments. Further analysis shows that our method can generate more salient and relevant summaries.

2   Related Work

In recent years, abstractive summarization [4,14,19,20,25] has received increasing attention for its promising commercial values. Different from extractive methods that directly selects salient sentences from the original documents, abstractive summarization aims to generate summaries word-by-word from the final vocabulary distribution, which is more consistent with the way of human beings.

The abstractive methods are more challenging for the following conspicuous problems: Out-of-vocabulary (OOV), repetition, and saliency. Therefore, some previous works [14,16,20] pay attention to tackle the OOV problem by introducing the pointer network. To eliminate repetitions, See et al. [20] propose a coverage mechanism that is a variant of the coverage vector from Machine Translation. However, the most difficult and concerning problem is how to improve the saliency.

To tackle this problem, some studies attempt to introduce the template discovered from the training dataset to guide the summary generation.For example, Cao et al. [2] employ the IR platform to retrieve proper summaries as candidate templates and then jointly conduct template reranking as well as template-aware summary generation. Wang et al. [22] propose a novel bi-directional selection mechanism with two gates to extract salient information from the source document and executes a multi-stage process to extract the high-quality template from the training corpus. Moreover, You et al. [24] extend the basic encoder-decoder framework with an information selection layer, which can explicitly model and optimize the information selection process. However, they are difficult to optimize for retrieving template first. And the wrong template will introduce noise to the model, which significantly hurt the generation performance.

Different from the works that incorporate templates, many researchers improve saliency by introducing topics. For example, Krishna et al. [11] take an article along with a topic of interest as input and generates a summary tuned to the target topic of interest. Li et al. [23] improve the coherence, diversity, and informativeness of generated summaries through jointly attending to topics and word-level alignment. To incorporate the important information, Li et al. [12] utilize keywords extracted by the extractive model to guide the generation process. Furthermore, Perez-Beltrachini et al. [18] train a Latent Dirichlet Allocation model [1] to obtain sentence-level topic distributions. Although these methods

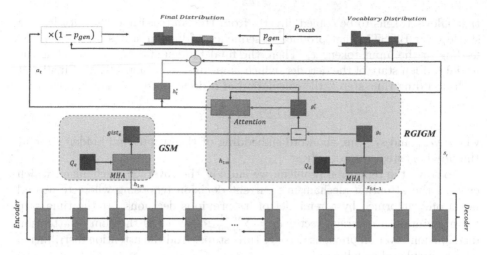

Fig. 1. The architecture of GIGN. It extends the pointer generator network with a gist summary module (GSM) and a remaining gist information guided Module (RGIGM).

have made some progress, they still fail to ignore the fact that the decoder should focus on different contents at different time steps and the information contains in topics is not enough to support the entire abstract.

In this paper, we propose a new framework, namely GIGN, to tackle these problems. We first introduce a gist summary module to obtain a global representation of the gist that contains several pieces of salient information. Then, the remaining gist information guided module is employed to attend to different contents during the decoding process.

3   Model

In this section, we introduce our Gist Information Guided Neural Network (GIGN) in detail. Given the source document $X = (x_1, x_2, \cdots, x_n)$, our model aims to generate the corresponding summary $Y = (y_1, y_2, \cdots, y_m)$. As shown in Fig. 1, our GIGN mainly includes the Gist Summary Module (GSM) and the Remaining Gist Information Guided Module (RGIGM). We briefly describe the pointer generator network in Sect. 3.1 firstly. Then, Sect. 3.2 introduces our gist summary module. And the remaining gist information guided module is described in Sect. 3.3. Finally, we introduce our training objective in the Sect. 3.4, which includes the gist consistency loss and the reinforcement learning objective. Notably, all W and b are learnable parameters.

3.1   Pointer Generator Network

The Pointer-Generator Network (PGN) aims to solve the out-of-vocabulary (OOV) problem, which extends ses2seq networks by adding a copy mechanism that allows tokens to be copied directly from the source. First, the encoder is a single layer BiLSTM that produces a sequence of hidden states $h_i, i \in [1, n]$ by feeding in the input tokens x_i. Then, the final hidden state h_n is served as the initial hidden state of the decoder, which is an un-directional LSTM. Finally, at each decoding time step t, the calculation of hidden state s_t is formulated as:

$$s_t = BiLSTM(s_{t-1}, y_{t-1}) \tag{1}$$

where y_{t-1} and s_{t-1} are the word embedding of the output and hidden state at the previous step, respectively.

To solve the repetition problem, we employ the coverage mechanism, which ensures the attention mechanism's current decision (choosing where to attend next) only informed by a reminder of its previous decisions. At the time step t, we maintain a coverage vector $c_t = \sum_{j=0}^{t-1} a^j$, which is the sum of attention distributions over all previous decoder time steps. And the attention distribution a_t are calculated as follows:

$$a_i^t = softmax(V^T tanh(W_h h_i + W_s s_t + W_c c_i^t + b_{attn})) \tag{2}$$

Based on the coverage vector and the previous hidden state, the vocabulary distribution of the next token is computed as follows:

$$P_{vocab} = softmax(W_{v'}(W_v[s_t; c_t] + b_v) + b_{v'}) \tag{3}$$

To solve the OOV problem, the pointer mechanism aims to copy rare or unknown words from the original document via pointing. Thus, the model first computes a generation probability:

$$p_{gen} = \sigma(W_{h^*}^T h_t^* + W_s^T s_t + W_x^T x_t + b_{ptr}) \tag{4}$$

where σ is the sigmoid function and $h_t^* = \sum_{j=1}^{n} a_j^t h_j$ is the context vector. Moreover, p_{gen} acts as a soft switch to decide whether to generate a word from the vocabulary by sampling from P_{vocab} or copy a word from the input sequence by sampling from the attention distribution a^t. Therefore, the final probability distribution is computed as follows:

$$P(w) = p_{gen} P_{vocab}(w) + (1 - p_{gen}) \sum_{i:w_i=w} a_i^t \tag{5}$$

3.2   Gist Summary Module

Inspired by the interaction between Question and Passage in Machine Reading Comprehension task [9], we introduce a Gist Summary Module (GSM) to obtain the gist of input document. As illustrated in Fig. 1, the GSM consists of a self-adjust query Q^e and a Multi-Head Attention (MHA) mechanism [21]. In our model, the self -adjust query Q^e is fed as the question, while the representation of source document $H = [h_1, \cdots, h_n]$ serves as the passage. The query vector seems to question the model "What is the document gist?". From the interaction between Q^e and H, we can get a question-aware global representation $gist_e$ that attends to several pieces of salient information for the document. The gist vector can be formulated as follows:

$$gist_e = MHA(Q^e, H)$$
$$MHA = [head_1; head_2 \ldots; head_k]W^O \tag{6}$$
$$head_j = softmax(\frac{Q^e W_j^Q (HW_j^K)^T}{\sqrt{d_k}})HW_j^V$$

where T represents transpose function and k is the number of heads. Notably, the Q^e is a learnable vector.

3.3 Remaining Gist Information Guided Module

The gist of a document contains several pieces of salient information, which needs to be distributed in different parts of the generated abstract. Therefore, we introduce a Remaining Gist Information Guide Module(RGIGM), which obtains the remaining gist information by calculating the difference between the generated semantics and the global gist. In this way, the decoder can attend to the most needed information of the gist for the current timestamp. As the continuation of the generation process, the information contained in the gist is constantly expressed, which can prevent semantic duplication and improve the conciseness of the summary. At time step t, we utilize the Gist Summary Module to obtain the generated information based on the previous hidden states $s_{1:t-1}$:

$$gist_t = MHA(Q^d, s_{1:t-1}) \tag{7}$$

where Q^d is the self-adjust question vector. Intuitively, $gist_t$ represents the expressed gist information contained in the generated sequence $y_{1:t-1}$. Then, we obtain the current remaining gist information $gist_t^r$ by subtracting $gist_t$ from the global $gist_e$:

$$gist_t^r = gist_e - gist_t \tag{8}$$

Furthermore, we utilize the remaining gist information $gist_t^r$ to guide the whole process of sequence generation. First, we propose the gist-aware attention distribution, which incorporates the context vector into the generation process. Then, Eq. 2 is modified as follows:

$$a_i^t = softmax(V^T tanh(W_h h_i + W_{sg}[s_t; gist_t^r] + W_c c_i^t + b_{attn})) \tag{9}$$

In this way, this distribution is not only related to the current hidden state s_t but also affected by the remaining gist information. Then, we also apply the remaining gist to the calculation of vocabulary distribution. And Eq. 3 can be modified as follows:

$$P_{vocab} = softmax(W_{v'}(W_v[s_t; gist_t^r; c_t] + b_v) + b_{v'}) \tag{10}$$

We further introduce the $gist_t^r$ into the pointer mechanism, which enables the pointer to identify the words that are relevant to the remaining salient information. And Eq. 4 is modified as follows:

$$p_{gen} = \sigma(W_{h*}^T h_t^* + W_{sg}^T[s_t; gist_t^r] + W_x^T x_t + b_{ptr}) \tag{11}$$

Finally, the whole RGIGM mechanism allows the decoder can generate unexpressed gist semantic information, which not only effectively prevents semantic repetition but also significantly enhances the saliency of the generated abstracts.

3.4   Model Training Objective

To train the model, we use a mixed training objective that jointly optimizes four loss functions, including the negative log-likelihood loss, the coverage loss, the gist consistency loss and the reinforcement learning loss.

Negative Log-Likelihood (NLL) Loss. The model is first pre-trained to optimize NLL loss, which is widely used in sequence generation tasks. We define (X, Y) is a document-summary pair in training set. The function is formulated as follows:

$$\mathcal{L}_{NLL} = -\sum_{t=1}^{m} \log p(y_t|y_1...y_{t-1}, X; \theta) \tag{12}$$

Coverage Loss. We utilize the coverage loss to alleviate the repetition problem, which aims to penalize the attention mechanism to focus on the same locations frequently. The formula can be described as follows:

$$\mathcal{L}_{Coverage} = \sum_i \min(a_i^t, c_i^t) \tag{13}$$

Gist Consistency Loss. To further ensure the consistency of the original document and the generated abstracts, we propose a gist consistency loss, which maximizes the similarity between the gist of source document and the salient information of generated results. At the time step t, we get the current salient semantics $gist_t$ for the generated tokens $y_1, \cdots, y_{t-1}$. Thus, we obtain the salient information of the entire generated summary $gist_m$ when decoding to the last word:

$$\mathcal{L}_{GCL} = cos(gist_m, gist_e) \tag{14}$$

where $cos(\cdot)$ represents the cosine similarity. By maximizing the similarity between the global gist and the generated gist, the salient information of documents can be expressed completely.

Reinforcement Learning (RL) Loss. In order to improve the naturalness of the generated sequence and alleviate the exposure bias problem, we utilize reinforcement learning [16] to directly optimize the ROUGE evaluation metric [13] of the discrete target, which is non-differentiable. For each training example X, two output sequences are generated: $\hat{y}_t$ is sampled from the probability distribution $p(\hat{y}_t|\hat{y}_1\cdots\hat{y}_{t-1}, X; \theta)$ at each time step and $\tilde{y}_t$ is the baseline output that is greedily generated by decoding from $p(\tilde{y}_t|\tilde{y}_1\cdots\tilde{y}_{t-1}, X; \theta)$. The training objective can be formulated as follows:

$$\mathcal{L}_{RL} = (r(\tilde{y}) - r(\hat{y}))\sum_{t=1}^{M} \log p(\hat{y}_t|\hat{y}_1...\hat{y}_{t-1}, X; \theta) \tag{15}$$

where $r(\cdot)$ denotes the reward score calculated by ROUGE-L. Intuitively, minimizing $\mathcal{L}_{RL}$ is equivalent to maximize the conditional likelihood of the sampled sequence $\hat{y}_t$ if it obtains a higher reward than the baseline $\tilde{y}_t$, thus increasing the reward expectation of our model.

Mixed Loss. In the training process, we combine all loss functions described above. The composite training objective is as follows:

$$\mathcal{L}_{MIXED} = (1 - \gamma)(\mathcal{L}_{NLL} + \lambda_1\mathcal{L}_{Coverage} + \lambda_2\mathcal{L}_{GCL}) + \gamma\mathcal{L}_{RL} \tag{16}$$

where λ_1, λ_2 and γ are tunable hyper parameters.

4 Experiments

4.1 Dataset

We perform experiments on the large-scale dataset CNN/Daily Mail [9], which is widely used in abstractive document summarization with multi-sentence summaries. For the data prepossessing, we utilize the scripts provided by [20] to obtain the non-anonymized dataset version[1], which contains 287,226 training pairs, 12,368 validation pairs and 11,490 test pairs. In addition, the average number of sentences in documents and summaries are 42.1 and 3.8, respectively.

4.2 Baselines

To validate the correctness and effectiveness of our model, we choose the following representative and competitive frameworks for comparison. **PGN+Cov** [20] proposes a novel architecture that extends the standard seq2seq attention model with pointer mechanism and coverage loss. **ML+RL** [6] allows users to define attributes of generated summaries and applies the copy mechanism for source entities. **Fast-Abs** [4] selects salient sentences and then rewrites them abstractively (i.e., compresses and paraphrases) to generate a concise summary. **DCA** [3] divides the hard task of encoding a long text across multiple collaborating encoder agents. **GPG** [8] promotes the standard attention model from both local and global aspects to reproduce most salient information and avoid repetitions. **Bottom-Up** [7] equips the seq2seq model with a data-efficient bottom-up

---

[1] https://github.com/abisee/cnn-dailymail.

Table 1. Automatic evaluation of our proposed model against recently released summarization systems on CNN/DailyMail dataset. The best performance is highlighted in bold and the results of all baselines are taken from the corresponding papers.

Models	ROUGE-1	ROUGE-2	ROUGE-L
PGN + Cov	39.53	17.28	36.38
ML + RL	39.87	15.82	36.90
Fast-Abs	40.88	17.80	38.54
GPG	40.95	18.05	37.19
ROUGESal + Ent	40.43	18.00	37.10
Bottom-Up	41.22	18.68	38.34
DCA	41.11	18.21	36.03
Our model (GIGN)	**42.04**	**19.08**	**39.15**

content selector. **ROUGESal+Ent** [15] utilizes the reinforcement learning approach with two novel reward functions: ROUGESal and Entail. Notably, due to the limited computational resource, we don't apply the pre-trained contextualized encoder (i.e. BERT [5]) to our model. Thus, we only compare with the models without BERT for the sake of fairness.

4.3   Hyper-parameters Settings

For a fair comparison, we limit the vocabulary size to 50k and initialize the tokens with 128-dimensional Glove embeddings [17]. The dimensions of hidden units are all set to 256 same as [20]. And the number of heads in attention mechanism is 8. During training, we set the batch size to 16 and optimize the model with Adam [10] method that the initial learning rate is 0.1. At test time, we utilize the beam search algorithm to generate summaries and the beam size is set to 5. Moreover, trigram avoidance [16] is used to avoid trigram-level repetition as previous methods. We implement our model on a Tesla V100 GPU.

4.4   Evaluation Metrics

To evaluate our model comprehensively, we adopt both automatic metrics and human judgments in our experiments. For automatic metrics, We evaluate our models with the standard ROUGE metric, measuring the unigram, bigram and longest common subsequence overlap between the generated and reference summaries as ROUGE-1, ROUGE-2 and ROUGE-L, respectively.

Moreover, human judgments can further evaluate the quality of the generated summaries accurately, which has been widely applied in previous works. We invite six volunteers (all CS majored students) as human annotators. For the fair comparison, given 100 randomly sampled source-target pairs from the CNN/Daily Mail test dataset, volunteers are required to score the results of all models from 1 to 5 based on the following indicators: *Relevance* (**C1**) represents

Table 2. Ablation study. The token "+" indicates that we add the corresponding module to the model.

Models	R-1	R-2	R-L
PGN + Cov	39.60	17.47	36.31
+ GSM	40.17	17.86	36.92
+ RGIGM	41.29	18.37	37.91
+ Consistency loss	41.72	**19.14**	38.78
+ RL	**42.04**	19.08	**39.15**

Table 3. The human evaluation results on Relevance (C1), Non-Redundancy (C2) and Readability (C3).

Models	C1	C2	C3
Reference	5.00	5.00	5.00
PGN + Cov	3.78	3.85	4.02
Fast-Abs	3.74	3.48	3.72
Our model (GIGN)	**4.18**	**4.32**	**4.34**

the correlation between the generated summaries and the ground truth. *Non-Redundancy* (**C2**) measures the diversity and informativeness of outputs. And *Readability* (**C3**) mainly evaluates whether the output is grammatically fluent.

5 Results

5.1 Automatic Evaluation

Table 1 presents the results of automatic evaluation on the CNN/DailyMail dataset. Obviously, our model significantly outperforms all the baselines on all metrics, which indicates the gist guided method can effectively generate more fluent summary. For the benchmark model PGN+Cov, our results are improved by 2.51, 1.80 and 2.77 in terms of ROUGE-1, ROUGE-2 and ROUGE-L, respectively. Particularly, the ROUGE-2 brings a 10.4% boost compared with PGN+Cov. Moreover, our model just utilizes end-to-end training instead of the two-step method like [7] to achieve the best results. In summary, we only add a few parameters to the baseline model, but we get a great improvement.

5.2 Ablation Study

We conduct the ablation study to evaluate the correctness and effectiveness of different modules. On the basic model PGN + Cov, we gradually add the GSM module, RGIGM mechanism, gist consistency loss and reinforcement learning. As shown in Table 2, we first add the GSM module to distill the salient gist information to guide the whole generation process, which achieves better results than the baseline. Taking the ROUGE-L for an example, the score exceeds the basic model by 0.61 points. Hence, the model can generate more coherent and fluent summaries by introducing the GSM module. Then, we introduce the RGIGM for the decoder, which brings a great performance improvement. It proves the remaining gist guided method makes the decoder concern about the information to be expressed next. Finally, we equip the model with a gist consistency loss to further improve the consistency between original documents and generated summaries that ensures the salient information is expressed completely. It is worth noting that the model has achieved state-of-the-art results at this time. We also verify reinforcement learning can further promote the performance of our model.

Table 4. The bold words in Article are salient parts contained in Reference Summary. The blue words in generated summaries are salient information and the red words are uncorrelated or error.

Article: **A video that was played during a preliminary hearing in a california courtroom on friday** showed a san diego police officer being hit with his own cruiser. **Officer Jeffrey Swett was allegedly run over by William Bogard in january after the suspect stole his car while it was running, according to prosecutors. Swett suffered two broken arms, a broken leg and severe head and neck trauma**, while Bogard has pleaded not guilty. Scroll down for video. A video from a hearing in a court on friday showed a san diego police officer being hit with his own cruiser. **William Bogard has pleaded not guilty after being charged with attempted murder, assault and vehicle theft**. Officer jeffrey ...

Reference Summary: Officer Jeffrey Swett was allegedly run over by William Bogard in january. Suspect stole officer 's car while it was running, according to prosecutors. Swett suffered broken arms, broken leg and severe head and neck trauma. Video of incident was played during preliminary hearing in court on friday. Bogard pleaded not guilty to charges including murder, assault and theft.

PGN+Cov: Officer jeffrey swett was charged with attempted murder, assault and vehicle theft. Swett suffered two broken arms, a broken leg and severe head and neck trauma. Bogard has pleaded not guilty after being charged with attempted murder.

Fast-Abs: Officer jeffrey swett was allegedly run over by William Bogard in january. Swett suffered two broken arms, a broken leg and severe head and neck trauma. William bogard has pleaded not guilty after being charged with attempted murder. Video shows san diego police officer being hit with his own cruiser. Bogard was smiling behind the wheel while running him down.

Our model: Officer Jeffrey Swett was allegedly run over by Billiam Bogard in january after the suspect stole his car while it was running, according to prosecutors. Swett suffered two broken arms, a broken leg and severe head and neck trauma, while Bogard has pleaded not guilty. The suspect was charged with attempted murder, assault and vehicle theft. The video was played during preliminary hearing in courtroom on friday.

5.3   Human Evaluation

The human evaluation results are calculated by averaging all scores from six annotators and the scores of reference summaries are set to 5. As shown in Table 3, our model significantly outperforms all baseline models we have implemented, especially in terms of the C1 (Relevance) and C2 (Non-Redundancy). Moreover, C1 measures the correlation between the generated summaries and the ground truth, while C2 evaluates the diversity and informativeness of outputs. Therefore, the high scores of C1 and C2 suggest that the gist information can make the model pay more attention to salient information of the input documents and the RGIGM mechanism improves the diversity of results by reducing semantic duplication. Furthermore, all scores of our model are very close to the ground truth, which indicates that our model can generate relevant, diverse and fluent summaries as human beings.

5.4 Case Study

To verify whether the performance improvements are owing to the gist information, we show a sample of summaries generated by our model and baseline models. As shown in the Table 4, without the guidance of remaining gist information, PGN+Cov fails to obtain some pieces of salient information and even generates false facts (officer jeffrey swett is a victim, not a suspect). Moreover, Fast-Abs not only losts the salient information, but also generates a number of trivial facts. By contrast, our model, with the guidance of gist, can avoid redundancy and generate summaries containing most pieces of salient information.

6 Conclusion

In this paper, we propose a novel framework that first introduces the gist concept in abstractive summarization. We propose the self-adjust query in multi-head attention mechanism to distill the salient semantics as global gist and calculate the remaining gist to guide the generation process dynamically, which can effectively reduce the redundancy and improve the readability. And the gist consistency loss further improves the consistency between documents and summaries. We conduct experiments on the CNN/Daily Mail dataset and the results indicate that our method significantly outperforms all baselines.

In the future, we can extend the gist guided method in many directions. An appealing direction is to investigate the abstractive method on the multi-document summarization, which is more challenging and lacks training data.

References

1. Blei, D.M., Ng, A.Y., Jordan, M.I.: Latent dirichlet allocation. J. Mach. Learn. Res. **3**, 993–1022 (2003). https://doi.org/10.1162/jmlr.2003.3.4-5.993, http://portal. acm.org/citation.cfm?id=944937
2. Cao, Z., Li, W., Li, S., Wei, F.: Retrieve, rerank and rewrite: soft template based neural summarization. In: Proceedings of the 56th Annual Meeting of the Association for Computational Linguistics (Volume 1: Long Papers), pp. 152–161 (2018)
3. Celikyilmaz, A., Bosselut, A., He, X., Choi, Y.: Deep communicating agents for abstractive summarization. In: Proceedings of the 2018 Conference of the North American Chapter of the Association for Computational Linguistics: Human Language Technologies, Volume 1 (Long Papers), pp. 1662–1675. Association for Computational Linguistics, New Orleans, Louisiana, June 2018. https://doi.org/10. 18653/v1/N18-1150, https://www.aclweb.org/anthology/N18-1150

4. Chen, Y.C., Bansal, M.: Fast abstractive summarization with reinforce-selected sentence rewriting. In: Proceedings of the 56th Annual Meeting of the Association for Computational Linguistics (Volume 1: Long Papers), pp. 675–686. Association for Computational Linguistics, Melbourne, Australia, July 2018. https://doi.org/10.18653/v1/P18-1063, https://www.aclweb.org/anthology/P18-1063

5. Devlin, J., Chang, M.W., Lee, K., Toutanova, K.: BERT: pre-training of deep bidirectional transformers for language understanding. In: Proceedings of the 2019 Conference of the North American Chapter of the Association for Computational Linguistics: Human Language Technologies, Volume 1 (Long and Short Papers), pp. 4171–4186. Association for Computational Linguistics, Minneapolis, Minnesota, June 2019. https://doi.org/10.18653/v1/N19-1423, https://www.aclweb.org/anthology/N19-1423

6. Fan, A., Grangier, D., Auli, M.: Controllable abstractive summarization. In: Proceedings of the 2nd Workshop on Neural Machine Translation and Generation. pp. 45–54. Association for Computational Linguistics, Melbourne, Australia, July 2018. https://doi.org/10.18653/v1/W18-2706, https://www.aclweb.org/anthology/W18-2706

7. Gehrmann, S., Deng, Y., Rush, A.: Bottom-up abstractive summarization. In: Proceedings of the 2018 Conference on Empirical Methods in Natural Language Processing, pp. 4098–4109. Association for Computational Linguistics, Brussels, Belgium, October–November 2018. https://doi.org/10.18653/v1/D18-1443, https://www.aclweb.org/anthology/D18-1443

8. Gui, M., Tian, J., Wang, R., Yang, Z.: Attention optimization for abstractive document summarization. In: Proceedings of the 2019 Conference on Empirical Methods in Natural Language Processing and the 9th International Joint Conference on Natural Language Processing, EMNLP-IJCNLP 2019, Hong Kong, China, 3–7 November 2019, pp. 1222–1228 (2019). https://doi.org/10.18653/v1/D19-1117, https://doi.org/10.18653/v1/D19-1117

9. Hermann, K.M., Kocisky, T., Grefenstette, E., Espeholt, L., Kay, W., Suleyman, M., Blunsom, P.: Teaching machines to read and comprehend. In: Advances in Neural Information Processing Systems, pp. 1693–1701 (2015)

10. Kingma, D.P., Ba, J.: Adam: a method for stochastic optimization. In: Bengio, Y., LeCun, Y. (eds.) 3rd International Conference on Learning Representations, ICLR 2015, Conference Track Proceedings , San Diego, CA, USA, 7–9 May 2015 (2015)

11. Krishna, K., Srinivasan, B.V.: Generating topic-oriented summaries using neural attention. In: Proceedings of the 2018 Conference of the North American Chapter of the Association for Computational Linguistics: Human Language Technologies, Volume 1 (Long Papers), pp. 1697–1705. Association for Computational Linguistics, New Orleans, Louisiana, June 2018. https://doi.org/10.18653/v1/N18-1153, https://www.aclweb.org/anthology/N18-1153

12. Li, C., Xu, W., Li, S., Gao, S.: Guiding generation for abstractive text summarization based on key information guide network. In: Proceedings of the 2018 Conference of the North American Chapter of the Association for Computational Linguistics: Human Language Technologies, Volume 2 (Short Papers), pp. 55–60. Association for Computational Linguistics, New Orleans, Louisiana, June 2018. https://doi.org/10.18653/v1/N18-2009, https://www.aclweb.org/anthology/N18-2009

13. Lin, C.Y.: Rouge: a package for automatic evaluation of summaries. In: Workshop on Text Summarization Branches Out, Post-Conference Workshop of ACL 2004, Barcelona, Spain, July 2004. https://www.microsoft.com/en-us/research/publication/rouge-a-package-for-automatic-evaluation-of-summaries/

14. Nallapati, R., Xiang, B., Zhou, B.: Sequence-to-sequence RNNs for text summarization. ArXiv abs/1602.06023 (2016)
15. Pasunuru, R., Bansal, M.: Multi-reward reinforced summarization with saliency and entailment. In: Proceedings of the 2018 Conference of the North American Chapter of the Association for Computational Linguistics: Human Language Technologies, Volume 2 (Short Papers), pp. 646–653. Association for Computational Linguistics, New Orleans, Louisiana, June 2018. https://doi.org/10.18653/v1/N18-2102, https://www.aclweb.org/anthology/N18-2102
16. Paulus, R., Xiong, C., Socher, R.: A deep reinforced model for abstractive summarization. In: 6th International Conference on Learning Representations, ICLR 2018, Conference Track Proceedings, Vancouver, BC, Canada, 30 April - 3 May 2018, OpenReview.net (2018). https://openreview.net/forum?id=HkAClQgA-
17. Pennington, J., Socher, R., Manning, C.: Glove: global vectors for word representation. In: Proceedings of the 2014 Conference on Empirical Methods in Natural Language Processing (EMNLP), pp. 1532–1543 (2014)
18. Perez-Beltrachini, L., Liu, Y., Lapata, M.: Generating summaries with topic templates and structured convolutional decoders. In: Proceedings of the 57th Annual Meeting of the Association for Computational Linguistics, pp. 5107–5116. Association for Computational Linguistics, Florence, Italy, July 2019. https://doi.org/10.18653/v1/P19-1504, https://www.aclweb.org/anthology/P19-1504
19. Rush, A.M., Chopra, S., Weston, J.: A neural attention model for abstractive sentence summarization. In: Proceedings of the 2015 Conference on Empirical Methods in Natural Language Processing, pp. 379–389. Association for Computational Linguistics, Lisbon, Portugal, September 2015. https://doi.org/10.18653/v1/D15-1044, https://www.aclweb.org/anthology/D15-1044
20. See, A., Liu, P.J., Manning, C.D.: Get to the point: summarization with pointer-generator networks. In: Proceedings of the 55th Annual Meeting of the Association for Computational Linguistics (Volume 1: Long Papers), pp. 1073–1083. Association for Computational Linguistics, Vancouver, Canada, July 2017
21. Vaswani, A., et al.: Attention is all you need. In: Advances in Neural Information Processing Systems, pp. 5998–6008 (2017)
22. Wang, K., Quan, X., Wang, R.: Biset: bi-directional selective encoding with template for abstractive summarization. arXiv preprint arXiv:1906.05012 (2019)
23. Wang, L., Yao, J., Tao, Y., Zhong, L., Liu, W., Du, Q.: A reinforced topic-aware convolutional sequence-to-sequence model for abstractive text summarization. In: Proceedings of the Twenty-Seventh International Joint Conference on Artificial Intelligence, IJCAI-18, pp. 4453–4460. International Joint Conferences on Artificial Intelligence Organization, July 2018. https://doi.org/10.24963/ijcai.2018/619, https://doi.org/10.24963/ijcai.2018/619
24. You, Y., Jia, W., Liu, T., Yang, W.: Improving abstractive document summarization with salient information modeling. In: Proceedings of the 57th Annual Meeting of the Association for Computational Linguistics, pp. 2132–2141 (2019)
25. Zheng, C., Zhang, K., Wang, H.J., Fan, L.: Topic-aware abstractive text summarization (2020)

Quality of Recommendations
and Cold-Start Problem in Recommender
Systems Based on Multi-clusters

Urszula Kużelewska[(✉)] [iD]

Faculty of Computer Science, Bialystok University of Technology,
Wiejska 45a, 15-351 Bialystok, Poland
u.kuzelewska@pb.edu.pl

Abstract. This article presents a new approach to collaborative filtering recommender systems that focuses on the problem of an active user's (a user to whom recommendations are generated) neighbourhood modelling. Precise identification of the neighbours has a direct impact on the quality of the generated recommendation lists. Clustering techniques are the solution that is often used for neighbourhood calculation, however, they negatively affect the quality (precision) of recommendations.

In this article, a new version of the algorithm based on multi-clustering, $M - CCF$, is proposed. Instead of one clustering scheme, it works on a set of multi-clusters, therefore it selects the most appropriate one that models the neighbourhood most precisely. This article presents the results of the experiments validating the advantage of multi-clustering approach, $M - CCF$, over the traditional methods based on single-scheme clustering. The experiments focus on the overall recommendation performance including accuracy and coverage as well as a cold-start problem.

Keywords: Multi-clustering · Collaborative filtering · Recommender systems · Cold-start problem

1 Introduction

With the rapid development of the Internet, a large expansion of data is observed. To help users to cope with the information overload, Recommender Systems (RSs) were designed. They are computer applications with the purpose to provide relevant information to a user and as a consequence reduce his/her time spent on searching and increase personal customer's satisfaction. The form of such relevant information is a list (usually ranked) of items that are interesting and useful to the user [8,16].

Collaborative filtering methods (CF) are the most popular type of RSs [4,8]. They are based on users' past behaviour data: search history, visited web sites, and rated items, and use them for similarity searching, with an assumption that users with corresponding interests prefer the same items. As a result, they

© Springer Nature Switzerland AG 2021
M. Paszynski et al. (Eds.): ICCS 2021, LNCS 12743, pp. 72–86, 2021.
https://doi.org/10.1007/978-3-030-77964-1_6

predict the level of interest of those users on new, never seen items [4,19]. Collaborative filtering approach has been very successful due to its precise prediction ability [18].

Although many complex algorithms to generate recommendations were proposed by scientists, it is still an open research challenge to build a universal system which is accurate, scalable, and time efficient [19].

During recommendation generation, a great amount of data is analysed and processed, whereas the generation outcome in real time is an issue. Ideally, an algorithm should produce entirely accurate propositions, that is, suggest items that are highly rated by users. At the same time, the method should be both vertically and horizontally scalable. Vertical scalability is related to the remaining real time of recommendation generation regardless of data size, whereas the horizontal scalability problem occurs when the data is sparse (when few items are connected by users, e.g. rated by them) [16].

The article is organised as follows: the first section presents the background of the neighbourhood identification problem in the field of Recommender Systems. This section discusses common solutions with their advantages and disadvantages as well. Next section describes the proposed multi-clustering algorithm, $M - CCF$ on the background of alternative clustering techniques, whereas the following section contains the results of the performed experiments to compare multi-clustering and single-clustering approaches. The algorithm $M - CCF$ is executed on different types of multi-clusters: when they come from the algorithm with different values of input parameters. The last section concludes the paper.

2   Background and Related Work

Generation of recommendation lists is connected with processing a large amount of input data. The input data is usually the ratings of users on a set of items. If a set of users is denoted as $X = \{x_1, \ldots, x_n\}$ and a set of items as $A = \{a_1, \ldots, a_k\}$, the matrix of the input data can be represented by a matrix $U = (X, A, V)$, where $V = \{v_1, \ldots, v_c\}$ and is a set of ratings values.

The main part of the processing of data is a similarity calculation of every pair of users, and with awareness of the fact that the number of ratings can reach millions values, the real time of recommendation generation appears as a challenge. A common solution to this problem is to reduce the search space around an active user to its closest neighbours [4]. A domain of CF focused on neighbourhood identification is still under intensive research [10,23].

2.1   Neighbourhood Identification Techniques

The traditional method for neighbourhood calculation is k Nearest Neighbours (kNN) [18]. It calculates all user-user or item-item similarities and identifies the most k similar objects (users or items) to the target object as its neighbourhood. Then, further calculations are performed only on the objects from the neighbourhood, improving the time of processing. The kNN algorithm is a reference method used to determine the neighbourhood of an active user for the collaborative filtering recommendation process [4].

The neighbourhood calculated by kNN technique can be noted as follows:

$$N_{knn}(y_i) = \forall_{y \in Y} \underset{p}{sim}(y_i, y) \tag{1}$$

where p is a number of the neighbours determined by k factor in kNN algorithm. This formula can be related to both users (X) or items (A), therefore the set is denoted generally as Y. Every object in N set is different from y_i.

An example similarity (between items a_i and a_j) formula based on Pearson correlation is as follows:

$$sim_P(a_i, a_j) = \frac{\sum_{k \in V_{ij}} (r(a_{ik}) - \mu_{a_i}) \cdot (r(a_{jk}) - \mu_{a_j})}{\sqrt{\sum_{k \in V_{ij}} (r(a_{ik}) - \mu_{a_i})^2} \cdot \sqrt{\sum_{k \in V_{ij}} (r(a_{jk}) - \mu_{a_j})^2}} \tag{2}$$

where $r(a_{ik})$ is a rating of the item a_i given by the user x_k, μ_{a_i} is an average rating of the item a_i given by all users who rated this item, V is a vector of possible ratings $V = \{v_1, \ldots, v_c\}$ and $V_{ij} = V(a_i) \cap V(a_j)$ - a set of ratings present in both item's vectors: i and j.

This equation can be used in item-item CF systems, however, it is possible to build an analogous equation for the calculation of a similarity between users in the case of a user-user recommender. Other similarity measures are: Euclidean-, and CityBlock-based Similarity Measures, Cosine Index, or Tanimoto Similarity [16]. They can be applied in both types of collaborative filtering recommender systems: item-item as well as user-user.

Simplicity and reasonably accurate results are the advantages of kNN approach; its disadvantages are low scalability and vulnerability to sparsity in data [19].

Clustering algorithms can be an efficient solution to the disadvantages of kNN approach due to the neighbourhood being shared by all cluster members. The neighbourhood calculated by clustering techniques can be described by (3).

$$N_{cl}(y_i) = C_j, \Rightarrow C_j = \{y_1, \ldots, y_{cj}\}, C_j \in C \tag{3}$$

where C_j is j-th cluster from one clustering scheme C and c is the number of objects in this cluster. Note that the object y_i is a member of the j-th cluster, as well. In this case, the metric of classification a particular object into a particular cluster is different from the similarity used in recommender systems - usually it is Euclidean distance.

The following problems may arise when one applies clustering algorithms to neighbourhood identification: significant loss of prediction accuracy and different every recommendation outcome. The diversity of results is related to the fact that most of the clustering methods are non-deterministic and therefore several runs of the algorithms can effect obtaining various clustering schemes. The following section, Sect. 2.2 is devoted to the clustering domain and methods.

Multi-clustering approach, instead of one clustering scheme, works on a set of partitions, therefore it selects the most appropriate one that models the neighbourhood precisely, thus reducing the negative impact of non-determinism. Section 2.3 presents a background of multi-clustering and describes selected solutions and applications.

2.2 Clustering Methods Used in RS Domain

Clustering is a part of Machine Learning domain. The aim of clustering methods is to organize data into separate groups without any external information about their membership, such as class labels. They analyse only the relationship among the data, therefore clustering belongs to Unsupervised Learning techniques [9].

Due to the independent *á priori* clusters identification, clustering algorithms are an efficient solution to the problem of RSs scalability, providing a predefined neighbourhood for the recommendation process [17]. The efficiency of clustering techniques is related to the fact that a cluster is a neighbourhood that is shared by all cluster members, in contrast to kNN approach determining neighbours for every object separately [17]. The disadvantage of this approach is usually the loss of prediction accuracy.

There are two major problems related to the quality of clustering. The first is the clustering results depend on the input algorithm parameters, and additionally, there is no reliable technique to evaluate clusters before on-line recommendation process. Moreover, some clustering schemes may better suit to some particular applications [22]. The other issue addressed to decreasing prediction accuracy is the imprecise neighbourhood modelling of the data located on the borders of clusters [11, 12].

Popular clustering technique is $k - means$ due to its simplicity and high scalability [9]. It is often used in CF approach [17]. A variant of $k - means$ clustering, bisecting $k - means$, was proposed for privacy-preserving applications [3] and web-based movie RS [17]. Another solution, ClustKNN [15] was used to cope with large-scale RS applications. However, the $k - means$ approach, as well as many other clustering methods, do not always result in clustering convergence. Moreover, they require input parameters, e.g., a number of clusters, as well.

2.3 Multi-clustering Approach to Recommendations

The disadvantages described above can be solved by techniques called alternate clustering, multi-view clustering, multi-clustering, or co-clustering. They include a wide range of methods which are based on widely understood multiple runs of clustering algorithms or multiple applications of a clustering process on different input data [2].

Multi-clustering or co-clustering has been applied to improve scalability in the domain of RSs. Co-clustering discovers samples that are similar to one another with respect to a subset of features. As a result, interesting patterns (co-clusters) are identified unable to be found by traditional one-way clustering [22]. Multiple clustering approaches discover various partitioning schemes, each capturing different aspects of the data [1]. They can apply one clustering algorithm changing the values of input parameters or distance metrics, as well as they can use different clustering techniques to generate a complementary result [22].

The role of multi-clustering in the recommendation generation process that is applied in the approach described in this article, is to determine the most accurate neighbourhood for an active user. The algorithm selects the best cluster from a set of clusters prepared in advance (see the following Section).

The method described in [14] uses a multi-clustering method, however, it is interpreted as clustering of a single scheme for both techniques. It groups the ratings to create an item group-rating matrix and a user group-rating matrix. As a clustering algorithm, it uses $k - means$ combined with a fuzzy set theory. In the last step of the pre-recommendation process, $k - means$ is used again on the new rating matrix to find groups of similar users that represent their neighbourhood with the goal to limit the search space for a collaborative filtering method. It is difficult to compare this approach with other techniques including single-clustering ones, because the article [14] describes the experiments on the unknown dataset containing only 1675 ratings.

The other solution is presented in [20]. The method $CCCF$ (Co-Clustering For Collaborative Filtering) first clusters users and items into several subgroups, where each subgroup includes a set of like-minded users and a set of items in which these users share their interests. The groups are analysed by collaborative filtering methods and the result recommendations are aggregated over all subgroups. This approach has advantages like scalability, flexibility, interpretability, and extensibility.

Other applications are: accurate recommendation of tourist attractions based on a co-clustering and bipartite graph theory [21] and $OCuLaR$ (Overlapping co-CLuster Recommendation) [7] - an algorithm for processing very large databases, detecting co-clusters among users and items as well as providing interpretable recommendations.

There are some other methods, which can be generally called as multi-view clustering, that find partitioning schemes on different data (e.g., ratings and text description) combining results after all ([2,13]). The main objective of a multi-view partitioning is to provide more information about the data in order to understand them better by generating distinct aspects of the data and searching for the mutual link information among the various views [6]. It is stated that single-view data may contain incomplete knowledge while multi-view data fill this gap by complementary and redundant information [5].

2.4   Contribution of Proposed Work

A novel recommender system with neighbourhood identification based on multi-clustering - $M - CCF$ - is described in this paper. The following are the major contributions of $M - CCF$:

1. Neighbourhood of an active user is modelled more precisely due to the fact that the system's overall neighbourhood is formed by a set of cluster schemes and the most similar cluster can be selected in every case, thereby improving the recommendation accuracy.

2. Precise neighbourhood increases the system's horizontal scalabililty, therefore a cold-start problem occurs rarely.
3. Clustering schemes obtained from different runs of a clustering algorithm with different values of input parameters model the neighbourhood better than the schemes obtained from clustering algorithms with the same parameter's values on their input, thereby improving the recommendation accuracy.

The last statement refers to a version of $M-CCF$ described in [12], in which the input data come from multi-clustering approach, however a value of an input parameter in $k-means$ was the same when building one $M-CCF$ RS system.

3 Description of M-CCF Algorithm

The novel solution consists of multiple types of clustering schemes that are provided for the method's input. It is implemented in the following way (for the original version, with one type of a clustering scheme, check in [11,12]).

Step I. Multiple Clustering
The first step of the $M-CCF$ is to perform clustering on the input data. The process is conducted several times and all results are stored in order to deliver them to the algorithm. In the experiments described in this paper, $k-means$ was selected as a clustering method, which was executed for $k = 10, 20, 50$ to generate input schemes (denoted by C set) for one $M-CCF$ RS system. This is illustrated in Fig. 1.

Step II. Building M-CCF RS System
It is a vital issue to have precise neighbourhood modelling for all input data. In $M-CCF$ it is performed by iterating every input object and selection of the best cluster from C set for it. The term *best* refers to the cluster which center is the most similar to the particular input object. Then, when all input data have their connected clusters, a traditional CF systems are built on these clusters. As a result, $M-CCF$ algorithm is created - a complex of recommender systems formed on their clusters as recommender data.

A general formula of a neighbourhood calculated by $M-CCF$ method can be described by (4).

$$N_{mcl}(y_i) = C_j(t), \Rightarrow C_j(t) \in C, C == \{C_1(1), \ldots, C_j(1), C_1(2), \ldots, C_g(h)\} \tag{4}$$

where $C_j(t)$ is j-th cluster from t-th clustering scheme, and C are all clustering schemes generated by a clustering algorithm in several runs of different values of its input parameters. In this case, the metric of classification a particular object into a particular cluster is different than the similarity used in recommender systems, as well.

Step III. Recommendation Generation
When generating recommendations for an active user, first of all, a relevant RS from $M-CCF$ is selected. It is also based on the similarity between the

active user's and cluster centers' ratings. Then, the process of recommendation generation is performed as it is implemented in the traditional collaborative filtering approach, however, searching for similar objects is limited to the cluster connected to the particular recommender in $M - CCF$ algorithm.

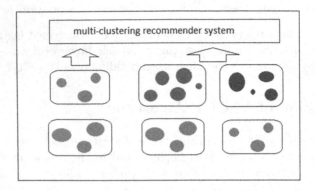

Fig. 1. Comparison of different inputs in $M - CCF$ algorithm

When a neighbourhood is modelled by a single-clustering method, the border objects have fewer neighbours in their closest area than the objects located in the middle of a cluster. The multi-clustering prevents such situations, as it identifies clusters in which particular users are very close to its center. A major advantage of $M - CCF$ algorithm is a better quality of an active user's neighbourhood modelling, therefore resulting in high precision of recommendations, including highly sparse cases.

4   Experiments

Evaluation of the performance of $M - CCF$ algorithm was conducted on two MovieLens datasets: a small one containing 534 users, 11 109 items and 100 415 ratings ($100k$), and a big dataset consisting of 4537 users, 16767 items and 1 000 794 ratings ($10M$) [24]. Note, that the small set is more sparse (contains fewer ratings per user and per item) than the big one.

The results obtained with $M - CCF$ were compared with the recommender system whose neighbourhood modelling is based on a single-clustering ($SCCF$). Attention was paid to the precision and completeness of recommendation lists generated by the systems. The evaluation criteria were related to the following baselines: Root Mean Squared Error ($RMSE$) described by (5) and *Coverage* described by (6). The symbols in the equations, as well as the method of calculation are characterised in detail below.

$$RMSE = \sqrt{\frac{1}{n \cdot k} \sum_{i=1}^{nk} (r_{real}(v_i) - r_{est}(v_i))^2}, r_{real} \in [2,3,4,5], r_{est} \in \mathbb{R}_+ \quad (5)$$

$$Coverage = \frac{\sum_{i=1}^{N} r_{est}(x_i) \in \mathbb{R}_+}{N} \cdot 100\% \tag{6}$$

where $\mathbb{R}_+$ stands for the set of positive real numbers. The performance of both approaches was evaluated in the following way. Before the clustering step, the whole input dataset was split into two parts: training and testing. In the case of $100k$ set, the parameters of the testing part were as follows: 393 ratings, 48 users, 354 items, whereas the case of $10M$: 432 ratings, 44 users and383 items. This step provides the same testing data during all the experiments presented in this paper, therefore making the comparison more objective.

In the evaluation process, the values of ratings from the testing part were removed and estimated by the recommender system. The difference between the original and the calculated value (represented respectively, as $r_{real}(x_i)$ and $r_{est}(x_i)$ for user x_i and a particular item i) was taken for $RMSE$ calculation. The number of ratings is denoted as N in the equations. The lower value of $RMSE$ stands for a better prediction ability.

During the evaluation process, there were cases in which estimation of ratings was not possible. It occurs when the item for which the calculations are performed, is not present in the clusters which the items with existing ratings belong to. It is considered in $Coverage$ index (6). In every experiment, it was assumed that $RMSE$ is significant if the value of $Coverage$ is greater than 90%. It means that if the number of users for whom the recommendations were calculated was 48 and for each of them it was expected to estimate 5 ratings, therefore at least 192 values should be present in the recommendation lists.

The experiments started from the precision evaluation of RS in which neighbourhood was determined by single-scheme $k-means$ algorithm. Table 1 contains evaluation results on $100k$ dataset, whereas Table 2 - on $10M$ data. In both cases, data were clustered independently six times into a particular number of groups to examine the influence of a non-determinism of $k-means$ results.

A clustering measure was one of the following: Euclidean, Cosine-based and Chebyshev. Although Euclidean and Chebyshev distances usually generate accurate partitions, the measure that is selected most often is Cosine-based due to its low complexity, especially for sparse vectors. It needs to be admitted that formally it is not a proper distance metric as it does not have the triangle inequality property. The number of groups was equal 10, 20, or 50. Finally, this experiment was performed 54 times (and repeated 5 times to decrease randomness) per input dataset.

The results concern similarity measures as well. The following indices were taken: $Cosine-based$, $LogLikelihood$, $Pearson\ correlation$, both $Euclidean$ and $CityBlock$ distance-based and $Tanimoto$ coefficient. There are minimal and maximal values of $RMSE$ to present a wide range of precision values which are a result of the non-determinism mentioned above. It means that there is no guarantee that the scheme selected for the recommendation process is optimal. The values are presented with a reference value in brackets that stands for $Coverage$.

Recommendation quality is definitely worse for $10M$ dataset. The best $RMSE$ values are 0.92 with $Coverage = 98\%$, whereas for $100k$ input data - 0.81

Table 1. RMSE of SCCF evaluated on 100k dataset. The best values are in bold.

Similarity measure	Clustering distance measure					
	Euclidean		Cosine-based		Chebyshev	
	Min	Max	Min	Max	Min	Max
Cosine-based	**0.83(99%)**	0.86(99%)	0.85(97%)	0.90(99%)	0.88(85%)	0.91(89%)
LogLikelihood	0.84(99%)	0.86(99%)	0.86(97%)	0.90(98%)	0.88(90%)	0.91(92%)
Pearson corr.	1.18(97%)	5.97(97%)	–	–	–	–
Euclidean	**0.81(99%)**	0.86(99%)	**0.84(94%)**	0.89(95%)	**0.87(86%)**	0.90(91%)
CityBlock	0.84(99%)	0.87(99%)	**0.84(92%)**	0.89(97%)	**0.87(86%)**	0.97(92%)
Tanimoto	**0.83(99%)**	0.85(99%)	**0.84(92%)**	0.89(97%)	0.88(91%)	0.91(95%)

Table 2. RMSE of SCCF evaluated on 10M dataset. The best values are in bold.

Similarity measure	Clustering distance measure					
	Euclidean		Cosine-based		Chebyshev	
	Min	Max	Min	Max	Min	Max
Cosine-based	**0.94(99%)**	0.98(84%)	0.95(97%)	0.99(97%)	0.95(95%)	1.00(99%)
LogLikelihood	**0.94(99%)**	1.00(98%)	0.95(97%)	0.99(98%)	0.93(95%)	1.00(98%)
Pearson corr.	1.02(77%)	2.52(99%)	0.96(94%)	2.52(98%)	0.98(94%)	2.52(98%)
Euclidean	0.95(99%)	0.99(99%)	**0.94(99%)**	0.97(99%)	0.92(91%)	0.99(98%)
CityBlock	0.95(99%)	0.99(99%)	**0.94(97%)**	0.99(99%)	**0.92(96%)**	0.99(98%)
Tanimoto	**0.93(99%)**	0.99(92%)	**0.93(97%)**	0.95(99%)	**0.92(98%)**	0.97(99%)

with $Coverage = 99\%$. However, for the big set, the range of values is smaller, regardless of a similarity or distance measure. The $Coverage$ is higher when the number of ratings increases as well.

The following experiments were performed on both input data described above. However, as a recommender, $M - CCF$ method was taken. Table 3 contains the results, containing $RMSE$ and $Coverage$ (they are average values from 5 different runs of $M - CCF$). Similarity and distance measures were the same as in the previous tests. Both input datasets were prepared as follows. All $k - means$ clustering schemes, regardless of a number of clusters, that were obtained in the previous experiments were placed as input data for $M - CCF$ algorithm.

Recommendation precision of $M - CCF$ was also worse in the case of 10M dataset. In comparison to the best $RMSE$ values for the single clustering algorithm, they were comparable or slightly worse. However, note that there are not any value ranges, but explicit numbers in every case. It means that the multi-clustering approach has eliminated the ambiguity of clustering scheme selection. Additionally, $Coverage$ was higher for both datasets. It means that $M - CCF$ is able to generate recommendations more often than the recommender system with neighbourhood strategy based on single-scheme clustering.

In the experiments described above, a big impact on $RMSE$ has a similarity measure, however only the best values are discussed there. For similarity as well as distance effect evaluation, see [12].

Table 3. RMSE of $M - CCF$ algorithm evaluated on both datasets: 100k and 10M. The best values are in bold.

Similarity measure	Euclidean clustering distance measure	
	100K dataset	10M dataset
Cosine-based	0.84(95%)	0.99(98%)
LogLikelihood	0.87(98%)	0.99(99%)
Pearson corr.	–	6.73(98%)
Euclidean	**0.84**(99%)	0.98(99
CityBlock	0.89(99%)	**0.97**(99%)
Tanimoto	0.86(99%)	**0.97**(99%)
Similarity measure	Cosine clustering distance measure	
	100K dataset	10M dataset
Cosine-based	**0.83(95%)**	0.94(97%)
LogLikelihood	0.89(98%)	0.96(99%)
Pearson corr.	4.98(96%)	1.14(99%)
Euclidean	**0.82(98%)**	**0.93**(99%)
CityBlock	**0.84**(98%)	**0.94**(99%)
Tanimoto	**0.82(98%)**	**0.93**(99%)
Similarity measure	Chebyshev clustering distance measure	
	100K dataset	10M dataset
Cosine-based	0.87(93%)	**0.98**(97%)
LogLikelihood	0.88(98%)	**0.98**(99%)
Pearson corr.	2.41(96%)	1.08(99%)
Euclidean	**0.84**(98%)	**0.97**(99%)
CityBlock	0.87(98%)	0.99(99%)
Tanimoto	0.87(98%)	**0.98**(99%)

Finally, the last experiment concerned a cold-start problem occurrence. The results are in Tables 4, 5, 6 and 7. A precision of recommendations generated for sparse data was tested, that is, for users who rated at most 1, 2, or 3 items. There users were taken from the test set, however, if a particular user had more ratings present, they were removed from the vectors. The ratings for the purge were selected randomly. All previously mentioned aspects (similarity, distance measure, the size of dataset) were taken into consideration.

Analysing the results, it can be noted that it is a common situation, when the recommendation precision increases if users rate more items. However, it is not

Table 4. Results of cold-start testing (RMSE) - SCCF evaluated on $100k$ dataset. The best values are in bold.

Similarity measure	Clustering distance measure					
	Euclidean			Cosine-based		
	1 rating	2 ratings	3 ratings	1 rating	2 ratings	3 ratings
Cosine	0.97–1.03	0.92–0.97	0.87–0.97	**0.90–0.97**	0.90–0.96	**0.91–0.97**
LogLike	**0.96–1.03**	**0.91–0.97**	0.87–0.97	**0.90–0.98**	0.90–0.96	**0.91–0.96**
Pearson	1.02–2.92	1.04–1.99	1.01–3.07	0.99–2.12	1.01–2.44	0.99–8.83
Euclidean	0.97–1.03	**0.91–0.97**	**0.87–0.96**	**0.90–0.97**	**0.89–0.95**	**0.91–0.96**
CityBlock	1.00–1.03	0.94–0.97	0.90–0.96	0.93–0.97	0.92–0.97	0.93–0.96
Tanimoto	**0.95–1.03**	**0.91–0.97**	**0.86–0.96**	**0.90–0.97**	**0.89–0.95**	**0.91–0.96**

always a rule. There were a few cases in which the generation of propositions was more difficult. If they appeared in the test set, the final precision was affected. Better quality of recommendations generated for sparse data is visible in the case of big dataset $10M$. This situation is common for both examined recommendation algorithms. However, when the input data were not sparse, the high number of ratings affected the precision negatively. The explanation is related to sparsity of data. The $10M$ data, although bigger, were more dense.

Table 5. Results of cold-start testing (RMSE) - SCCF evaluated on $10M$ dataset. The best values are in bold.

Similarity measure	Clustering distance measure					
	Euclidean			Cosine-based		
	1 rating	2 ratings	3 ratings	1 rating	2 ratings	3 ratings
Cosine	0.90–0.93	**0.89–0.93**	0.88–0.94	**0.89–0.94**	0.91–0.93	0.89–0.92
LogLike	0.90–0.93	**0.89–0.93**	0.88–0.95	**0.89–0.94**	**0.90–0.93**	**0.89–0.91**
Pearson	2.17–5.72	1.08–2.98	2.16–5.31	1.07–1.81	2.11–2.86	2.29–2.56
Euclidean	**0.89–0.92**	**0.88–0.93**	0.88–0.95	**0.89–0.93**	0.91–0.93	**0.89–0.91**
CityBlock	0.93–0.97	0.93–0.98	0.92–0.98	0.94–0.98	0.95–0.96	0.93–0.96
Tanimoto	**0.89–0.92**	**0.89–0.93**	**0.86–0.95**	**0.89–0.93**	**0.90–0.93**	**0.89–0.91**

The goal of the last experiment was to evaluate which algorithm better succeeded in managing a cold-start problem. In the case of recommendations with single-clustering based on neighbourhood modelling, the best values obtained for $10M$ dataset equal 0.89–0.92 - in the case of 1 rating, 0.87–0.91 - in the case of 2 ratings, and 0.86–0.95 - in the case of 3 ratings present in users' vectors. For $100k$ dataset, the values were more scattered - 0.90–0.97 - in the case of 1 rating, 0.89–0.95 - in the case of 2 ratings present, and 0.86–0.96 - in the case of 3 ratings present in users' vectors.

Table 6. Results of cold-start testing (RMSE) - SCCF evaluated on both datasets. The best values are in bold.

Similarity measure	Chebyshev clustering distance measure					
	100k dataset			10M dataset		
	1 rating	2 ratings	3 ratings	1 rating	2 ratings	3 ratings
Cosine	0.90–1.15	0.89–1.03	0.93–1.01	**0.90–0.93**	**0.87–0.91**	0.88–0.93
LogLike	**0.90–1.12**	0.88–1.01	0.92–1.00	**0.90–0.93**	**0.87–0.91**	**0.87–0.93**
Pearson	1.27–5.47	1.18–2.86	2.17–3.82	2.64–5.58	2.34–5.59	2.05–3.04
Euclidean	0.90–1.17	0.88–1.04	0.92–1.02	0.91–0.94	**0.87–0.91**	0.88–0.93
CityBlock	0.95–1.18	0.91–1.05	0.94–1.02	0.95–0.98	0.92–0.95	0.92–0.96
Tanimoto	**0.91–1.07**	**0.88–0.97**	**0.91–0.94**	**0.90–0.93**	**0.88–0.90**	**0.87–0.92**

Table 7. Results of data sparsity testing (RMSE) - $M - CCF$ evaluated on both datasets. The best values are in bold.

Similarity measure	Euclidean clustering distance measure					
	100k dataset			10M dataset		
	1 rating	2 ratings	3 ratings	1 rating	2 ratings	3 ratings
LogLike	**0.91**	0.97	0.98	**0.87**	0.84	0.97
Euclidean	1.04	0.94	0.92	1.06	0.96	1.01
CityBlock	1.06	1.04	0.97	0.94	**0.89**	**0.96**
Tanimoto	0.96	0.99	0.93	0.95	0.90	0.97
Similarity measure	Cosine clustering distance measure					
	100k dataset			10M dataset		
	1 rating	2 ratings	3 ratings	1 rating	2 ratings	3 ratings
LogLike	0.94	0.95	1.02	0.84	0.94	0.97
Euclidean	0.91	**0.92**	1.02	**0.78**	0.88	**0.94**
CityBlock	**0.88**	**0.92**	**0.92**	0.83	0.92	**0.94**
Tanimoto	**0.86**	**0.91**	0.94	**0.81**	**0.90**	**0.93**
Similarity measure	Chebyshev clustering distance measure					
	100k dataset			10M dataset		
	1 rating	2 ratings	3 ratings	1 rating	2 ratings	3 ratings
LogLike	1.01	0.93	1.01	**0.89**	**0.87**	**0.91**
Euclidean	**0.92**	**0.85**	**0.89**	0.92	0.95	1.00
CityBlock	0.95	0.89	0.93	0.91	0.96	1.03
Tanimoto	**0.93**	**0.87**	0.93	0.91	0.97	1.03

In the case of $M - CCF$ recommender which was given on its input the clustering schemes obtained using $k - means$ algorithm for $k = 10, 20, 50$, the best values were obtained for $10M$ dataset equal 0.81 - in the case of 1 rating, 0.84 - in the case of 2 ratings, and 0.91 - in the case of 3 ratings present in users' vectors. For $100k$ dataset, the values were as follows - 0.86 - in the case of 1 rating, 0.85 - in the case of 2 ratings, and 0.89 - in the case of 3 ratings present in users' vectors. The values are more advantageous, that is, $M - CCF$ outperforms the compared method. Moreover, the final value is always unambiguous, the recommender system is able to generate the most optimal propositions. The results for Cosine and Pearson coefficients were not presented due to low *Coverage*.

Taking into consideration all the experiments presented in this article, it can be observed, that the method based on multi-clustering used for neighbourhood modelling in RS is more successful in precise recommendation generation. Moreover, the recommendation lists are more covered, and the results are free from the ambiguousity present in the case of RS in which the neighbourhood is modelled by single-clustering approach.

5 Conclusions

A new developed version of a collaborative filtering recommender system, $M - CCF$, was described in this paper. To improve RS scalability, a search space is limited to variously defined users' neighbourhoods. The presented method models the neighbourhood using a multi-clustering algorithm. It works as follows: $M - CCF$ dynamically selects the most appropriate cluster for every user to whom recommendations are generated. Properly adjusted neighbourhood leads to more accurate recommendations generated by a recommender system. The algorithm eliminates a disadvantage appearing when the neighbourhood is modelled by a single-clustering method - dependence of the final performance of a recommender system on a clustering scheme selected for the recommendation process. Additionally, the preparation of input data was improved. Data come from many clustering schemes obtained from $k - means$ algorithm, however, its input parameters (k) were diversified.

The experiments which are described in this paper confirmed the better performance of $M - CCF$ over the traditional method based on single-clustering. The recommendations have greater precision (lower $RMSE$ values) and are not ambiguous due to working with a mixture of clustering schemes instead of a single one, which may appear not optimal. $M - CCF$ improved better accuracy in the case of a cold-start problem as well.

Next experiments will be performed to prepare a mixture of clustering schemes that is more adjusted to the input data. The clusters will be evaluated and only the best ones will be given to RS input. Finally, it should improve the overall quality of recommendation: precision as well as scalability.

Acknowledgment. The work was supported by the grant from Bialystok University of Technology WZ/WI-IIT/2/2020 and funded with resources for research by the Ministry of Science and Higher Education in Poland.

References

1. Dan, A., Guo, L.: Evolutionary parameter setting of multi-clustering. In: Proceedings of the 2007 IEEE Symposium on Computational Intelligence in Bioinformatics and Computational Biology, pp. 25–31 (2007)
2. Bailey, J.: Alternative Clustering Analysis: a Review. Intelligent Decision Technologies: Data Clustering: Algorithms and Applications, pp. 533–548. Chapman and Hall/CRC (2014)
3. Bilge, A., Polat, H.: A scalable privacy-preserving recommendation scheme via bisecting K-means clustering. Inf. Process Manage. **49**(4), 912–927 (2013)
4. Bobadilla, J., Ortega, F., Hernando, A., Gutiérrez, A.: Recommender systems survey. Knowl.-Based Sys. **46**, 109–132 (2013)
5. Ye, Z., Hui, Ch., Qian, H., Li, R., Chen, Ch., Zheng, Z.: New Approaches in Multi-View Clustering. Recent Applications in Data Clustering. InTechOpen (2018)
6. Guang-Yu, Z., Chang-Dong, W., Dong, H., Wei-Shi, Z.: Multi-view collaborative locally adaptive clustering with Minkowski metric. Exp. Syst. Appl. **86**, 307–320 (2017)
7. Heckel, R., Vlachos, M., Parnell, T., Duenner, C.: Scalable and interpretable product recommendations via overlapping co-clustering. In: IEEE 33rd International Conference on Data Engineering, pp. 1033–1044 (2017)
8. Jannach, D.: Recommender Systems: An Introduction. Cambridge University Press (2010)
9. Kaufman, L.: Finding Groups in Data: An Introduction to Cluster Analysis. John Wiley (2009)
10. Kumar, R., Bala, P.K., Mukherjee, S.: A new neighbourhood formation approach for solving cold-start user problem in collaborative filtering. Int. J. Appl. Manage. Sci. (IJAMS) **12**(2) (2020)
11. Kużelewska, U.: Dynamic neighbourhood identification based on multi-clustering in collaborative filtering recommender systems. In: Zamojski, W., Mazurkiewicz, J., Sugier, J., Walkowiak, T., Kacprzyk, J. (eds.) DepCoS-RELCOMEX 2020. AISC, vol. 1173, pp. 410–419. Springer, Cham (2020). https://doi.org/10.1007/978-3-030-48256-5_40
12. Kużelewska, U.: Effect of dataset size on efficiency of collaborative filtering recommender systems with multi-clustering as a neighbourhood identification strategy. In: Krzhizhanovskaya, V.V., et al. (eds.) ICCS 2020. LNCS, vol. 12139, pp. 342–354. Springer, Cham (2020). https://doi.org/10.1007/978-3-030-50420-5_25
13. Mitra, S., Banka, H., Pedrycz, W.: Rough-fuzzy collaborative clustering. IEEE Trans. Syst. Man Cybern. Part B (Cybern.) **36**(4), 795–805 (2006)
14. Puntheeranurak, S., Tsuji, H.: A Multi-clustering hybrid recommender system. In: Proceedings of the 7th IEEE International Conference on Computer and Information Technology, pp. 223–238 (2007)
15. Rashid, M., Shyong, K.L., Karypis, G., Riedl, J.: ClustKNN: a highly scalable hybrid model - & memory-based CF algorithm. In: Proceeding of WebKDD (2006)
16. Ricci, F., Rokach, L., Shapira, B.: Recommender systems: introduction and challenges. In: Ricci, F., Rokach, L., Shapira, B. (eds.) Recommender Systems Handbook, pp. 1–34. Springer, Boston, MA (2015). https://doi.org/10.1007/978-1-4899-7637-6_1
17. Sarwar, B.: Recommender systems for large-scale e-commerce: scalable neighborhood formation using clustering. In: Proceedings of the 5th International Conference on Computer and Information Technology (2002)

18. Schafer, J.B., Frankowski, D., Herlocker, J., Sen, S.: Collaborative Filtering Recommender Systems, pp. 291–324. The Adaptive Web (2007)
19. Singh, M.: Scalability and sparsity issues in recommender datasets: a survey. Knowl. Inf. Syst. **62**(1), 1–43 (2018). https://doi.org/10.1007/s10115-018-1254-2
20. Wu, Y., Liu, X., Xie, M., Ester, M., Yang, Q.: CCCF: improving collaborative filtering via scalable user-item co-clustering. In: Proceedings of the Ninth ACM International Conference on Web Search and Data Mining, pp. 73–82 (2016)
21. Xiong, H., Zhou, Y., Hu, C., Wei, X., Li, L.: A novel recommendation algorithm frame for tourist spots based on multi - clustering bipartite graphs. In: Proceedings of the 2nd IEEE International Conference on Cloud Computing and Big Data Analysis, pp. 276–282 (2017)
22. Yaoy, S., Yuy, G., Wangy, X., Wangy, J., Domeniconiz, C., Guox, M.: Discovering multiple co-clusterings in subspaces. In: Proceedings of the 2019 SIAM International Conference on Data Mining, pp. 423–431 (2019)
23. Zhang, L., Li, Z., Sun, X.: Iterative rating prediction for neighborhood-based collaborative filtering. Appl. Intell. 1–13 (2021). https://doi.org/10.1007/s10489-021-02237-1
24. MovieLens Datasets. https://grouplens.org/datasets/movielens/25m/. Accessed 10 Oct 2020

Model of the Cold-Start Recommender System Based on the Petri-Markov Nets

Mihail Chipchagov$^{(\boxtimes)}$ (iD) and Evgeniy Kublik (iD)

Financial University Under the Government of the Russian Federation, Moscow, Russia

Abstract. The article describes a model for constructing a cold-start recommendation system based on the mathematical apparatus of Petri-Markov nets. The model combines stochastic and structural approaches to building recommendations. This solution allows you to differentiate recommendation objects by their popularity and impose restrictions on the available latent data about the user.

Keywords: Cold-start · Recommendation system · Petri-markov nets

1 Introduction

A recommender system is understood as a set of algorithms and programs that aim to predict the user's interest in objects (goods, products, services) based on user data, data about the object, and the history of the relationship between objects and users of the system.

Recommender systems are now of great importance in everyday life. Before the advent of the Internet, social networks, or online commerce, people in their preferences were based on their own experience or recommendations of friends and relatives. Now, an opinion about a product or service can be formed based on the reviews of millions of people around the world.

Recommender systems are actively penetrating many areas of human activity. Their role as a defining vector in the system of modern trade and content search for information can hardly be overestimated.

Most of the recommendation systems are based on the analysis of accumulated statistical information about users and objects. The collection of information about user preferences can be done explicitly or implicitly. An example of an explicit collection of information is product reviews and recommendations from users themselves. The implicit collection of information includes a statistical analysis of a typical consumer basket.

Recommender algorithms are divided into two groups:

- Memory-based. Recommendations are created based on processing the entire available data set. Such algorithms are considered more accurate, but they require significant resource costs.

M. Paszynski et al. (Eds.): ICCS 2021, LNCS 12743, pp. 87–91, 2021.
https://doi.org/10.1007/978-3-030-77964-1_7

– Model-based. A model of relationships between users and objects is created, and recommendations for a specific user are already formed based on this model. This approach is considered less accurate due to the impossibility of real-time processing of new data, but it allows calculating recommendations in real-time.

2 Related Work

Most of the recommendation algorithms are based on processing existing user data. One of the main problems of such algorithms is the creation of recommendations for new users, not yet known to the system. This problem is also referred to in some literature as the cold-start problem. It is associated with the lack of explicit information on new users to predict recommendations.

To somehow personalize the user and achieve relevant recommendations, a latent classification of the user takes place based on cookies, demographic and geographic data, Internet activity, and data from social networks [5–9].

A fairly common approach to constructing recommender systems is the stochastic approach to describing the user-object relationship. Many researchers in their works use Markov chains to build a probabilistic model of recommendations. Thus, in their work, Shudong Liu and Lei Wang [1] presented a self-adaptive recommendation algorithm based on the Markov model, which gives recommendations to users depending on their geographic location. In the article by Mehdi Hosseinzadeh Aghdam [2], a hierarchical hidden Markov model is considered for revealing changes in user preferences over time by modeling the latent context of users. Fatma Mlika and Wafa Karoui [3] proposed an intelligent recommendation system model based on Markov chains and genre groupings. Yijia Zhang et al. [4] integrated social networks and Markov chains in their work to create a recommendation in a cold-start.

3 Proposed Model

In our work, we propose to use the mathematical apparatus of Petri-Markov nets to build a model of a recommender system in a cold start. This will combine the stochastic and structured approach to building recommendations. The Markov process allows one to differentiate the popularity of recommendation objects. Petri net makes it possible to impose restrictions on the recommendation net in appliance with the available latent data of users.

Petri-Markov nets (PMN) is a structural-parametric model defined by the set [10]:

$$Y = \{P, M\} \tag{1}$$

where P is a description of the structure of a bipartite graph, which is a Petri net; M is a description of the parameters imposed on the structure of P and determining the probabilistic and logical characteristics of the PMN.

The structure of the PMN allows combining the Markov probabilistic model of recommendations and the Petri net, which will allow formulating the conditions for developing recommendations for a specific user.

The PMN structure is characterized by the set:

$$P = \{X, Z, I_X(Z), O_X(Z)\} \tag{2}$$

where $X = \{x_{1(x)}, \ldots, x_{j(x)}, \ldots, x_{J(x)}\}$ is a finite set of places of the Petri net that simulate the initial conditions, categories, and objects of recommendations;

$Z = \{z_{1(z)}, \ldots, z_{j(z)}, \ldots, z_{J(z)}\}$ is a finite set of Petri net transitions that simulate the conditions for choosing a recommendation object;

$I_x(Z) = \{I_x(z_{1(z)}), \ldots, I_x(z_{j(z)}), \ldots, I_x(z_{J(z)})\}$ is the input transition function;

$O_x(Z) = \{O_x(z_{1(z)}), \ldots, O_x(z_{j(z)}), \ldots, O_x(z_{J(z)})\}$ is the output function of transitions;

$J(x)$ is the total number of places;

$J(z)$ is the total number of transitions.

In the context of the task of building a model of a recommender system:

- set places $X = \{x_{1(x)}, \ldots, x_{j(x)}, \ldots, x_{J(x)}\}$ can be represented by the mathematical similarity of the category, subcategory, or object of the recommendation;
- set transitions $Z = \{z_{1(z)}, \ldots, z_{j(z)}, \ldots, z_{J(z)}\}$ simulate a refined choice of a category or a recommendation object.

The Petri net P determines the structure of the PMN, and the random process M is superimposed on the structure of P and determines the probabilistic characteristics of the PMN.

The parametric aspects Petri-Markov nets are described by the following set:

$$M = \{q, p, \Lambda\} \tag{3}$$

where $q = \{q_{1(z)}, \ldots, q_{j(z)}, \ldots, q_{J(z)}\}$ is the vector of transition triggering probabilities;
$p = [p_{j(x)j(z)}]$ is the probability matrix;
$\Lambda = [\lambda_{j(Z)j(X)}]$ is a matrix of logical conditions, the elements of which are equal to

$$\lambda_{j(X)j(Z)} = \begin{cases} L\{\sigma[x_{j(x)} \in I_x(z_{j(z)}), z_{j(z)}]\}, & \text{if } x_{j(x)} \in O_x(z_{j(z)}); \\ 0, & \text{if } x_{j(x)} \notin O_x(z_{j(z)}); \end{cases} \tag{4}$$

The function L is a logical function that allows the execution of half-steps from transitions to states by the structure of the Petri net. $\sigma[x_{j(x)} \in I_x(z_{j(z)}), z_{j(z)}]$ is the half-step, which is defined as a logical variable that takes the values

$$\sigma[x_{j(x)} \in I_x(z_{j(z)}), z_{j(z)}] = \begin{cases} 1, & \text{if a half - step from place j(x) to transition j(z)} \\ & \text{is performed} \\ 0, & \text{if a half - step from place j(x) to transition j(z)} \\ & \text{is not performed} \end{cases} \tag{5}$$

Half-step $\sigma_{i(x)j(z)} = \sigma[x_{i(x)} \in I_x(z_{j(z)}), z_{j(z)}]$ or $\sigma_{j(z)i(x)} = \sigma[z_{i(z)}, x_{i(x)} \in O_x(z_{j(z)})]$ is called switching the state of PMN, in which from place $x_{i(x)} \in I_x(z_{j(z)})$ they get to the transition zj(z), or from the transition zj(z) they get to place $x_{i(x)} \in O_x(z_{j(z)})$. Two consecutive half-steps form a step.

Graphically, PMN are depicted in the form of oriented weighted digraph. Places are indicated by circles. Transitions are indicated by a bold line. The possibility of performing a half-step is indicated by an arrow.

Let's explain the approach to modeling using a small example of an online toy store. In Fig. 1 shows the model of recommendations based on the application of the Petri-Markov nets.

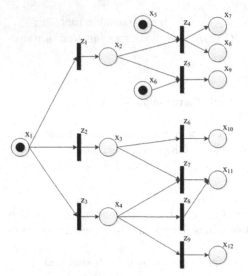

X_1 – starting position of the model
X_2 – category "Games"
X_3 – category "Model Trains & Railway Sets"
X_4 – category "Characters & Brands"
X_5 – condition "Region"
X_6 – condition "Age"
X_7 – product "Casino (Bingo) Equipment"
X_8 – product "Poker Chips – 120 Deluxe Gaming Chips Jaques of London"
X_9 – product "Drinking Games"
X_{10} – product "Bachmann 31-588 Freightliner Class 70 005 Powerhaul Diesel Weathered"
X_{11} – product "Thomas & Friends Trackmaster Paxton Motorised Engine"
X_{12} – product "Star Wars Black Series Figures Wave 4 – Luke Skywalker Dagobah"

Fig. 1. An example of a recommendation model based on a Petri-Markov nets.

The place X_1 defines the starting place of the model with the marker set. Let places X_2, X_3, X_4 represent product categories ("Games", "Model Trains & Railway Sets", "Characters & Brands"), and places X_7, X_8, X_9, X_{10}, X_{11}, X_{12} are products from the corresponding categories. Z transitions determine the probability of transition from one place to another. The probability of a transition being triggered is determined by the popularity of the category, subcategory, or the product itself among other users of the system.

Places X_5 and X_6 model additional conditions for triggering transitions. Markers in these places are set based on the analysis of latent information about the user. For example, place X_5 can be a geographic or regional condition for selecting items from places X_7 and X_8. And the marker in place X_6 can be an age restriction for triggering the transition Z_5. Then the objects corresponding to the X_7, X_8 and X_9 places will be offered as recommendations only if there is a marker in the place X_5 and X_6.

The proposed model combines various product classifications. The product corresponding to position X_{11} can be recommended both as a character of the popular cartoon "Thomas & Friends" and as a functional model of the steam locomotive "Thomas & Friends Trackmaster Paxton Motorized Engine".

4 Conclusion

The proposed model of the recommender system differs from the existing ones in that it combines the structural and probabilistic approaches to the construction of cold-start recommendations by using the mathematical apparatus of Petri-Markov nets. The model allows one to describe the complex structure of the classification relationships of recommendation objects and combine it with a probabilistic model of object selection.

The model allows:

- take into account a priori user data for making recommendations, which allows you to personalize the search model based on latent data from cookie files, data from social networks, etc.;
- to issue different recommendations to users each time due to the probabilistic model of triggering the transitions of the Petri-Markov nets;
- describe the complex structure of the classification relationships of the recommendation objects;
- to form a choice of recommendation objects by their popularity among other users of the system.

References

1. Liu, S., Wang, L.: A self-adaptive point-of-interest recommendation algorithm based on a multi-order Markov model. Futur. Gener. Comput. Syst. **89**, 506–514 (2018)
2. Aghdam, M.: Context-aware recommender systems using hierarchical hidden Markov model. Phys. A **518**, 89–98 (2019)
3. Mlika, F., Karoui, W.: Proposed model to intelligent recommendation system based on Markov chains and grouping of genres. Procedia Comput. Sci. **176**, 868–877 (2020)
4. Zhang, Y., Shi, Z., Zuo, W., Yue, L., Liang, S., Li, X.: Joint personalized Markov chains with social network embedding for cold-start recommendation. Neurocomputing **386**, 208–220 (2020)
5. Herce-Zelaya, J., Porcel, C., Bernabé-Moreno, J., Tejeda-Lorente, A., Herrera-Viedma, E.: New technique to alleviate the cold start problem in recommender systems using information from social media and random decision forests. Inf. Sci. **536**, 156–170 (2020)
6. Natarajan, S., Vairavasundaram, S., Natarajan, S., Gandomi, A.: Resolving data sparsity and cold start problem in collaborative filtering recommender system using Linked Open Data. Expert Syst. Appl. **149**, 52–61 (2020)
7. Silva, N., Carvalho, D., Pereira, A.C.M., Mourão, F., Rocha, L.: The Pure Cold-Start Problem: a deep study about how to conquer first-time users in recommendations domains. Inf. Syst. **80**, 1–12 (2019)
8. Pliakos, K., Joo, S., Park, J., Cornillie, F., Vens, C., Noortgate, W.: Integrating machine learning into item response theory for addressing the cold start problem in adaptive learning systems. Comput. Educ. **137**, 91–103 (2019)
9. Peng, F., Lu, J., Wang, Y., Xu, R., Ma, C., Yang, J.: N-dimensional Markov random field prior for cold-start recommendation. Neurocomputing **191**, 187–199 (2016)
10. Ignatev, V., Larkin, E.: Seti Petri-Markova. Uchebnoe posobie. TGU, Tula (1997)

Text-Based Product Matching with Incomplete and Inconsistent Items Descriptions

Szymon Łukasik[1,2,3]([✉]) [iD], Andrzej Michałowski[3], Piotr A. Kowalski[1,2][iD], and Amir H. Gandomi[4][iD]

[1] Faculty of Physics and Applied Computer Science, AGH University of Science and Technology, al. Mickiewicza 30, 30-059 Kraków, Poland
{slukasik,pkowal}@agh.edu.pl
[2] Systems Research Institute, Polish Academy of Sciences, ul. Newelska 6, 01-447 Warsaw, Poland
{slukasik,pakowal}@ibspan.waw.pl
[3] Synerise PLC, ul. Lubostroń 1, 30-383 Kraków, Poland
{andrzej.michalowski,szymon.lukasik}@synerise.com
[4] Faculty of Engineering and Information Technology, University of Technology Sydney, Ultimo, NSW 2007, Australia
gandomi@uts.edu.au

Abstract. In recent years Machine Learning and Artificial Intelligence are reshaping the landscape of e-commerce and retail. Using advanced analytics, behavioral modeling, and inference, representatives of these industries can leverage collected data and increase their market performance. To perform assortment optimization – one of the most fundamentals problems in retail – one has to identify products that are present in the competitors' portfolios. It is not possible without effective product matching. The paper deals with finding identical products in the offer of different retailers. The task is performed using a text-mining approach, assuming that the data may contain incomplete information. Besides the description of the algorithm, the results for real-world data fetched from the offers of two consumer electronics retailers are being demonstrated.

Keywords: Product matching · Assortment · Missing data · Classification

1 Introduction

Over the last ten years, the share of people ordering goods or services online increased significantly. In the United States between 2010 and 2019, e-commerce sales as a percent of total retail sales tripled and reached 12% [24]. The impact of machine learning and artificial intelligence in this context cannot be neglected. Over 70% of surveyed tech executives see the positive impact of AI on the retail industry, and 80% state that the use of AI allows maximizing profits [8]. Typically

© Springer Nature Switzerland AG 2021
M. Paszynski et al. (Eds.): ICCS 2021, LNCS 12743, pp. 92–103, 2021.
https://doi.org/10.1007/978-3-030-77964-1_8

machine learning is currently being used for personalization of services, managing relationships with customers and enterprise resource planning [12, 23, 26].

Optimizing assortment corresponds to the problem of selecting a portfolio of products to offer to customers to maximize the revenue [3]. Similarly, price optimization represents a problem of finding the best pricing strategy that maximizes revenue or profit based on demand forecasting models [13]. Both are crucial for gaining a competitive advantage in the market. To solve the aforementioned issues, it is important to match products present in the offers of competitors to the ones sold by the analyzed retailer. Such product matching is not a trivial problem as it involves using the information provided by different retailers. Its accuracy and completeness varies. Therefore designed algorithms have to deal with different data types, missing values, imprecise/erroneous entries.

The task of this paper is to present a new algorithm of product matching. It is based on analyzing product descriptions with text-mining tools. The matching problem is formulated here as an instance of classification. Our approach uses a decision-tree ensemble with gradient boosting as the learning algorithm. Together with chosen text similarity measures, it allows us to successfully deal with missing values and imprecise product descriptions. It is important to note that we assume here that no structured data regarding technical product attributes are available for both retailers – which is frequently the case.

The paper is organized as follows. In the next Section, an overview of existing approaches to product matching is being provided. It is followed by a brief summary of text similarity measures and methods of word embeddings. These elements serve as the methodological preliminary to the description of the proposed algorithm presented in Sect. 3. It is followed by the results of the first studies on the algorithm's performance. Finally, the last Section of the paper contains a conclusion and plans for further research.

2   Related Work

2.1   Product Matching

Product matching, in essence, can be perceived as an instance of a much broader problem of entity matching. Given two sets of products A and B – also known as product feeds – coming from two different sources (retailers) S_A and S_B, the task of product matching is to identify all correspondences between products in A and B. Such correspondence means that two matched products listed in A and B represent the same real-world object. The match result is typically represented by a set of correspondences together with similarity values $s_{ij} \in [0, 1]$ indicating the similarity or strength of the correspondence between the two objects [17].

There are not many machine learning papers dealing strictly with product matching. In [16] authors perform sophisticated feature engineering along with identifying product codes to improve matching performance. In [22] an approach using Siamese networks standard neural network using fastText (on concatenated product descriptions) is being tested. It also uses global product identifiers (if available). In [1] one can find an interesting application of fuzzy matching – with

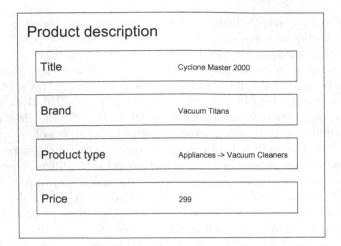

Fig. 1. Elements of product description

basic similarity measures (e.g. Q-grams and exact string matching). Study [20] is the most complete one of the analyzed here. It uses both textual information as well as images of products (feeding convolutional neural network). It uses standard Word2Vec embeddings and offers average matching performance. Finally, when overviewing literature on product matching it is worth noting here that there are no publicly available datasets, allowing us to test and compare various approaches to product matching. Most of the studies are using customized datasets – obtained or generated by authors themselves.

Product matching can be reformulated as a task of binary classification – its goal is to assign element x_{ij} to one of the two classes/categories with a known set of representative elements (known as the training set T) [5]. For product matching x_{ij} should represent a set of matching descriptors calculated for product i from feed A and product j from feed B. Such descriptors should essentially correspond to the similarity of products – calculated for different product attributes. Classification outcome for a given pair should be either crisp decision – match or no match, or matching probability value $p_{ij} \in [0, 1]$ corresponding to product similarity mentioned above.

There exists a variety of classification approaches that might be used in the aforementioned problem. Among others Support Vector Machines (SVM) [10], neural networks [9] or decision trees [18] could be named here.

As one of the typical features of product feeds is the incompleteness of product descriptions, the algorithms that are designed to deal with missing values are strongly preferred. Here we propose to use the XGBoost algorithm [6] based on decision-tree ensemble with gradient boosting. One of its essential features is treating the non-presence of attribute value as a missing value and learning the best direction to handle such a situation.

In the subsequent part of the paper, we will treat product description as a minimal set of features shown in Fig. 1 along with the exemplary product. The feature set contains the product name, brand, product type (or in other words category it belongs to). Besides these textual features, the product price is also available. Problems with the accuracy of such description can manifest in the form of:

1. errors in product's titles and overusing it to list product's attributes;
2. different structures of categories between two considered feeds;
3. missing brand;
4. hidden components of the price (e.g. costs of delivery).

To overcome these difficulties, product matching descriptors have to employ soft similarity measures allowing to compare not only the strict set of words used in the descriptions but also their semantic context. Possible solutions with this respect will be covered in the next subsection of the paper.

2.2 Text Similarity Measures

Finding similarities between texts is an essential task for a variety of text-mining applications. It can be perceived as a hierarchical procedure – with the similarity of individual words situated at the bottom level of this structure. It is fundamental for analyzing higher-level similarities: of sentences, paragraphs and, documents. Words can be similar in two aspects: lexical and semantic. The first corresponds to the occurrence of similar character sequences. On the other hand, words are similar semantically if they are associated with the same/opposing concepts or representing different kinds of objects of the same type. Lexical similarity can be analyzed with string-based approaches, while semantic one is usually tackled with corpus-based and knowledge-based algorithms. For a more detailed outlook on the text similarity measures, one could refer to the survey paper [11]. Here we will concentrate on lexical similarity as semantic context will be captured with the additional tool – covered in the next subsection.

The string-based similarity is calculated either on a character basis or using terms forming a string. The first approach tries to quantify differences between the sequences of letters. One of the well-known technique of this type is Levenshtein distance. Levenshtein distance between two words corresponds to the minimum number of single-character edits (insertions, deletions or substitutions) required to change one word into the other [27]. Damerau-Levenshtein distance represents a generalization of this concept – by including transpositions among its allowable operations [7].

As one of the most popular techniques of term-based similarity evaluation Jaccard similarity could be named [14]. It is calculated by:

$$J(A, B) = \frac{|A \cap B|}{|A \cup B|} = \frac{|A \cap B|}{|A| + |B| + |A \cap B|}. \tag{1}$$

It means that to obtain its value for text matching, one has to divide the number of words present in both words by the number of words present in either of the documents.

2.3   Word Embeddings

The aim of word embedding is to find a mapping from a set of words to a set of multidimensional – typically real-valued – vectors of numbers. A useful embedding provides vector representations of words such that the relationship between two vectors mirrors the linguistic relationship between the two words [21]. In that sense, word embeddings can be particularly useful for comparing the semantics of short texts.

The most popular approach to word embeddings is Word2Vec technique [19], which can build embeddings using two methods: CBOW model, which takes the context of each word as the input and tries to predict the word corresponding to the context and skip-gram model. The latter takes every word in large corpora and also takes one-by-one the words that surround it within a defined window to then feed a neural network that after training will predict the probability for each word to actually appear in the window around the focus word [25].

fastText is a more recent approach based on the skipgram model, where each word is represented as a bag of character n-grams. A vector representation is associated with each character n-gram, with words being represented as the sum of these representations. What is important from the point of product matching fastText allows us to compute word representations for words that did not appear in the training data [4]. fastText, by default, allows us to obtain an embedding with 300 real-valued features, with the models for a variety of languages already built.

3   Proposed Approach

The algorithm introduced here employs only textual features listed in Fig. 1 – which are available for almost all product feeds. Matching descriptors are formed of real-valued vectors consisting of 6 features:

1. The first element of the feature vector corresponds to the Jaccard similarity obtained for titles of tested products from Feed A and Feed B. It is aimed at grasping the simplest difference between products – in their names.
2. The second is calculated as the cosine similarity between 300-dimensional fastText embeddings of product types, with prebuilt language models corresponding to the retailer's business geographic location. In this way, synonyms used by different retailers in product categories are being properly handled.
3. The third feature corresponds to the Damerau-Levenshtein similarities calculated for the brand of two products under investigation. It allows us to ignore any typos or misspellings.
4. Fourth element is calculated as a relative price difference:

$$\frac{|priceA - priceB|}{\max(priceA, priceB)} \tag{2}$$

and it is inspired by the fact that the same product sold by two different retailers would be priced similarly for both of them.

5. Finally, two last features are built using 2-dimensional embedding with Principal Components Analysis [15] for the product type. It is obtained from the 300-dimensional fastText embeddings. It allows us to differentiate the impact of other features between different product categories.

Constructed features serve as an input to classifier – pretrained using manually processed dataset containing feature values and corresponding matching labels. As a classifier extreme gradient boosting ensemble (XGboost) is being used. Thanks to the sparsity-aware split finding algorithm, it is able to accurately handle missing values (denoted as NaN), which might be present in the dataset [2].

Figure 2 summarizes the detailed scheme of the algorithm.

4   Experimental Results

We have preliminarily tested the proposed approach on the real-world instance of product matching. To achieve this goal, we have been using two product feeds – obtained by web-scraping – from two EU-based consumer electronics retailers.

Feed A consists of 130 000 products, Feed B - of 30 000 items. Such an imbalance of feeds sizes makes the problem even more challenging. In addition to that, their category trees are not consistent. Examples of observed inconsistencies were shown in Table 1. It can be seen that the same group of products are not located for Feed A and Feed B under the same path in their category trees.

The matching model was trained using a set of 500 randomly chosen and manually labeled (as match or no match) pairs of products. The testing dataset consisted of 100 randomly chosen products from feed A. For the purpose of subsequent analysis, matchings obtained from the classifier for this set of items were again analyzed manually.

Table 1. Examples of inconsistencies in category trees

Feed A	Feed B
Small appliances > Kitchen - cooking and food preparation > Kitchen utensils > Breadboxes	Small appliances > Kitchen accessories > Breadboxes
Audio-video devices and TV sets > Car Audio and Navigation > Sound > Speakers	Audio-video devices and TV sets > Car audio > Speakers
Parent Zone > Feeding > Breast pumps	Small household appliances > For children> Breast pumps
Audio-video devices and TV sets > HiFi Audio > DJ controllers	Audio-video devices and TV sets > Party > DJ controllers
Small household appliances > Home - cleanliness and order > Vacuum cleaners > Standard vacuum cleaners	Small household appliances > Cleaning > Standard vacuum cleaners

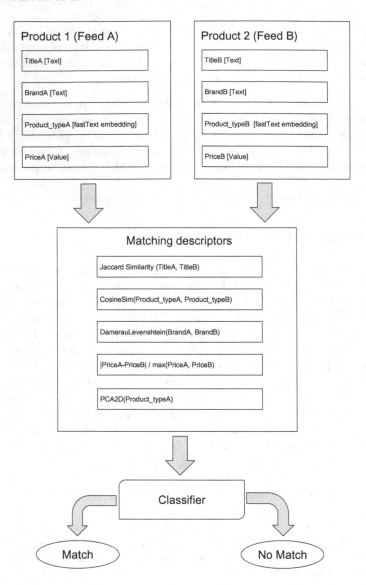

Fig. 2. Proposed product matching algorithm scheme

Figure 3 demonstrates feature importances during the training phase reported by XGBoost. It can be seen that naturally, product title is a very important factor to take into account while matching products. Still, other descriptors, such as price ratio or category of product being matched, seem to have a non-negligible impact.

Table 2 contains selected pairs with the highest matching probabilities calculated with respect to the product from the main feed A. The first four rows

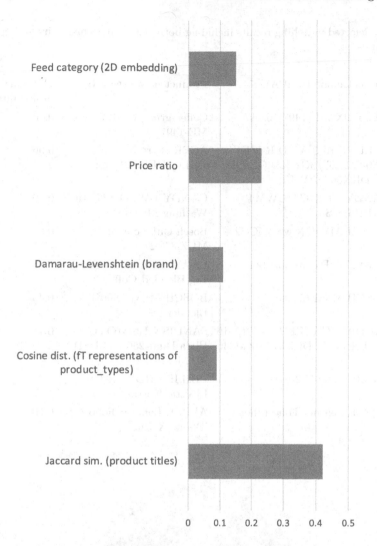

Fig. 3. Importance of features obtained while fitting model to the training dataset

demonstrate examples of a perfect match – when the algorithm is able to recover product similarities very well, even though in some cases, the compatibility for product names is not perfect. The rest of the table presents cases where the algorithm (as a matching probability threshold, a natural value of $p_{ij} > 0.5$ could be chosen) was able to detect a no-match situation. In all of these cases, rejection of matching was not so evident as product descriptions contained a lot of similar elements.

Table 2. Selected matching results including both closest matched pairs and matching probabilities

Product name (feed A)	Product name (feed B)	Matching probability
METROX ME-1497 Black	Coffee grinder METROX ME-1497	0.86
ACER Nitro 5 AMD RYZEN 5 2500U/15.6"/8GB/256GBSSD/ AMDRX560X/W10	ACER Nitro 5 (NH.Q3REP.021) Laptop	0.96
CANDY BIANCA BWM4 137PH6/1-S	CANDY BWM4 137PH6 Washing Machine	0.90
BOSCH ADDON MUZ5CC2	Bosch Cube cutter MUZ5CC2	0.83
LAVAZZA Pienaroma 1kg	LAVAZZA Qualita Oro 1kg Blended Coffee	0.28
BOSCH MSM 67190	BOSCH MSM 6S90B Blender	0.45
SANDISK CRUZER ULTRA 16 GB USB 3.0 SDCZ73-016G-G46	SANDISK USB/OTG Ultra Dual 256 GB USB 3.0 Memory	0.11
PHILIPS HD 9126/00	PHILIPS HD 4646/70 Electric Kettle	0.34
APPLE Leather Folio iPhone XS, Peony Pink	APPLE Leather Folio for iPhone X Black	0.50

■ Matched properly   ■ No match found   ■ Wrong matching

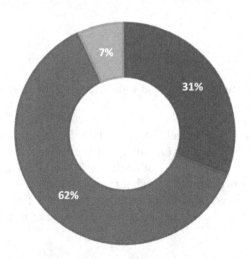

Fig. 4. Statistics of matching results

Figure 4 illustrates the results of matching quantitatively. It can be seen that the majority of products from feed A have not found a match from a corresponding feed B. It is a natural consequence of inequality in the number of products between those feeds. At the same time, of 38 obtained matches (with probability > 0.5), 31 of them were correct (which accounts for 81% of all matches). It demonstrates the robustness of the method and its reasonable performance also when compared with other studies in the field of product matching (e.g. see [20]).

5   Conclusion

The paper examined the possibility of using text-based machine learning algorithms on sets of product feeds to obtain product matchings. Our goal was to introduce the approach using the minimal amount of information from the product description, at the same time tolerating inaccurate and missing information provided in the product feeds.

Through the first round of experiments performed on the real-world data the algorithm proved to be effective and robust. Further studies will involve using additional, optional matching descriptors such as product images and their technical features. Experiences from the field of product matching will be used for building new category matching algorithms – which are also important from the data presentation standpoint. The existing framework of obtaining product descriptors could also be used for other tasks such as product grouping and categorization.

Acknowledgment. The work was supported by the Faculty of Physics and Applied Computer Science AGH UST statutory tasks within the subsidy of MEiN.

References

1. Amshakala, K., Nedunchezhian, R.: Using fuzzy logic for product matching. In: Krishnan, G.S.S., Anitha, R., Lekshmi, R.S., Kumar, M.S., Bonato, A., Graña, M. (eds.) Computational Intelligence, Cyber Security and Computational Models. AISC, vol. 246, pp. 171–179. Springer, New Delhi (2014). https://doi.org/10.1007/978-81-322-1680-3_20
2. Rusdah, D.A., Murfi, H.: XGBoost in handling missing values for life insurance risk prediction. SN Appl. Sci. **2**(8), 1–10 (2020). https://doi.org/10.1007/s42452-020-3128-y
3. Bernstein, F., Kök, A.G., Xie, L.: Dynamic assortment customization with limited inventories. Manuf. Serv. Oper. Manag. **17**, 538–553 (2015)
4. Bojanowski, P., Grave, E., Joulin, A., Mikolov, T.: Enriching word vectors with subword information. CoRR abs/1607.04606 (2016). http://arxiv.org/abs/1607.04606
5. Charytanowicz, M., Niewczas, J., Kulczycki, P., Kowalski, P.A., Łukasik, S.: Discrimination of wheat grain varieties using x-ray images. In: Piętka, E., Badura, P., Kawa, J., Wieclawek, W. (eds.) Information Technologies in Medicine. AISC, vol. 471, pp. 39–50. Springer, Cham (2016). https://doi.org/10.1007/978-3-319-39796-2_4

6. Chen, T., Guestrin, C.: Xgboost: a scalable tree boosting system. In: Proceedings of the 22nd ACM SIGKDD International Conference on Knowledge Discovery and Data Mining, pp. 785–794. KDD 2016, Association for Computing Machinery, New York, NY, USA (2016). https://doi.org/10.1145/2939672.2939785, https://doi.org/10.1145/2939672.2939785

7. Damerau, F.J.: A technique for computer detection and correction of spellingerrors. Commun. ACM **7**(3), 171–176 (1964). https://doi.org/10.1145/363958.363994

8. Edelman: 2019 Edelman AI Survey. Whitepaper, Edelman (2019)

9. Faris, H., Aljarah, I., Mirjalili, S.: Training feedforward neural networks using multi-verse optimizer for binary classification problems. Appl. Intell. **45**(2), 322–332 (2016). https://doi.org/10.1007/s10489-016-0767-1

10. Gaspar, P., Carbonell, J., Oliveira, J.: On the parameter optimization of support vector machines for binary classification. J. Integr. Bioinform. **9**(3), 201 (2012). https://doi.org/10.2390/biecoll-jib-2012-201

11. Gomaa, W., Fahmy, A.: A survey of text similarity approaches. Int. J. Comput. Appl. **68**(13), 13–18 (2013). https://doi.org/10.5120/11638-7118

12. Ismail, M., Ibrahim, M., Sanusi, Z., Cemal Nat, M.: Data mining in electronic commerce: benefits and challenges. Int. J. Commun. Netw. Syst. Sci. **8**, 501–509 (2015). https://doi.org/10.4236/ijcns.2015.812045

13. Ito, S., Fujimaki, R.: Large-scale price optimization via network flow. In: Lee, D.D., Sugiyama, M., Luxburg, U.V., Guyon, I., Garnett, R. (eds.) Advances in Neural Information Processing Systems 29, pp. 3855–3863. Curran Associates, Inc. (2016). http://papers.nips.cc/paper/6301-large-scale-price-optimization-via-network-flow.pdf

14. Ivchenko, G., Honov, S.: On the jaccard similarity test. J. Math. Sci. **88**(6), 789–794 (1998)

15. Jolliffe, I.: Principal Component Analysis. Springer Verlag, New York (2002)

16. Köpcke, H., Thor, A., Thomas, S., Rahm, E.: Tailoring entity resolution for matching product offers. In: Proceedings of the 15th International Conference on Extending Database Technology, pp. 545–550. EDBT 2012, Association for Computing Machinery, New York, NY, USA (2012). https://doi.org/10.1145/2247596.2247662, https://doi.org/10.1145/2247596.2247662

17. Köpcke, H., Rahm, E.: Frameworks for entity matching: a comparison. Data Knowl. Eng. **69**(2), 197 – 210 (2010).https://doi.org/10.1016/j.datak.2009.10.003,http://www.sciencedirect.com/science/article/pii/S0169023X09001451

18. Liu, L., Anlong Ming, Ma, H., Zhang, X.: A binary-classification-tree based framework for distributed target classification in multimedia sensor networks. In: 2012 Proceedings IEEE INFOCOM, pp. 594–602 (March 2012). https://doi.org/10.1109/INFCOM.2012.6195802

19. Mikolov, T., Sutskever, I., Chen, K., Corrado, G., Dean, J.: Distributed representations of words and phrases and their compositionality (2013)

20. Ristoski, P., Petrovski, P., Mika, P., Paulheim, H.: A machine learning approach for product matching and categorization: use case: enriching product ads with semantic structured data. Semant. Web **9**, 1–22 (2018). https://doi.org/10.3233/SW-180300

21. Schnabel, T., Labutov, I., Mimno, D., Joachims, T.: Evaluation methods for unsupervised word embeddings. In: Proceedings of the 2015 conference on empirical methods in natural language processing, pp. 298–307 (2015)

22. Shah, K., Kopru, S., Ruvini, J.D.: Neural network based extreme classification and similarity models for product matching. In: Proceedings of the 2018 Conference of the North American Chapter of the Association for Computational Linguistics: Human Language Technologies, vol. 3 (Industry Papers), pp. 8–15. Association for Computational Linguistics, New Orleans - Louisiana (Jun 2018). https://doi.org/ 10.18653/v1/N18-3002, https://www.aclweb.org/anthology/N18-3002
23. Srinivasa Raghavan, N.R.: Data mining in e-commerce: a survey. Sadhana **30**(2), 275–289 (2005). https://doi.org/10.1007/BF02706248
24. US Census Bureau: quarterly retail e-commerce sales. News report CB19-170, US Census Bureau,19 November 2019
25. Vieira, A., Ribeiro, B.: Introduction to deep learning business applications for developers: from Conversational Bots in Customer Service to Medical Image Processing. Apress (2018). https://books.google.pl/books?id=K3ZZDwAAQBAJ
26. Yu, G., Xia, C., Guo, X.: Research on web data mining and its application in electronic commerce. In: 2009 International Conference on Computational Intelligence and Software Engineering, pp. 1–3 (December 2009). https://doi.org/10. 1109/CISE.2009.5363366
27. Yujian, L., Bo, L.: A normalized levenshtein distance metric. IEEE Trans. Pattern Anal. Mach. Intell. **29**(6), 1091–1095 (2007). https://doi.org/10.1109/TPAMI. 2007.1078

Unsupervised Text Style Transfer via an Enhanced Operation Pipeline

Wanhui Qian[1,2], Jinzhu Yang[1,2], Fuqing Zhu[1(✉)], Yipeng Su[1],
and Songlin Hu[1,2]

[1] Institute of Information Engineering, Chinese Academy of Sciences,
Guangzhou, China
{qianwanhui,yangjinzhu,zhufuqing,suyipeng,husonglin}@iie.ac.cn
[2] School of Cyber Security, University of Chinese Academy of Sciences,
Beijing, China

Abstract. Unsupervised text style transfer aims to change the style attribute of the given unpaired texts while preserving the style-independent semantic content. In order to preserve the content, some methods directly remove the style-related words in texts. The remaining content, together with target stylized words, are fused to produce target samples with transferred style. In such a mechanism, two main challenges should be well addressed. First, due to the style-related words are not given explicitly in the original dataset, a detection algorithm is required to recognize the words in an unsupervised paradigm. Second, the compatibility between the remaining content and target stylized words should be guaranteed to produce valid samples. In this paper, we propose a multi-stage method following the working pipeline – *Detection*, *Matching*, and *Generation*. In the *Detection* stage, the style-related words are recognized by an effective joint method and replaced by mask tokens. Then, in the *Matching* stage, the contexts of the masks are employed as queries to retrieve target stylized tokens from candidates. Finally, in the *Generation* stage, the masked texts and retrieved style tokens are transformed to the target results by attentive decoding. On two public sentimental style datasets, experimental results demonstrate that our proposed method addresses the challenges mentioned above and achieves competitive performance compared with several state-of-the-art methods.

Keywords: Unsupervised learning · Natural language generation · Sentiment transfer

1 Introduction

Text style transfer task focuses on alternating the style attribute of given texts while preserving the original semantic content and has drawn much attention from natural language generation community, especially with the inspiration of transfer learning. Due to the ability to modify attributes of texts in a fine-grained manner, text style transfer has potential applications on some specific tasks, such

© Springer Nature Switzerland AG 2021
M. Paszynski et al. (Eds.): ICCS 2021, LNCS 12743, pp. 104–117, 2021.
https://doi.org/10.1007/978-3-030-77964-1_9

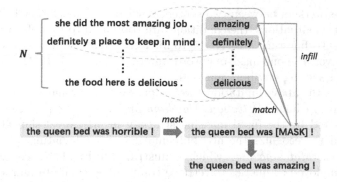

Fig. 1. An illustration of the context matching operation. The context of the '[MASK]' looks up the table of attribute markers (for simplicity, attribute markers are all set as words) extracted from N samples sentences. The word '*amazing*' is selected to infill the mask slot.

as dialogue system [19], authorship obfuscation [25]. However, pair-wised style-to-style translation corpora are hard to obtain. Therefore, searching an effective unsupervised method to conduct style transfer on unpaired texts has become a primary research direction.

Majority of the existing methods focus on separating style information and content apart at the first step. According to the separation manner, these methods fall into two groups – implicit disentanglement and explicit disentanglement. Specifically, implicit disentanglement methods [6,17,24] usually leverage the adversarial training strategy [7] to formulate the consistency of content distribution disentangled with various styles. Explicit disentanglement methods [31,32,34] recognize the *attribute markers* (words or phrases in the sentence, indicating a particular attribute[1]) in the original sentences[2], and replace the attribute markers with the expected ones. Compared to the implicit disentanglement operation, explicit replacement improves the model interpretability and enhances the ability of content preservation. However, the existing explicit disentanglement methods have apparent limitations in handling two problems.

First, selecting an effective attribute marker detection algorithm is critical for subsequent processing. Frequency-based [16] and Attention-based methods [32] are designed for detecting attribute markers. However, due to the inherent flaws, both of the two methods can hardly maintain the detection precision at a relatively high level. Second, the compatibility of the target attribute markers and style-independent content is the guarantee for producing valid sentences. For example, if we want to transfer the sentiment style of "*I love the movie*" from positive to negative, the word '*love*' should be replaced with '*hate*', but not '*disappoint*'. *DeleteAndRetrieve* [16] achieves the style transfer by exchanging

---

[1] The concept of *attribute marker* is borrowed from the work in [16], we use this term to indicate the style information.

[2] In this work, the 'text' and 'sentence' terms are exchangeable.

the attribute markers between two sentences with similar content but different styles. In actual situations, the conditions are too harsh to be well satisfied. Wu et al. (2019) [31] fine-tune a BERT [5] to infer the attribute markers to infill the masks. However, the one mask-one word infilling manner of BERT restricts the flexibility of selecting attribute markers.

To address the above limitations, we propose a multi-stage method following the enhanced pipeline – *Detection*, *Matching* and *Generation*. In the *Detection* stage, motivated by the fused masking strategy in [31], we design an effective joint method to recognize the attribute markers for alleviating the detection problem. In the *Matching* stage which is illustrated in Fig. 1, the contexts of the masks are employed as queries to retrieve compatible attribute markers from a set of sampled attribute tokens. In the last *Generation* stage, the masked texts and retrieved attribute markers are fed into a decoder to produce the transferred samples. To ensure the style is correctly transferred, we introduce an auxiliary style classifier pre-trained on the non-parallel corpus.

We conduct experiments on public two sentiment style datasets. Three aspects indicators are utilized to evaluate the transferred samples, i.e., target style accuracy, content preservation and language fluency. The main contributions of this paper are summarized as follows:

– A multi-stage (including *Detection*, *Matching* and *Generation*) method is proposed for unsupervised text style transfer task, transferring the given unpaired texts by explicitly manipulation.
– In the *Detection* and *Matching* stages, the attribute marker detection and content-marker compatibility problems are well alleviated by the joint detection method and the context-matching operation.
– Experimental results demonstrate that the proposed model achieves competitive performance compared with several state-of-the-art methods. Besides, the model gains an excellent trade-off between style accuracy and content preservation.

2   Related Work

Recently, unsupervised text style transfer task has attracted broad interest. Shen et al. (2017); Fu et al. (2018); and John et al. (2019) [6,9,24] assume that the style and content can be separated via generative models (e.g. GAN [7], VAE [12]). Follow the above assumption, Yang et al. (2018) [33] adopt a pre-trained language model to improve the fluency of the generated sentences. Considering the content preservation, Prabhumoye et al. (2018) [22] design a dual language translation model. The semantic content in the source and target sentences of translation remains unchanged. Logeswaran et al. (2018) [17] discard the language translation process and back-translate the transferred sentence to the original sentence directly. Despite the developments of those methods, some work [4,15,29] suspects the efficiency of the separation in latent space and proposes methods with end-to-end translation fashion. Some work attempts to pre-build a parallel dataset from the original non-parallel corpus. Zhang et al. (2018a) [35]

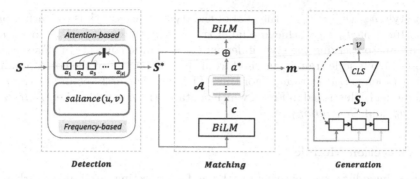

Fig. 2. Overview structure of the proposed model. First, the original sentence s is fed into the *Detection* module, the attribute markers are recognize and replaced by masks. Then the remaining content s^* is passed to the *Matching* module. The bidirectional language model BiLM extracts the contexts c from s^*, and perform a matching operation between mask context in c and a sampled attribute token set $\mathcal{A}$. The retrieved results a^* and s^* are combined and encoded into memories m. Finally, the *Generation* module decodes the m to the target sentence s_v. Additionally, a pre-trained style classifier CLS is appended to the decoder to strengthen the style control ability of the decoder.

initialize the dataset by the unsupervised word mapping techniques [1, 14] which are widely used in the unsupervised machine translation task. Jin et al. (2019) [8] construct sentence pairs according to the distance measurement between sentences. More straightforwardly, Luo et al. (2019) [18] utilize the transferred results of [16] as the initial target samples.

In our work, the style of sentences is transferred through word-level manipulation. The content of the original sentences is preserved effectively, and the manipulation process is interpretable. Previous methods based on explicit word manipulation usually improve the word detection process or the generation process. Zhang et al. (2018b)[34] propose a self-attention method to detect attribute words. Li et al. (2018) [16] present a frequency-based method and four generating strategies. [32] adopt a similar attention-based method with that in [34] and propose a cycle-reinforcement learning algorithm. Wu et al. (2019) [31] fuse the detection methods in [16] and [32], and generate target sentences with a fine-tuned BERT [5]. Sudhakar et al. (2019) [26] improve each operation step in [16] and gain better performance. Compared to the above methods, our model overcomes the inherent weakness of previous detection methods and a well-designed joint method is utilized to locate attribute markers. Additionally, our method retrieves target attribute markers by a matching operation, and the final results are generated through attentive decoding.

3   Proposed Method

In this paper, we employ the corpus with a style set $\mathcal{V}$. Each collection $\mathcal{D}_v$ represents the sentences with the style $v \in \mathcal{V}$. Given any source style v_{src} and any target style v_{tgt} ($v_{\text{src}} \neq v_{\text{tgt}}$), the goal of the style transfer task is to learn a projection function $f_{\text{src}\rightarrow\text{tgt}}$ achieving label transfer of sentences in $\mathcal{D}_{v_{\text{src}}}$ from v_{src} to v_{tgt} while preserving the style-independent semantic content. In this section, we describe our style transfer method from the perspective of the corresponding working pipeline and learning algorithm. The overview architecture of the proposed model is illustrated in Fig. 2, which contains three modules (i.e., *Detection*, *Matching*, and *Generation*).

3.1   Detection Module

We first introduce two existing methods for detecting attribute markers – Frequency-based method [16] and Attention-based method [32]. Then, a joint detection method is proposed for better identifying attribute markers.

Frequency-based Method. If the frequency of an given n-gram u appears in $\mathcal{D}_v$ is much higher than that in other datasets, u has a higher probability of being a v-style attribute marker. Specifically, for a given n-gram u and a chosen style v from the style set $\mathcal{V}$, a quantity $s(u,v)$ called *salience* is defined for the statement as:

$$s(u,v) = \frac{\text{count}(u, \mathcal{D}_v) + \lambda}{\sum_{v' \in \mathcal{V}, v' \neq v} \text{count}(u, \mathcal{D}_{v'}) + \lambda}, \tag{1}$$

where $\text{count}(u, \mathcal{D}_v)$ is the number of times that an n-gram u appears in $\mathcal{D}_v$, and λ is the smoothing parameter. If $s(u,v)$ is larger than a predefined threshold, u will be identified as an attribute marker for the style v.

Attention-based Method. To apply this method, a pre-trained attention-based LSTM classifier is required. Given a sentence consists of a sequence of tokens $s = [t_1, t_2, \ldots, t_{|s|}]$ with style $v \in \mathcal{V}$, a LSTM module first encodes the tokens into a sequence of hidden states $h = [h_1, h_2, \ldots, h_{|s|}]$. Then, an attention operation is conducted between v and h to obtain a sequence of normalized weights $\alpha = [\alpha_1, \alpha_2, \ldots, \alpha_{|s|}]$:

$$\alpha = \text{softmax}(\text{attention}(v, h)), \tag{2}$$

Finally, an accumulated feature vector $\alpha \cdot h$ is utilized to predict the style label v.

In such a process, the weights α can be regarded as the contributions of corresponding tokens for style prediction. Therefore, we use α_i denoting the score of style for token t_i, $i \in \{1, 2, \ldots, |s|\}$. To identify the attribute markers, we define the attribute identifier $\hat{\alpha}_i$ as follows:

$$\hat{\alpha}_i = \begin{cases} 0, & \text{if } \alpha_i < \bar{\alpha}; \\ 1, & \text{if } \alpha_i \geq \bar{\alpha}, \end{cases} \tag{3}$$

where the $\bar{\alpha}_i$ is the mean of weights α. If the identifier $\hat{\alpha}_i$ is equals to 1, the corresponding token will be recognized as a single-token attribute marker, otherwise, the token will be ignored.

Joint Detection Method. Even though the above two methods implement the extraction of attribute markers, they both have some limitations. Occasionally, as mentioned in [31], the Frequency-based method recognizes style-independent grams as attribute markers. This misdetection ruins both style transfer and content preservation. As for the Attention-based method, the LSTM module affects the precision of detecting the attribute markers. Because the last output of the LSTM accumulates the full sequential information, resulting in that the classifier tends to pay more attention to the end of outputs. To alleviate the inherent problems, we design a joint heuristic method combining the Frequency-based method and Attention-based method.

Given style v, the joint method assign a score α_u^* for a gram u according to following equation:

$$\alpha_u^* = \alpha_u * \log s(u, v). \tag{4}$$

The criterion of decision making has the same formulation with Eq. 3. We explain the formulation of the joint detection method (Eq. 4) from three points:

1. The 'misdetection' problem in Frequency-based method is alleviated by multiplying the score α_u acquired from Attention-based method.
2. The LSTM module in Attention-based method is removed to prevent the 'focusing-on-last' problem.
3. The logarithm over $s(u, v)$ prevents the extremely large values affecting the detection accuracy.

Once the attribute markers detected, we replace them with mask tokens (one mask for one attribute marker).

3.2   Matching Module

In this module, we construct compatible attribute markers through matching operation. The overall process consists of determining queries, constructing candidates, and matching.

In our implementation, the queries are the contexts of corresponding masks. Ubiquitous bidirectional language models BiLM (such as BiLSTM [21], Transformer Encoder [5]) are effective tools for extracting contexts. Given a masked sentence, $s^* = [t_1, t_2, \ldots, t_{|s^*|}]$. Without loss of generality, we/ suppose that the k-th token in s^* is a mask[3]. The context information $c = [c_1, c_2, \ldots, c_{|s^*|}]$ is obtained by:

$$c = \text{BiLM}(s^*). \tag{5}$$

Obviously, the k-th element c_k is the context for mask t_k. Now the query c_k is prepared, the next problem is how to construct the candidates to be matched.

By investigating the corpus of interest, we observe that the frequencies of attribute tokens (Words that consist attribute markers) have the long-tail phenomena. Most of the attribute tokens appear only a few times, and the vast

---

[3] To simplify the demonstration, only one mask t_k is considered. The strategy of multi-masks is the parallel situation for single mask.

majority of sentences share a small number of attribute tokens. Therefore, we randomly samples N sentences and extract the attribute tokens from the sentences as the candidates, represented by $\mathcal{A} = [a_1, a_2, \ldots, a_{|\mathcal{A}|}]$.

The matching operation indicates an attention between c_i and $\mathcal{A}$, which produces the a compatible attribute marker a^*:

$$\beta = \text{softmax}(\text{attention}(c_i, \mathcal{A})), \tag{6}$$

$$a^* = \beta \cdot \mathcal{A}, \tag{7}$$

where the β is normalized attention weights. The attentive result a^* is the weighted sum of candidates in $\mathcal{A}$, which can be viewed as a composition of attribute tokens. A new sentence representation $\hat{s} = [t_1, t_2, \ldots, a^*, \ldots, t_{|s^*|}]$ is acquired by replacing t_k with a^* in s^*. At last, we encode $\hat{s}$ to an external memories m for further processing:

$$m = \text{BiLM}(\hat{s}). \tag{8}$$

3.3 Generation Module

Compared to the vanilla auto-encoder framework, decoding with attention enables to stabilize the generation of sentences. The generation module adopts the attention structure from [2].

A sentence s_v with style v is generated by recurrent decoding:

$$s_v = \text{Decoder}(I_v, m), \tag{9}$$

where the given style indicator I_v for style v is set as the initial hidden state of the recurrent decoder, and m is the external memory from the matching module to be attended.

Previous methods for text style transfer usually adopt vanilla auto-encoder as the backbone. The attention mechanism is disabled because the sentence reconstruction tends to degenerate to a copy operation. Inspired by the solution in [13], we add noises to original sentences to prevent the corruption. Each token in content has an equivalent probability p_{noise} to be removed, replaced or appended with a sampled token.

3.4 Learning Algorithm

The learning process has two steps. The first step attempts to reconstruct the original sentence s with style v_{src}. The second step evaluates the style accuracy of the transferred samples s_{tsf} with target style v_{tgt}. Two steps produce two losses – reconstruction loss and classification loss, respectively.

Reconstruction Loss. Due to the non-parallel nature of datasets, we follow a self-transfer routine $s \rightarrow s^* \rightarrow s$. To recover the original sentence s, the attribute token candidates $\mathcal{A}_{\text{src}}$ is aggregated with those extracted from sampled sentences in $\mathcal{D}_{v_{\text{src}}}$ (s is included in the sampled sentences). At generation step, the style indicator I_v is set as $I_{v_{\text{src}}}$. The reconstruction loss is formulated as:

$$\mathcal{L}_{\text{rec}} = -\log p(s | s^*, \mathcal{A}_{\text{src}}, I_{v_{\text{src}}}). \tag{10}$$

Classification Loss. As previous methods done, we append an auxiliary pre-trained style classifier to the end of decoder. The classifier aims to enhance the control over style transferring. The prediction routine is $s \to s^* \to s_{\mathrm{tsf}} \to p(v_{\mathrm{tgt}}|s_{\mathrm{tsf}})$, the transferred sample s_{tsf} is expected to possess the target style v_{tgt}. To enable the style transfer, the attribute token candidates $\mathcal{A}_{\mathrm{tgt}}$ consists of those extracted from sampled sentences in $\mathcal{D}_{v_{\mathrm{tgt}}}$. The generation indicator is set as $I_{v_{\mathrm{tgt}}}$. The classification loss is:

$$\mathcal{L}_{\mathrm{cls}} = -\log p(v|s^*, \mathcal{A}_{\mathrm{tgt}}, I_{v_{\mathrm{tgt}}}). \tag{11}$$

To resolve the discreteness problem of texts, we adopt the same strategy in [31].

The final optimization target is:

$$\min_{\phi} \mathcal{L} = \mathcal{L}_{\mathrm{rec}} + \eta \mathcal{L}_{\mathrm{cls}}, \tag{12}$$

where ϕ represents the trainable parameter set of the model, and η is a predefined parameter for scaling the classification loss.

4   Experiments

In this section, we first describe the experimental settings, including datasets, baselines, evaluation metrics and experimental details. Then we show the experimental results and analysis, where some comparisons will be provided to demonstrate the effectiveness of the proposed method.

4.1   Datasets

This paper evaluates the performance of the proposed method on two public sentimental style datasets released by [16], i.e., *Yelp* and *Amazon*, including reviews with binary polarity. The above datasets are pre-processed and divided into three sets for training, developing and testing. Additionally, crowd-workers are hired on Amazon Mechanical Turk to write references for all testing sentences [16], ensuring each of the references hold opposite sentiment and similar content with the original sentence. The references can be regarded as the standard gold outputs to evaluate the performances of the proposed model.

4.2   Baselines

In this paper, the following state-of-the-art methods are employed as baselines for comparison, including **CrossAE** [24], three strategies (**Template, DeleteOnly, Del&Retr**) in [16], **CycleS2S** [32], **C-BERT** [31], **DualRL** [18], **PTO** [30].

4.3   Evaluation Metrics

Automatic Evaluation. Following the work [16,24], we estimate the **style accuracy** (ACC) of the transferred sentences with a pre-trained classifier of fastText[4] [10]. The validation accuracy on development set of *Yelp* and *Amazon*

---

[4] https://github.com/facebookresearch/fastText.

achieves 97% and 80.4%, respectively. Similar to the machine translation task, the BLEU [20] score between generated result and the human reference is the measurement of **content preservation**. In our implementation, the BLEU score is calculated through the moses script[5]. To evaluate the **fluency** of sentences, we pre-train two language models on *Yelp* and *Amazon* datasets by fine-tuning two distinct GPT-2[6] [23]. The perplexity (PPL) of transferred sentences indicates the fluency rate.

Human Evaluation. For either *Yelp* or *Amazon*, we sample and annotate 100 transferred sentences (50 for each sentiment) randomly. Without any knowledge about the model which produces the sentences, three annotators are required to evaluate every sentence from the aspect of style control, content preservation, and language fluency. For each target sentence, the annotator should give answers for three questions: 1) Does the sentence hold the correct style? 2) Is the content preserved in the target sentence? 3) Is the expression fluent? For any question with answer 'yes', the target sentence is labeled with '1', otherwise labeled with '0'.

4.4 Experimental Details

During the data pre-processing, the sentence length on *Yelp* and *Amazon* datasets is limited to 23 and 30, respectively. To alleviate the word sparsity problem, we set the word as 'unk' if the corresponding frequency is below 5. The noise rate p_{noise} is set to 0.05 to stabilize the training process. In terms of model setting, the style detection module recognizes n-grams with up to 4 tokens, the smoothing parameter λ is 1.0. The BiLM and Decoder are implemented as recurrent networks, which both adopt GRU [3] as the recurrent unit. The dimension of word embedding and the hidden size of GRU are both set to 512. All the attention operations in the proposed model are employed in the *Scaled Dot-Product* schema [28]. At training stage, the scale coefficient η and batch size is set to 0.05 and 100 respectively. Then, the training is done after 30K iterations, optimized by an Adam [11] optimizer with a fixed learning rate 0.0003.

4.5 Experimental Results and Analysis

The automatic evaluation results are shown in Table 1. To avoid the margin problem described in [27], each of the results is an averaged value from 5 single running models initialized with different random seeds. Compared to state-of-the-art systems, the proposed model achieves a competitive performance and get a more excellent style accuracy on both *Yelp* and *Amazon*. However, some methods, such as C-BERT, DualRL, PTO, preserve more style-independent content

---

Table 1. Automatic evaluation results on the *Yelp* and *Amazon* datasets. ↑ denotes the higher the better and ↓ denotes the lower the better.

Models	Yelp			Amazon		
	Accuracy(%)↑	BLEU↑	Perplexity↓	Accuracy(%)↑	BLEU↑	Perplexity↓
CrossAE [24]	84.4	5.4	53.3	60.7	1.7	45.2
Template [16]	84.2	22.2	111.7	69.8	31.8	95.9
DeleteOnly [16]	85.6	15.1	68.8	47.2	27.9	69.9
Del& Retr [16]	89.1	15.5	46.4	47.9	27.9	54.8
CycleS2S [32]	52.8	18.5	161.7	49.9	14.2	N/A
C-BERT [31]	95.2	26.0	54.9	88.4	33.9	114.0
PTO [30]	86.3	**29.3**	46.0	45.9	**36.6**	76.6
DualRL [18]	88.4	27.5	48.2	45.8	34.8	**43.9**
Ours	**95.5**	25.8	**41.4**	**94.0**	23.8	47.4

Table 2. Human evaluation results (%) on the *Yelp* and *Amazon* datasets. The evaluation includes three aspects: style accuracy (denoted as Sty), content preservation (denoted as Con), and fluency (denoted as Flu). Each cell indicates the proportion of sentences that passed the human test.

Models	Yelp			Amazon		
	Sty	Con	Flu	Sty	Con	Flu
Del& Retr	19.3	25.8	36.8	7.0	29.1	26.8
C-BERT	32.8	**43.1**	41.3	**16.8**	**33.1**	30.1
Ours	**34.6**	42.8	**45.5**	10.1	21.1	**36.1**

than the proposed model. The reason maybe that the detection module in the proposed model tends to boost the recall rate of recognizing attribute markers. As a result, more tokens in original sentences are removed, then the less contents are preserved. In terms of language fluency (i.e., Perplexity), the proposed model is superior to most baselines. Because the matching module guarantees the compatibility of target attribute markers and contents, the attentive generation module keeps the stability of the generation process.

Table 2 shows the human evaluation results of the two well-performed models (both have the similar training process of our model), Del&Retr and C-BERT, in automatic evaluation. We find that the results on *Yelp* are generally consistent with that in automatic evaluation. However, the most confusing part is the style accuracy of the proposed model on *Amazon* (i.e., our model performs 6.7% points lower than C-BERT. However in the automatic evaluation, our model perform 5.6% points higher than C-BERT.) By investigating the *Amazon* dataset, we find that the reason is the imbalance phenomena in positive and negative product reviews. For example, the word 'game' appears 14,301 times in negative training set, while it appears only 217 times in positive set. Therefore, the detection module tends to recognize the 'game' as a negative attribute marker. This phenomenon interferes the detection of attribute markers severely, and more

Table 3. Examples of generated sentences from CrossAE [24], Del&Retr [16], PTO [30] and our model. Words with different colors have different meanings, specifically: blue → sentiment words; green → correct transferred part; red → errors (i.e. sentiment error, grammar error, content changed and etc.).

	Yelp: negative → positive	*Yelp*: positive → negative
Source	i ca n't believe how inconsiderate this pharmacy is.	portions are very generous and food is fantastically flavorful.
CrossAE	i do n't know this store is great.	people are huge and the food was dry and dry.
Del&Retr	this pharmacy is a great place to go with.	portions are bland and food is fantastically not at all flavorful.
PTO	i delightfully n't believe how great this is.	portions are very bland and food is not flavorful.
Ours	i always believe how good this pharmacy is.	portions are very weak and food is deeply bland.
	Amazon: negative → positive	*Amazon*: positive → negative
Source	it crashed for no reason, saves got corrupted.	exactly what i need for my phone and at the best price possible.
CrossAE	it works for me for # years, etc.	i don t believe the price for # months and i am using it.
Del&Retr	it works flawlessly, works, and does easy to use and clean, saves got.	was really excited to get this for my phone and at the best price possible.
PTO	it crashed for no reason, got delicious.	exactly what i need for my phone and at the worst price possible.
Ours	it worked great for no reason, got perfect results.	not what i needed for my game and at the same price possible.

style-independent content is misidentified. We believe that leveraging external knowledge could alleviate the above imbalance problem, this is a potential candidate for further exploring.

Besides the formal evaluations, some transferred sentences are presented in Table 3 for further qualitative analysis. The grammar of sentences generated by our model is generally correct. The semantic is more consistent than some baselines. The presented results demonstrates that the cooperation of the *Detection* and *Matching* indeed make a stable improvement of our model. We observe the results in *Yelp* and *Amazon* datasets, respectively. Most of the models struggle in transferring sentiment of sentences on *Amazon*. Our model has transferred the the last example (*Amazon*: positive → negative) by replacing the word 'phone' with 'game'. This phenomena is consistent with the observation of imbalance problem mentioned in human evaluation part.

4.6   Ablation Study

To estimate the influence of different components on the overall performance, we remove the components individually and check the model performance on *Yelp* dataset. The results are reported in Table 4.

Table 4. Automatic evaluation results of ablation study on *Yelp* dataset. 'Joint → Freq' indicates replacing Joint Detection with Frequency-based method. Similarly, 'Joint → Attn' indicates replacing Joint Detection with the Attention-based method.

Models	Accuracy(%)	BLEU	Perplexity
Joint → Freq	76.6	25.8	35.4
Joint → Attn	94.6	25.8	44.0
- match	95.0	24.8	41.8
- noise	88.0	26.4	42.4
- $\mathcal{L}_{cls}$	86.1	26.4	40.6
Full model	95.5	25.8	41.4

First, we replace the Joint-Detection method with Frequency-based method. As a result, the style accuracy reduces drastically (18.9% below the full model) while the fluency is increased. The Frequency-based method tends to recognize style-independent tokens as attribute markers, and the undetected attribute markers deeply affect the style accuracy. If we replace the Joint-Detection with Attention-based method, the style accuracy and language fluency decrease slightly. The above results show that the proposed Joint-Detection method is superior to the Frequency-based and Attention-based methods in terms of the detection accuracy.

Then, we discard the matching operation in the matching module. The rest model is similar to the *DeleteOnly* in [16], which infers the removed attribute markers based on style-independent content. The overall performance reduces in all three aspects. This result further supports our claim that the matching operation tends to select compatible attribute markers.

Finally, two training tricks – denoising mechanism and classification loss, are taken into consideration. Without noises, the model can preserve more content as the BLEU score increases. However, the style accuracy dropped by 7.5% at the same time. Moreover, due to the lack of denoising ability, the model cannot generate sentences smoothly (the corresponding perplexity rises). If we remove the classification loss $\mathcal{L}_{cls}$, the style accuracy decreases from 95.5% to 86.1%. Therefore, the classification loss is critical for boosting the style transfer strength. Due to the model would corrupt if the $\mathcal{L}_{rec}$ is disabled, we ignore the ablation study on $\mathcal{L}_{rec}$.

In summary, the studies on the detection and matching modules have proved that the issues of detection accuracy and compatibility could be resolved. The studies on denoising mechanism and classification loss demonstrate their crucial rule in improving the style transfer strength.

5 Conclusion

In this paper, we propose a multi-stage method to address the detection accuracy and compatibility issues for unsupervised text style transfer. The joint detection method is designed to combine the Frequency-based and Attention-based methods for recognizing attribute markers. The matching operation is presented to seek the compatible tokens for retrieving the target style information. Both the automatic and human evaluation results show that the proposed model achieves competitive performance compared with several state-of-the-art systems. The ablation study confirms that the designed joint detection method enhances the style transfer strength, and the matching operation improves the fluency of generated sentences. In *Amazon*, we observe the data imbalance problem which severely reduces the model performance. Therefore, achieving unsupervised text style transfer in imbalanced scenario is the topic for our future exploration.

Acknowledgement. This research is supported in part by the Beijing Municipal Science and Technology Project under Grant Z191100007119008.

References

1. Artetxe, M., Labaka, G., Agirre, E.: Learning bilingual word embeddings with (almost) no bilingual data. In: Proceedings of the ACL, pp. 451–462 (2017)
2. Bahdanau, D., Cho, K., Bengio, Y.: Neural machine translation by jointly learning to align and translate. In: Proceedings of the ICLR (2015)
3. Cho, K., van Merrienboer, B., Gülçehre, Ç., Bahdanau, D., Bougares, F., Schwenk, H., Bengio, Y.: Learning phrase representations using RNN encoder-decoder for statistical machine translation. In: Proceedings of the EMNLP, pp. 1724–1734 (2014)
4. Dai, N., Liang, J., Qiu, X., Huang, X.: Style transformer: unpaired text style transfer without disentangled latent representation. In: Proceedings of the ACL, pp. 5997–6007 (2019)
5. Devlin, J., Chang, M., Lee, K., Toutanova, K.: BERT: pre-training of deep bidirectional transformers for language understanding. In: Proceedings of the NAACL, pp. 4171–4186 (2019)
6. Fu, Z., Tan, X., Peng, N., Zhao, D., Yan, R.: Style transfer in text: exploration and evaluation. In: Proceedings of the AAAI, pp. 663–670 (2018)
7. Goodfellow, I., et al.: Generative adversarial nets. In: Proceedings of the NeurIPS, pp. 2672–2680 (2014)
8. Jin, Z., Jin, D., Mueller, J., Matthews, N., Santus, E.: Unsupervised text style transfer via iterative matching and translation. In: Proceedings of the EMNLP, pp. 3097–3109 (2019)
9. John, V., Mou, L., Bahuleyan, H., Vechtomova, O.: Disentangled representation learning for non-parallel text style transfer. In: Proceedings of the ACL, pp. 424–434 (2019)
10. Joulin, A., Grave, E., Bojanowski, P., Mikolov, T.: Bag of tricks for efficient text classification. In: Proceedings of the EACL, pp. 427–431 (2017)
11. Kingma, D.P., Ba, J.: Adam: a method for stochastic optimization. In: Proceedings of the ICLR (2015)
12. Kingma, D.P., Welling, M.: Auto-encoding variational bayes. In: Proceedings of the ICLR (2014)
13. Lample, G., Conneau, A., Denoyer, L., Ranzato, M.: Unsupervised machine translation using monolingual corpora only. In: Proceedings of the ICLR (2018)
14. Lample, G., Conneau, A., Ranzato, M., Denoyer, L., Jégou, H.: Word translation without parallel data. In: Proceedings of the ICLR (2018)

15. Lample, G., Subramanian, S., Smith, E.M., Denoyer, L., Ranzato, M., Boureau, Y.: Multiple-attribute text rewriting. In: Proceedings of the ICLR (2019)
16. Li, J., Jia, R., He, H., Liang, P.: Delete, retrieve, generate: a simple approach to sentiment and style transfer. In: Proceedings of the NAACL, pp. 1865–1874 (2018)
17. Logeswaran, L., Lee, H., Bengio, S.: Content preserving text generation with attribute controls. In: Proceedings of the NeurIPS, pp. 5103–5113 (2018)
18. Luo, F., et al.: A dual reinforcement learning framework for unsupervised text style transfer. In: Proceedings of the IJCAI, pp. 5116–5122 (2019)
19. Oraby, S., Reed, L., Tandon, S., Sharath, T. S., Lukin, S.M., Walker, M.A.: Controlling personality-based stylistic variation with neural natural language generators. In: Proceedings of the SIGDIAL, pp. 180–190 (2018)
20. Papineni, K., Roukos, S., Ward, T., Zhu, W.J.: Bleu: a method for automatic evaluation of machine translation. In: Proceedings of the ACL, pp. 311–318 (2002)
21. Peters, M.E., et al.: Deep contextualized word representations. In: Proceedings of the NAACL, pp. 2227–2237 (2018)
22. Prabhumoye, S., Tsvetkov, Y., Salakhutdinov, R., Black, A.W.: Style transfer through back-translation. In: Proceedings of the ACL, pp. 866–876 (2018)
23. Radford, A., Wu, J., Child, R., Luan, D., Amodei, D., Sutskever, I.: Language models are unsupervised multitask learners (2019)
24. Shen, T., Lei, T., Barzilay, R., Jaakkola, T.: Style transfer from non-parallel text by cross-alignment. In: Proceedings of the NeurIPS, pp. 6830–6841 (2017)
25. Shetty, R., Schiele, B., Fritz, M.: A4NT: author attribute anonymity by adversarial training of neural machine translation. In: Proceedings of the USENIX, pp. 1633–1650 (2018)
26. Sudhakar, A., Upadhyay, B., Maheswaran, A.: "transforming" delete, retrieve, generate approach for controlled text style transfer. In: Proceedings of the EMNLP, pp. 3260–3270 (2019)
27. Tikhonov, A., Shibaev, V., Nagaev, A., Nugmanova, A., Yamshchikov, I.P.: Style transfer for texts: retrain, report errors, compare with rewrites. In: Proceedings of the EMNLP, pp. 3927–3936 (2019)
28. Vaswani, A., et al.: Attention is all you need. In: Proceedings of the NeurIPS, pp. 5998–6008 (2017)
29. Wang, K., Hua, H., Wan, X.: Controllable unsupervised text attribute transfer via editing entangled latent representation (2019). arXiv preprint arXiv:1905.12926
30. Wu, C., Ren, X., Luo, F., Sun, X.: A hierarchical reinforced sequence operation method for unsupervised text style transfer. In: Proceedings of the ACL, pp. 4873–4883 (2019)
31. Wu, X., Zhang, T., Zang, L., Han, J., Hu, S.: Mask and infill: applying masked language model for sentiment transfer. In: Proceedings of the IJCAI, pp. 5271–5277 (2019)
32. Xu, J., et al.: Unpaired sentiment-to-sentiment translation: a cycled reinforcement learning approach. In: Proceedings of the ACL, pp. 979–988 (2018)
33. Yang, Z., Hu, Z., Dyer, C., Xing, E.P., Berg-Kirkpatrick, T.: Unsupervised text style transfer using language models as discriminators. In: Proceedings of the NIPS, pp. 7287–7298 (2018)
34. Zhang, Y., Xu, J., Yang, P., Sun, X.: Learning sentiment memories for sentiment modification without parallel data. In: Proceedings of the EMNLP, pp. 1103–1108 (2018)
35. Zhang, Z., et al.: Style transfer as unsupervised machine translation (2018). arXiv preprint arXiv:1808.07894

Exemplar Guided Latent Pre-trained Dialogue Generation

Miaojin Li[1,2], Peng Fu[1(✉)], Zheng Lin[1(✉)], Weiping Wang[1], and Wenyu Zang[3]

[1] Institute of Information Engineering, Chinese Academy of Sciences, Beijing, China
{limiaojin,fupeng,linzheng,wangweiping}@iie.ac.cn
[2] School of Cyber Security, University of Chinese Academy of Sciences,
Beijing, China
[3] China Electronics Corporation, Beijing, China
wenyuzang@sina.com

Abstract. Pre-trained models with latent variables have been proved to be an effective method in the diverse dialogue generation. However, the latent variables in current models are finite and uninformative, making the generated responses lack diversity and informativeness. In order to address this problem, we propose an *exemplar guided latent pre-trained dialogue generation* model to sample the latent variables from a continuous sentence embedding space, which can be controlled by the exemplar sentences. The proposed model contains two parts: exemplar seeking and response generation. First, the exemplar seeking builds a sentence graph based on the given dataset and seeks an enlightened exemplar from the graph. Next, the response generation constructs informative latent variables based on the exemplar and generates diverse responses with latent variables. Experiments show that the model can effectively improve the propriety and diversity of responses and achieve state-of-the-art performance.

Keywords: Multiple responses generation · Pre-trained generation · Dialogue generation

1 Introduction

Text generation, which is a challenging task due to the limited dataset and complex background knowledge, is one of the most popular branches of machine learning. By pre-trained on large-scale text corpora and finetuning on downstream tasks, self-supervised pre-trained models such as GPT [1], UNILM [2], and ERNIE [3,4] achieve prominent improvement in text generation. Dialogue generation is one of the text generation tasks, so the pre-trained models also gain remarkable success in dialogue generation [5,6]. However, these researches treat the dialogue dataset as the general text and ignore the unique linguistic pattern in conversations.

Different from general text generation, dialogue generation requires the model to deal with the one-to-many relationship. That is, different informative replies

© Springer Nature Switzerland AG 2021
M. Paszynski et al. (Eds.): ICCS 2021, LNCS 12743, pp. 118–132, 2021.
https://doi.org/10.1007/978-3-030-77964-1_10

can answer the same query in human conversation. To better address this relationship, some studies [7–12] focus on improving pre-trained conversation generation with the help of external knowledge. For example, incorporating additional texts [8], multi-modal information [13], and personal characters [7,11,12]. They have shown external knowledge can effectively improve response generation. However, the defect of these methods is obvious since they all heavily rely on high-quality knowledge beyond the dialogue text. In addition, the external knowledge is unavailable in a real conversation scenario. To model the one-to-many relationship without external knowledge, inducing latent variables into a pre-trained model is a valid way. With discrete latent variables corresponding to latent speech acts, PLATO [14] and PLATO-2 [15] can generate diverse responses and outperform other state-of-the-art methods. However, the discrete latent variables in their work are finite and contain insufficient information, so that the diversity and informativeness of generated responses are limited.

In order to allow latent variables to be more diverse and more informative, we propose the exemplar guided latent pre-trained dialogue generation model. The model treats latent variables as the continuous embeddings of sentences instead of discrete ones. Besides, it has been proved that the usage of exemplar can bring more beneficial and diverse information [16–19] for dialogue generation, so we utilize exemplars to guide the model to construct more informative latent variables. The exemplars contain explicit referable exemplary information from the given dataset. Different from the previous work, which directly searches exemplars with similar semantics to the current conversation, we first constructed a sentence graph and then find the relevant exemplars in the graph. By altering different exemplars, the model can generate more diverse and informative responses.

To achieve our goal, the proposed model is designed with two main parts: exemplar seeking and response generation. In the exemplar seeking, we construct a sentence graph according to the given dataset offline, and then find the proper exemplar in the sentence graph for the current dialogue. Next, in the response generation, we construct the response variable based on the dialogue context and the given exemplar and use a transformer structure to generate a response based on the response variable.

Experimental results show that this approach can achieve state-of-the-art performance. Our contributions can be summarized as follows:

- To generate diverse and informative responses, we propose an exemplar guided latent pre-trained dialogue generation model, in which latent variables are continuous sentence embeddings.
- To obtain referable exemplars, we build a sentence graph and search for an enlightened exemplar in the graph. By altering exemplars and sampling variables, we can get various latent variables to generate diverse and informative responses.
- We conduct our model on three dialogue datasets and achieve state-of-the-art performance.

2 Related Work

Pre-trained dialogue generation models. DialoGPT [6] exhibits a compelling performance in generating responses. But it pays not enough attention to the one-to-many phenomenon in the dialogue. And the one-to-many relationship can be effectively built with extra knowledge. The researcher [7] uses conditional labels to specify the type of target response. The study [8] provides related information from wiki articles to the pre-trained language model. The work [9] learns different features of arXiv-style and Holmes-style from two separate datasets to fine-tune the pre-trained model to generate responses towards the target style. TOD-BERT [10] incorporates two special tokens for the user and the system to model the corresponding dialogue behavior. The researches [11] and [12] both take the usage of personality attributes to improve the diversity of pre-trained models. However, those methods train models based on the characteristic datasets with specific and additional labels, which is not easy to collect. Without additional information besides dialogue text, PLATO [14,15] uses discrete latent variables, which are dialogue acts, to improve the diversity of generation. However, the diversity and informativeness of dialogue acts are limited, so their model's improvement is limited.

Hybrid neural conversation models that combine the benefits of both retrieval and generation methods can promote dialogue generation effectiveness. Studies [20] and [21] apply retrieved sentences to assist the generation network in producing more informative responses. Still, retrieved responses may better than generated ones, so studies [22,23] rerank all responses achieved by these two methods to find the best response. To further improve the quality of generated responses, Researchers [24] and [25] induce the reinforcement learning process. Researcher [25] employs the top N retrieved exemplars to estimate the reward for the generator, and researcher [24] proposes a generative adversarial framework based on reinforcement learning. However, they ignore the phenomenon that irrelevant information in retrieved sentences may mislead the generation. Therefore, researchers [26–28] focus on refining the retrieved sentence into a useful skeleton and utilize this skeleton instead of the whole sentence to guide the generation. Besides, the work [29] applies a two-stage exemplar retrieval model to retrieves exemplar responses in terms of both the text-similarity and the topic proximity. In recent, the transformer structure becomes a popular method in text generation because of its effectiveness. Studies [16–19] conduct this structure to improve the quality of dialogue generation. However, those works ignore the shift phonemes of content in human dialogue [30]. To better capture the direction of this shifting, we form a sentence graph and search for a guiding exemplar in it to direct the generation.

3   Methodology

3.1   Framework

To generate diverse and informative responses, we propose an exemplar guided latent pre-trained dialogue generation model that composed of two main parts: exemplar seeking and response generation.

The exemplar seeking is designed to get the exemplar sentence. This module uses the dataset to construct the sentence graph at first. In this graph, nodes and edges represent keywords and sentences respectively. Then the module starts from the keyword nodes of the query, wandering along the edge in the graph, to find the end keyword nodes of the response. Finally, it selects the decent sentence to be the exemplar sentence in the edges between start and end.

To generate responses according to the dialogue context and intrusive exemplary, we invent the responses generation module. It first encodes the dialogue context to form a Gaussian distribution and samples a context variable from it. At the same time, this module encodes the exemplar to form another Gaussian distribution and samples an exemplar variable from it. Next, this module calculates the response variable based on the context variable and exemplar variable. Finally, this module generates a response based on the response variable.

3.2   Exemplar Seeking

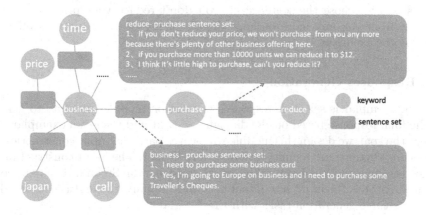

Fig. 1. An example of sentence graph. The blue nodes in the graph represent keywords, while the gray edge between the two nodes represents the set of sentences with these two nodes as keywords. (Color figure online)

This module tries to get an enlightened exemplar to guide the response generation stage. Instead of directly retrieving a sentence just by literally-similarity methods in endeavors [16–19], we hope to find an exemplar that can reflect the content shift [30,31] in conversation, and use this exemplar to guide the generation process. For this purpose, we design a graph-based method to capture the semantic transfer phenomenon, which consists of two parts: graph construction and exemplar selection. The former first constructs a sentence graph based on the given dataset, and the latter selects a suitable exemplar from the sentence graph for current dialogue.

Graph Construction. This section purposes to form a sentence graph, from which we can quickly locate a suitable exemplar for the current conversation. To this end, we treat the sentences as the edges and the keywords in the sentence as the nodes. What's more, we define words with the Top-k TF-IDF values in each sentence as the keywords of this sentence. After that, we extract tuples like (keywords1, sentence set, keywords2), where keywords1 and keywords2 are two keywords that exist in the same sentence, and the sentence set is the set of sentences that contains keywords1 and keywords2. As shown in the Fig. 1, we next construct the graph from those tuples.

Exemplar Selection. After preparing the whole graph, exemplar selection searches for an exact exemplar in the graph. We set keywords in the query as start nodes, and collect keywords from n hops. If the keywords in the response are reachable in n hops, we random choose a sentence from the edge of the last hop as the exemplar sentence. For example, as shown in the Fig. 1, the keyword "price" is the keyword in the query, and the "reduce" is the keyword in the response, so we chose the sentence "If you don't reduce your price, we won't purchase from you any more because there's plenty of other business offering here." as the exemplar.

3.3 Responses Generation

After an exemplar is selected, responses generation module attempts to generate the final response based on the dialogue context and the given exemplar. To achieve the goal, we design this module with three parts: an input construction, a latent construction, and a multi-task decoder. At first, The input construction is assigned to construct the input of the latent construction. Second, the latent construction encodes the input into latent variables. Finally, the multi-task decoder completes three generation tasks with these variables.

Input Construction. Input construction aims to construct a comprehensive input of generation stage. To enrich input knowledge, we summarize four kinds of embeddings as the final input embeddings for each token, including the vocab, role, type, turn, and position embeddings. Vocab embedding are intialized with UNILM [2]. Role embedding is managed to distinguish the replier and the interlocutor in a conversation. Type embedding mainly separates dialogue utterances and knowledge information since persona and DSTC dataset contains external

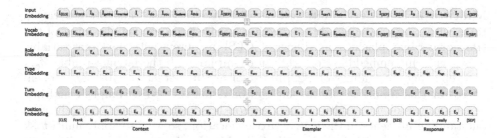

Fig. 2. Input construction. The input embedding of each token is the embedding sum of the role, turn, position, segment, and vocab embeddings.

knowledge about current dialogue, such as summary and personal profile. Turn embedding is numbered from reply to the beginning of the conversation, the reply is numbered 0, the last statement is numbered 1, and so on. Position embeddings are computed according to the token position in each utterance. As shown in the Fig. 2, we concatenate the embeddings of special tags, context, exemplar, and response as the final input series.

Latent Construction. The latent construction module works to produce the latent variables based on the input. Since a response is related to the dialogue context and guided by the examplar, the latent response variable can be measured according to the context variable and the examplar variable:

$$Z_{res} = Z_{con} + (Z_{exe} - Z_{con}) * L \tag{1}$$

Where Z_{con}, Z_{exe}, and Z_{res} are latent variables of the dialogue context, the exemplar, and the response separately. L represents the distance of the Z_{res} in the direction from Z_{con} to Z_{exe}. As shown in Fig. 3, we use Multi-Layer Transformer [32] as the backbone network and apply a special attention mask for the latent construction. This structure encodes contextual information from both directions and can encode better contextual representations of text than its unidirectional counterpart.

We further induce Gaussian distribution to approximate the ideal distribution of latent variables and use the reparameterization method [33] to get samples of the latent variable. The latent variable of context can be constructed as:

$$Z_{con} = \mu_{con} + \sigma_{con} * \varepsilon \tag{2}$$

where $\varepsilon \sim N(0,1)$ is a random sampling error, the μ_{con} and the σ_{con} are mean and standard deviation of the distribution, which derived by a linear layer:

$$\begin{bmatrix} \mu_{con} \\ \log\left(\sigma_{con}^2\right) \end{bmatrix} = W_{con} O_{con} + b_{con} \tag{3}$$

where W_{con} and b_{con} are training parameters, O_{con} is the output of last transformer block, at the position of [CLS] before the context as shown in Fig. 3. Also, we conduct the same method to get the Z_{exe}.

Multi-task Decoder. Multi-task decoder strives to do multiple generation tasks with latent variables. In this section, we first reconstruct the input of the transformer by replacing embeddings of [CLS] in context with the latent variable Z_{con}, replacing embeddings of [CLS] in exemplar with the latent variable Z_{exe} and replacing embeddings of [S2S] in response with latent variable Z_{res}. Then, based on this input, we use a unidirectional network to accomplish multiple generation tasks. The multiple tasks method has been confirmed to produce an outstanding influence in the field of text generation, so we also apply multi-task to improve our model, including Masked LM, Unidirectional LM, and bag-of-words prediction. The three tasks are all generation tasks, and the bidirectional structure of Bert will bring future knowledge to the current period. Therefore, as shown in the Fig. 3, we use the unidirectional attention mask so that the generation only receive the data ahead of the current time sequence.

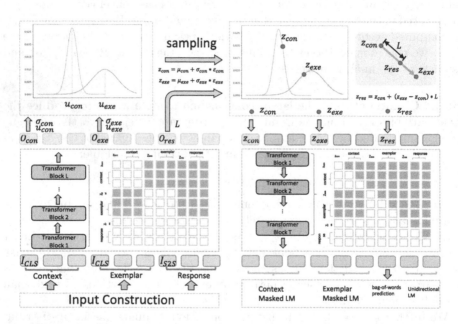

Fig. 3. The architecture of responses generation with latent construction and multiple tasks decode. The self-attentions of latent construction and multi-task decoder are shown at the right of transformer blocks. Grey dots represent preventing attention and blank ones represent allowing attention.

For Masked LM, the model could trivially predict the target masked word in a multi-layered network and the prediction objective of masked words in context can be formed as:

$$\mathcal{L}_{MASKC} = -\mathbb{E}_{z \sim p(\mathbf{z}_{con}|c)} \log p(w|c) = -\mathbb{E}_{z \sim p(\mathbf{z}_{con}|c)} \sum_{i \in I} \log p(w_i|z, w_{<i}) \quad (4)$$

where w_i is the masked word at i-th position and the I is the set of positions of all masked words in context. At the same time, $\mathcal{L}_{MASKE}$ is the prediction goal of masked words in exemplar and is obtained similarly to $\mathcal{L}_{MASKC}$.

For Unidirectional LM, we use the Unidirectional BERT [2] without token mask operation to generate all tokens in response. That is, when generating the t th token in the response r, former tokens are always given to the model as the input. Then generation probability of response is:

$$\mathcal{L}_{NLL} = -\mathbb{E}_{z \sim p(\mathbf{z_{res}}|c,e,r)} \log p(r|c,e,z) = -\mathbb{E}_{z \sim p(\mathbf{z_{res}}|c,e,r)} \sum_{t=1}^{T} \log p(r_t|z,c,e,r_{<t})$$

(5)

where z_{res} is estimated by Latent Construction module given a tuple of (c,e,r), and T is the length of response sequence.

For bag-of-words prediction, we use the bag-of-words loss [34] to facilitate the training process of responses' latent variables and tackle the vanishing latent variable problem:

$$\mathcal{L}_{BOW} = -\mathbb{E}_{z \sim p(\mathbf{z_{res}}|c,e,r)} \sum_{t=1}^{T} \log p(r_t|c,e,r) = -\mathbb{E}_{z \sim p(\mathbf{z}|c,r)} \sum_{t=1}^{T} \log \frac{e^{f_{r_t}}}{\sum_{v \in V} e^{f_v}}$$

(6)

where V is vocabulary size and f if a network layer to generate the words in the target response in a non-autoregressive way:

$$f = softmax(W_{bow} h_{res} + b_{bow})$$

(7)

where h_{res} is the output of z_{res} decoded by the T-th transfomer block in multi-task decoder as shown in Fig. 3. W_{bow} and b_{bow} are the training parameters.

Overall, the total objective of our model is to jointly minimize the integrated loss:

$$\mathcal{L} = \mathcal{L}_{MASKC} + \mathcal{L}_{MASKE} + \mathcal{L}_{NLL} + \mathcal{L}_{BOW}$$

(8)

3.4 Post-training and Fine-Tuning

We employ the pre-trained parameters of the UNILM [2] to initialize our network. Though UNILM has been proven to be an effective language model, it can not directly adapt to the task of dialogue generation. Since UNILM is trained on a large non-conversational corpus, and there is a huge natural gap between the dialogic corpus and other corpora. In order to make up for the reduced effect caused by different corpus, we carry out post-training in dialogue corpus. For each dataset, we first post-train the UNILM structure with masked LM task on the dialogue corpus without exemplar sentences. After that, we implement the proposed model to fine-tune corpus with exemplar with all three tasks.

4 Experiments

4.1 Datasets

In order to evaluate the performance of our proposed method, we conducted comprehensive experiments on three different datasets: Persona chat, Daily dialog, and Dstc7-avsd. The daily dialog contains only the dialogue text, but Persona chat and Dstc7-avsd involve knowledge beyond dialogue. To avoid changing the structure of our model, we concatenate knowledge text with dialogue context and treat this combination text as the dialogue context in training.

- Persona chat [35] is a knowledge-based dialogue dataset consisting of 164,356 utterances between crowdworkers who were randomly paired and asked to act the part of a given provided persona (randomly assigned, and created by another set of crowdworkers).
- Daily dialog [36] is a high-quality multi-turn dialogue dataset containing conversations about our daily life, in which human communicate with others for two main reasons: exchanging information and enhancing social bonding.
- Dstc7-avsd [37] is an abbreviation of dstc7 challenged audio-visual scene aware dialogue. It is a conversational QA dataset. Given dialogue context and background knowledge, the system tries to generate answers in this challenge. In our experiment, we utilize the unimodal information of text, which includes the title and abstract of video.

4.2 Baseline and Evaluation

The following models have been compared in the experiments.

- **Seq2Seq:** Sequence to sequence with attention is employed as the baseline for the experiments.
- **LIC:** LIC obtains the well-known performance [38] in the ConvAI2 challenge [39], in which Persona-Chat dataset is utilized.
- **iVAE MI:** iVAE MI [40] generates diverse responses with sample-based latent representation and achieves state-of-the-art performance on the dataset of Daily Dialog.
- **CMU:** CMU [41] gains excellent achievement in all the evaluation metrics of DSTC7-AVSD.
- **PLATO:** PLATO [14] outperforms other methods in all datasets for now.

We use three metrics to evaluate our proposed model, including automatic evaluation BLEU, DISTINCT, and MSCOCO platform.

- **BLEU:** We adopt BLEU-(1-4) to measure the overlap of candidates and references at the character level.
- **DISTINCT:** This metric is proposed by [42] to evaluate the diversity of the responses. In the generated responses, the number of distinct unigrams and bigrams are divided by the total number of generated unigrams and bigrams in the test set.

- **MSCOCO:** We employ the MSCOCO platform [43] to evaluate the performance in DSTC7-AVSD, including metrics of BLEU, METEOR, ROUGH-L and CIDEr.

The evaluation methods and results of the baselines are from the project of PLATO [14].

4.3 Results and Analysis

Table 1. The automatic evaluation results of DSTC7-AVSD

Dataset	Model	BLEU-1	BLEU-2	BLEU-3	BLEU-4	METEOR	ROUGH-L	CIDEr
DSTC7- AVSD	Baseline	0.626	0.485	0.383	0.309	0.215	0.487	0.746
	CMU	0.718	0.584	0.478	0.394	0.267	0.563	1.094
	PLATO	**0.784**	0.637	0.525	0.435	0.286	0.596	1.209
	Our method	0.736	**0.644**	**0.607**	**0.522**	**0.297**	**0.610**	**1.219**

Comparison of BLEU. As shown in Table 2 and Table 3, the large-scale pre-trained model achieves better performance than the seq2seq on three datasets, which shows the effectiveness of the large-scale pre-trained model in dialogue generation tasks. Our method achieves the best results on Persona-Chat, Daily Dialog datasets. Because our approach introduces a suitable exemplar, providing better guidance for the generation process, and improving the accuracy. Our model's score is lower than PLATO in the BLEU-1 with DSTC7-AVSD dataset but higher than PLATO in BLEU-2/3/4, which indicates that our model has more advantages in generating more extended responses. It is because much information, which our latent variables provide, may confuse the model when generating short text but be beneficial when generating long text (Table 1).

Table 2. The automatic evaluation results of Persona-Chat

Dataset	Model	BLEU-1	BLEU-2	Distinct-1	Distinct-2
Persona-Chat	Seq2Seq	0.448	0.353	0.004	0.016
	PLATO w/o Latent	0.405	0.320	0.019	0.113
	LIC	0.458	0.357	0.012	0.064
	PLATO	0.406	0.315	**0.021**	0.121
	Our method	**0.459**	**0.362**	0.020	**0.123**

Table 3. The automatic evaluation results of Daily Dialog

Dataset	Model	BLEU-1	BLEU-2	Distinct-1	Distinct-2
Daily-Dialog	Seq2Seq	0.336	0.268	0.030	0.128
	iVAE MI	0.309	0.249	0.029	0.250
	PLATO w/o Latent	0.405	0.322	0.046	0.246
	PLATO	0.397	0.311	0.053	0.291
	Our method	**0.417**	**0.334**	**0.056**	**0.295**

Comparison of Distinct. Our model gets the highest score in distinct-1/2 on the daily dataset, but it is slightly lower than that of PLATO on the distinct-1 on the persona dataset. The most likely reason is that the data in these two datasets are considerably different. The data of daily dialogue only contains dialogue text. The information in conversation without the exemplar is limited, so that information is insufficient to generate various responses. In this case, our method of introducing more information is effective. However, the corpus of the persona dataset contains enough additional information about the dialogue. In this case, the exemplar cannot bring much more useful knowledge about the current conversation, so our method's improvement is limited. However, in most conversation scenes, external knowledge does not exist, so that our model can be more practical than PLATO.

4.4   Case Study

We show some generation examples in the Table 4. According to the same query, we search for different exemplars in the graph to guide the generation process. Combined with different exemplars, the model generates different informative responses. Some of the generated responses have obvious overlap with the exemplar, such as "I'm going to work late". And some others have a weak connection with the exemplar, such as "You have to be quick". This phenomenon demonstrates that both exemplars can guide the generation effectively. Also, most exemplars contain knowledge beyond the query so that exemplars can provide more extra information for model generation.

4.5   The Influence of Exemplar Quality

In order to verify the effectiveness of the exemplar seeking module, we use different quality sentences as the exemplar in our method. We consider the quality of ground truth is the best, and use random sentences and blank sentences to replace exemplars in our model. As shown in Table 5, Random sentences get the worst performance in BLEU because they do not contain accurate information and bring misleading noise to the generation. Blank sentences get the worst score in Distinct owning to they convey the least information to the generation process. Our method gains the closest performance to the ground truth, which shows that the exemplar seeking module is valid.

Table 4. Some responses generated by our method.

Query	Exemplar	Response
The taxi drivers are on strike again	I'm going to work	I'm going to work late
	OK, let's go and ask	Let's go and try to find one
	We can have a company car take you there	You have to take a bus
	What? Are you serious?	Are you serious?
	Why don't we go for a walk?	You have to be quick
I really need to start eating healthier	Anyway health is the most important thing	I really need to lose weight
	Oh I see, okay	I think so, too
	In fact your body will only store fat if you miss meals	I really need to get some exercise
	Do you know of another good restaurant?	Do you have any plans?
	That's a good choice	That's a good idea

Table 5. The influence of exemplar quality

Exemplar position	BLEU-1	BLEU-2	Distinct-1	Distinct-2
Ground truth	0.470	0.361	0.058	0.310
Exemplar in graph (our method)	0.417	0.334	0.055	0.295
Random	0.351	0.326	0.048	0.241
NULL (our method w/o latent)	0.356	0.329	0.039	0.234

5   Conclusion

In this paper, we propose the exemplar guided latent pre-trained dialogue generation model to generate diverse and informative responses. In the proposed model, we treat the latent variable as the continuous sentence embedding and induce an enlightened exemplar to guide the generation. Results of experiments prove the proposed model improves the diversity and informativeness of responses.

Acknowledgment. This work was supported by National Natural Science Foundation of China (No. 61906187, No. 61976207, No. 61902394).

References

1. Radford, A., Narasimhan, K., Salimans, T., Sutskever, I.: Improving language understanding by generative pre-training (2018)

2. Dong, L., et al.: Unified language model pre-training for natural language understanding and generation. In: NeurIPS 2019, 8–14 December 2019, Vancouver, BC, Canada (2019)
3. Sun, Y., et al.: ERNIE 2.0: a continual pre-training framework for language understanding. In: AAAI, 7–12 February 2020, New York, NY, USA (2020)
4. Xiao, D., et al.: ERNIE-GEN: an enhanced multi-flow pre-training and fine-tuning framework for natural language generation. In: Proceedings of IJCAI 2020 (2020)
5. Rashkin, H., Smith, E.M., Li, M., Boureau, Y.-L.: Towards empathetic open-domain conversation models: a new benchmark and dataset. In: Proceedings of ACL 2019, 28 July–2 August 2019, Florence, Italy (2019)
6. Zhang, Y., et al.: DIALOGPT: large-scale generative pre-training for conversational response generation. In: Proceedings of ACL 2020, Online, 5–10 July 2020 (2020)
7. Zeng, Y., Nie, J.: Generalized conditioned dialogue generation based on pre-trained language model. CoRR, vol. abs/2010.11140 (2020)
8. Zhao, X., Wu, W., Xu, C., Tao, C., Zhao, D., Yan, R.: Knowledge-grounded dialogue generation with pre-trained language models. In: Proceedings of EMNLP 2020, Online, 16–20 November 2020 (2020)
9. Yang, Z., et al.: StyleDGPT: stylized response generation with pre-trained language models. In: Proceedings of Findings, EMNLP 2020, Online Event, 16–20 November 2020 (2020)
10. Wu, C.-S., Hoi, S.C., Socher, R., Xiong, C.: TOD-BERT: pre-trained natural language understanding for task-oriented dialogue. In: Proceedings of EMNLP, Online (2020)
11. Zheng, Y., Zhang, R., Huang, M., Mao, X.: A pre-training based personalized dialogue generation model with persona-sparse data. In: Proceedings of AAAI 2020, 7–12 February 2020, New York, NY, USA (2020)
12. Cao, Y., Bi, W., Fang, M., Tao, D.: Pretrained language models for dialogue generation with multiple input sources. In: Proceedings of EMNLP 2020, Online Event, 16–20 November 2020 (2020)
13. Le, H., Hoi, S.C.H.: Video-grounded dialogues with pretrained generation language models. In: Jurafsky, D., Chai, J., Schluter, N., Tetreault, J.R. (eds.) Proceedings of ACL 2020, Online, 5–10 July 2020 (2020)
14. Bao, S., He, H., Wang, F., Wu, H., Wang, H.: PLATO: pre-trained dialogue generation model with discrete latent variable. In: Proceedings of ACL 2020, Online, 5–10 July 2020 (2020)
15. Bao, S., et al.: PLATO-2: towards building an open-domain chatbot via curriculum learning. CoRR, vol. abs/2006.16779 (2020)
16. Zhang, L., Yang, Y., Zhou, J., Chen, C., He, L.: Retrieval-polished response generation for chatbot. IEEE Access **8**, 123882–123890 (2020)
17. Shalyminov, I., Sordoni, A., Atkinson, A., Schulz, H.: Hybrid generative-retrieval transformers for dialogue domain adaptation. CoRR, vol. abs/2003.01680 (2020)
18. Gupta, P., Bigham, J.P., Tsvetkov, Y., Pavel, A.: Controlling dialogue generation with semantic exemplars. CoRR, vol. abs/2008.09075 (2020)
19. Ma, T., Yang, H., Tian, Q., Tian, Y., Al-Nabhan, N.: A hybrid Chinese conversation model based on retrieval and generation. Future Gener. Comput. Syst. **114**, 481–490 (2021)
20. Weston, J., Dinan, E., Miller, A.H.: Retrieve and refine: improved sequence generation models for dialogue. In: Proceedings of SCAI@EMNLP 2018, 31 October 2018, Brussels, Belgium (2018)

21. Pandey, G., Contractor, D., Kumar, V., Joshi, S.: Exemplar encoder-decoder for neural conversation generation. In: Proceedings of ACL 2018, Melbourne, Australia, 15–20 July 2018, Volume 1: Long Papers (2018)

22. Song, Y., Li, C., Nie, J., Zhang, M., Zhao, D., Yan, R.: An ensemble of retrieval-based and generation-based human-computer conversation systems. In: Proceedings of IJCAI 2018, 13–19 July 2018, Stockholm, Sweden (2018)

23. Yang, L., et al.: A hybrid retrieval-generation neural conversation model. In: Proceedings of CIKM 2019, 3–7 November 2019, Beijing, China (2019)

24. Zhang, J., Tao, C., Xu, Z., Xie, Q., Chen, W., Yan, R.: EnsembleGAN: adversarial learning for retrieval-generation ensemble model on short-text conversation. In: Proceedings of SIGIR 2019, 21–25 July 2019, Paris, France (2019)

25. Zhu, Q., Cui, L., Zhang, W., Wei, F., Liu, T.: Retrieval-enhanced adversarial training for neural response generation. In: Proceedings of ACL 2019, 28 July–2 August 2019, Florence, Italy (2019)

26. Wu, Y., Wei, F., Huang, S., Wang, Y., Li, Z., Zhou, M.: Response generation by context-aware prototype editing. In: Proceedings of AAAI 2019, 27 January–1 February 2019, Honolulu, Hawaii, USA (2019)

27. Cai, Y.D., et al.: Skeleton-to-response: dialogue generation guided by retrieval memory. In: Proceedings of NAACL-HLT 2019, Minneapolis, MN, USA, 2–7 June 2019, Volume 1 (Long and Short Papers) (2019)

28. Cai, D., Wang, Y., Bi, W., Tu, Z., Liu, X., Shi, S.: Retrieval-guided dialogue response generation via a matching-to-generation framework. In: Proceedings of EMNLP-IJCNLP 2019, 3–7 November 2019, Hong Kong, China (2019)

29. Cai, H., Chen, H., Song, Y., Zhao, X., Yin, D.: Exemplar guided neural dialogue generation. In: Proceedings of IJCAI 2020 (2020)

30. Wu, S., Li, Y., Zhang, D., Zhou, Y., Wu, Z.: TopicKA: generating commonsense knowledge-aware dialogue responses towards the recommended topic fact. In: Proceedings of IJCAI 2020 (2020)

31. Zhang, H., Liu, Z., Xiong, C., Liu, Z.: Grounded conversation generation as guided traverses in commonsense knowledge graphs. In: Proceedings of ACL 2020, Online, 5–10 July 2020 (2020)

32. Devlin, J., Chang, M., Lee, K., Toutanova, K.: BERT: pre-training of deep bidirectional transformers for language understanding. In: Proceedings of NAACL-HLT 2019, Minneapolis, MN, USA, 2–7 June 2019, Volume 1 (Long and Short Papers) (2019)

33. Kingma, D.P., Welling, M.: Auto-encoding variational bayes. In: Bengio, Y., LeCun, Y. (eds.) 2nd International Conference on Learning Representations, ICLR 2014, Banff, AB, Canada, 14–16 April 2014, Conference Track Proceedings (2014). http://arxiv.org/abs/1312.6114

34. Zhao, T., Zhao, R., Eskénazi, M.: Learning discourse-level diversity for neural dialog models using conditional variational autoencoders. In: Proceedings of ACL 2017, Vancouver, Canada, 30 July–4 August 2017, Volume 1: Long Papers (2017)

35. Zhang, S., Dinan, E., Urbanek, J., Szlam, A., Kiela, D., Weston, J.: Personalizing dialogue agents: i have a dog, do you have pets too? In: Proceedings of ACL 2018, Melbourne, Australia, 15–20 July 2018, Volume 1: Long Papers (2018)

36. Li, Y., Su, H., Shen, X., Li, W., Cao, Z., Niu, S.: DailyDialog: a manually labelled multi-turn dialogue dataset In: Proceedings of IJCNLP 2017, Taipei, Taiwan, 27 November–1 December 2017 - Volume 1: Long Papers (2017)

37. AlAmri, H., et al.: Audio visual scene-aware dialog. In: CVPR 2019, 16–20 June 2019, Long Beach, CA, USA (2019)

38. Golovanov, S., Kurbanov, R., Nikolenko, S.I., Truskovskyi, K., Tselousov, A., Wolf, T.: Large-scale transfer learning for natural language generation. In: Proceedings of ACL 2019, Florence, Italy, 28 July–2 August 2019, Volume 1: Long Papers (2019)
39. Dinan, E., et al.: The second conversational intelligence challenge (ConvAI2). CoRR, vol. abs/1902.00098 (2019)
40. Fang, L., Li, C., Gao, J. Dong, W. Chen, C.: Implicit deep latent variable models for text generation. In: Proceedings of EMNLP-IJCNLP 2019, 3–7 November 2019, Hong Kong, China (2019)
41. Ramon Sanabria, S.P., Metze, F.: CMU sinbads submission for the DSTC7 AVSD challenge. In: AAAI Dialog System Technology Challenge Workshop (2019)
42. Li, J., Galley, M., Brockett, C., Gao, J., Dolan, B.: A diversity-promoting objective function for neural conversation models. In: Proceedings of NAACL HLT 2016, 12–17 June 2016, San Diego California, USA (2016)
43. Chen, X., et al.: Microsoft COCO captions: Data collection and evaluation server. CoRR, vol. abs/1504.00325 (2015)

Monte Carlo Winning Tickets

Rafał Grzeszczuk(iD) and Marcin Kurdziel(✉)(iD)

AGH University of Science and Technology,
al. A. Mickiewicza 30, 30-059 Krakow, Poland
{grzeszcz,kurdziel}@agh.edu.pl

Abstract. Recent research on sparse neural networks demonstrates that densely-connected models contain sparse subnetworks that are trainable from a random initialization. Existence of these so called *winning tickets* suggests that we may possibly forego extensive training-and-pruning procedures, and train sparse neural networks from scratch. Unfortunately, winning tickets are data-derived models. That is, while they can be trained from scratch, their architecture is discovered via iterative pruning. In this work we propose Monte Carlo Winning Tickets (MCTWs) – random, sparse neural architectures that resemble winning tickets with respect to certain statistics over weights and activations. We show that MCTWs can match performance of standard winning tickets. This opens a route to constructing random but trainable sparse neural networks.

Keywords: Lottery tickets hypothesis · Neural network initialization · Sampling

1 Introduction

Contemporary neural network architectures tend to employ a large number of trainable parameters. In computer vision, for example, convolutional nets frequently have from over a million (as is the case in deep residual models) to tens of millions of parameters (e.g. in certain DenseNet architectures [6]). Natural language processing applications employ even larger models, with the current record held by a transformer network with 175 *billion* parameters [2]. However, available empirical evidence suggests that such large numbers of parameters are *not* a necessary prerequisite for strong performance in learned tasks. On the contrary: even though final performance in training usually increases with the number of parameters, a trained network can often be pruned of unimportant weights with virtually no performance loss. The fraction of parameters that can be pruned from a trained model is significant: performance of a dense network can often be matched by a pruned model with a tenth of the original number

Research presented in this paper was supported by the funds assigned to AGH University of Science and Technology by the Polish Ministry of Science and Higher Education. This research was supported in part by PL-Grid Infrastructure.

© Springer Nature Switzerland AG 2021
M. Paszynski et al. (Eds.): ICCS 2021, LNCS 12743, pp. 133–139, 2021.
https://doi.org/10.1007/978-3-030-77964-1_11

of weights [5]. This may, in turn, translate to major computational gains in a dedicated inference hardware, especially in energy-limited applications.

An immediate question arising from the empirical evidence for strong performance of pruned networks is whether we can avoid training dense models altogether. Surprisingly, at least up until recently the answer was no. Frankle and Carbin [3] show that randomly sampled sparse networks cannot be trained to match the performance of pruned networks. Previously, Han et al. [5] observed that pruned networks trained from scratch (i.e. from a random initialization) also do not converge to good solutions. Crucially, however, Frankle and Carbin also demonstrated that randomly initialized dense networks *do* contain sparse subnetworks that train well – what they call *winning tickets*. To find such subnetworks they start from dense models, which are iteratively trained, pruned and rewound to original parameter values. Each iteration remove only a small fraction of trainable weights. Ultimately, after a number of pruning iterations they uncover sparse subnetworks that train well starting from original initializations.

Winning tickets, by their construction, are data-derived subnetworks – the connectivity retained in the sparse model arises from pruning networks trained on the given data set. Furthermore, the specific initial parameter values – which can be seen as indirectly chosen via data-dependent pruning – also play an important role. In particular, randomly reinitialized winning tickets do not train as well as their counterparts with original initialization. Frankle and Carbin therefore suggest that winning tickets can possibly be seen as networks whose structures are adapted to the solved learning task. If so, then winning ticket-like networks could be uncovered only by training large, dense models, usually across many pruning iterations – a procedure with large computational cost.

In this work we explore an alternative hypothesis. Specifically, we investigate sparse neural networks with random architectures. However, we do not sample these architectures from a uniform distribution, but from a distribution that approximates certain statistics over weights and activations in an untrained winning ticket. In other words, we investigate networks that resemble untrained winning tickets with respect to certain statistics, but are otherwise randomly sampled. Our goal is to see whether such random architectures – which we call *Monte Carlo Winning Tickets* (MCWTs) – could achieve performance level close to the original, data-derived winning tickets. The main outcome from our experiments is that this may indeed be the case: we demonstrate Monte Carlo winning tickets with performance close to the iteratively pruned winning tickets. The main implication from this finding is that sparse, trainable neural networks can possibly be constructed without expensive retraining and pruning of dense models.

2   Monte Carlo Winning Tickets

We hypothesise that a sparse trainable neural networks can be constructed by sampling architectures (more precisely: network connections) from a distribution that replicates certain statistics over weights and activations in an untrained winning ticket. In this work we focus on two such statistics: magnitudes of weights and *connectivity*, which we describe below in more details.

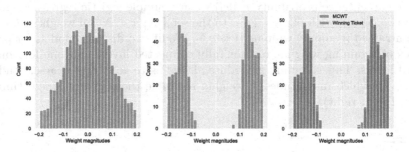

Fig. 1. Weight magnitudes in an untrained dense layer (left) and a winning ticket (center). Right: match between weight magnitudes in a winning ticket and an MCWT.

Let Φ_N be a set of parameters (i.e. weights) in neural network N, e.g. a densely connected convolutional network with certain number of layers and channels. Further, let $w_{cij}^{kl} \in \Phi_N$ be the (c, i, j) element of the k-th kernel in the l-th convolutional layer, where c enumerates input channels and (i, j) are indices over the kernel width and height, respectively. Similarly, for a fully connected layer l we will write $w_{ij}^l \in \Phi_N$ to denote the weight between the i-th neuron in layer $l - 1$ and j-th neuron in layer l. We will also write $w \in \Phi_N$ to denote a network parameter irrespective of its location in the architecture. Consider an untrained network N. If N is densely connected, or is a sparse network sampled from a uniform distribution over all possible subsets of parameters, the distribution $p(w \mid w \in \Phi_N)$ will simply match the density from which we sample initial parameter values. However, this will generally not be the case if N is a sparse network chosen in some non-uniform way from all possible subsets of parameters. In particular, if N is an untrained winning ticket, the parameters $w \in \Phi_N$ will be chosen by iterative training, pruning and rewinding to initial values. Thus, the conditional density $p(w \mid w \in \Phi_N)$ may be quite different from the density used to sample initial parameter values. Indeed, in Fig. 1 (center) we report an empirical distribution of initial parameter values retained by a winning ticket in a fully connected layer. Under uniform pruning we would expect a Gaussian distribution, which was used to initialize the layer (Fig. 1, left). However, winning tickets appears to preferentially prune weights with small initial values. We observed similar tendency in weights retained in convolutional layers.

Our first goal is to construct a randomly sampled sparse network S (Monte Carlo winning ticket) which replicates empirical distribution of weights in a genuine winning ticket. To approximately replicate this distribution, we begin with a densely connected model and sample pruning masks for its parameters from Bernoulli distributions parametrized by probabilities that depend on magnitudes of weights. That is, let $z_w \in \{0, 1\}$ be an indicator variable such that $z_w = 1$ if w is retained in S, and $z_w = 0$ otherwise. Then:

$$z_w \sim \text{Bernoulli}\left(f\left(|w|\right)\right). \tag{1}$$

To fit the acceptance probability $f\left(|w|\right)$ we fit a simple logistic regression model $f\left(|w|\right) = \sigma\left(u|w|\right)$, with parameter u, to the set $\{w \mid w \in N\}$ of weights retained in an untrained (genuine) winning ticket N. We fit two distinct models for probabilities of retaining weights: one for fully connected layers and one for convolutional layers. This sampling procedure allows us to construct sparse, random networks in which magnitudes of weights resemble those in untrained winning tickets (Fig. 1, right).

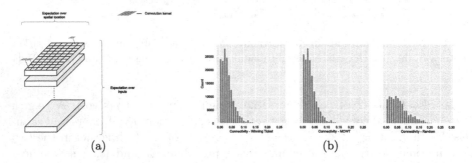

(a)   (b)

Fig. 2. (a) Connectivity measure in a sparse neural network. (b) Connectivity in a standard winning ticket (left), Monte Carlo winning ticket (center) and a uniformly sampled sparse network (right).

Sampling that is driven purely by weight magnitudes disregards any connectivity structure in the sampled network. To remedy this, we introduce a simple connectivity measure illustrated in Fig. 2a. Let $\mathbf{x} \sim D$ be an input observation chosen randomly from the train set D. Further, let $a_c^{l-1}\left(m, n; \mathbf{x}\right)$ be an input activation map (channel) connected to the parameter w_{cij}^{kl} and evaluated for input $\mathbf{x}$. Indices m, n enumerate spatial locations in the input channel. We define the connectivity of w_{cij}^{kl} as an expected input to that parameter in an untrained network:

$$g\left(w_{cij}^{kl}\right) = \mathbb{E}_{\substack{x \sim D \\ m,n}}\left[a_c^{l-1}\left(m, n; \mathbf{x}\right)\right]. \tag{2}$$

Note that expectation is taken with respect to input examples and spatial locations in the channel connected to w_{cij}^{kl}. In practice, we approximate this expectation by averaging across all inputs in the training set. A similar connectivity measure can be defined for fully connected layers:

$$g\left(w_{ij}^{l}\right) = \mathbb{E}_{x \sim D}\left[a_{l-1}\left(i; \mathbf{x}\right)\right], \tag{3}$$

where $a_{l-1}\left(i; \mathbf{x}\right)$ is the i-th input to the fully connected layer l in an untrained network evaluated for $\mathbf{x}$.

Connectivity measures introduced above express the expected input to a given network parameter, which in turn reflect contributions to that input from parameters in proceeding layers. In Fig. 2b we compare distributions of connectivity measure in an untrained standard winning ticket (left) and a uniformly sampled sparse network (right). Note a substantial difference between the two distributions. We leverage this discrepancy to incorporate connectivity information into the sampling model for Monte Carlo winning tickets. That is, we extend the magnitude-base sampling (Eq. 1) to account for $g(w)$:

$$z_w \sim \text{Bernoulli}\left(f\left(|w|, g(w)\right)\right),$$
$$f\left(|w|, g(w)\right) = \sigma\left(u_1|w| + u_2 g(w)\right). \tag{4}$$

Again, we use a simple logistic regression model for the acceptance probability (with parameters u_1, u_2), and fit it to the magnitudes of weights and connectivities observed in a genuine winning ticket. The resultant connectivity distribution in a Monte Carlo winning ticket is pictures in Fig. 2b (center). To estimate connectivity in a layer l we must known which parameters from the proceeding layers are retained. Thus, we sample Monte Carlo wining tickets in a greedy layer-wise way, by moving from the network input towards the last layer. In the first layer we account only for weight magnitudes. In subsequent layers we also estimate connectivity values.

3   Results

Experiments in this work follow the setup used in Frankle and Carbin [3]. Specifically, we evaluate MCWTs on *Conv2*, *Conv4* and *Conv6* architectures used therein, and use hyper-parameter values reported in that work. In each case, we prune 80% of weights in every network layer using either random pruning, iterative pruning described by Frankle and Carbin [3] (i.e. winning tickets) or Monte Carlo method described in the previous section.

We use CIFAR-10 dataset [7] to evaluate MCWTs. It consists of 60,000 images from 10 classes. Of these images, 50,000 are in the train set and the remaining 10,000 in the test set. All images are 32×32 pixels in size. To follow the experimental setup from [3], we use only the 50,000 train samples. Specifically, we train networks on 45,000 examples and evaluate on the remaining 5,000. Even though we train for up to 300–500 epochs, MCWTs learn faster, achieving near-final performance after 10–20 epochs.

Results from our experiments are reported in Table 1. The main finding here is that Monte Carlo winning tickets allow for low-computational-cost discovery of sparse trainable neural network architectures. Specifically, we observe similar performance for original winning tickets and our Monte Carlo winning tickets. These results are slightly above unpruned networks and significantly outperform baseline, i.e. random pruning. In Monte Carlo winning tickets most of the improvement comes from fitting distributions of initial weights, with connectivity playing a less important role. For *Conv4* and *Conv6* models our approach

Table 1. Experimental results for networks with 80% pruning. To facilitate comparison, we reproduced the original Winning Ticket results in our code. We also compare against random pruning and unpruned networks.

Experiment	Conv2	Conv4	Conv6
Unpruned network	67.20%	74.62%	77.40%
Randomly pruned network	66.90%	73.20%	73.14%
Original winning ticket	**70.42%**	74.92%	78.04%
Monte Carlo winning ticket - w/o connectivity	69.47%	**76.39%**	78.10%
Monte Carlo winning ticket - connectivity	68.40%	76.21%	**78.17%**

shows modest improvement over original winning tickets, while for smaller *Conv2* architecture it exhibit slightly worse, but still similar performance. Note, however, that it is often impossible to exactly replicate experimental conditions in deep learning, due to missing data on some hyper-parameters or differences in software versions. As a result, we observed higher performance for unpruned networks and slightly better results for the original winning tickets, than those reported in [3].

4 Related Work

The starting point for the research presented in this work was the Lottery Tickets Hypothesis formulated by Frankle and Carbin [3]. From a practical perspective it presents an iterative pruning mechanism for discovering sparse, trainable neural networks. The main disadvantage of this approach is the necessity to train the network several times, each time increasing the number of pruned parameters by a certain, architecture-dependant factor. Zhou et al. [9] conduct further research on this subject and introduce the concept of *supermasks*. They show that choosing the pruning masks with a specific, carefully designed criteria can lead to significantly better-than-chance performance of the randomly-initialized sparse network. Our research shows that for good results, we can simply sample the connections in the sparse model in such a way that they resemble winning tickets with respect to certain statistics over magnitudes of weights and network connectivity. Frankle et al. [4] further develop their technique for finding winning tickets, thereby enabling its use on more complex datasets, such as ImageNet. Most importantly they show that it is easier to start pruning networks that are trained for a few epochs, rather than rewinding them all the way to the initial parameter values. This suggest an avenue for further research on Monte Carlo Winning tickets, namely investigation of performance of models sampled using our approach on ImageNet-scale datasets. Concurrently to work on Monte Carlo winning tickets, Blalock et al. [1] suggested alternative ways to assess neural network pruning. However, their findings cannot be applied directly to our work, because we compare two networks with the same architecture and pruning ratio. Finally, Xie et al. [8] investigated randomly wired neural networks constructed using random graph models. They obtained competitive performance in computer vision tasks.

5 Conclusions

In this work we presented Monte Carlo winning tickets. These sparse neural networks are constructed by sampling connections from a probability distribution that replicates statistics over weights and connectivity in standard winning tickets. We demonstrated that Monte Carlo winning tickets are trainable from scratch, i.e. they train to a performance level matching standard winning tickets and densely connected model. Typically, this level of performance used to be achieved by training and then pruning a densely connected model. Thus, Monte Carlo winning tickets open an avenue to lower the computational cost of deploying sparse neural nets.

References

1. Blalock, D., Ortiz, J.J.G., Frankle, J., Guttag, J.: What is the state of neural network pruning? arXiv preprint arXiv:2003.03033 (2020)
2. Brown, T.B., et al.: Language models are few-shot learners. In: Advances in Neural Information Processing Systems, vol. 33 (2020)
3. Frankle, J., Carbin, M.: The lottery ticket hypothesis: finding sparse, trainable neural networks. In: 7th International Conference on Learning Representations, ICLR 2019 (2019). https://arxiv.org/pdf/1803.03635.pdf
4. Frankle, J., Dziugaite, G.K., Roy, D.M., Carbin, M.: Stabilizing the lottery ticket hypothesis. arXiv preprint arXiv:1903.01611 (2019)
5. Han, S., Pool, J., Tran, J., Dally, W.J.: Learning both weights and connections for efficient neural networks. In: Advances in Neural Information Processing Systems, vol. 28, pp. 1135–1143 (2015)
6. Huang, G., Liu, Z., Van Der Maaten, L., Weinberger, K.Q.: Densely connected convolutional networks. In: 2017 IEEE Conference on Computer Vision and Pattern Recognition (CVPR), pp. 2261–2269. IEEE Computer Society (2017)
7. Krizhevsky, A., Hinton, G., et al.: Learning multiple layers of features from tiny images (2009)
8. Xie, S., Kirillov, A., Girshick, R.B., He, K.: Exploring randomly wired neural networks for image recognition. In: 2019 IEEE/CVF International Conference on Computer Vision, ICCV, pp. 1284–1293. IEEE (2019)
9. Zhou, H., Lan, J., Liu, R., Yosinski, J.: Deconstructing lottery tickets: zeros, signs, and the supermask. arXiv preprint arXiv:1905.01067 (2019)

Interpreting Neural Networks Prediction for a Single Instance via Random Forest Feature Contributions

Anna Palczewska[1]($^{(\boxtimes)}$) and Urszula Markowska-Kaczmar[2]($^{(\boxtimes)}$)

[1] School of Built Environment, Engineering and Computing, Leeds Beckett
University, Leeds, UK
a.palczewska@leedsbeckett.ac.uk
[2] Wroclaw University of Science and Technology, Wroclaw, Poland
urszula.markowska-kaczmar@pwr.edu.pl

Abstract. In this paper, we are focusing on the problem of interpreting Neural Networks on the instance level. The proposed approach uses the Feature Contributions, numerical values that domain experts further interpret to reveal some phenomena about a particular instance or model behaviour. In our method, Feature Contributions are calculated from the Random Forest model trained to mimic the Artificial Neural Network's classification as close as possible. We assume that we can trust the Feature Contributions results when both predictions are the same, i.e., Neural Network and Feature Contributions give the same results. The results show that this highly depends on the level the Neural Network is trained because the error is then propagated to the Random Forest model. For good trained ANNs, we can trust in interpretation based on Feature Contributions on average in 80%.

Keywords: Model interpretation · Artificial Neural Network · Feature contributions

1 Introduction

Neural Networks (NNs) are widely accepted machine learning technique to learn complex relationships for classification and prediction problems. Their pattern-matching and learning capabilities allowed them to address many difficult problems, impossible to solve by other computational methods. Unfortunately, they lack transparency. It is hard to see how the network arrives at a particular conclusion due to the network architecture's complexity. Therefore, ANN (Artificial Neural Network) is often called a black-box model [12]. The interpretation of the model (why the model makes a particular decision) is important [20], but for non-linear models' extraction of such knowledge is difficult to achieve.

There are two approaches to ANN models interpretation: methods based on rule extraction and variable importance.

© Springer Nature Switzerland AG 2021
M. Paszynski et al. (Eds.): ICCS 2021, LNCS 12743, pp. 140–153, 2021.
https://doi.org/10.1007/978-3-030-77964-1_12

Rule extraction methods, that try to interpret trained neural networks or opaque models, have a long track record in machine learning and its applications. The definition of the problem can be found in [5]. The taxonomy of rule extraction from neural networks distinguishes the following: decompositional (local methods), pedagogical (global methods) [3] and eclectic methods. The main disadvantage of this approach is a limited interpretation of a model for data with a large number of variables. Models built for datasets that contain thousands of variables (e.g., codes DNA, chemical compounds or binary data) are not readily interpretable by rules.

Estimation of variable importance for ANN models explains the relative contribution of each variable to the prediction result. In [14] authors presented the interquartile range (IQR) method to rank variables based on their importance. This method was used to rank variables but does not explain the influence of a variable on predicted value. In [4,6,11] methods, based on partial derivatives in ANN sensitivity analysis were proposed to calculate variable importance. In [12] the relative importance of variable, calculated using various methods, was averaged to handle the instability problem of variable importance.

The variable importance is applicable to datasets with a large number of input variables as feature selection method. The variables with the most significant importance are further used to build more accurate models [21]. The need for interpretation and difficulties connected with this problem grow when we consider deep networks. A survey paper [2] and two latest methods [22] used flip points to explain the boundary between two classes and [6] proposed enhanced integrated gradients. Using principal component analysis (PCA) and rank-revealing QR factorization (RR-QR), the set of directions from each training input variable to its closest flip point provides explanations of how a trained neural network processes a dataset.

In some cases, we would like to interpret the model behaviour on the instance level. As an example, let us consider two toxic chemicals (class toxic) with similar structures. We would like to know which part of the structures are the most toxic by extracting contributions of chemical substructures toward the toxicity. Applying the rules approach we could find that they share the same conditions that classify them toxic, but when we look at variable contributions, we may see differences in substructures toxicity. In [18] authors presented a method for colouring molecule using a heat map for interpretation of support vector machine models. Another method called Feature Contributions was proposed by Kuzmin et al. in 2011 [8]. It was designed to extract feature contributions for random forest models for regression problems. It has been extended to random forest classification models in [13] and used in work [10], where authors compared the predictions' chemical interpretability based on scoring schemes for assessing heat map images of substructural contributions. Another example of feature contribution was presented in [17] where authors propose the novel explanation technique LIME (Local Interpretable Model-agnostic Explanations) that approximates an interpretable model locally. Also, in [9], authors presented SHAP (SHapley Additive exPlanations) values allowing interpretation of predictive models based on a game theory approach.

Feature contributions are numerical values that allow extraction of a relationship between a particular feature value and a model's decision. For each instance, we calculate how much a given variable/feature contributed to the predicted outcome. We can see which features have a positive/negative impact on a predicted value and which of them have a more decisive influence.

There are no methods (that analyse a network structure) to interpret neural networks prediction on an instance level. Currently, in the era of the application of deep neural network in almost all areas with large data availability, the interpretation of feature influence on the model decision would be beneficial for many decision-making models. Many methods for extracting feature/variable contributions are on a model level that is not sufficient for more detailed analysis. Unfortunately, the structure of the neural network does not allow extraction of such information, because it is distributed in the network.

In this paper, we address the problem of interpretation of neural networks on an instance level. To achieve a solution, we propose the use of the neural network as an oracle within a pedagogical approach (similar to rule extraction). This oracle could be any opaque model. Within that approach, we use a Random Forest (RF) model together with its Feature Contribution (FC) method described in [13]. In the presented research, we assume that the FCs are acquired from RF mimicking the activity of ANN, we have to check whether we can trust the result offered by FCs. We use feature contributions to build a classifier. If the ANN responds with the same class as FCs for a new input vector, then this is an indication that we can trust the interpretation delivered by FCs.

The paper is organised as follows. Section 2 describes the proposed method for the ANN model interpretation. It provides the formal problem statement and includes the definition of a random forest model, feature contributions, and their analysis. Section 3 describes the experimental study and discusses the obtained results. Section 4 concludes our work.

2   Methodology

Although extraction of feature contributions is not new, as we are borrowing from existing methods, feature contributions in the context of ANN model interpretation require the development of some methodology. In this section, we recall the definition for feature contributions, and we describe how the feature contributions can be used to interpret neural networks.

2.1   ANN Model Interpretation for a Single Instance

We assume that the ANN model trained for a specific classification problem is given. Our idea is to train the Random Forest model to mimic the behaviour of ANN then to calculate Feature Contributions.

The workflow of the ANN model interpretation is presented in Fig. 1. In step 1 we build Random Forest (RF) model using input data $\mathbf{x}$ and output $\mathbf{y} \in Y_{RF}$ produced by the ANN model. Thus, the training data set for RF is composed of pairs $<\mathbf{x}, \mathbf{y}>$, where $\mathbf{x} \in D_{RF}$ and $\mathbf{y} \in Y_{RF}$. In step 2, we extract Feature Contributions from a Random Forest model. When, for a new input instance vector $\mathbf{x}_{new}$ we want to interpret the ANN classification result, we calculate Feature Contributions (FCs) for this instance. They show the influence (negative or positive) of each feature (input variable) on a predicted class. To assess whether we can trust the result we perform classification based on FCs, and the evaluation is positive if the class predicted by ANN and by FCs are the same.

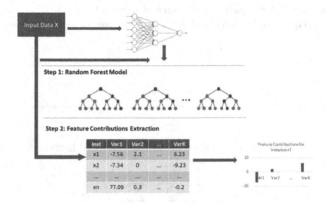

Fig. 1. A schema for the ANN model interpretation method via Random Forest model and Feature Contributions

2.2 Random Forest Feature Contributions

Firstly, we recall the definition of the feature contributions proposed in [8,13]. Feature contributions calculated for a given instance represent the influence (negative or positive) of each feature (input variable) on a predicted target. They are computed in two steps. Firstly, local increments are calculated for each node in the forest's trees using the trees training datasets:

$$LI_{fc} = \begin{cases} Y_{mean}^c - Y_{mean}^p, & \text{if the split in the parent is performed over the feature} f, \\ 0, & \text{otherwise,} \end{cases}$$

where Y_{mean} is a fraction of the training instances in a given node c, where c - is a child node and p - is a parent node, belonging to a selected class (for details see [13]) or an average over the instances within the node for regression models. A local increment for feature f represents the change of the probability of being in a given class between the child node and its parent node in a tree.

Algorithm 1. The method (in pseudocode) of ANN interpretation using feature contributions

---

Require: ANN, D_{RF}, Y_{RF} and D_{New}, Y_{New}
1: Train a random forest model RF on D_{RF}, Y_{RF} datasets
2: Calculate feature contributions FC from the trained RF model
3: Find the class representative FC_{rep}^c for feature contributions (medians or cluster centres)
4: **for** each instance $\mathbf{x}_i$ in D_{New} **do**
5:    calculate feature contribution FC_i for an instance $\mathbf{x}_i$)
6:    **for** each class c in datasets classes C **do**
7:       Calculate Euclidean distance between feature contributions FC_i for the instance $\mathbf{x}_i$ and class representative: $d_E(FC_i, FC_{rep}^c)$
8:    **end for**
9:    Select the class c for which the distance is minimal.
10:   **if** class c is equal to the predicted ANN model class $\mathbf{y}_i$ for the instance $\mathbf{x}_i$ **then**
11:       $p_i = 1$
12:   **else**
13:       $p_i = 0$
14:   **end if**
15: **end for**

Secondly, for any instance and a variable f these local increments are summed on tree paths:

$$FC_{if} = \frac{1}{ntree} \sum_{k=1}^{ntree} \sum_{l=1}^{knode} LI_{if_{kl}}, \tag{1}$$

where the value $LI_{if_{kl}}$ is a local increment for the instance i, feature f in k tree and its l node. The values $ntree$ and $knode$ represent the number of trees in the forest and the number of nodes from the k tree, which split over a feature f, respectively.

Feature Contribution values estimate a contribution of feature values to the difference between the actual prediction and the mean prediction for the current set of feature values. As reported in [13], $Y' = Y^r + \sum_j FC_j$ where Y' denotes a predicted value and Y^r averages of Y_{mean} overall root nodes in the forest with the assumption of unanimity (all elements in trees nodes belonging to the same class). The magnitude represents how strongly the feature contributes and the sign represents the direction (such as toward the model decision or against).

2.3   Interpretation of ANN Prediction Based on Feature Contributions

Once feature contributions are extracted from a Random Forest model they can be interpreted by domain expert reviling the model decision process. As these values were calculated within the pedagogical approach we need to assess the certainty of such interpretation. This procedure is shown in Algorithm 1. As the input, the algorithm requires trained ANN, datasets D_{RF}, Y_{RF} for RF training.

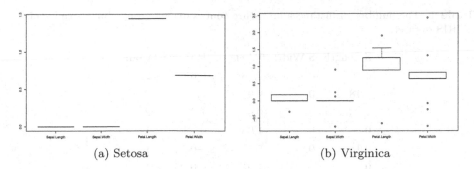

(a) Setosa   (b) Virginica

Fig. 2. An example of feature contribution variations. Box plots of feature contributions for two classes of IRIS dataset [1]. The axes x and y represent the IRIS features and values of their contributions, respectively.

To test if we can trust the interpretation of ANN prediction for a new instance $\mathbf{x}_{new}$ we use a distance between feature contributions of the new data and feature contributions *representatives* for the Random Forest training dataset (line 2–3 in Algorithm 1). As described in [13] we can consider two feature contribution representatives: median and cluster centroids, computed for each class separately. To calculate representatives, we used all instances from the random forest training dataset that were correctly classified. Then:

- if there is no variation within feature contributions which means that all values are distributed around the FC mean (see for example Fig. 2a) then as feature contributions *representative* we use a median.
  To classify a new i-th instance $\mathbf{x}_{new}^{i}$ based on its feature contributions, we calculate feature contributions first. Then, the Euclidean distance d_E (Eq. 2) is computed for all class representative's medians (line 5–9 in Algorithm 1), and minimal distance is selected:

$$d_E^i = \min_l \sqrt{\sum_{f=1}^{nvar} (FC_{if} - m_{fl})^2}, \qquad (2)$$

  where FC_{if} is calculated using Eq. (1), $nvar$ is a number of features (variables) in the input vector and m_{fl} is a feature contributions median of f-th feature and l-th class. The smallest distance indicates the class of the new data i predicted by the feature contributions method.
- otherwise, there is a variation within feature contributions (see Fig. 2b for *Virginica* class as an example).
  Many instances have values close to FC mean, and there are few elements with different values. These few elements can produce a small group with another feature contribution that differs from the majority group created. The group with the smallest variance is called a core cluster [13] and its centre is used as the class *representative*. If clusters have the same variance (e.g. equal to zero) we can have more than one representative for a class. For each class,

Table 1. The number of instances in feature contributions groups for Virginica class of IRIS dataset.

S.Length	S.Width	P.Length	P.Width	Count
0	0	1.26	0.65	25
0.18	0	0.9	0.83	9
−0.32	0	0.9	1.3	2
0	0	1.08	0.83	1
0.18	0.92	5.48	−0.74	1
0	−0.75	1.91	−0.25	1
0.18	0.25	1.55	−0.07	1
0	0.25	1.91	−0.25	1
0	0.3	−0.66	2.43	1

the best number of clusters is obtained using the elbow method [15].

Training instances $\mathbf{x} \in D_{RF}$ are assigned to these clusters. To classify a new instance $\mathbf{x}_{new}^i$, the Euclidean distance (Eq. 3) is calculated to all cluster centres and the smallest distance is selected (line 5–9):

$$d_E^i = \min_l \sqrt{\sum_{f=1}^{nvar} (FC_{if} - c_{fj_l})^2} \qquad (3)$$

where c_{fj_l} is a centroid of a cluster j of class l, $nvar$ is the number of variables f. The smallest distance indicates which cluster a new data $\mathbf{x}_{new}^i$ belongs to and defines a class for the new instance.

To illustrate how to use centroids as representatives, let us consider the example of *Virginica* class in detail. Table 1 shows examples of patterns in feature contributions for the Virginica class from the IRIS dataset. There were 42 elements in the training dataset for the random forest model that were correctly classified. We can notice that there are two main groups with cardinality 25 and 9 elements. The clusters that have the smallest variance become core clusters and core clusters are further used to evaluate whether we can trust in the interpretation of ANN offered by FC.

If the class assessed by the use of feature contributions is the same as the class predicted by the ANN, we trust the FCs interpretation result ($p_i = 1$), in another case, it is not possible ($p_i = 0$) (lines 10–14 in Algorithm 1).

3   Experimental Study

The experimental research goal is to test whether the feature contributions method can be used to interpret a trained ANN model. In this research, we focused on the shallow ANN, but it could also be a more complex model. The process of training ANN for a given training dataset, developing the Random Forest model, extracting Feature Contributions, identifying the model FCs representatives, and testing ANN model reliability using FCs was repeated fifty times. The averaged results are presented in this section.

3.1 Datasets

Eight datasets from the UCL Machine Learning Dataset Repository [1] were used. We selected the datasets that were often used as benchmark sets in rule extractions for ANN models [7]: Breast Cancer Wisconsin (Diagnostic) Dataset (BCWD), COX2 [19], German Credit Scoring, IRIS, SEEDS, Teaching Assistant Evaluation (TEACHING), WAVEFORM, Database Generator (Version 1), WINE.

3.2 Training the Artificial Neural Network Model

The multi-layer perceptron (MLP) network has been used as ANN model. Training is performed by the backpropagation method. We used the default settings for the *MLP* model from the RSNNS package in R. We only set a parameter *size* (describing the number of hidden neurons) to be equal to the averaged sum of input and output variables, learning coefficient - *learnFuncParams* equals 0.1 and the maximal number of iteration equals 50. The ANN model had only one hidden layer. The number of ANN's output neurons was equal to the number of classes in a given dataset because we used 1 of n encoding for the output layer. We did not focus on the *MLP* model accuracy, so we did not optimize the model parameters to get the most accurate model (the model accuracy was not the subject of this study).

Table 2 presents the averaged results from building the *MLP* model. First four columns show the cardinality of each dataset and the split for training (D_{train}), testing (D_{test}) and validating (D_{new}) datasets. Testing D_{test} and D_{new} datasets were randomly selected taking 20% of data for both datasets. To have an equally represented set of elements in each class this selection was conducted for each class separately. The fifth and sixth columns in the table represent the number of attributes and classes for each dataset, respectively. The last two columns show the averaged accuracies for the *MLP* models for training and testing datasets obtained from the repeated procedure of 50 runs, each time splitting the dataset and generating a new ANN model.

In the training *MLP* procedure, we do not focus on high-quality results, therefore one can see that the *MLP* model gives for some datasets (BCWD, IRIS, SEEDS, WAVEFORM and WINE) high averaged accuracy around 0.9, but for some (TEACHING and COX2 dataset) they are less satisfying.

Table 2. Characteristics of datasets and average accuracy (ACC) of ANN over 50 runs of the ANN model development procedure. The columns represent: the number of instances in the dataset (Inst), the number of instances for the training and testing dataset for the ANN model ($\#D_{train}$, $\#D_{test}$), the number of instances for a validating dataset ($\#D_{new}$), the number of dataset's attributes and classes (#Attr, #Class), average accuracy for the training dataset (ACC_{train}) and accuracy for the testing dataset ACC_{train} for the ANN model

Name	#Inst	$\#D_{train}$	$\#D_{test}$	$\#D_{new}$	#Attr	#Class	ACC_{train}	ACC_{test}
BCWD	683	409	137	137	9	2	0.981	0.966
COX2	190	114	38	38	255	2	1.000	0.677
German_CS	1000	600	200	200	20	2	0.878	0.737
IRIS	150	90	30	30	4	3	0.958	0.941
SEEDS	210	126	42	42	7	3	0.953	0.916
TEACHING	151	90	30	31	5	3	0.576	0.491
WAVEFORM	5000	2998	1000	1002	21	3	0.904	0.856
WINE	178	105	36	37	13	3	1.000	0.980

3.3 Training Random Forest Model and Calculating Feature Contributions

Random Forest model was trained on a combined dataset D_{train} and D_{test} called D_{RF} and Y_{RF} - an output of ANN for D_{RF} as described in Sect. 2.1. We used *randomForest* package in R. The number of trees was set to the number of input variable for each dataset separately. The reason lies in avoiding the overfitting for datasets like IRIS with a small number of variables. We used default settings for this method. We set the parameter *replace* = False to avoid selection with a replacement for training trees. We also keep information on records that were used to train a tree in a forest by setting the parameter *keep.inbag* = True. This is needed to calculate Feature Contributions.

Table 3 shows the averaged results for Random Forest models and for the *MLP* models. Column $\#D_{new}$ informs how many instances contains the D_{new} dataset. The averaged accuracy of *MLP* model for D_{new} is included in the column (ANN_{new}). The column RF_{new} describes average accuracy for the Random Forest models. The table also shows the average accuracy of the Random Forest models for training data (column RF_{train}) achieved on the D_{RF} dataset.

To test how well the Random Forest model mimics the ANN model, we calculated the average Area Under Curve for each RF model. Table 3 presents averaged AUCs for D_{new}. The higher the AUC value – closed to one, the less noise/error was introduced by the Random Forest model, and the better interpretability of the ANN model we can expect. As the ground truth, the instances from Y_{RF} and Y_{new} were considered, respectively.

Table 3. Average accuracy for ANN (ANN_{new}) and RF (RF_{new}) models for validation dataset D_{new} and for RF training D_{RF} dataset (RF_{train} column), AUC for ANN and RF models for D_{new}

Name	#D_{new}	ANN_{new}	$AUC_{ANN_{new}}$	RF_{train}	RF_{new}	$AUC_{RF_{new}}$
BCWD	137	0.97	0.96	0.99	0.98	0.97
COX2	38	0.68	0.67	0.95	0.77	0.74
German_CS	200	0.74	0.66	0.96	0.81	0.72
IRIS	30	0.95	0.94	0.99	0.97	0.96
SEEDS	42	0.93	0.84	0.99	0.92	0.90
TEACHING	31	0.64	0.53	0.98	0.93	0.90
WAVEFORM	1002	0.86	0.84	0.97	0.84	0.83
WINE	37	0.98	0.98	0.99	0.94	0.94

3.4 Certainty Assessment of ANN Interpretation

Feature Contributions calculated for the instance $\mathbf{x}_{new}$ give information about the relation between predicted class and input features for an RF model. Because the RF model only mimics the ANN model we are interested in evaluating how much we can rely on this interpretation. To decide whether the extracted Feature Contributions for an instance $\mathbf{x}_{new}$ give certain interpretation we test them against ANN model prediction for this instance. The verification of the ANN model prediction is based on the comparison of the classification of $\mathbf{x}_{new}$ data made with the ANN model and the class found by the Feature Contribution analysis. If the prediction from FC agrees with the prediction from ANN for an instance $\mathbf{x}_{new}$, we say that interpretation is *certain* for this instance. If the predicted class from ANN agrees with FC prediction and with the original class for this instance, we say that prediction is *correct*.

Following the Algorithm 1 we calculated Feature Contributions for instances from the Random Forest training dataset D_{RF} and the validation dataset D_{new}. We used the *rfFC* R package [16]. We selected these instances from D_{RF} for which predictions from RF and ANN models agree with the original value of the output variable. Then for each class, we calculated Feature Contributions medians. In the second step, we applied *k-means* to cluster Feature Contributions within each class. For each Feature Contribution subset with non zero variance, the number of clusters was assessed using the *MClust* R package. Finally, we extract the Feature Contributions representatives for each class. In Fig. 3 we present medians representatives of Feature Contributions for two datasets and all classes (for each dataset). Contributions can be positive as well as negative values and representatives differ between classes.

Having the Feature Contributions representatives for each class, we calculate Feature Contributions for each $\mathbf{x}_{new}$ instance from D_{new}. Then, to find the class for the new instance, we compute distances between representatives and Feature Contributions for instances from D_{new} using Eq. (2) and (3). The smallest distance assigns the class.

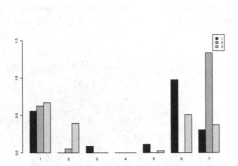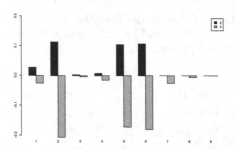

(a) Feature contributions medians for SEED dataset. The variable numbers represent: area, perimeter, compactness, length of kernel, width of kernel, asymmetry coefficient and length of kernel groove, respectively.

(b) Feature contributions medians for BCWD. The variable numbers represent: Clum Thickness, Uniformity of Cell Size, Uniformity of Cell Shape, Marginal Adhesion, Single Epithelial Cell Size, Bare Nuclei, Bland Chromatin, Normal Nucleoli, Mitoses, respectively.

Fig. 3. Example of Feature Contributions median for selected datasets

Interpretability Method Evaluation for All Datasets. In this section, we repeat the procedure described in Algorithm 1 for all eight chosen datasets. Table 4 shows averaged results from repeated runs of the method for each dataset. The values were rounded to the nearest integer. The first column in this table shows the number of elements in the new dataset D_{new}. The second (Med_Certain) column shows the number of instances that were marked certain with the median approach. The third (Med_Correct) column shows how many instances were correctly classified by the median approach concerning the original class value. The last two columns show the number of interpretations that were marked as certain based on the clustering approach (Clust_Certain) and the number of correctly classified instances for the original class (Clust_Correct). Also, Table 5 presents detailed results from the certainty assessment of the interpretability method. For each dataset, columns represent instances for which ANN interpretation was marked as certain and uncertain for both median and clustering methods. In rows, we have ANN prediction expressed by instances that were classified wrongly by the ANN model.

The aggregated results confirm that the presented method is suitable to interpret the ANN model for new data. For ANN models with good predictive accuracy such as for IRIS, BCWD, WINE, SEEDS, the certainty of ANN interpretation is greater than 80%. This means that Feature Contributions represent the true importance of the ANN model. For weak models (TEACHING and German_CS datasets), the certainty is greater than 60%. It is worth noticing that models for these two datasets had a low predictive accuracy. This shows that the proposed approach of assessment of the ANN model interpretability can filter instances with correct ANN prediction and with certain Feature Contribution values. Also, the results show that the use of clustering seems to work better than the use of the median approach.

Table 4. Number of elements from the D_{new} dataset marked as a correctly predicted by ANN model via median and clustering methods in respect to their original class label

Name	#D_{new}	Med_Certain	Med_Correct vs Orig	Clust_Certain	Clust_Correct
BCWD	137	130 (94,8%)	127	133 (97%)	130
COX2	38	28 (73,6%)	21	29 (76,3%)	23
German_CS	200	161 (80,5%)	127	180 (90%)	145
IRIS	30	27 (90%)	26	28 (93,3%)	27
SEEDS	42	35 (83,3%)	33	39 (92,8%)	37
TEACHING	31	22 (70,0%)	19	27 (87%)	22
WAVEFORM	1002	674 (67,2%)	585	745 (74,3%)	663
WINE	37	31 (83,7,4%)	29	34 (91,8%)	33

Table 5. Certain/Uncertain vs correct/non correct prediction for elements of D_{new} dataset.

Name	#Valid	ANN Pred	Median		Cluster	
			Certain	Uncertain	Certain	Uncertain
BCWD	137	correct	127	2	130	1
		Incorrect	3	5	3	3
COX2	38	correct	21	3	23	3
		Incorrect	7	7	6	6
German_CS	200	correct	127	32	145	19
		Incorrect	34	7	35	1
IRIS	30	correct	26	2	27	2
		Incorrect	1	1	1	0
SEEDS	42	correct	33	4	37	2
		Incorrect	2	3	2	1
TEACHING	31	correct	19	1	22	0
		Incorrect	3	8	5	4
WAVEFORM	1002	correct	585	107	663	98
		Incorrect	89	221	82	159
WINE	37	correct	29	4	33	2
		Incorrect	2	2	1	2

4   Conclusions

In this paper, we showed that Feature Contributions could be used to interpret an ANN model for a before unseen data (instance) to find relationships between instance variables and the predicted outcome. We used shallow ANN models as the example of a non-transparent model. This approach offers interpretation for any opaque model and does not limit its architecture. The idea of method interpretation lies in building a forest of trees that with high accuracy emulates the behaviour of the opaque model and then Features Contributions calculation allow us interpretation on an instance level.

To test the certainty of Feature Contribution for the ANN model interpretation, we proposed the procedure for the classification of instances based on their feature contribution values. Using a distance measure between a new instance feature contribution and the model representatives Feature Contributions we can decide wherever to trust the interpretation of the ANN model. The representatives in this work were defined by a median or by cluster centres defined on the model training dataset. The averaged results showed that for the best ANN models in 80% of new instances we were able to tell whether the interpretation was certain. The experiment was carried on eight datasets from the UCI Machine Learning repository.

A study on the threshold level for the Euclidean distances used in median and clustering methods and its influence on the ability of ANN interpretation is the next step of our research in this area. Further research, focusing on the distance metrics choice will be an essential enhancement of the study presented here. Comparison of the proposed method with other available methods to test the agreement on the explained model decision will be the next interesting research problem to address.

References

1. Bache, K., Lichman, M.: UCI machine learning repository. University of California, Irvine, School of Information and Computer Sciences (2013). http://archive.ics.uci.edu/ml/datasets. Accessed 28 Aug 2016
2. Fan, F.-L., Xiong, J., Li, M., Wang, G.: On interpretability of artificial neural networks: a survey (2020). https://arxiv.org/ftp/arxiv/papers/2001/2001.02522.pdf
3. de Fortuny, E., Martens, D.: Active learning-based pedagogical rule extraction. IEEE Trans. Neural Netw. Learn. Syst. **26**(11), 2664–2677 (2015)
4. Gevrey, M., Dimopoulos, I., Lek, S.: Two-way interaction of input variables in the sensitivity analysis of neural network models. Ecol. Modell. **195**(1–2), 43–50 (2006). Selected Papers from the Third Conference of the International Society for Ecological Informatics (ISEI), 26–30 August 2002, Grottaferrata, Rome, Italy
5. Huysmans, J., Baesens, B., Vanthienen, J.: Using rule extraction to improve the comprehensibility of predictive models In: Research 0612, K.U.Leuven (2006)
6. Jha, A., Aicher, J.K., Gazzara, M.R., Singh, D., Barash, Y.: Enhanced integrated gradients: improving interpretability of deep learning models using splicing codes as a case study. Genome Biol. **149**(21) (2020, online)
7. Kamruzzaman, S.M., Islam, M.M.: An algorithm to extract rules from artificial neural networks for medical diagnosis problems. CoRR abs/1009.4566 (2010)
8. Kuz'min, V.E., Polishchuk, P.G., Artemenko, A.G., Andronati, S.A.: Interpretation of QSAR models based on random forest methods. Mol. Inf. **30**(6–7), 593–603 (2011)

9. Lundberg, S.M., Lee, S.I.: A unified approach to interpreting model predictions. In: Guyon, I., et al. (eds.) Advances in Neural Information Processing Systems, vol. 30. Curran Associates, Inc. (2017). https://proceedings.neurips.cc/paper/2017/file/8a20a8621978632d76c43dfd28b67767-Paper.pdf

10. Marchese Robinson, R.L., Palczewska, A., Palczewski, J., Kidley, N.: Comparison of the predictive performance and interpretability of random forest and linear models on benchmark data sets. J. Chem. Inf. Model. **57**(8), 1773–1792 (2017). https://doi.org/10.1021/acs.jcim.6b00753. pMID: 28715209

11. Olden, J.D., Joy, M.K., Death, R.G.: An accurate comparison of methods for quantifying variable importance in artificial neural networks using simulated data. Ecol. Model. **178**(3–4), 389–397 (2004)

12. de Oña, J., Garrido, C.: Extracting the contribution of independent variables in neural network models: a new approach to handle instability. Neural Comput. Appl. **25**(3), 859–869 (2014)

13. Palczewska, A., Palczewski, J., Marchese Robinson, R., Neagu, D.: Interpreting random forest classification models using a feature contribution method. In: Bouabana-Tebibel, T., Rubin, S.H. (eds.) Integration of Reusable Systems. AISC, vol. 263, pp. 193–218. Springer, Cham (2014). https://doi.org/10.1007/978-3-319-04717-1_9

14. Paliwal, M., Kumar, U.A.: Assessing the contribution of variables in feed forward neural network. Appl. Soft Comput. **11**(4), 3690–3696 (2011)

15. Qin, L.X., Self, S.G.: The clustering of regression models method with applications in gene expression data. Biometrics **62**(2), 526–533 (2006)

16. rfFC: Random forest feature Contrubutions. https://r-forge.r-project.org/R/?group_id=1725. Accessed 28 Aug 2016

17. Ribeiro, M.T., Singh, S., Guestrin, C.: "Why should i trust you?" Explaining the predictions of any classifier. In: KDD 2016, San Francisco, CA, USA (2016)

18. Rosenbaum, L., Hinselmann, G., Jahn, A., Zell, A.: Interpreting linear support vector machine models with heat map molecule coloring. J. Cheminf. **3**(1), 1–12 (2011)

19. Sutherland, J., O'Brien, L., Weaver, D.: A comparison of methods for modeling quantitative structure activity relationships. J. Med. Chem. **47**(22), 5541–5554 (2004). pMID: 15481990

20. Tropsha, A., Gramatica, P., Gombar, V.: The importance of being earnest: validation is the absolute essential for successful application and interpretation of QSPR models. Mol. Inf. **22**(1), 69–77 (2003)

21. Wang, T., Guan, S.-U., Ma, J., Liu, F.: Linear feature sensibility for output partitioning in ordered neural incremental attribute learning. In: He, X., et al. (eds.) IScIDE 2015. LNCS, vol. 9243, pp. 373–383. Springer, Cham (2015). https://doi.org/10.1007/978-3-319-23862-3_37

22. Yousefzadeh, R., O'Leary, D.P.: Proceedings of The First Mathematical and Scientific Machine Learning Conference, PMLR, vol. 107, pp. 1–26 (2020). http://proceedings.mlr.press/v107/yousefzadeh20a.html

A Higher-Order Adaptive Network Model to Simulate Development of and Recovery from PTSD

Laila van Ments[1] and Jan Treur[2(✉)]

[1] AutoLeadStar, Jerusalem, Israel
laila@autoleadstar.com
[2] Social AI Group, Vrije Universiteit Amsterdam, Amsterdam, The Netherlands
j.treur@vu.nl

Abstract. In this paper, a second-order adaptive network model is introduced for a number of phenomena that occur in the context of PTSD. First of all the model covers simulation of the formation of a mental model of a traumatic course of events and its emotional responses that make replay of flashback movies happen. Secondly, it addresses learning processes of how a stimulus can become a trigger to activate this acquired mental model. Furthermore, the influence of therapy on the ability of an individual to learn to control the emotional responses to the traumatic mental model was modeled. Finally, a form of second-order adaptation was covered to unblock and activate this learning ability.

Keywords: PTSD · Higher-order adaptive · Mental model · Flashback movie

1 Introduction

A Post Traumatic Stress Disorder (PTSD) is usually developed after experiencing one or a course of events that trigger strong negative emotions like fear; e.g., [7, 20]. One of the symptoms is a recurring re-experiencing of the course of events that led to the trauma and that are played again and again in the mind as a kind of flashback movie and thereby trigger the strong negative emotions again. In the literature such as [2, 3, 13, 27] strong evidence can be found for relations to amygdala, dorsal anterior cingulated cortex, ventromedial prefrontal cortex and hippocampus. One of the reported issues here is a reduction of the connections to regions of the prefrontal cortex, which makes it difficult to apply emotion regulation. The role of the amygdala in activating fear and of the relation between amygdala and the pre-frontal cortex areas in suppressing fear was found to be crucial; e.g., [2, 19]. If the emotion regulation strategy based on suppression is strengthened, this leads to a decrease in physiological and experiential effects of negative emotions; e.g., [9, 18, 26].

Multiple forms of adaptivity play a crucial role in both the development of PTSD and therapies to recover from it. During the development, an important role is played by the learning of a form of mental model of the course of events leading to the trauma.

© Springer Nature Switzerland AG 2021
M. Paszynski et al. (Eds.): ICCS 2021, LNCS 12743, pp. 154–166, 2021.
https://doi.org/10.1007/978-3-030-77964-1_13

This is a form of observational learning; e.g., [4, 25]. It is this learnt mental model that is the basis of the flashback symptoms. Moreover, during development also learning takes place to connect different stimuli (by themselves irrelevant but just co-occurring with the traumatic events) to the traumatic stimuli which makes them triggers for the flashbacks; this is a form of sensory preconditioning; e.g., [5, 11]. To recover from PTSD another form of learning is required: learning to strengthen the connections to the relevant prefrontal cortex areas to improve emotion regulation; e.g., [18, 26]. However, this learning capability is impaired by the stress itself, which prevents the learning from taking place in a natural manner. This effect is called metaplasticity; e.g., [10]. Metaplasticity [1] is a form of second-order adaptation, as it exerts a form of control over adaptation. In contrast, the other forms of adaptation mentioned above are called first-order adaptation.

The focus in the current paper is to introduce a computational network model addressing all these forms of adaptivity pointed out above. This leads to a second-order adaptive network model in which during development of PTSD a mental model for the flashbacks is learnt and also an association of a trigger to the traumatic events (both first-order adaptation). As an additional effect of the development phase, a negative effect of metaplasticity occurs that impairs the plasticity of the emotion regulation (second-order adaptation). For recovery, a therapy is applied to resolve the impairment of the plasticity of the emotion regulation which is a positive effect of metaplasticity (second-order adaptation). After this, the learning to strengthen the emotion regulation takes place which then leads to recovery (first-order adaptation).

In Sect. 2 some background knowledge is discussed for the different types of adaptation. Section 3 introduces the second-order adaptive network model to address these forms of adaptation. In Sect. 4 some example simulations for this network model are discussed. Finally, Sect. 5 is a discussion.

2 Background Knowledge on Adaptation Principles Used

As discussed above, different forms of adaptation play a role in development of and recovery from traumas. The more specific adaptation principles for these forms of adaptation are discussed in this section.

2.1 First-Order Adaptation Principle: Hebbian Learning

In neuroscientific literature such as [6], two types of first-order adaptation principles are discussed: synaptic and non-synaptic. An example of the latter type is intrinsic excitability adaptation, which will not be used here. Hebbian learning is a well-known first-order adaptation principle of the first type; it addresses adaptive connectivity [12]. It can be explained by:

'When an axon of cell A is near enough to excite B and repeatedly or persistently takes part in firing it, some growth process or metabolic change takes place in one or both cells such that A's efficiency, as one of the cells firing B, is increased.' [12], p. 62 (1)

This is sometimes simplified (neglecting the phrase 'one of the cells firing B') to:

'What fires together, wires together' [14, 21] (2)

This first-order adaptation principle will be used to model adaptation for the following.

- Development of the trauma:

  – Learning of a connection of a trigger stimulus to the traumatic course of events based on sensory preconditioning [5, 11]
  – Learning the connections in the mental model of the traumatic course of events based on observational learning, also using sensory preconditioning [4, 25]

- Recovery from the trauma:

  – Strengthening emotion regulation for recovery by learning the connections to the prefrontal cortex areas [18, 26].

2.2 Second-Order Adaptation Principle: Stress Reduces Adaptation Speed

In [10] the focus is on the role of stress in reducing or blocking plasticity. Many mental and physical disorders are stress-related, and are hard to overcome due to poor or even blocked plasticity that comes with the stress. Garcia [10] describes the negative role of stress-related metaplasticity for this, which often leans to a situation that a patient is locked in his or her disorder by that negative pattern. However, he also shows that by some form of therapy this negative cycle might be broken:

> 'At the cellular level, evidence has emerged indicating neuronal atrophy and cell loss in response to stress and in depression. At the molecular level, it has been suggested that these cellular deficiencies, mostly detected in the hippocampus, result from a decrease in the expression of brain-derived neurotrophic factor (BDNF) associated with elevation of glucocorticoids.' [10], p. 629

> '…modifications in the threshold for synaptic plasticity that enhances cognitive function is referred here to as 'positive' metaplasticity. In contrast, changes in the threshold for synaptic plasticity that yield impairment of cognitive functions, for example (..) in response to stress (..), is referred to as 'negative' metaplasticity.' [10], pp. 630–631

> 'In summary, depressive-like behavior in animals and human depression are associated with high plasma levels of glucocorticoids that produce 'negative' metaplasticity in limbic structures (…). This stress-related metaplasticity impairs performance on certain hippocampal-dependent tasks. Antidepressant treatments act by increasing expression of BDNF in the hippocampus. This antidepressant effect can trigger, in turn, the suppression of stress-related metaplasticity in hippocampal-hypothalamic pathways thus restoring physiological levels of glucocorticoids.' [10], p. 634

This second-order adaptation principle will be used to model adaptation for the following.

- Development of the trauma:

  - Reducing the adaptation speed for the learning of the emotion regulation connections to the prefrontal cortex areas due to the high stress levels [10]

- Recovery from the trauma:

  - Increasing the adaptation speed for the learning of the emotion regulation connections to the prefrontal cortex areas due to a therapy that (temporarily) reduces the stress levels [10].

In Sect. 3 it will be discussed how these have been modeled by using a so-called self-modeling network model.

3 The Second-Order Network Model

In this section, a detailed overview is presented of the designed second-order adaptive network model for modeling the learning of PTSD trauma and the influence of therapy on recovery. For the modeling, we use the Network-Oriented Modeling approach introduced in [22] and further developed to cover higher-order adaptive networks in [23, 24], where also the supporting dedicated software environment is presented.

3.1 The General Format

This approach can be broken down in the following steps:

- Translating the domain into a conceptual causal network model in terms of network characteristics
- Transcribing the conceptual causal network model into a standard table format called *role matrix format*. These role matrices break down the network characteristics for all the different types of causal influences on a state in the model
- The network characteristics are grouped into the following types:

  1. **Connectivity characteristics**
     What *states X, Y* and *connections $X \rightarrow Y$* are there in the model and what are the *weights $\omega_{X,Y}$* of the connections? These are specified in role matrix **mb** (for the states and their connections) and **mcw** (for the connection weights $\omega_{X,Y}$)
  2. **Aggregation characteristics**
     How are different impacts from other states on a state Y aggregated by a *combination function $c_Y(..)$* and what are the values of the *parameters* for these combination functions? The combination functions are chosen from a library by assigning weights $\gamma_{i,Y}$ to them and values for the parameters $\pi_{i,j,Y}$ are set. These characteristics are specified in role matrix **mcfw** (for combination function weights $\gamma_{i,Y}$) and **mcfp** (for the combination function parameters $\pi_{i,j,Y}$)

3. **Timing characteristics**
   How fast do the states Y change upon the received impact, due to their *speed factor* η_Y? These speed factors η_Y are specified in role matrix **ms**.

- Providing the above network characteristics as tables in role matrix format as input for the available dedicated software environment. Based on these received tables, the software environment runs simulations.

3.2 Translating the Domain Knowledge into a Conceptual Causal Model

Based on a domain study, the first step towards building a computational model is translating the processes and brain mechanisms discussed in the literature into a conceptual causal network model. To accommodate for the forms of adaptation of different orders order for the model, the conceptual model uses so-called *self-modeling networks* that include self-models, in this case leading to three levels (see Fig. 1):

1. **The Base Level**
   This level includes all *basic* (non-adaptive/non-learning) *processes* of the conceptual model.
2. **The First-Order Self-Model Level (or First Reification Level)**
   On this level, states are added that represent (adaptive) network characteristics of the base level. For example, a *self-model state* $\mathbf{W}_{X,Y}$ can be added to represent an adaptive connection weight $\omega_{X,Y}$, or a *self-model state* $\mathbf{H}_Y$ can be added to represent a speed factor η_Y. In the model in this way the learning of several connections in the base level takes place through Hebbian learning. These learning connections are represented by the dynamics of the $\mathbf{W}$-states in the blue middle plane. This first-order self-model enables adaptation of the connections of the mental model in the base level.
3. **The Second-Order Self-Model Level (or Second Reification Level)**
   Because the learning itself is adaptive as well, another level is added on top of the first-order self-model level: the second-order self-model level. This level allows to *control the learning speed* of the states $\mathbf{W}_{X,Y}$ for the learning connections by adding state $\mathbf{H}_{\mathbf{W}_{X,Y}}$ here representing the speed factor of $\mathbf{W}_{X,Y}$.

See for the connectivity of the network model Fig. 1; Table 1 shows the states and brief explanations of them. Within the network model, the first-order adaptation based on the Hebbian learning principle has been modeled by using a *connectivity self-model* (in the blue plane) based on self-model states $\mathbf{W}_{X,Y}$ representing connection weights $\omega_{X,Y}$. These self-model states need incoming and outgoing connections to let them function within the network. To incorporate the 'firing together' part of (2) from Sect. 2, for the self-model's connectivity, incoming connections from X and Y to $\mathbf{W}_{X,Y}$ are used; see Fig. 1 (upward arrows in blue). These upward connections have weight 1. Also a connection from $\mathbf{W}_{X,Y}$ to itself with weight 1 is used to model persistence of the learnt effect; in pictures they are usually left out. In addition, an outgoing connection from $\mathbf{W}_{X,Y}$ to state Y is used to indicate where this self-model state $\mathbf{W}_{X,Y}$ has its effect; again see Fig. 1 (pink downward arrow). The downward connection indicates that the value of $\mathbf{W}_{X,Y}$ is actually used for the connection weight of the connection from X to Y. For the *aggregation characteristics* of the first-order self-model, the Hebbian learning rule is defined by the combination function $\mathbf{hebb}_\mu(V_1, V_2, W)$ for self-model state $\mathbf{W}_{X,Y}$ from Table 4.

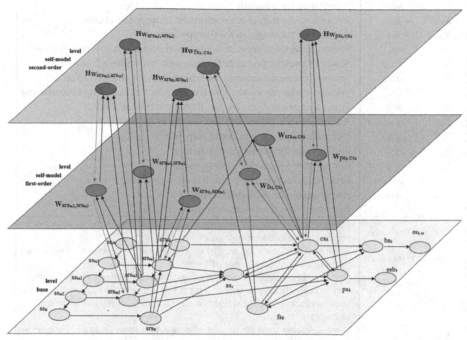

Fig. 1. Connectivity of the introduced second-order adaptive network model (Color figure online)

The sensing of an example of a traumatic course of events is modeled by the sensor states ss_{te1}, ss_{te2}, ss_{te3}. For example, te1 or traumatic event 1, is a potentially dangerous situation for a child you observe, te2 is an action from your side with the intention to save the child from that situation and te3 is an unfortunate failure of your action such that the child actually gets hurt. During this traumatic course of affairs, sensory representations srs_{te1}, srs_{te2}, srs_{te3} are activated for these events te1, te2 and te3, and by sensory preconditioning the connections between these sensory representations are learned. By this observational learning process, the mental model of the traumatic course of events is formed and represented by first-order self-model states $\mathbf{W}_{srs_{te1},srs_{te2}}$ and $\mathbf{W}_{srs_{te2},srs_{te3}}$. Similarly, the connection between the sensory representations of the trigger tr and the traumatic events is learnt based on sensory preconditioning, represented by $\mathbf{W}_{srs_{tr},srs_{te1}}$. These newly formed connections activate the mental model every time the trigger is sensed. For the traumatized person this shows as an internal flashback movie of the traumatic course of events. In turn, this flashback movie activates the related negative emotions experienced at the original traumatic events.

In contrast to what was believed earlier, such learnt connections usually do not show any form of natural extinction; e.g., [15], p. 507. Therefore, to make their effect more

Table 1. The states in the network model and their explanation

state		explanation
X_1	ss_{te1}	Sensor state for traumatic event phase 1: observation te1
X_2	ss_{te2}	Sensor state for traumatic event phase 2: action te2
X_3	ss_{te3}	Sensor state for traumatic event phase 3: effect te3
X_4	ss_{tr}	Sensor state for trigger tr for traumatic sequence of events
X_5	ss_{th}	Sensor state for trigger th for therapy input
X_6	srs_{te1}	Sensory representation state for traumatic event phase 1: observation te1
X_7	srs_{te2}	Sensory representation state for traumatic event phase 2: action te2
X_8	srs_{te3}	Sensory representation state for traumatic event phase 3: effect te3
X_9	srs_{tr}	Sensory representation state for trigger tr for traumatic sequence of events
X_{10}	srs_{th}	Sensory representation state for therapy th from therapy
X_{11}	as_{te}	Awareness state for traumatic sequence of events te
X_{12}	ps_b	Preparation state for emotional response b
X_{13}	fs_b	Feeling state for emotional response b
X_{14}	cs_b	Control state for emotional response b
X_{15}	$bs_{b,te}$	Belief that emotional response b is from traumatic event te
X_{16}	es_b	Bodily expressed emotional response b
X_{17}	$es_{b,te}$	Expressing that emotional response b is from te
X_{18}	$W_{srs_{te1},srs_{te2}}$	Representation state for weight of connection from srs_{te1} to srs_{te2} for imprinting traumatic sequence of events
X_{19}	$W_{srs_{te2},srs_{te3}}$	Representation state for weight of connection from srs_{te2} to srs_{te3} for imprinting traumatic sequence of events
X_{20}	$W_{srs_{tr},srs_{te1}}$	Representation state for weight of connection from srs_{tr} to srs_{te1} for sensory preconditioning to link trigger tr to the traumatic sequence of events
X_{21}	W_{ps_b,cs_b}	Representation state for weight of connection from ps_b to cs_b for learning of emotion regulation
X_{22}	W_{fs_b,cs_b}	Representation state for weight of connection from fs_b to cs_b for learning of emotion regulation
X_{23}	W_{th,cs_b}	Representation state for weight of connection from th to cs_b for learning of emotion regulation from therapy
X_{24}	$HW_{srs_{te1},srs_{te2}}$	Control state for adaptation speed for weight of connection from srs_{te1} to srs_{te2}
X_{25}	$HW_{srs_{te2},srs_{te3}}$	Control state for adaptation speed for weight of connection from srs_{te2} to srs_{te3}
X_{26}	$HW_{srs_{tr},srs_{te1}}$	Control state for adaptation speed for weight of connection from srs_{tr} to srs_{te1}
X_{27}	HW_{ps_b,cs_b}	Control state for adaptation speed for weight of connection from ps_b to cs_b
X_{28}	HW_{fs_b,cs_b}	Control state for adaptation speed for weight of connection from fs_b to cs_b

bearable, the only option is to suppress the emotional consequences related to the trauma by activating the emotion regulation control state cs_b. However, due to the high negative emotion levels the learning process for the activation of cs_b is impaired: learning speeds $\mathbf{HW}_{ps_b,cs_b}$ and $\mathbf{HW}_{fs_b,cs_b}$ are very low. Therefore, without any additional help the situation will stay as it is. But, following [10] the therapy *th* is able to temporarily reduce the level of negative emotions, so that $\mathbf{HW}_{ps_b,cs_b}$ and $\mathbf{HW}_{fs_b,cs_b}$ get higher values. Due to this, learning of the connections to the control state takes place: $\mathbf{W}_{ps_b,cs_b}$ and $\mathbf{W}_{fs_b,cs_b}$ get higher values.

3.3 Transcribing the Conceptual Model into Role Matrices

To allow for easy formalization of the conceptual model into role matrices and an executable computational model, we use generic ways to describe the states, intra-level connections and interlevel connections. See an abstracted overview of all types of states and connections used in the model in Tables 2 and 3.

Table 2. Overview of types of states

State name	Representation
ss_y	Sensor state for state y in the world
srs_y	Sensory representation state for y
as_s	Awareness state for s
fs_b	Feeling state for feeling b
ps_b	Preparation state for feeling b
cs_b	Control state for feeling b
bs_b	Belief state for feeling b
es_b	Execution state for feeling b
$\mathbf{W}_{X,Y}$	Connection weight representation state for connection $X \rightarrow Y$
$\mathbf{H}\mathbf{W}_{X,Y}$	Learning control state for the connection weight state for connection $X \rightarrow Y$

The model with connectivity shown in Fig. 1 was then specified by tables in role matrix format: Connectivity characteristics (1), aggregation characteristics (2) and timing characteristics (3). See the Appendix at URL https://www.researchgate.net/public ation/350159052. Four different combination functions from the library are used that each serve a different purpose; see Table 4.

The advanced logistic sum combination function combines influences of multiple states by adding them but makes sure they stay between 0 and 1, with parameters steepness σ and threshold τ. The Hebbian learning combination function is used for learning of a connection weight. The stepmod function allows for an activation of states with a predefined length and frequency (here, that is used for the recurring trigger state). The steponce function allows for the activation of states with predefined length and start time (here, that is used for the therapy and trauma states).

The following standard generic difference equation is used for simulation purposes and also for analysis. It incorporates the network characteristics $\omega_{X,Y}$, $\mathbf{c}_Y(..)$, η_Y in a numerical difference equation format:

$$Y(t + \Delta t) = Y(t) + \eta_Y[\mathbf{c}_Y(\omega_{X_1,Y}X_1(t), \ldots, \omega_{X_k,Y}X_k(t)) - Y(t)]\Delta t \qquad (1)$$

Table 3. Overview of types of connections

Connection	Representation	Connection type
$X \rightarrow Y$	Connection between base states X and Y	Intra-level (horizontal) connection
$X \rightarrow \mathbf{W}_{X.Y}$ $Y \rightarrow \mathbf{W}_{X.Y}$	Connections from base level states X and Y to connection adaptation state $\mathbf{W}_{X.Y}$ to support the Hebbian learning formation	Interlevel connection, upward from the base level to the first-order self-model level
$\mathbf{W}_{X.Y} \rightarrow Y$	Connection from connection adaptation state $\mathbf{W}_{X.Y}$ to base state Y; these connections effectuate the learnt connection	Interlevel connection, downward from the first-order self-model level to the base level
$\mathbf{W}_{X.Y} \rightarrow \mathbf{H}_{\mathbf{W}_{X.Y}}$ $X \rightarrow \mathbf{H}_{\mathbf{W}_{X.Y}}$ $Y \rightarrow \mathbf{H}_{\mathbf{W}_{X.Y}}$	Connections from connection adaptation state $\mathbf{W}_{X.Y}$ and base level states X and Y to learning control state $\mathbf{H}_{\mathbf{W}_{X.Y}}$	Interlevel connections, upward from the base level to the second-order self-model level, and upward from the first-order self-model level to the second-order self-model level
$\mathbf{H}_{\mathbf{W}_{X.Y}} \rightarrow \mathbf{W}_{X.Y}$	Connection from learning control state $\mathbf{H}_{\mathbf{W}_{X.Y}}$ to adaptive connection adaptation state $\mathbf{W}_{X.Y}$ to effectuate learning control	Interlevel connection, downward from the third level to the second level

Table 4. The combination functions used from the library

Combination function	Notation	Formula	Parameters
Advanced logistic sum	$\mathbf{alogistic}_{\sigma,\tau}(V_1, \dots,V_k)$	$\left[\frac{1}{1+e^{-\sigma(V_1+\dots+V_k-\tau)}} - \frac{1}{1+e^{\sigma\tau}} \right](1 + e^{-\sigma\tau}))$	Steepness $\sigma > 0$ Excitability threshold τ
Hebbian learning	$\mathbf{hebb}_{\mu}(V_1, V_2, W)$	$V_1 V_2 (1 - W) + \mu W$	Persistence factor $\mu \geq 0$
Steponce	$\mathbf{steponce}(V)$	1 if $\alpha \leq t \leq \beta$, else 0	$\alpha \geq 0$ begin, $\beta \geq \alpha$ end time
Stepmod	$\mathbf{stepmod}_{\rho,\delta}(V_1, \dots,V_k)$	0 if $t \bmod \rho < \delta$, else 1	Repetition $\rho \geq 0$ Duration $\delta \geq 0$

for any state Y and where X_1 to X_k are the states from which Y gets its incoming connections. Based on the role matrices as input, his generic difference equation is automatically applied to all network states (including the self-model states) within the dedicated software environment used to perform simulation experiments.

4 Example Simulations

The role matrices can easily be transferred to the dedicated softare environment for simulations. Running the software loops over a chosen time period (in this case a time interval from 0 to 1400 with step size $\Delta t = 0.5$) and provides as output a simulation graph for the model. In Fig. 2 the development of PTSD is shown based on traumatic events te1 to te3 in time period from 100 to 200 without applying therapy. The trigger also occurs from 100 to 200 and after that regularly recurs from 300 to 400, from 500 to 600, et cetera. In Fig. 3 the same is shown but this time therapy is taking place from time 400 to time 800 where the therapy leads to recovery. In both Fig. 2 and 3 in the time period from 100 to 200 the sequence of traumatic events te1 to te3 in the world are sensed (via sensor states ss_{te1}, …, ss_{te3}) of which internal representations srs_{te1}, …, srs_{te3} are made. Due to sensory preconditioning (first-order adaptation based on Hebbian learning), the connections between them are developed (thus forming a mental model of the traumatic course of events) and also a connection from the trigger representation srs_{tr} to srs_{te1}. Moreover, they trigger the negative emotional response preparation ps_b and feeling state fs_b, and these in turn reduce the adaptation speed (represented by the **H**-states) of the learning of the connections to the control state cs_b (second-order adaptation

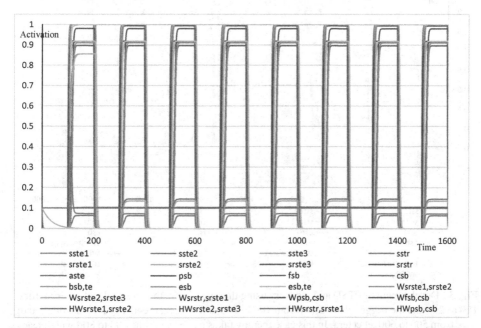

Fig. 2. Development of PTSD without using therapy. The trauma develops from time 100 to 200. The trigger also occurs from 100 to 200 and after that regularly recurs from 300 to 400, from 500 to 600, et cetera. No recovery from PTSD takes place.

for metaplasticity). Therefore, no strengthening of the emotion regulation takes place, what would be needed to get rid of the negative feelings. Every time period that the trigger recurs, due to the connection from srs_{tr} to srs_{te1} and the connections between $srs_{te1},..,srs_{te3}$, the flasback movie is replayed (as a form of internal simulation) and because of that the negative emotion and feeling are activated to high values again.

5 Discussion

In this work, a second-order adaptive model was developed to allow for simulation of the formation of a mental model of a trauma that is built up over time and its emotional responses, and neurological processes of how a stimulus can become a trigger to activate this mental model. Furthermore, the influence of therapy on the ability of an individual to control the emotional response to the trauma mental model was explored. The computational model was developed following the approach described in [24], using the following steps:

- A conceptual causal network model was designed based on literature on patients with PTSD and existing theories and models about PTSD and emotion regulation
- The conceptual causal network model was translated into role matrices format

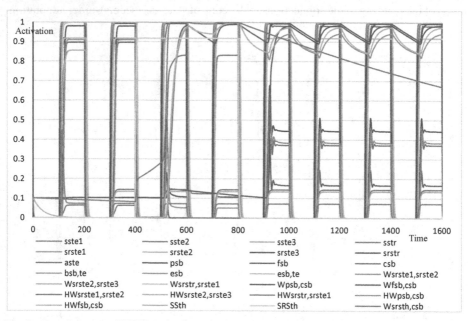

Fig. 3. Development of PTSD and recovery using therapy. Again, the trauma develops from time 100 to 200 and the trigger also occurs from 100 to 200 and after that regularly recurs from 300 to 400, from 500 to 600, et cetera. In this case therapy takes place from time 400 to 800 which leads to recovery.

- The role matrices were used in the dedicated software environment to obtain simulations; this software environment is available at
- https://www.researchgate.net/project/Network-Oriented-Modeling-Software.

Different simulation experiments were done, for individuals developing a trigger response, individuals not developing a trigger response, and individuals receiving therapy.

Other work addressing computational modelling for trauma development and recovery can be found in [8, 16, 17]. However, none of these previous works allowed for the adaptation of the learnt connections of the mental model and therapy. In addition, in [16, 17] it is assumed that already built-in upward connections for the emotion regulation exist and are static, while in the model presented here an important part of the development of a trauma is the learning for the mental model of the traumatic course of affairs. In another comparison, [8] addresses social support instead of the type of therapy suggested by Garcia [10] and used in the current paper. Moreover, the underlying second-order adaptation process as explained extensively by [10] is fully addressed here while it is ignored in [8, 16, 17]. Finally, in the current paper the source of the trauma can be a process taking place over a longer time period with a successive course of events over time, and modeled in the form of an internal mental model that can be replayed as a flashback movie, while in [8, 16, 17] only one traumatic state at one time point is assumed where a flashback is only one static image, which is not quite realistic.

The second-order adaptive model described in this paper can be used as a basis for development of integrated computing applications to support PTSD therapy or to develop virtual characters illustrating the processes involved in patients with PTSD. In such contexts, also possibilities may be exploited for further validation of the model.

References

1. Abraham, W.C., Bear, M.F.: Metaplasticity: the plasticity of synaptic plasticity. Trends Neurosci. **19**(4), 126–130 (1996)
2. Admon, R., Milad, M.R., Hendler, T.: A causal model of post-traumatic stress disorder: disentangling predisposed from acquired neural abnormalities. Trends Cogn. Sci. **17**(7), 337–347 (2013)
3. Akiki, T.J., Averill, C.L., Abdallah, C.G.: A network-based neurobiological model of PTSD: evidence from structural and functional neuroimaging studies. Curr. Psychiatry Rep. **19**(11), 1–10 (2017). https://doi.org/10.1007/s11920-017-0840-4
4. Benbassat, J.: Role modeling in medical education: the importance of a reflective imitation. Acad. Med. **89**(4), 550–554 (2014)
5. Brogden, W.J.: Sensory preconditioning of human subjects. J. Exp. Psychol. **37**, 527–539 (1947)
6. Chandra, N., Barkai, E.: A non-synaptic mechanism of complex learning: modulation of intrinsic neuronal excitability. Neurobiol. Learn. Mem. **154**, 30–36 (2018)
7. Duvarci, S., Pare, D.: Amygdala microcircuits controlling learned fear. Neuron **82**, 966–980 (2014)
8. Formolo, D., Van Ments, L., Treur, J.: A computational model to simulate development and recovery of traumatised patients. Biol. Inspired Cogn. Archit. **21**, 26–36 (2017)

9. Fitzgerald, J.M., DiGangi, J.A., Phan, K.L.: Functional neuroanatomy of emotion and its regulation in PTSD. Harv. Rev. Psychiatry **26**(3), 116–128 (2018)
10. Garcia, R.: Stress, metaplasticity, and antidepressants. Curr. Mol. Med. **2**, 629–638 (2002)
11. Hall, G.: Learning about associatively activated stimulus representations: implications for acquired equivalence and perceptual learning. Animal Learn. Behav. **24**(3), 233–255 (1996). https://doi.org/10.3758/BF03198973
12. Hebb, D.O.: The Organization of Behavior: A Neuropsychological Theory. Wiley, Hoboken (1949)
13. Holmes, S.E., et al.: Cerebellar and prefrontal cortical alterations in PTSD: structural and functional evidence. Chronic Stress **2**, 1–11 (2018). https://doi.org/10.1177/2470547018786390
14. Keysers, C., Gazzola, V.: Hebbian learning and predictive mirror neurons for actions, sensations and emotions. Philos. Trans. R. Soc. Lond. B Biol. Sci. **369**, 20130175 (2014)
15. Levin, R., Nielsen, T.A.: Disturbed dreaming, posttraumatic stress disorder, and affect distress: a review and neurocognitive model. Psychol. Bull. **133**, 482–528 (2007)
16. Naze, S., Treur, J.: A computational agent model for post-traumatic stress disorders. In: Samsonovich, A.V., Johannsdottir, K.R. (eds.) Proceedings of the Second International Conference on Biologically Inspired Cognitive Architectures, BICA 2011, pp. 249–261. IOS Press (2011)
17. Naze, S., Treur, J.: A computational model for development of post-traumatic stress disorders by hebbian learning. In: Huang, T., Zeng, Z., Li, C., Leung, C.S. (eds.) ICONIP 2012. LNCS, vol. 7664, pp. 141–151. Springer, Heidelberg (2012). https://doi.org/10.1007/978-3-642-34481-7_18
18. Ochsner, K.N., Gross, J.J.: The neural bases of emotion and emotion regulation: a valuation perspective. In: Handbook of Emotional Regulation, 2nd edn., pp. 23–41. Guilford, New York (2014)
19. Panksepp, J., Biven, L.: The Archaeology of Mind: Neuroevolutionary Origins of Human Emotions. Chap. 1. W.W. Norton, New York (2012)
20. Parsons, R.G., Ressler, K.J.: Implications of memory modulation for post-traumatic stress and fear disorders. Nat. Neurosci. **16**(2), 146–153 (2013)
21. Shatz, C.J.: The developing brain. Sci. Am. **267**, 60–67 (1992). https://doi.org/10.1038/scientificamerican0992-60
22. Treur, J.: Network-Oriented Modeling: Addressing Complexity of Cognitive, Affective and Social Interactions. Springer, Cham (2016). https://doi.org/10.1007/978-3-319-45213-5
23. Treur, J.: Modeling higher-order adaptivity of a network by multilevel network reification. Netw. Sci. **8**, S110–S144 (2020)
24. Treur, J.: Network-Oriented Modeling For Adaptive Networks: Designing Higher-Order Adaptive Biological, Mental and Social Network Models. Springer, Cham (2020). https://doi.org/10.1007/978-3-030-31445-3
25. Van Gog, T., Paas, F., Marcus, N., Ayres, P., Sweller, J.: The mirror neuron system and observational learning: implications for the effectiveness of dynamic visualizations. Educ. Psychol. Rev. **21**(1), 21–30 (2009)
26. Webb, T.L., Miles, E., Sheeran, P.: Dealing with feeling: a meta-analysis of the effectiveness of strategies derived from the process model of emotion regulation. Psychol. Bull. **138**(4), 775 (2012)
27. Zandvakili, A., et al.: Mapping PTSD symptoms to brain networks: a machine learning study. Transl. Psychiatry **10**, e195 (2020)

Trojan Traffic Detection Based on Meta-learning

Zijian Jia[1,2], Yepeng Yao[1,2], Qiuyun Wang[1], Xuren Wang[1,3], Baoxu Liu[1,2], and Zhengwei Jiang[1,2(✉)]

[1] Institute of Information Engineering, Chinese Academy of Sciences, Beijing, China
{jiazijian,yaoyepeng,wangqiuyun,liubaoxu,
jiangzhengwei}@iie.ac.cn
[2] School of Cyber Security, University of Chinese Academy of Sciences,
Beijing, China
[3] College of Information Engineering, Capital Normal University, Beijing, China
wangxuren@cnu.edu.cn

Abstract. At present, Trojan traffic detection technology based on machine learning generally needs a large number of traffic samples as the training set. In the real network environment, in the face of Zero-Day attack and Trojan variant technology, we may only get a small number of traffic samples in a short time, which can not meet the training requirements of the model. To solve this problem, this paper proposes a method of Trojan traffic detection using meta-learning for the first time, which mainly includes the embedded part and the relation part. In the embedding part, we design a neural network combining ResNet and BiLSTM to transform the original traffic into eigenvectors and allocate the meta tasks of each round of training in the form of a C-way K-shot. In the relation part, we design a relationship network improved by dynamic routing algorithm to calculate the relationship score between samples and categories in the meta-task. The model can learn the ability to calculate the difference between different types of samples on multiple meta-tasks. The model can use a small number of samples to complete training and classify quickly according to prior knowledge. In few-shot, our method has better results in Trojan traffic classification than the traditional deep learning method.

Keywords: Network security · Trojan traffic detection · Deep learning · Meta learning

1 Introduction

In recent years, the number of network attacks is gradually increasing with the development of the Internet. Trojan is one of the means of attack. Attackers usually implant the Trojan virus into the victim's host to control the victim's host through this program. Because there must be communication behavior between

© Springer Nature Switzerland AG 2021
M. Paszynski et al. (Eds.): ICCS 2021, LNCS 12743, pp. 167–180, 2021.
https://doi.org/10.1007/978-3-030-77964-1_14

the attacker and the victim host, detecting the traffic generated in it has become an extremely effective method to detect the Trojan virus.

At present, many researchers use machine learning method for traffic detection. They extract many features from the original traffic and use machine learning methods to detect and classify malicious traffic [2,3]. But the feature extraction is very subjective. With the continuous development of deep learning, more and more people begin to build various neural network models to detect Trojan traffic behavior and content. For example, the researcher transformed the payload of traffic into a gray image and used CNN to train the recognition model. Or the payload is directly regarded as semantic information for numerical calculation. They use RNN and other networks to train the recognition model directly. All these works show that the use of deep learning in traffic classification can achieve better results.

Compared with traditional machine learning, deep learning has significant advantages because it does not need to extract features manually. It mostly avoid the influence of human subjective factors on the model. But the premise of this model training is to have a large number of data as training samples so that the model can obtain enough knowledge from the iteration and adjust the parameters to achieve a better classification effect. According to the current real cyberspace situation, such a premise is difficult to achieve. Trojan attacks often change their behavior, leading to the problem of slow sample capture and a large number of sample data for training cannot be obtained quickly.

In this paper, we propose a meta-learning method based on the metric to solve this problem. We transform the original network traffic into feature vectors. An improved relational network is constructed to calculate the difference between different vectors. Our model obtains prior knowledge through a large number of meta-task and learns to calculate the differences between different samples so that the Trojan traffic detection can be implemented when the dataset is imbalance. The main contributions of this method are the following three aspects:

- We propose a network combining ResNet and BiLSTM as the embedded part of meta-learning. On a large number of dataset, this network's result is better than that of a general neural network, which can better learn and extract the features of the original traffic.
- We use a dynamic routing algorithm to replace the superposition part of vectors in a traditional relational network. This algorithm can make the vector fusion simultaneously retain more information, such as the direction of the vector. As the representative vector of category in meta task is more accurate.
- Based on the CTU13 dataset, we build an unbalanced dataset and compare the standard deep learning model's classification effect and the proposed meta-learning model. The results show that our model is better than the general deep learning model when the dataset is imbalance.

2   The Description of Problems

The general deep learning to find a mapping relationship between sample x and label y such that $f(x) = y$. If the dataset is imbalance, it is easy to overfit the sample recognition. To solve this problem, we propose a meta-learning method based on metric for Trojan traffic detection. This method contains a task set $T = \{T_i, i \in N^*\}$ for training. This setting can make the model through the training of n tasks to use the prior knowledge obtained from the existing tasks to complete training quickly. The representation of meta-learning architecture is shown in Fig. 1. It contains Meta-training set D_{train} and Meta-testing set D_{test}. Each item in the two sets is a small meta-task T which containing the Support Set and the Query Set.

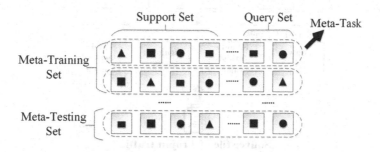

Fig. 1. The representation of meta learning

The task allocation can be summarized as a C-way K-shot problem. Each task randomly selects C classes from the dataset, and K samples from each classes. The support set contains different Trojan sample traffic classes and the query set includes the same kind of Trojan sample traffic as the support. The model is trained by meta-task like Fig. 1. When the training is completed, the model can calculate the score between the unknown sample and each category in new task to achieve samples' classification.

3 Trojan Traffic Detection Based on Meta-learning Model

3.1 Transform Trojan Traffic into Feature Vector

At present, many researchers use deep learning to solve the problem of network traffic detection. No matter what the network structure design of deep learning is, they all need a fixed input mode. In this chapter, a neural network based on ResNet-LSTM is proposed to extract the original traffic feature vectors as the embedded part of the meta-learning framework. The primary extraction process is shown in Fig. 2.

Data Preprocessing of Pcap Package. The model extracts the two-way TCP flow from the packet as the sample. After extracting the data stream, we need to filter some of the information.

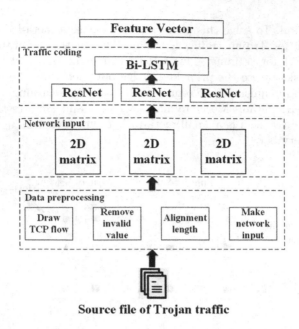

Fig. 2. The extraction of feature vector

The traffic in the dataset is captured by a trojan virus running in the virtual machine environment. The MAC address and other header's information in the packet header will interfere with the model learning. So we should remove this information. According to Rezaei's research [4], the characteristics of the first few packets can represent the whole data flow. In the experiment, we focus on the initial stage of the whole TCP flow. According to the experimental results, the first three to five packets of the data stream are needed. Data cleaning is needed after extracting the data stream. The header information such as five tuples in each packet is cleared and the payload part in TCP flow is directly extracted. This part extracts 784 bytes. If the length is not enough, we fill zero at the end. Then a 28 × 28 two-dimensional matrix is formed as the input of the embedded part of the model.

Make Feature Vector by ResNet-LSTM. Most of researchers build 3 to 5 layers of CNN to detect traffic and don't explore more deep network. A large number of experiments show that the network's recognition efficiency decreases when the depth of the general CNN network increases gradually. He K et al. proposed the ResNet model, which can deepen the network level and make the whole network have better learning ability [1]. In this paper, this model is combined with bidirectional LSTM to extract the original traffic characteristics in space and time. The ResNet model designed in this experiment is shown in Table 1.

Table 1. The structure of ResNet

Network name	Network output	Network structure	
Conv1	$128 \times 28 \times 28$	$1 \times 1, 128$	
Conv2_Res	$128 \times 14 \times 14$	$\begin{bmatrix} 3 \times 3, 128 \\ 3 \times 3, 128 \end{bmatrix}$	$\times 4$
Conv3_Res	$256 \times 7 \times 7$	$\begin{bmatrix} 3 \times 3, 256 \\ 3 \times 3, 256 \end{bmatrix}$	$\times 6$
Conv4_Res	$512 \times 4 \times 4$	$\begin{bmatrix} 3 \times 3, 512 \\ 3 \times 3, 512 \end{bmatrix}$	$\times 3$
Ave_pool	$512 \times 1 \times 1$	4×4	
Full_connected	512	512	

The Conv1 layer transform the sample into one hundred and twenty-eight 28×28 characteristic matrices. Then there are three blocks of ResNet. This network is made up of CNN with a jump layer. When the depth of the network deepens, it is equivalent to mapping the current network to the next layer. If there is an identity mapping between the network and the next layer, the model will degenerate into an external network, and there is no degradation problem. This is what the residual network looks like, as shown in Fig. 3.

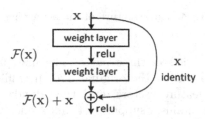

Fig. 3. Neuron structure of residual network [1]

The three residual blocks contain different convolution layers. In Table 1, network structure represents the shape of every network layer. At last, the generated characteristic matrix is input into a fully connected layer as the next layer's input to bidirectional LSTM.

A convolutional neural network extracts the spatial characteristics of the content in the traffic. The traffic is continuous in terms of time and the payload in the communication about Trojan traffic has semantic properties. In the paper, the bidirectional LSTM model is used to encode the features of the ResNet. It mainly processes the two unidirectional LSTM networks' output and obtains the feature vector used to represent the sample, which provides a unified input for the meta-learning model.

3.2   The Structure of Meta-learning Model

As we all know, neural network iterates and optimizes by learning many labeled dataset. So when the dataset is imbalance, the model will be overfitted. In this model, the neural network can learn how to calculate the difference between a sample and the class in different tasks. In this way, the model has the ability of generalization. Even if the samples of a single class are unbalanced, the model can also help classify the samples through the knowledge of comparison. The overall framework is shown in Fig. 4.

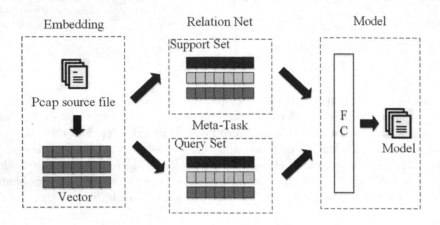

Fig. 4. The structure of meta-learning

The whole model consists of three parts. The first part is the embedding part. This part aims to extract the original Trojan traffic samples, transform them into computable feature vectors, and then divide the model into meta tasks. Each meta task contains a support set and a query set. The second part is the relation network based on metric. Facing each meta task, the model will fuse the samples contained in all categories in the support set so that all classes have their representative vectors. The vector and all samples in the query set are input into the full join layer to calculate the relationship score. The network is optimized iteratively through the score. The third part is the verification part. The samples to be tested are input into the trained model in the form of tasks, and the classification results are obtained according to the scores between samples and classes.

Task Generation in Meta-learning. The training set and test set contain meta-tasks in this model, including samples selected from different classes. The specific algorithm is shown in Algorithm 1.

Algorithm 1. Generate One Meta-Task For Training

Input: Labelset $L = \{1, 2, ..., n\}$, Dataset$D = \{(X_1, L_1), ..., (X_n, L_n)\}$, $L_i \in L$, X_i is the i kind of sample, $X_i = \{x_1, ..., x_n\}$, select K_s samples from Support Set, select K_q samples from Query Set

Output: Meta learning task$T = \{Support, Query, L_{Query}\}$, S is Support Set, Q is Query Set, L_Q is Labsels of Query Set.

Require: $RandomSample(A, K)$ Indicates that k samples are randomly selected from the A set, $Remove(A, B)$, Indicates that the B sample is removed from the A set

```
 1: procedure GENERATETASK(D, K_s, K_q)
 2:     GenerateSet:
 3:     for i = 1 to n do                          ▷ Traverse all categories in the dataset
 4:         S_{L_i} ← RandomSample(X_i, K_s)
 5:         Other ← Remove(X_i, S_{L_i})
 6:         Q_{L_i} ← RandomSample(Other, K_q)
 7:     end for
 8:     GenerateTask:
 9:     Support ← ∑_1^n S_i                        ▷ Splice the selected samples
10:     Query ← ∑_1^n Q_i
11:     L_{Query} ← ∑_1^n L_i xlen(Q_i)
12:     return T ← {Support, Query, L_{Query}}
13: end procedure
```

Meta-learning Models Based on Metric. After meta-task assignment, each task will contain different Trojan class feature vectors and each class has K samples. When K is greater than 1, Sung et al. used an embedding module to sum each class's elements step by step in early study [14]. This embedding module consists of a convolutional neural network. The combined class level mapping feature is combined with the mapping feature in the query set to calculate the relationship score. The number of samples in the support set is greater than 1 in our experiment. We use the dynamic routing algorithm from the capsule network to fuse the vectors.

Hinton et al. first proposed this algorithm [17]. Using this method is that when the traditional convolutional neural network does the dimension mapping from the bottom to the top [18,19], the relative relations such as direction and space information in the vector cannot be learned in the pooling layer. In our experiment, we need to carry out high-dimensional mapping on multiple vectors of the same kind and different vectors will also have deviations in the direction. Such information can not be transmitted upward using simple superposition. So the dynamic routing method is adopted, which encapsulates the space state of the underlying vector and passes it to the next neuron to complete the expression of different Trojan class vectors. The algorithm is shown in Algorithm 2.

This kind of neuron calculates vector. Its essence is to update the initial weight value of b_{ij} through vector iteration. This step's update is equivalent to using the high-level output vector and the low-level input vector for dot product operation update. When the two vectors' directions are the same, the point product operation will increase the weight update value and vice versa. The

---

Algorithm 2. Dynamic routing algorithm

Input: Vector of samples v_{ij} in Support Set S,Initialization vector $b_{ij} = 0$, Number of Route iterations r

Output: The Vector represent a Class V_k in Support Set S

1: **procedure** DYNAMIC(v_{ij}, b_{ij}, r)
2:     **for** r iterations **do**
3:         **for** i *in layer* l **do**
4:             $c_i \leftarrow softmax(b_i)$
5:         **end for**
6:         **for** j *in layer* $(l+1)$ **do**
7:             $s_j \leftarrow \sum_{j=1}^{n} c_{ij} v_{ij}$
8:             $v_j \leftarrow squash(s_j)$ ▷ Squeezing function
9:         **end for**
10:        **for** i *in layer* l *and* j *in layer* $(l+1)$ **do**
11:            $b_{ij} \leftarrow b_{ij} + v_i * v_j$
12:        **end for**
13:    **end for**
14:    **return** v_j
15: **end procedure**

---

algorithm repeats the iteration process until r times. Hinton also pointed out that a large number of iterations will lead to overfitting [1]. They recommended using three iterations. So we use the same super parameters for training.

After fusion, we can obtain the representative eigenvectors of different classes of Trojan traffic in meta-task. They are used as the support set to calculating the relation score with other classes of samples in the query set. Two vectors are input into two fully connected layers, calculate relation score by *softmax* and obtain a classification result. The score is also used to calculate the loss value of the task. As the whole training is multitasking, the model uses the MSE as the loss value and the formula is shown in (1).

$$L = \sum_{i=1}^{C} \sum_{j=1}^{K} (r_{ij} - 1 * (y_j == y_i))^2 \tag{1}$$

In formula 1, C stands for the selected C classes, K stands for K samples, the final output of the model is the relation score between classes and samples. The label of classes judged by the score. The *Adam* and *SGD* optimizers are compared on the optimizer. And *Adam* is selected as the optimization algorithm of the model according to the experimental results. The purpose of this meta-learning model after optimization is to get the metric function G_Φ, as shown in (2).

$$r_{ij} = g_\phi(C(f_\phi(x_i), f_\phi(x_i))) \qquad i = 1, ..., C \tag{2}$$

Through the neural network's nonlinear optimization process, a function g_ϕ which can express the metric between vector is obtained. Each task generated in training is to generalize this form of metric calculation so that the network can learn the ability of comparison. When new tasks appear, the network can calculate the differences between samples and different classes appropriately through the comparison function to achieve classification.

4   Evaluation

4.1   Dataset

To test the performance of the model, we use the CTU13 dataset. It's from the statosphere lab project of Czech Polytechnic University. It's often used in the industry which contains the malicious traffic generated by many classic Trojans. Ten Trojans are selected for experiment and analysis as shown in Table 2.

Table 2. The scale of dataset

Types of Trojan horse	The size of original data (M)	Samples (TCP)
Andromeda	620	29343
Emotet	500	23444
Geodo	2560	22938
Locky	372	55207
Sathurbot	1510	68820
Tinba	1100	34450
Trojan.Rasftuby	1210	41447
Variant	1460	32756
Yakes	1350	41725
Zeus	782	64161

Table 2 shows the classes of Trojans that build the experimental dataset. The original pcap packet samples are cleaned and each bidirectional TCP flow is extracted as a sample. The samples that do not contain any payload information are removed. Due to network traffic characteristics, each byte is between 0 and 255, just in the gray values range. Each sample is saved in the form of a gray image as the input part of the meta-learning model.

4.2   The Results of Experiments

Firstly, we evaluate the embedded part of the model. In the experiment, we construct different neural networks to learn the original TCP stream's content and check the classification effect. We have constructed four kinds of neural networks, including CNN, BiLSTM, ResNet and ResLSTM. The detection results of these four kinds of neural networks are shown in Table 3. This experiment uses all

Table 3. The results of embedding

	Precision	Recall	F1-score
CNN	94.7%	94.5%	94.6%
BiLSTM	96.4%	98.9%	97.6%
ResNet	99.0%	96.3%	97.6%
ResLSTM	99.9%	99.9%	99.9%

the data sets, the data sets are balanced, each class contains a large number of samples, mainly used to verify the ability of the embedded part of the network to identify the original traffic.

The experimental results are the average of the recognition accuracy and recall rate of different Trojan traffic in the dataset. It can be seen from the experimental results that the method of combining ResNet and BiLSTM can obtain a better recognition effect in a large number of the dataset. Therefore, this method can better extract features from the original traffic. So in the experiment, we use the model proposed in the third section to encode the original traffic and transform the original traffic into a computable numerical vector in space.

After determining the embedded model of meta-learning, we need to verify the whole meta-learning model's detection ability in the case of few-shot. This small sample setting is that when the unknown Trojan traffic appears, the model only uses a small amount of data collected quickly for training and then detects and classifies the subsequent traffic. To verify this case, we first set the traffic generated by any Trojan horse in the dataset as unknown attack traffic and reduce the data volume to 0.25% of the original. The reserved dataset is used as a training set and test set. It together with other Trojan traffic to form an unbalanced dataset. The remaining data of this category is used as the validation set after model training.

Table 4. Experimental results of few-shot

	Precision	Recall	F1-score
CNN	75.4%	75.9%	75.6%
BiLSTM	74.1%	76.8%	75.4%
ResLSTM	78.8%	75.3%	77.0%
Relation Net	81.7%	76.1%	78.8%
Method in Paper	85.7%	82.9%	84.2%

The data in Table 4 shows the comparison of different deep learning models. There are five models, including CNN, BiLSTM, ResLSTM, the original relational network and our the model of this paper. The first three are standard deep learning models, and the last two are meta-learning models. We use the method proposed by Wang et al. [5]. ResLSTM uses the model we put forward in the embedded layer and input the model's output directly into the softmax function for classification operation. In the case of few-shot, although the recall of this model is not as good as the CNN and BiLSTM, the precision becomes higher, and the F1 is higher than them. The latter two models adopt the measurement-based meta-learning model. Relation Net uses the relational network proposed by Sung. In the meta-learning model, we set super parameters in advance when assigning meta tasks. Each time we select all ten categories, the number of samples K in each type is set to 10, half of which is divided into a support set, and half is placed as the query set for training.

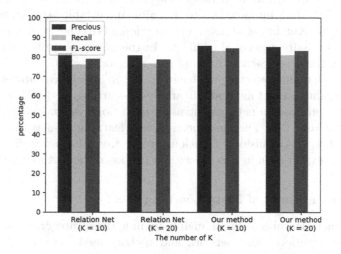

Fig. 5. The results of experiments about different K

At the same time, we also tested the influence of the sample size of K on the results. In the C-way K-shot problem, the number of samples selected for each class is generally less than 20. In the experiment's task allocation phase, we test the impact of K = 10 and 20 on the experimental results. This allocation method can ensure that the number of samples in the support set and query set is 5 and 10. The results are shown in Fig. 5. The selection of K has little effect on the results. This may be due to the significant difference between different traffic. The model can be used as the representative vector of each class only using a small number of samples for fusion.

5   Related Work

5.1   Trojan Traffic Detection

It is challenging to prevent Trojan's attack, so we need to detect according to the communication behavior after intrusion. Li et al. analyzed Trojan's actual network behavior, extracted features from the TCP handshake process and detected them [20]. D. Jiang et al. extracted the number and length of packets in the early communication stage Trojan [21]. They established a machine learning model for detection. This kind of research extract features from Trojan communication and builds models.

5.2 Traffic Detection Based on Deep-Learning

In recent years, researchers gradually apply deep learning methods to traffic detection. Wang et al. proposed a variety of malicious traffic detection methods using deep learning. They used 1D-CNN, 2D-CNN and other methods to model the payload information in malicious traffic [5–7]. Ren Hwang et al. used LSTM to transform TCP flow into word vector training and established a traffic classification model [8]. Xue Liu et al. used a bidirectional GRU network and attention mechanism to classify encrypted traffic [9]. In the aspect of traffic detection and classification, the deep learning method can get good results.

The research about detecting malicious traffic on imbalance dataset generally use the data enhancement method [16] and meta-learning method is rarely used. Xu et al. first proposed a network intrusion detection system based on a meta-learning framework [15]. They use metric-based learning models to construct a binary classifier, distinguishing malicious traffic from normal traffic. We have carried on the exploration in this aspect and proposed the method in this paper.

5.3 Research Status of Meta-Learning

At present, meta-learning network mainly include three categories: model-based network, optimization-based network and metric-based network. The model-based network method is to quickly update parameters on a small number of samples through the model structure's design [10,11]. The optimization-based network completes the task of small sample classification by adjusting the optimization method [12].

The metric-based network calculates the distance between the samples in the training set and the test set. Koch et al. proposed the Siamese Network [13]. Different pairs of samples are constructed by combination and input into the network for training. The distance between two samples is used to judge whether they belong to the same class. Sung et al. proposed the relational network, which considers that the measurement method needs to be modeled [14]. They built a neural network (CNN) to learn how to calculate distance between two samples.

6 Conclusion

We propose a Trojan traffic detection method based on meta-learning, which can classify Trojan traffic. In the classification task, the model preprocesses the original pcap packet, uses ResLSTM network as the embedded part to encode the feature vector of the sample, and then calculates the association score between the sample and each category, and judges whether the sample belongs to the class through the association score. To evaluate the method's effectiveness, we select different Trojan traffic types from CTU13 and construct a data set with the balanced and unbalanced sample number. The experimental results show that the imbalanced data set makes the traditional deep learning model overfit, and the detection result is lower than the sample balanced data set. The meta-learning method can achieve better detection results on many verification sets after training on unbalanced data sets.

In future research, we will continue to explore the use of other meta-learning methods to detect Trojan traffic. And we will analyze the non-standard encrypted Trojan traffic to study the model's detection ability for this traffic.

Acknowledgement. This research is supported by the National Key Research and Development Program of China (Grant No. 2018YFC0824801). It is also partially supported by Key Laboratory of Network Assessment Technology, Chinese Academy of Sciences and Beijing Key Laboratory of Network Security and Protection Technology.

References

1. He, K., Zhang, X., Ren, S., et al.: Deep residual learning for image recognition. In: Proceedings of the IEEE Conference on Computer Vision and Pattern Recognition, pp. 770–778 (2016)
2. Moore, A.W., Zuev, D.: Discriminators for use in flow-based classification. Technical report, Intel Research, Cambridge (2005)
3. Feizollah, A., Anuar, N.B., Salleh, R., et al.: Comparative study of k-means and mini batch k-means clustering algorithms in android malware detection using network traffic analysis. In: 2014 International Symposium on Biometrics and Security Technologies (ISBAST), pp. 193–197. IEEE (2014)
4. Rezaei, S., Liu, X.: Deep learning for encrypted traffic classification: an overview. IEEE Commun. Mag. **57**(5), 76–81 (2019)
5. Wang, W., Zhu, M., Zeng, X., et al.: Malware traffic classification using convolutional neural network for representation learning. In: 2017 International Conference on Information Networking (ICOIN), pp. 712–717. IEEE (2017)
6. Wang, W., Zhu, M., Wang, J., et al.: End-to-end encrypted traffic classification with one-dimensional convolution neural networks. In: 2017 IEEE International Conference on Intelligence and Security Informatics (ISI), pp. 43–48. IEEE (2017)
7. Wang, W., Sheng, Y., Wang, J., et al.: HAST-IDS: learning hierarchical spatial-temporal features using deep neural networks to improve intrusion detection. IEEE Access **6**, 1792–1806 (2017)
8. Hwang, R.H., Peng, M.C., Nguyen, V.L., et al.: An LSTM-based deep learning approach for classifying malicious traffic at the packet level. Appl. Sci. **9**(16), 3414 (2019)
9. Liu, X., You, J., Wu, Y., et al.: Attention-based bidirectional GRU networks for efficient HTTPS traffic classification. Inf. Sci. **541**, 297–315 (2020)
10. Santoro, A., Bartunov, S., Botvinick, M., et al.: One-shot learning with memory-augmented neural networks. arXiv preprint arXiv:1605.06065 (2016)
11. Munkhdalai, T., Yu, H.: Meta networks. In: Proceedings of the 34th International Conference on Machine Learning, vol. 70, pp. 2554–2563. JMLR. org (2017)
12. Ravi, S., Larochelle, H.: Optimization as a model for few-shot learning. In: International Conference on Learning Representations (2017)

13. Koch, G., Zemel, R., Salakhutdinov, R.: Siamese neural networks for one-shot image recognition. In: ICML Deep Learning Workshop, p. 2 (2015)
14. Sung, F., Yang, Y., Zhang, L., et al.: Learning to compare: Relation network for few-shot learning. In: Proceedings of the IEEE Conference on Computer Vision and Pattern Recognition, pp. 1199–1208 (2018)
15. Xu, C., Shen, J., Du, X.: A method of few-shot network intrusion detection based on meta-learning framework. IEEE Trans. Inf. Forensics Secur. **PP**(99), 1 (2020)
16. Zhenyan, L., Yifei, Z., Pengfei, Z., et al.: An imbalanced malicious domains detection method based on passive DNS traffic analysis. Secur. Commun. . **2018**, 1–7 (2018)
17. Sabour, S., Frosst, N., Hinton, G.E.: Dynamic routing between capsules. In: Advances in neural Information Processing Systems, pp. 3856–3866 (2017)
18. Mandal, B., Ghosh, S., Sarkhel, R., et al.: Using dynamic routing to extract intermediate features for developing scalable capsule networks. In: 2019 Second International Conference on Advanced Computational and Communication Paradigms (ICACCP), pp. 1–6. IEEE (2019)
19. Lin, A., Li, J., Ma, Z.: On learning and learned data representation by capsule networks. IEEE Access **7**, 50808–50822 (2019)
20. Li, S., Yun, X., Zhang, Y., Xiao, J., Wang, Y.: A general framework of trojan communication detection based on network traces. In: IEEE Seventh International Conference on Networking Architecture and Storage, pp. 49–58 (2012)
21. Jiang, D., Omote, K.: An approach to detect remote access trojan in the early stage of communication. In: 2015 IEEE 29th International Conference on Advanced Information Networking and Applications, Gwangiu, pp. 706–713 (2015). https://doi.org/10.1109/AINA.2015.257

Grasp the Key: Towards Fast and Accurate Host-Based Intrusion Detection in Data Centers

Mengtian Gu[1,2], Biyu Zhou[1(✉)], Fengyang Du[1,2], Xuehai Tang[1,2], Wang Wang[1,2], Liangjun Zang[1], Jizhong Han[1], and Songlin Hu[1]

[1] Institute of Information Engineering, Chinese Academy of Sciences, Beijing, China
zhoubiyu@iie.ac.cn
[2] School of Cyber Security, University of Chinese Academy of Sciences, Beijing, China

Abstract. With the rapid development of data center facilities and technology, in addition to detection accuracy, detection speed has also become a concern for host-based intrusion detection. In this paper, we propose a DNN model to detect intrusion for host with high accuracy. Along with that, a data reduction method based on SHapley Additive exPlanations (SHAP) is incorperated to reduce the execution time of the DNN model. Extensive evaluation on two well-known public datasets in this field shows that our proposed method can achieve high-efficiency intrusion detection while ensuring high-precision.

Keywords: Intrusion detection · Explainable artificial intelligence · Neural networks

1 Introduction

In recent years, security threats against computers have become more and more serious. The Host-based Intrusion Detection System (HIDS), as a kind of active protection technology, has been gaining extensive attention in the field of computer security. However, despite considerable efforts in this field, HIDS has encountered the well-known big data challenge brought by the rapid development of data center facilities and technologies [13]. First of all, large data centers require high precision for intrusion detection. Once an exception occurs within the data center, it will spread quickly and finally affect the whole cluster. Furthermore, handling a large number of system call traces has become a basic requirement for modern data centers, which is a big challenge for detection efficiency due to the real-time requirement.

Deep learning has shown its ability to discover potential patterns of big data and achieve high-precision intrusion detection in recent years. However, the high complexity of the network also results in long execution time of the model. For instance, it may take several days to train a deep learning model in a large

© Springer Nature Switzerland AG 2021
M. Paszynski et al. (Eds.): ICCS 2021, LNCS 12743, pp. 181–194, 2021.
https://doi.org/10.1007/978-3-030-77964-1_15

data center. This is not a problem if the system only needs to be trained once. But in practice, because of new patches and modifications to the system, it is often necessary to train several times to accommodate the new changes. More importantly, the model needs to handle a large number of fine-grained system call traces at the same time when doing intrusion detection. Understandably, this causes a speed limit.

In this paper, we seek to develop a novel intrusion detection method and introduce specialized designs to solve the above challenges. In terms of intrusion detection model, we choose two neural networks: multi-filter CNN and attention-based BiLSTM, which can extract local and global feature representation well. We combine these two neural networks to achieve a high precision. Then we use a XAI method—SHapley Additive exPlanations (SHAP) to obtain the *important decision interval*, so that we can implement model acceleration by data reduction.

We experimented with two publicly available datasets: ADFA-LD and UNM. The results show that our proposed method can greatly reduce execution time of the intrusion detection model and has little impact on precision. To the best of our knowledge, this is the first intrusion detection technology to be customized for big data by using XAI. By accelerating the model execution process, our efforts can make a positive contribution to building a reliable deep learning intrusion detection system in the current big data and cloud computing environment.

The rest of this article is composed as follows. In Sect. 2, we present a brief background and the motivation; In Sect. 3, we give a literature review. In Sect. 4, we describe the proposed method; Two different system call datasets are evaluated in Sect. 5; Sect. 6 discusses our conclusion and summarizes the future work.

2   Background and Motivations

With the frequent occurrence of various network security issues, intrusion detection can actively defend against various attacks and has gradually become a research hotspot in the field of computer security. Intrusion Detection Systems (IDS) are devices which monitor systems to detect potential intrusions. They can be divided into Network-based Intrusion Detection Systems (NIDS) and Host-based Intrusion Detection Systems (HIDS). NIDS collects information from the packets of data, then monitors and analyzes network traffic to protect the system from network-based threats. Compared with NIDS, HIDS mainly collects information such as host system calls or logs. It focuses on using these indicators to determine whether the host system has been compromised.

System calls provide the basic interface between the process and the operating system. The system call is referenced when the running process requests the kernel service from the operating system. Thus, it constitutes a trace which describes the behavior of the monitored process. Compared with other data sources for HIDS, system call traces can better reflect differences between normal and abnormal behavior. At present, most HIDS use system call traces as their main source of information. This article also mainly discuss intrusion detection technology based on system calls.

The traditional HIDS mainly performs intrusion analysis on independent hosts installed with independent detection software. However, with the rapid development of data centers and other technologies, the application scenario of HIDS has changed. Processing a large number of system call traces of multiple hosts has become a basic requirement for modern data centers. Therefore, this presents some new challenges to HIDS.

Challenge 1: Detection accuracy—System call traces generated by a large data center are a kind of big data, which are huge and complex. On the other hand, once an internal server is abnormal, this exception will spread quickly and then affect the entire cluster. This may cause considerable damage to the data center. Therefore, we need an intrusion detection technology which can accurately identify each intrusion behavior. In other words, it requires high accuracy and low false alarm rate.

Challenge 2: Execution time—The execution time of HIDS is usually measured by the training time and detection time of the model. For the training time, it is very cumbersome to maintain or update large amounts of traditional HIDS software installed on each host or virtual host in the network. As for the detection time, Fig. 1 shows the detection time of a RNN model with the increase of data volume under a single GPU. With the increase of data volume, the detection time of HIDS will no longer meet the requirement of real-time. One solution is to use multiple GPUs deployed on physical hosts to speed up the detection process. However, it can be expensive and space-consuming.

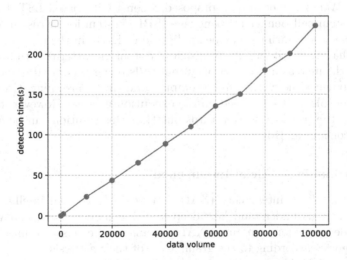

Fig. 1. Detection time of a RNN model with the increase of data volume under a single GPU

3 Related Work

This section briefly discusses the work closely related to this paper.

3.1 Host-Based Intrusion Detection

Numerous work and surveys have been published in the area of intrusion detection. In the field of traditional methods, Forrest et al. [6] first introduced a system-call-based method into intrusion detection research works, which is known as "sequence time-delay embedding (STIDE)". This method constructs normal databases to detect anomalies which would take a lot of time to update and maintain. Compared with STIDE, Wang et al. [21] implemented a method based on Bloom filter which takes up less memory. However, the limitation of Bloom Filter also exits as it allows false positives. Pierre-Francois Marteau [16] proposed a new sequence similarity measure to distinguish normal and abnormal sequences. However, it is difficult to obtain dependencies between sequences. Some well-known machine-learning models such as support vector machine (SVM) [1,9], the k-nearest neighbor algorithm (KNN) [5,12] and decision trees have been implemented on HIDS. Yet these methods are difficult to adapt to the current distributed computing environment.

Some recent works have used deep learning to improve the performance of HIDS, which has achieved remarkable success in many other fields. Staudemeyer and Trivedi [19] applied LSTM neural network to intrusion detection and achieved preferable results. Ghosh et al. [7] introduced the Elman neural network into HIDS. Wenqi Xie et al. [22] proposed a sensitivity-based LSTM model to design a System-call Behavioral Language (SBL) system for intrusion detection. In conclusion, deep learning may be applicable to HIDS in the big data environment as it has shown its ability to discover potential patterns within big data. But due to the increasing number of system calls being generated, the execution time of deep neural networks can be complex and time consuming. One solution is to use multiple GPUs to speed up the execution process. However, the result presented in the previous section turns out that this solution can be expensive and space-consuming [13].

3.2 Explainable Artificial Intelligence

Explainable artificial intelligence (XAI) is a field of artificial intelligence (AI) which can help us understand how deep learning models learn and why they make such decisions for each input. XAI systems can be classified into local and global categories according to the granularity of their analysis.

Locally explainable methods aim to help people understand the decision-making process of the learning model for the specific input sample. Some methods are based on back propagation [3,17,18], which are usually limited to convolutional neural networks (CNN). Ribeiro et al. introduced Local Interpretable Model Agnostic Explanations (LIME) [8]. Although LIME applies to any model, it assumes that the characteristics of input samples are independent of each other. A game theoretically optimal solution using Shapley values for model explainability was proposed by Lundberg et al. [15], which called SHapley Additive exPlanations (SHAP). This method effectively makes up for the deficiency of LIME by considering the influence of variable groups. Globally explainable methods try to understand the complex logic and the internal working mechanism behind the model as a whole. Classic globally explainable methods include rules extraction [20], Model Distillation [14], Concept Activation Vectors (CAVs) [10] and so on. Furthermore, LIME and SHAP can also provide a global understanding of the model.

4   Method

Figure 2 shows the overall flow of the method which is divided into two stages. In the first phase, we leverage the historical system call sequences of the data center and use them to train a high-precision intrusion detection model (in Sect. 4.1). Then, we choose SHAP (in Sect. 4.2) to obtain global explanations of the trained model and analyze an *important decision interval*. In the second stage, we set a period T to collect real-time system calls in the data center on a regular basis. Then, we can only detect the important fragments through data reduction which is described in Sect. 4.3.

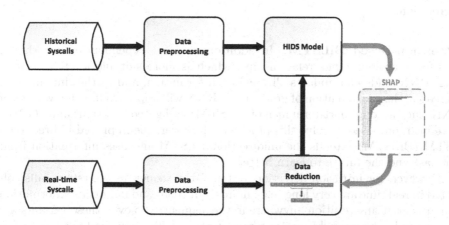

Fig. 2. Overview of the flow of the proposed method

4.1   Intrusion Detection Model

Host-based Intrusion Detection Systems (HIDS) always consider the traces of system calls generated by executing computer programs as input samples. If we treat these traces as sentences, then intrusion detection can be treated as a text classification task. Just like text classification, we need to consider both the local and global features of input sequences during detection. Therefore, our detection model is a combination of two kinds of neural networks: multi-filter convolutional neural network (CNN) and attention-based Bi-directional Long Short-Term Memory (BiLSTM).

Multi-filter CNN. Convolutional neural network (CNN) has become one of the most popular neural networks by virtue of its expertise in capturing local dependencies. In CNN, local receptive field is regarded as the input of the bottom layer and the information is transmitted forward through each layer which is composed of filters in order to obtain relevant features of input data. This processing way, which is closer to the visual system of human brain, makes CNN have unique advantages in processing image data. As for 1D data such as system call traces and texts, we can use 1D CNN to generate representations by sliding over sequences.

Nevertheless, traditional CNN only uses one type of convolution kernel for feature extraction which makes it difficult to capture all local dependencies of the entire sequence. In this work, we solve this problem by using multi-filter CNN [11]. We design filters with different sizes in order to get receptive fields of different widths. This allows to extract words of different lengths each time and capture local features at different levels just like n-gram with multi-window size. Therefore, multi-filter CNN can achieve a better extraction effect for the local correlation.

Attention-Based BiLSTM. In addition to local dependencies, we should also focus on long-range relationships, which is more suitable for RNN. However, RNN reads and updates all previous information, and as the time interval increases, the accumulation of gradients in RNN will approach 0. Some variant of RNN, such as long short-term memory (LSTM) and gated recurrent unit (GRU), have been proposed to solve this problem. Moreover, the improved bidirectional LSTM (BiLSTM) extends the unidirectional LSTM and uses information from the past and the future to learn better.

However, for intrusion detection systems, the sequences of system calls collected in real time are very long (e.g., more than 2000) and important information can appear at any position anywhere in the sequence. So even these variants are difficult to learn reasonable vector representation, which may lead to the decrease of detection accuracy. To tackle these problems, the method which is proposed to do relation classification tasks in the natural language processing domain is transferred to our model [23], which acts the attention mechanism on the output vector produced by the BiLSTM. The representation r of the sequence is actually a weighted sum of the output of the hidden layer in BiLSTM at each time as follows:

$$M = \tanh(H) \tag{1}$$

$$\alpha = softmax(w^T M) \tag{2}$$

$$r = H\alpha^T \tag{3}$$

where:

- H is a matrix consisting of output vectors of the hidden layer.
- w is a trained parameter vector.
- α is the corresponding weights of each output vector of the hidden layer.

In this way, the attention-based BiLSTM can learn different weight coefficient α at different moments by introducing the attention mechanism. This allows to selectively learn the input sequence and capture the long-range dependence relationship better.

Finally, we combine these two neural networks into the intrusion detection model as shown in Fig. 3. Given a trace of system calls, our proposed model generates word-embedding vectors via an embedding layer. Then, we feed these embedding vectors to both two neural networks to create local and global feature vectors. The final vector is the combination of these two vectors. In the end, a fully connected layer is used for classification. In this way, our proposed model can achieve a good detection performance by combining the advantages of multi-filter CNN and attention-based BiLSTM.

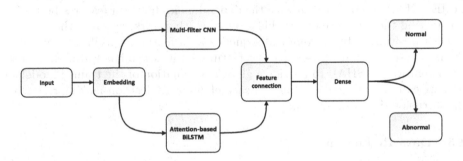

Fig. 3. Overview of the structure of the proposed intrusion detection model

4.2 Using SHAP for Model Analysis

Although our intrusion detection model mentioned above can achieve a good performance, the detection efficiency cannot meet the standards of real-time detection, especially under the circumstance of big data. Therefore, we propose a method to improve detection efficiency by using eXplainable Artificial Intelligence (XAI) to reduce the input data. In the first place, we use SHapley Additive exPlanations (SHAP) to analyze the trained detection model.

SHAP is a general method proposed by [15] to explain model predictions. Inspired by cooperative game theory, SHAP constructed an additive explanatory model that treated all features as "contributors". For each sample, the model generates a predicted value, and each SHAP value measures how much each feature contributes. SHAP specifies the explanation for an instance x as:

$$y_i = y_{base} + f(x_{i1}) + f(x_{i2}) + \ldots + f(x_{ik}) \tag{4}$$

where:

- y_i is the predicted value of the model.
- y_{base} is the baseline of the entire model (usually the mean value of the target variable of all samples).
- x_i is the ith sample and x_{ij} is the jth feature of it.
- f is the SHAP value. $f(x_i, 1)$ means the contribution of the first feature to the final predicted value y_i in the ith sample. The larger $f(x_i, 1)$ is, the more this feature affects the final decision.

SHAP provides three significant advantages compared to other XAI methods. Firstly, SHAP can work under a black box setting. SHAP does not require any knowledge of the model internals and analyze the model by sending inputs and observing outputs, which effectively support the application environment of intrusion detection. Secondly, SHAP can provide not only explanations for the prediction results of each sample, but also the global explanation for the model, which is what we need. Thirdly, compared with other blackbox methods like LIME, SHAP can better deal with the dependencies between features better by considering the influence of variable groups. In this work, we treat the element at each position in the system call sequence as a feature. Specifically, we label the first system call in the sequence as feature 1, the second as feature 2, and so on. Then, we use SHAP to provide a global explanation of the trained model so that we can obtain the importance of each element to decision-making results of the intrusion detection model.

4.3   Data Reduction

In deep learning, the influence of input features on model is different. Therefore, system calls from different locations have different effects on results of intrusion detection. And because the system call sequence has behavior continuity, there will be some sequence fragments that have little influence on decision-making results of the model. Our aim is to reduce this part of data and retain the fragments that have a large impact on model decisions.

Therefore, after obtaining the global explanation provided by SHAP, we can know which elements in the detection sequence are important for the model and which are not. We select the range of important elements (features) as the *important decision interval* of the input sequence. This allows us to subtract the system call sequences retrieved next based on this interval. In other words, we only use the sequence within the *important decision interval* for model retraining and intrusion detection, so that the efficiency of intrusion detection can be improved on the premise of ensuring the accuracy is not greatly affected. It is worth noting that the length of the *important decision interval s* can be manually set according to the actual situation.

Table 1. The specific composition of datasets

	Normal	Abnormal	Training	Detecting	max_len
ADFA-LD	746	746	998	494	3143
UNM (live lpr)	1196	1001	1460	737	7720

5 Evaluation

In this section, we first present the datasets and simulation settings of our experiments. Then, we evaluate the performance of the proposed method and conduct a series of comparisons.

5.1 Datasets and Experimental Configuration

We use two publicly available datasets to conduct our experiments, i.e. ADFA-LD and UNM [2,4]. Both of them are commonly used in this field. As different types of programs use different mapping files in UNM, we chose the live lpr set which has larger data volume than others. We selected traces with equal proportion in order to balance the normal and abnormal data and divided them into training data and detecting data according to the ratio of 2:1 (see Table 1).

The experimental configuration related to the model operation is described below. The embedding layer transforms one-hot encoding of integers in the call sequence into a dense vector of size 128. Regarding to multi-channel CNN model, we use filter windows of 4, 6, 8, and 10 with 128 feature maps each. The size of memory cells used in BiLSTM stage accounts for 128. We applied the dropout technique with the dropout rate of 0.1 on BiLSTM and the fully connected neural network. In addition, our proposed model is trained by using mini-batch stochastic gradient descent (SGD)for higher accuracy. The size of each mini-batch is 50. And we used Adam optimizers with a learning rate of 0.004. The length of the important decision interval is set to 100. All experiments in this paper are performed under Ubuntu 16.04.6 LTS system, the kernel version is GNU/Linux 3.10.0-1062.12.1.el7.x86_64, the CPU is Intel (R) Xeon (R) with a frequency of 2.40 GHz, a NVIDIA TESLA M40 GPU with 24 G memory, the version of CUDA is 10.2.

5.2 Experimental Results

Detection Accuracy. Since most of the traditional methods are relatively old and can achieve low accuracy, we only compare with Support Vector Machines (SVM), which is one of the most common classification methods in machine learning. As shown in Table 2, The precision of our model is 75.6% and 0.2% higher than that of SVM on the two datasets respectively, and the false positives rate (FAR) is reduced by 93.5% and 100%, where a high false-alarm rate can affect the performance of HIDS. This proves that our method performs well on detection accuracy.

Table 2. Intrusion detection performance comparison by two evaluation metrics.

Method	ADFA-LD		UNM	
	Precision	FAR	Precision	FAR
SVM	0.557	0.437	0.995	0.00215
Our proposed	**0.978**	**0.0283**	**0.997**	**0**

Table 3. Ablation experiments

	Multi_filter CNN	BiLSTM	CNN+BiLSTM	Our proposed
ADFA-LD	0.929	0.900	0.965	**0.978**
UNM	0.935	0.983	0.997	**0.997**

We also built three other models for Ablation experiments: (1) Multi-filter CNN (2) BiLSTM (3) Multi-filter CNN and BiLSTM. The detailed parameters of these models are same as our proposed model. As shown in Table 3, our model performs best on both two datasets. This result confirms that giving consideration to both the local and global features is better than simply focusing on one type of features. In addition, the accuracy of our model (0.978) is higher than the model with multi-channel CNN and BiLSTM on the ADFA-LD dataset (0.965). As for the UNM dataset, both of their accuracy are reach 0.997. As can be seen from these results, the attention mechanism has a certain improvement in the model effects.

Global Explanation. Figure 4 shows the top 30 important features extracted by SHAP for the ADFA-LD and UNM datasets. Specifically, the y-axis shows features, and the x-axis shows average values of the absolute SHAP values, which reflect average impacts on model output magnitude. Different colors correspond to different classes. The impacts on the Normal and Abnormal classes are marked in blue and red, respectively. The features are ordered according to their importance. Labels of important features extracted for the ADFA-LD dataset are almost less than 100, which means the important pieces of input sequences for the intrusion detection model are in the interval [0, 100]. Similarly, as for the UNM dataset, the critical detection segments are between [100, 200]. In this way, we can reduce the input sequences based on these two *important decision intervals*.

Fidelity Tests. In order to validate the correctness of the explanation, we design fidelity tests to show whether the features in the selected important decision intervals are the major contributors to the detection results. We denote the selected features as F_x. There are two intuitions:

- If features F_x are accurately selected, then removing F_x from input x will greatly affect the detection accuracy.
- If features F_x are accurately selected, only keeping F_x as input has little impact on accuracy.

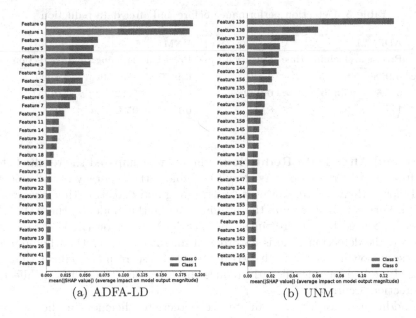

(a) ADFA-LD   (b) UNM

Fig. 4. The top 30 features extracted by SHAP

Table 4. Fidelity tests by two evaluation metrics.

	ADFA-LD		UNM	
	Precision	FAR	Precision	FAR
x	0.978	0.0283	0.997	0.0
$t(x)_1$	0.862	0.215	0.634	0.0
$t(x)_2$	0.960	0.0769	0.997	0.0

Using these intuitions, we constructed two new types of input samples $t(x)_1$ and $t(x)_2$ for feature verification. To be more specific, we construct samples $t(x)_1$ by nullifying the selected feature F_x from the original data x and construct $t(x)_2$ by only preserving the feature values of the selected features F_x.

Table 4 shows the results of the fidelity tests. By only nullifying the features of important decision intervals, the accuracy on two datasets decreased by 11% and 36% respectively, and FAR on the ADFA-LD dataset increased from 2% to 21%. This drastic decrease of performance indicates that this small set of features are highly important to the classification. In addition, using only the features in the important decision intervals, the accuracy only decrease by 1% on the ADFA-LD dataset and have no effect on the UNM, indicating that the core patterns have been successfully captured. Therefore, both sets of tests verifies that the features in the important decision intervals we obtained are indeed the major contributors to the detection results.

Table 5. Detection performance before and after data reduction

	ADFA-LD			UNM		
	Precision	Training_time	Detection_time	Precision	Training_time	Detection_time
Before	**0.978**	86.595	2.140	**0.997**	180.350	4.344
After	0.966	**6.306**	**0.232**	0.996	**9.139**	**0.251**
Gain	1%	93%	89%	0.1%	95%	94%

Before and After Data Reduction. Finally, we compared the results before and after the data reduction. As shown in Table 5, the accuracy of the model is slightly lower than before. However, the training and detection time are greatly reduced. For training duration, the training time of the model on the ADFA-LD dataset is reduced by over 10 times, and nearly 20 times on the UNM dataset. Moreover, the detection time is an order of magnitude lower than beforen and this observation holds for both two datasets. These results confirm that our proposed method can significantly reduce the execution time without affecting the detection accuracy.

In addition, in order to evaluate the detection efficiency in the big data environment, we expand the scale of the dataset. Figure 5 shows the variation of the detection time before and after data reduction. With the increase of the sample size, the detection time of the method without data reduction grows to a large scale while ours grows slowly, and the gap between them becomes broader and broader.

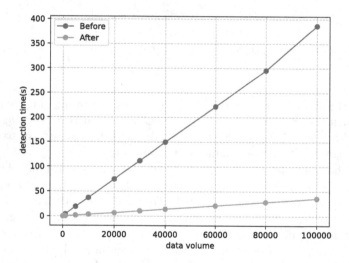

Fig. 5. The variation of detection time before and after data reduction

6   Conclusion

This paper introduces a method to realize high-precision and high-efficiency intrusion detection in the data center. We first design a high-precision intrusion detection model, which can better capture the local and global dependencies of sequence data through integrating multi-filter CNN and attention-based BiL-STM. Next, the important decision interval is proposed through explaining this DNN model by SHAP. Finally, the execution time is shortened by data reduction. The extensive experiments using ADFA-LD and UNM datasets which are public demonstrate that compared with traditional methods, this proposed DNN model can achieve high-precision detection. Furthermore, we also verify that the features in the important decision intervals we obtained are indeed the major contributors to the detection results. Moreover, the detection time only produce small increases with the growth of data volume, which verifies the strong adaptability of our method under the environment of big data. Based on the accurate and efficient intrusion detection, our future work is to build a whole set of intrusion detection system and apply it to physical environment.

References

1. Ambusaidi, M., He, X., Nanda, P., Tan, Z.: Building an intrusion detection system using a filter-based feature selection algorithm. IEEE Trans. Comput. **65**, 2986–2998 (2016)
2. Assem, N., Rachidi, T., Graini, M.T.E.: Intrusion detection using Bayesian classifier for arbitrarily long system call sequences. IADIS Int. J. Comput. Sci. Inf. Syst. **9**(1), 71–81 (2014)
3. Bach, S., Binder, A., Montavon, G., Klauschen, F., Müller, K.R., Samek, W.: On pixel-wise explanations for non-linear classifier decisions by layer-wise relevance propagation. Plos One **10**, e0130140 (2015)
4. Creech, G., Hu, J.: Generation of a new ids test dataset: time to retire the KDD collection. In: 2013 IEEE Wireless Communications and Networking Conference (WCNC) (2013)
5. Yuxin, D., Xuebing, Y., Di, Z., Li, D.: Feature representation and selection in malicious code detection methods based on static system calls. Comput. Secur. **30**, 514–524 (2011)
6. Forrest, S., Hofmeyr, S.A., Somayaji, A., Longstaff, T.A.: A sense of self for unix processes. In: SP (1996)
7. Ghosh, A.K., Schwartzbard, A., Schatz, M.: Learning program behavior profiles for intrusion detection. In: Workshop on Intrusion Detection and Network Monitoring, vol.51462, pp. 1–13 (1999)
8. Godsell, M.: Why should we trust you? Marketing (2007)
9. Khreich, W., Murtaza, S.S., Hamou-Lhadj, A., Talhi, C.: Combining heterogeneous anomaly detectors for improved software security. J. Syst. Softw. **137**(MAR.), 415–429 (2017)
10. Kim, B., et al.: Interpretability beyond feature attribution: quantitative testing with concept activation vectors (TCAV) (2017)
11. Kim, Y.: Convolutional neural networks for sentence classification. arXiv preprint arXiv:1408.5882 (2014)

12. Liao, Y., Vemuri, V.R.: Using text categorization techniques for intrusion detection. USENIX Security Symposium, vol. 12, pp. 51–59 (2002)
13. Liu, M., Xue, Z., Xu, X., Zhong, C., Chen, J.: Host-based intrusion detection system with system calls: review and future trends. ACM Comput. Surv. **51**(5), 98.1–98.36 (2019)
14. Liu, X., Wang, X., Matwin, S.: Improving the interpretability of deep neural networks with knowledge distillation. In: 2018 IEEE International Conference on Data Mining Workshops (ICDMW) (2018)
15. Lundberg, S., Lee, S.I.: A unified approach to interpreting model predictions (2017)
16. Marteau, P.F.: Sequence covering for efficient host-based intrusion detection. IEEE Trans. Inf. Forensics Secur. **14**(4), 994–1006 (2018)
17. Shrikumar, A., Greenside, P., Shcherbina, A., Kundaje, A.: Not just a black box: learning important features through propagating activation differences (2016)
18. Simonyan, K., Vedaldi, A., Zisserman, A.: Deep inside convolutional networks: visualising image classification models and saliency maps. arXiv preprint arXiv:1312.6034 (2013)
19. Staudemeyer, R.C.: Applying long short-term memory recurrent neural networks to intrusion detection. South Afr. Comput. J. **56**(1), 136–154 (2015)
20. Tickle, A.B., Andrews, R., Golea, M., Diederich, J.: The truth will come to light: directions and challenges in extracting the knowledge embedded within trained artificial neural networks. IEEE Trans. Neural Netw. **9**(6), 1057–1068 (1998)
21. Wang, K., Parekh, J.J., Stolfo, S.J.: Anagram: a content anomaly detector resistant to mimicry attack. In: Zamboni, D., Kruegel, C. (eds.) RAID 2006. LNCS, vol. 4219, pp. 226–248. Springer, Heidelberg (2006). https://doi.org/10.1007/11856214_12
22. Xie, W., Xu, S., Zou, S., Xi, J.: A system-call behavior language system for malware detection using a sensitivity-based LSTM model. In: Proceedings of the 2020 3rd International Conference on Computer Science and Software Engineering, pp. 112–118 (2020)
23. Zhou, P., Shi, W., Tian, J., Qi, Z., Xu, B.: Attention-based bidirectional long short-term memory networks for relation classification. In: Proceedings of the 54th Annual Meeting of the Association for Computational Linguistics (Volume 2: Short Papers) (2016)

MGEL: A Robust Malware Encrypted Traffic Detection Method Based on Ensemble Learning with Multi-grained Features

Juncheng Guo[1,2], Yafei Sang[1(✉)], Peng Chang[1], Xiaolin Xu[3],
and Yongzheng Zhang[1,2]

[1] Institute of Information Engineering, Chinese Academy of Sciences, Beijing, China
{guojuncheng,sangyafei,changpeng,zhangyongzheng}@iie.ac.cn
[2] School of Cyber Security, University of Chinese Academy of Sciences,
Beijing, China
[3] National Computer Network Emergency Response Technical Team/Coordination
Center of China, Beijing, China
xuxiaolin@cert.org.cn

Abstract. As the use of encryption protocols increase, so does the challenge of identifying malware encrypted traffic. One of the most significant challenges is the robustness of the model in different scenarios. In this paper, we propose an **ensemble learning** approach based on **multi-grained** features to address this problem which is called MGEL. The MGEL builds diverse base learners using multi-grained features and then identifies malware encrypted traffic in a stacking way. Moreover, we introduce the self-attention mechanism to process sequence features and solve the problem of long-term dependence. We verify the effectiveness of the MGEL on two public datasets and the experimental results show that the MGEL approach outperforms other state-of-the-art methods in four evaluation metrics.

Keywords: Malware encrypted traffic detection · Ensemble learning · Multi-grained features · Self-attention

1 Introduction

With the widespread use of encryption technology, the privacy, freedom, anonymity of Internet users have been greatly protected, but also it has allowed attackers to evade the anomaly detection system. For example, an attacker invades and attacks the system by encrypting malware traffic. Besides, criminals penetrate the darknet through tools such as Tor [12] to trade illegally. That

Supported by the National Key Research and Development Program of China (Grant No.2018YFB0804702), and the National Natural Science Foundation of China (Grant No.U1736218).

is to say, the abuse of encryption technology poses many challenges to network anomaly detection and secure management. Therefore, the identification of malware encrypted traffic has aroused great concern in academia and industry.

There are traditional rule-based methods such as port numbers [18] and deep packet inspection (DPI) [11]. Port numbers are a quick and easy approach. However, its accuracy has declined because of the use of well-known port numbers. Deep packet inspection (DPI) is to find patterns or keywords in the payloads. But this method can only handle plaintext information and the matching process is computationally costly.

Recently, some researchers have been using machine learning methods to solve such problems. They extract packet features from plaintext messages of the SSL/TLS handshake packets [1, 2] or use statistical features at the flow level [9]. However, TLS/SSL handshake information may be incomplete in real scenarios (e.g., the client reconnects to the server via session ID without the process of certificate exchange and negotiate the secret key again), resulting in a low accuracy rate of identification. The statistical features, such as byte distribution, transition probabilities of lengths and times are often used as features. However, statistical features require additional preprocessing, and the statistical method is based on experience, which weakens the generalization ability of the model.

Deep learning has achieved great success in the areas of image process and natural language process, and are also gradually being applied to the field of traffic identification. In some recent literature, researchers mainly use AutoEncoder [10], CNN [16, 17], LSTM [19] and other network structures to learn representations from raw data. Although deep learning methods have high accuracy, there are some limitations in practical use:

(1) Deep learning methods are overly dependent on data size, and when there is not enough data for the real scenario, deep learning methods cannot learn a good representation.
(2) Some research only uses encrypted payloads of a few packets in a flow for identification, which can fail in some application-level issues. Moreover, flow sequence features mainly face the challenge of the long-term dependence problem.
(3) Deep learning methods typically have millions of parameters, and model performance is greatly influenced by hyperparameters. Moreover, Deep learning methods cannot adapt well to the new scenario.

The ability to generalize in different scenarios is critical because there are many types of malware encrypted traffic. Drawing the idea of ensemble learning that two heads are better than one, we use diverse base learners to enhance the model's robustness in different scenarios. In order to obtain diverse base learners, we use multi-grained features to learn separately. We first define packet features and flow features. Packet features refer to key fields in certain packets, such as port number, validity time of certificate, cipher suite, etc. Using the discrepancy in these key fields, we can directly identify malware encrypted traffic. Flow features refer to sequence features that reflect the communication process between the client and the server, such as packet length sequence, message type sequence, and time interval sequence. By learning these sequence features, we can get the pattern of communication of different types of traffic, so as to distinguish between normal and malware encrypted traffic. After getting diverse base learners, we use the idea of stacking to ensemble each base learner to get the final identification result. This idea of ensemble learning based on multi-grained features enhances the robustness of the model in different scenarios.

In this paper, We propose an ensemble learning approach based on multi-grained features (MGEL) for malware encrypted traffic identification. The MGEL first obtains diverse base learners using models suitable for different grained features. Specifically, we use Xgboost to learn from packet features and use a model based on Bi-LSTM+Self-Attention to learn from flow features. Then, we introduce the idea of stacking to ensemble each base learner. We verify the effectiveness of the MGEL on two public datasets and the experimental results show that our MGEL approach outperforms other state-of-the-art methods in four evaluation metrics.

Our contributions can be briefly summarized as follows:

(1) We propose an ensemble learning approach based on multi-grained features for malware encrypted traffic identification. The attributes of multi-grained features and the characteristics of ensemble learning enhance the robustness of the model in different scenarios.
(2) For sequence features, we introduce a self-attention mechanism to solve the long-term dependency problem. Meanwhile, self-attention runs faster than RNN because its computation does not depend on the previous state.
(3) Our MGEL achieves excellent results on the two public datasets for malware traffic encrypted identification and outperforms other state-of-the-art methods.

2   Background and Related Work

2.1   SSL/TLS Encrypted Protocols

The Secure Sockets Layer (SSL) [6] and its successor Transport Layer Security (TLS) [5] protocol are popular encryption protocols used to protect client-server sessions. Figure 1 shows the process of an SSL/TLS session. It mainly includes the handshake process and communication process. The handshake process is used to negotiate secret key and verify identification. Specifically, the client and server first exchange Client Hello and Server Hello messages to establish the session. Then both sides negotiate the secret key through Certificate, Server Key Exchange, Client Key Exchange messages, and finally the client sends Change-CipherSpec and Finished messages to complete the handshake process. Then the subsequent communication process encrypts the communication payloads using the negotiated secret key and encryption algorithm.

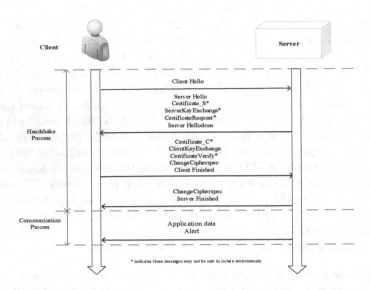

Fig. 1. A communication session of SSL/TLS protocol.

2.2   Mainstream Method

Conventional Methods. Conventional methods generally identify malware encrypted traffic by port number and DPI (Deep packet Inspect). The port-based approach is very efficient as it identifies the port number extracted from the packet and matches it with IANA TCP/UDP. However, the accuracy of port-based methods has dramatically decreased due to the widespread use of port obfuscation and dynamically assigned ports [4]. Deep Packet Inspect (DPI) technology is based on the analysis of information available in the payload of packets. This method can only handle plaintext information and the matching process is computationally costly.

Machine Learning Methods. The machine learning method is mainly based on the assumption that there are some distinguishable statistical features between normal and malware encrypted traffic. From millions of flows, Anderson *et al.* analyze the discrepancy in TLS key fields and summarize a number of distinguishing features such as Cipher Suite, Extension, Client's Public key Length, Validity of Certificate, *etc.* [2]. In addition, some flow-level statistical features such as byte distribution, transition probabilities of lengths and times [7,9,13] are often used as features. However, since the statistical features are designed empirically, they may be valid only for data in certain scenarios and have weak generalization ability.

Deep Learning Methods. Deep learning methods have been widely used in image process and natural language process, but are still new to the problem of encrypted traffic identification. Currently, the research of encrypted traffic identification based on deep learning mainly uses flow sequences or raw byte data as input to learning a good representation by AutoEncoder (AE), Convolutional Neural Neteork (CNN), Recurrent Neural Network (RNN), *etc.* Wang *et al.* transform the raw data into images and learn feature representations using 2D-CNN [17] and 1D-CNN [16]. Liu *et al.* proposed an AutoEncoder model based on flow sequences and improve performance by adding classification loss and reconstruction loss [10]. Li *et al.* proposed a byte segment neural network where payloads are divided into multiple segments [8]. They use the Attention mechanism to further select significant representations as to the input of the softmax layer.

3   The MGEL Framework

Our ensemble learning model with multi-grained features (MGEL) is shown in Fig. 2. The MGEL consists of three parts which are extracting multi-grained features, first stage of MGEL, second stage of MGEL. In the first part, we use the open source tool flowcontainer[1] to obtain multi-grained features from the raw flow data, and in the second part, we learn three base learners (MGEL-1, MGEL-2, MGEL-3) using multi-grained features. In the third part, we use the output of the three base learners to learn the meta learner. In the following section, we will show the details of the last two parts.

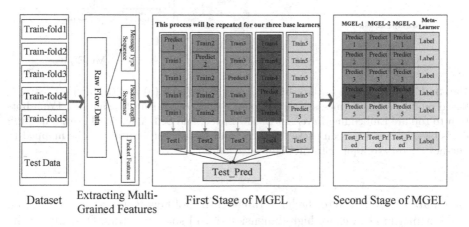

Fig. 2. The flow diagram of identifying malware encrypted traffic by MGEL. We perform a five-fold cross-validation on the model, which is represented by 5 colors. (Color figure online)

---

[1] https://github.com/jmhIcoding/flowcontainer.

3.1   The First Stage of MGEL

In the first stage, we use the training dataset to learn three base learners by multi-grained features. Three base learners are based on flow features and packet features as shown in Fig. 3. Suppose the size of the training dataset and test dataset are M_{train}, M_{test}, then after the first stage of learning, we will get the prediction matrix of $P_{train} \in \mathbb{R}^{M_{train} \times 3}$, $P_{test} \in \mathbb{R}^{M_{test} \times 3}$.

Base Learner Based on Flow Feature. Flow features reflect the communication process between the client and the server. As shown in Fig. 3(a) and Fig. 3(b), MGEL-1 and MGEL-2 are based on message type sequences and packet length sequences, respectively. They consist of three layers, which are Embedding Layer, Bidirectional LSTM Layer and Self-Attention Layer.

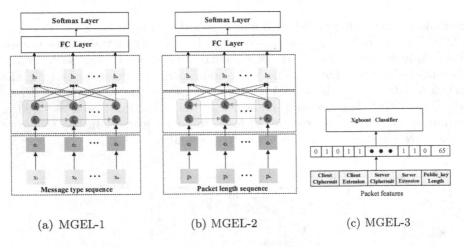

(a) MGEL-1	(b) MGEL-2	(c) MGEL-3

Fig. 3. The architecture of three base learners. Subgraph (a) and (b) are base learners based on flow features (message type sequence and packet length sequence). From bottom to top are Embedding Layer, Bi-LSTM Layer, and Self-Attention Layer. Subgraph (c) is base learner based on packet features, and the packet features will be mapped to one-hot vectors to input into the Xgboost Claffifer.

Embedding Layer. Due to the large range values of flow features, the traditional one-hot method can cause high-dimensional and sparse vectors. Drawing on the idea of embedding in natural language processing, we map each value of flow features to a fixed-length vector. We also introduce pad and unk tokens in the dictionary to deal with unknown values and sequence length inconsistency.

Assume that the size of the dictionary is V, and the embedded vector dimension is d. The embedding layer can be viewed as a matrix $E \in \mathbb{R}^{V \times d}$, where each row of the matrix represents a d-dimensional vector corresponding to a certain value. The matrix E is learnable so that we can get embedding vectors that are more context-sensitive.

Bi-directional LSTM Layer. A bi-directional *lstm* runs a forward and backward *lstm* on a sequence starting from the left and the right ends, respectively. A forward language model computes the probability of the sequence by modeling the probability of tokens s_t given the history $(s_1, ..., s_{t-1})$:

$$p(s_1, ..., s_N) = \prod_{t=1}^{N} p(s_t | s_1, ..., s_{t-1}) \tag{1}$$

A backward language model is similar to a forward language model, except it runs over the sequence in reverse, predicting the previous token given the future context:

$$p(s_1, ..., s_N) = \prod_{t=1}^{N} p(s_t | s_{t+1}, ..., s_N) \tag{2}$$

After going through bi-directional *lstm* Layer, we can get two context-dependent representations $\overrightarrow{h}_t$ and $\overleftarrow{h}_t$ at each position t. we concatenate them and get the final output for each position: $o_t = [\overrightarrow{h}_t, \overleftarrow{h}_t]$. With this setting, the output of bi-directional *lstm* can be expressed as $o = [o_1, o_2, ..., o_n]$, where o contains bi-directional information about the the whole sequence.

Self-Attention Layer. In order to improve the speed of the model and solve the problem of long-term dependencies, we introduce a self-attention mechanism [15], which has recently been widely used in natural language processing and computer vision. The self-attention mechanism generally adopts the form of query-key-value (Q, K, V), and its calculation equation can be expressed as follows:

$$H = softmax(\frac{K^T Q}{\sqrt{d_k}})V$$
$$Q = XW^Q, K = XW^K, V = XW^V \tag{3}$$

where $X = [x_1, x_2, ..., x_N]$ is the input sequence, $H = [h_1, h_2, ..., h_N]$ is the output sequence. d_k is the dimension of Q and K. W^Q, W^K, W^V are learnable parameters. Through the self-attention mechanism, we can learn the parts that should be focused on in the sequence. At the same time, because the calculation of each step does not need to depend on the previous step, the computational speed is faster than the RNN model.

Base Learner Based on Packet Feature. Packet features refer to some key fields in the packet such as cipher suite, extension, length of public key, validity time of certificate, *etc.*. According to the analysis in [2], there is a clear distinction between these key fields in normal and malware encrypted traffic, which can be learned by a machine learning model. Through a comprehensive comparison, we choose Xgboost as the base learner and the structure of the model is shown in Fig. 3(c).

Xgboost. Xgboost [3] is a scalable machine learning system for tree boosting methods, which achieves state-of-the-art results in many areas. We choose Xgboost as the base learner based on two main considerations. Firstly, gradient tree boosting combines the advantages of bagging and boosting and has been shown to give state-of-the-art results in many applications. Secondly, by proposing a novel tree learning algorithm, Xgboost can handle sparse data and the lack of packet features. For a given sample , a Xgboost model uses M additive functions to predict the output:

$$\hat{y}_i = \phi(x_i) = \sum_{k=1}^{M} f_k(x_i) \tag{4}$$

where f_k is a regression tree (also known as CART). Since the parameters of the model are functions, we cannot use optimization algorithms like SGD. Instead, the model is trained in an additive manner. Formally, let $\hat{y}_i^{(t-1)}$ be the prediction of the ith instance at the $(t-1)^{th}$ iteration, we will need to add f_t to minimize the following objective function.

$$\mathcal{L}^t(\phi) = \sum_{i=1}^{N} l(y_i, \hat{y}_i^{(t-1)} + f_t(x_i)) + \Omega(f_t) \tag{5}$$

$$\text{where } \Omega(f_t) = \gamma T + \frac{1}{2}\lambda \|w\|^2$$

where l is a convex loss function that measures the difference between the target y_i and prediction $\hat{y}_i^{(t-1)}$, regularizer $\Omega(f_t)$ is used to penalize the complexity of the tth tree model. Equation 4 means we should add the f_t that most minimizes $\mathcal{L}^t(\phi)$. In practice, Second-order approximation can be used to quickly optimize the objective. By performing a second-order Taylor approximation expansion of the loss function, we can approximate the objective function as following.

$$\mathcal{L}^t(\phi) \simeq \sum_{i=1}^{N}[g_i f_t(x_i) + \frac{1}{2}h_i f_t^2(x_i)] + \Omega(f_t) \tag{6}$$

where $g_i = \partial_{\hat{y}_i^{(t-1)}} l(y_i, \hat{y}_i^{(t-1)})$, $h_i = \partial_{\hat{y}_i^{(t-1)}}^2 l(y_i, \hat{y}_i^{(t-1)})$.

Therefore, we can transform packet features into one-hot vectors as the input of Xgboost and set the parameter M for training. Moreover, the Xgboost model is interpretable, and we can evaluate the importance of features based on the number of times they are selected as split points.

3.2   The Second Stage of MGEL

In the second stage, we first train the meta learner using the output $P_{train} \in \mathbb{R}^{M_{train} \times 3}$ of the three base learners on the training dataset. For simplicity, we choose logistic regression as the meta learner, which can be represented as:

$$Y = \sigma(f(P_{train})) \tag{7}$$

where $\sigma(\cdot)$ is activation function, f is the parameter to be learned, $Y \in \mathbb{R}^{M_{train} \times 1}$.

After the parameters are trained, we use $P_{test} \in \mathbb{R}^{M_{train} \times 3}$ as input to predict the performance of our MGEL model on the test dataset.

4 Experment and Results

4.1 Experiment Settings

Dataset. The first dataset (*CIC-InvesAndMal2019*) is regenerated from *CIC-InvesAndMal2019* [14] which includes 10 different families of Ransomware, such as Charger family, Jisut family, Koler family, *etc.* A total of 3,797,000 data packets and 63,953 flows are included in the malware sample. The benign sample comes from normal Android applications, including 13,474,342 data packets and 69,670 flows.

The second dataset (*MTA*) is regenerated from malware-traffic-analysis.net, which is a website that provides real-time updates on current malware prevalent in Europe and the United States. We collected malware samples from 2019 to 2020, totaling 10,793,979 data packets and 50,289 flows. Since the normal sample is not provided on the website, we captured 9,990,438 data packets and 50,243 flows on the normal applications of Android.

Evaluation Metrics. We evaluate and compare our model with the state-of-the-art methods using four metrics. Namely, Accuracy, Precision, Recall, and F-Measure:

$$Accuracy = \frac{TP + TN}{TP + FP + FN + TN} \tag{8}$$

$$Recall = \frac{TP}{TP + FN} \tag{9}$$

$$Precision = \frac{TP}{TP + FP} \tag{10}$$

$$F - Measure = \frac{2 Precision * Recall}{Precision + Recall} \tag{11}$$

where TP (True Positive) represents malware traffic is correctly identified as malware traffic; FP (False Positive) represents benign traffic is incorrectly identified as malware traffic; TN (True Negative) represents benign traffic is correctly identified as benign traffic; FN (False Negative) represents malware traffic is incorrectly identified as benign traffic.

Experimental Settings and Baselines. For the base learner based on flow features, We use the Adam optimizer with a batch size of 128, for training where the learning rate was set to 5e−3. For the base learner based on Xgboost, we choose the maximum depth of the tree to be 3 and the number of trees to be 160.

We compare our methods with other state-of-the-art methods as follows:

(1) FS-Net uses a multi-layer encoder-decoder structure and reconstruction mechanism to identify malware flows from flow sequences (message type sequences or packet length sequences) [10].
(2) MAMPF uses Random Forest to identify malware flows with features learned from message type and length block Markov models [9].
(3) FoSM constructs a first-order Markov model using message sequences to discriminate the class of traffic based on the maximum likelihood probability [7].
(4) SoSM is similar to FoSM, but utilizes a second-order Markov model [13].
(5) 2D-CNN converts raw traffic data into an image and use two-dimensional CNN for classification [17].
(6) 1D-CNN views raw traffic data as an article and extracts features using a 1D CNN and Max Pooling layer, and finally classifies them using a softmax layer [16].

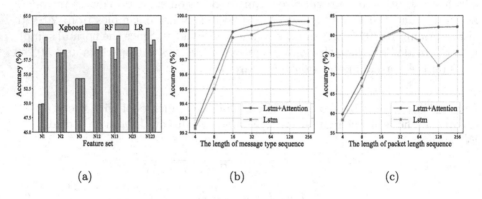

(a)   (b)   (c)

Fig. 4. The experimental results of MGEL key component

4.2   Analysis on MGEL Component

We explore several key points of each base learner in this section. For the base learner based on packet features, there are two main problems. The first one is which packet features should be selected, and the second one is which classifier should be chosen.

In Fig. 4(a), We compare the performance of Xgboost, Random Forest(RF) and Logistic Regression(LR) on the *CIC-InvesAndMal2019* with different combinations of features. Due to space constraints, we use *N1*, *N2*, *N3* accordingly to denote the client-side cipher suite and extension fields, the server-side cipher suite and extension fields, and the public key length. The combination of numbers indicates the combination of features, *e.g.*, *N12* means we use both client-side and server-side cipher suites and extension fields. According to the experimental results, we can see that the Xgboost method can obtain the highest accuracy. It can also be seen that as more features are added, the accuracy of the model is improved except for Logistic Regression. This is attributed to the fact that the tree-based model is more adept at dealing with missing features. Therefore, we choose Xgboost as the base learner and *N123* as the packet features.

For the base learner based on flow features, we focus on exploring the effect of sequence length and the self-attention mechanism on the accuracy of the model.

In Fig. 4(b) and Fig. 4(c), we take the length of the message type sequence and packet length sequence as [4, 8, 16, 32, 64, 128, 256], respectively, and compare the accuracy under the *CIC-InvesAndMal2019* dataset with and without attention layer. From the experimental results, it can be seen that the introduction of the self-attention layer can improve the accuracy of the model. This becomes more obvious as the sequence length increases, which is attributed to the fact that the self-attention mechanism is stronger than LSTM in solving the long-term dependence problem. Moreover, we can find that when the sequence length exceeds 128, the accuracy improvement of the model is no longer obvious. Therefore, we set the sequence length of both MGEL-1 and MGEL-2 to 128.

4.3   Comparison Results

We conducted comparative experiments on two public datasets using the six state-of-the-art methods mentioned in Sect. 4.1, and the experimental results are shown in Table 1 and Table 2. From Table 1 and Table 2, we can obtain the following conclusions:

(1) MGEL achieves the best performance and outperforms all the other methods on all the overall metrics. In addition, we can see that MGEL can reach more than 99% of F1 values on both two public datasets.
(2) Models using a single feature (FS-Net, FoSM, SoSM) have unstable performances on the two public datasets. For example, FS-Net has higher accuracy on *MTA* dataset, while FoSM and SoSM have better performance on *CIC* dataset. This is due to the different sensitivity of data to features, further confirming the importance of using multi-grained features.
(3) Deep learning models using raw byte data (1D-CNN, 2D-CNN) and multi-attribute features models(MAMPF) have a robust performance on both public datasets, but in general worse than MGEL. This is because of the advantage of self-attention and stacking ensemble method.

4.4   The Advanced of Ensemble Learning

In order to verify the advanced of the ensemble method, we compared the performance of a single model on two public datasets. Table 3 is our results on the *CIC-InvesAndMal2019* dataset, and Table 4 is our results on the *MTA* dataset. It can be seen from Table 3 and Table 4 that a base learner trained using only a single feature cannot achieve the best performance on different datasets. With the stacking ensemble method, the adaptability of features on different datasets can be adjusted adaptively, so that the best performance can be achieved on different datasets.

Table 1. Comparison results on *CIC-InvesAndMal2019* dataset

Method	Accuracy	Precision	Recall	F-Measure
MGEL	**0.9988**	**0.9998**	**0.9978**	**0.9988**
FS-Net	0.7953	0.7760	0.8859	0.8273
MAMPF	0.9332	0.9184	0.9613	0.9394
FoSM	0.933	0.9765	0.8976	0.9354
SoSM	0.9587	0.9558	0.9680	0.9619
2D-CNN	0.744	0.7429	0.7463	0.7446
1D-CNN	0.7424	0.694	0.7684	0.7293

Table 2. Comparison results on *MTA* dataset

Method	Accuracy	Precision	Recall	F-Measure
MGEL	**0.9999**	**0.9999**	**1.0**	**0.9999**
FS-Net	0.9998	0.9998	0.9997	0.9998
MAMPF	0.9984	0.9989	0.9977	0.9983
FoSM	0.9772	0.9904	0.9599	0.9749
SoSM	0.9780	0.9955	0.9566	0.9577
2D-CNN	0.9949	0.9958	0.9934	0.9946
1D-CNN	0.9970	0.9947	0.9991	0.9969

Table 3. The results of advanced analysis on *CIC-InvesAndMal2019* dataset

Method	Accuracy	Precision	Recall	F-Measure
MGEL	**0.9988**	**0.9998**	**0.9978**	0.9988
Base-learner 1	0.9988	0.9998	0.9978	**0.9998**
Base-learner 2	0.7403	0.7634	0.6964	0.7284
Base-learner 3	0.6221	0.6406	0.5502	0.5963

Table 4. The results of advanced analysis on *MTA* dataset

Method	Accuracy	Precision	Recall	F-Measure
MGEL	**0.9999**	0.9999	**1.0**	**0.9999**
Base-learner 1	0.9672	0.9661	0.9683	0.9672
Base-learner 2	0.9980	0.9984	0.9969	0.9977
Base-learner 3	0.9908	**1.0**	0.9817	0.9907

5   Conclusion and Future Work

In this paper, we propose an ensemble learning approach with multi-grained features for malware encrypted traffic identification. It jointly trains three base learners using flow features and packet features, and finally ensembles the three base learners by a stacking way. Since the three base learners use multi-grained features for learning, they can focus on different parts of the data. Moreover, our model is more robust on different scenarios by ensembling base learners through stacking. We validate the effectiveness of the MGEL on two public datasets, and the experimental results demonstrate that the MGEL can achieve an excellent identification performance and outperform other state-of-the-art methods. In the future, we would like to explore the performance of the model on multi-classification tasks. In addition, we would like to apply for the recent advances in the field of deep learning to improve the performance of traffic classification.

References

1. Anderson, B., McGrew, D.: Identifying encrypted malware traffic with contextual flow data. In: Proceedings of the 2016 ACM Workshop on Artificial Intelligence and Security, pp. 35–46 (2016)
2. Anderson, B., Paul, S., McGrew, D.: Deciphering malware's use of TLS (without decryption). J. Comput. Virol. Hack. Tech. **14**(3), 195–211 (2018)
3. Chen, T., Guestrin, C.: Xgboost: a scalable tree boosting system. In: Proceedings of the 22nd ACM SIGKDD International Conference on Knowledge Discovery and Data Mining, pp. 785–794 (2016)
4. Constantinou, F., Mavrommatis, P.: Identifying known and unknown peer-to-peer traffic. In: Fifth IEEE International Symposium on Network Computing and Applications (NCA 2006), pp. 93–102. IEEE (2006)
5. Dierks, T., Rescorla, E.: The transport layer security (TLS) protocol version 1.2 (2008)
6. Freier, A., Karlton, P., Kocher, P.: The secure sockets layer (SSL) protocol version 3.0. Technical report, RFC 6101 (2011)
7. Korczyński, M., Duda, A.: Markov chain fingerprinting to classify encrypted traffic. In: IEEE INFOCOM 2014-IEEE Conference on Computer Communications, pp. 781–789. IEEE (2014)
8. Li, R., Xiao, X., Ni, S., Zheng, H., Xia, S.: Byte segment neural network for network traffic classification. In: 2018 IEEE/ACM 26th International Symposium on Quality of Service (IWQoS), pp. 1–10. IEEE (2018)
9. Liu, C., Cao, Z., Xiong, G., Gou, G., Yiu, S.M., He, L.: MaMPF: encrypted traffic classification based on multi-attribute Markov probability fingerprints. In: 2018 IEEE/ACM 26th International Symposium on Quality of Service (IWQoS), pp. 1–10. IEEE (2018)
10. Liu, C., He, L., Xiong, G., Cao, Z., Li, Z.: FS-net: a flow sequence network for encrypted traffic classification. In: IEEE INFOCOM 2019-IEEE Conference on Computer Communications, pp. 1171–1179. IEEE (2019)
11. Park, J.S., Yoon, S.H., Kim, M.S.: Performance improvement of payload signature-based traffic classification system using application traffic temporal locality. In: 2013 15th Asia-Pacific Network Operations and Management Symposium (APNOMS), pp. 1–6. IEEE (2013)

12. Rezaei, S., Liu, X.: Deep learning for encrypted traffic classification: an overview. IEEE Commun. Mag. **57**(5), 76–81 (2019)
13. Shen, M., Wei, M., Zhu, L., Wang, M., Li, F.: Certificate-aware encrypted traffic classification using second-order Markov chain. In: 2016 IEEE/ACM 24th International Symposium on Quality of Service (IWQoS), pp. 1–10. IEEE (2016)
14. Taheri, L., Kadir, A.F.A., Lashkari, A.H.: Extensible android malware detection and family classification using network-flows and API-calls. In: 2019 International Carnahan Conference on Security Technology (ICCST), pp. 1–8. IEEE (2019)
15. Vaswani, A., et al.: Attention is all you need. arXiv preprint arXiv:1706.03762 (2017)
16. Wang, W., Zhu, M., Wang, J., Zeng, X., Yang, Z.: End-to-end encrypted traffic classification with one-dimensional convolution neural networks. In: 2017 IEEE International Conference on Intelligence and Security Informatics (ISI), pp. 43–48. IEEE (2017)
17. Wang, W., Zhu, M., Zeng, X., Ye, X., Sheng, Y.: Malware traffic classification using convolutional neural network for representation learning. In: 2017 International Conference on Information Networking (ICOIN), pp. 712–717. IEEE (2017)
18. Zejdl, P., Ubik, S., Macek, V., Oslebo, A.: Traffic classification for portable applications with hardware support. In: 2008 International Workshop on Intelligent Solutions in Embedded Systems, pp. 1–9. IEEE (2008)
19. Zou, Z., Ge, J., Zheng, H., Wu, Y., Han, C., Yao, Z.: Encrypted traffic classification with a convolutional long short-term memory neural network. In: 2018 IEEE 20th International Conference on High Performance Computing and Communications; IEEE 16th International Conference on Smart City; IEEE 4th International Conference on Data Science and Systems (HPCC/SmartCity/DSS), pp. 329–334. IEEE (2018)

TS-Bert: Time Series Anomaly Detection via Pre-training Model Bert

Weixia Dang[1,2], Biyu Zhou[1,2(✉)], Lingwei Wei[1,2], Weigang Zhang[1,2], Ziang Yang[1,2], and Songlin Hu[1,2]

[1] School of Cyber Security, University of Chinese Academy of Sciences, Beijing, China
[2] Institute of Information Engineering, Chinese Academy of Sciences, Beijing, China
zhoubiyu@iie.ac.cn

Abstract. Anomaly detection of time series is of great importance in data mining research. Current state of the art suffer from scalability, over reliance on labels and high false positives. To this end, a novel framework, named TS-Bert, is proposed in this paper. TS-Bert is based on pre-training model Bert and consists of two phases, accordingly. In the pre-training phase, the model learns the behavior features of the time series from massive unlabeled data. In the fine-tuning phase, the model is fine-tuned based on the target dataset. Since the Bert model is not designed for the time series anomaly detection task, we have made some modifications thus to improve the detection accuracy. Furthermore, we have removed the dependency of the model on labeled data so that TS-Bert is unsupervised. Experiments on the public data set KPI and yahoo demonstrate that TS-Bert has significantly improved the f1 value compared to the current state-of-the-art unsupervised learning models.

Keywords: Anomaly detection · Pre-training model · Time series analysis

1 Introduction

In large-scale distributed systems, monitoring is inevitable. Analyzing monitoring data is of great significance to ensure the quality of service of the system and protect corporations against malicious attacks. From the perspective of data analysis, this means that we need to perform real-time analysis on large volume of time series generated by monitoring systems in order to detect potential errors and anomalies.

However, anomaly detection on large volume of time series is not easy. Specifically, it has the following challenges:

Challenge 1: *Efficiency.* Due to the large scale of the problem, time series anomaly detection algorithms require rapid detection of anomalies. In industrial applications, time series anomaly detection systems need to process hundreds of millions of time series in real time.

© Springer Nature Switzerland AG 2021
M. Paszynski et al. (Eds.): ICCS 2021, LNCS 12743, pp. 209–223, 2021.
https://doi.org/10.1007/978-3-030-77964-1_17

Challenge 2: *Generalization*. The patterns of time series are diverse, it is time-consuming, laborious and unrealistic to build a model for each pattern of time series. The generalization ability of industrial anomaly detection services is important to work well on various patterns of time series.

Challenge 3: *Lack of labels*. In industrial applications, obtaining large-scale, high-quality time series anomaly detection label data requires a lot of manpower and material resources. This means that algorithms that overly rely on labels are practically infeasible.

We argue that a good time series anomaly detection algorithm should be efficient, unsupervised, and can be adapted to most scenarios with some simple adjustments. Unfortunately, existing proposals [5,8–10,15,17,20,23] fail to adequately meet all these requirements. According to our observations, feature extraction and long-distance dependent modeling have great influence on the accuracy of time series anomaly detection. However, the widely adopted feature extraction models for sequences such as CNN [22], RNN [4] and LSTM [12] do not deal well with long-distance dependencies. In this paper, we propose to model time series by referring to the Bert [7] model in natural language processing (NLP) field. The reasons are three folds: (1) The core algorithm of Bert is Transformer, which can solve the long-distance dependence issue by orchestrating self-attention modules. (2) Bert is a pre-training model [7,13], which can learn effectively from large-scale raw text to alleviate the dependence on supervised learning during the pre-training phase. This pattern can be borrowed to time series abnormal detection to solve the problem of missing tags. Besides, the pattern of pre-training is essentially transfer learning. By fine-tuning with target datasets, it can be applied to many scenarios, thus is more generic. (3) Bert is a thorough open source framework which can be used directly, thus saving development costs.

Based on the above analysis, we propose a pre-training model TS-Bert based on the Bert model in NLP to solve the time series anomaly detection problem in this paper. Technically, TS-Bert involves two phases: pre-training and fine-tuning. In the pre-training phase, the model learns the behavior features of the time series from a large amount of unlabeled data. Note that the pre-training task includes masked LM and next sentence prediction. In the fine-tuning phase, the parameters of the pre-training model is tuned according to the specific time series anomaly detection task. Since the Bert model is not designed for this task, we have made some modifications to the model to improve the detection accuracy. It is worth noting that although the original Bert model is supervised, TS-Bert is an unsupervised model. By using SR [15] method to generate labels, we remove the dependency of the model on labeled data. Moreover, the model can perform better when partial labeled data is provided.

The major contributions are highlighted as below:

- For the first time, we introduce the pattern of pre-training and fine-tuning to the field of time series anomaly detection.
- We propose to adopt the Bert model in NLP field to model time series thus can address the long-distance dependent modeling issue. Accordingly, we solve

the problems of mapping the Bert model to the time series anomaly detection task, such as large knowledge span, mismatched input format, etc.
- We conducted extensive experiments to verify the performance of TS-Bert. Simulation results on two widely used public datasets demonstrate that our method is 21% and 11% more accurate on KPI dataset and yahoo dataset than the state-of-the-art solution.

2   Related Work

This section summarizes the work closely relevant to this paper.

The traditional statistical model [5,11,16] such as ARIMA [23] and Holt-winter [5], typically utilize handcrafted features to model normal/anomaly patterns. These methods are only applicable to time series data that conform to certain characteristics, with poor universality.

Supervised anomaly detection [8–10,17] using labeled data to train anomaly detection classifiers. With the rapid development of machine learning and data mining, many time series anomaly detection methods based on supervised learning models [8–10,17] have been proposed. However, obtaining large-scale, high-quality time series anomaly detection label data requires a lot of manpower and material resources. This means that anomaly detection algorithms based on supervised learning are not feasible in large-scale industrial applications.

Unsupervised anomaly detection does not require labels during the training process which alleviates the dependence of supervised learning. In recent years, many unsupervised time series anomaly detection algorithms have been proposed. VAE [20] is a reconstruction models, which learns the representation for normal time series by reconstructing the original input based some latent variables. SR [15] combines the benefits of Spectral Residual and convolutional neural network to detect anomalies in time series. Current state of the art anomaly detection approaches based on unsupervised models suffer from scalability and a large number of false positive compromising user experience.

Pre-training model has received widespread attention in recent years [7,13]. It designs pre-tasks to learn rich semantic representations from a large amount of unlabeled data and the learned data representations can be transferred to downstream tasks. In this way, the problem of label data shortage can be solved. BERT is a pre-training learning model that obtains the the state-of-the-art results in various natural language processing tasks [6,18,19,21]. As far as we know, TS-Bert is the first pre-trained model for the field of time series anomaly detection.

3   Methodology

Time series anomaly detection (TSAD) is to check whether the current data has obviously deviated from the normal situation through historical data analysis. The TSAD problem can be defined as follows.

Problem Definition: The input of TSAD is denoted by $T \in R^n$, where n is the length of timestamps. The task of TSAD is to produce an output vector $Y \in R^n$, where $y_i \in \{0,1\}$, 0 means normal, and 1 means abnormal.

We propose TS-Bert to address the TSAD problem. The framework of TS-Bert is shown in Fig. 1. To improve the robustness of our model, all the time series datasets are normalized before use. In the pre-training phase, the model will learn the behavior features of the time series from massive unlabeled data. After that the parameters of the pre-trained model will be fine-tuned according to the target dataset. Since the Bert model is not designed for the TSAD task, we have made some modifications to the model thus to improve the detection accuracy. Furthermore, we removed the dependency of the model on labeled data so that TS-Bert is unsupervised. In the following, we will describe TS-Bert in detail. The notations used is summarized on Table 1.

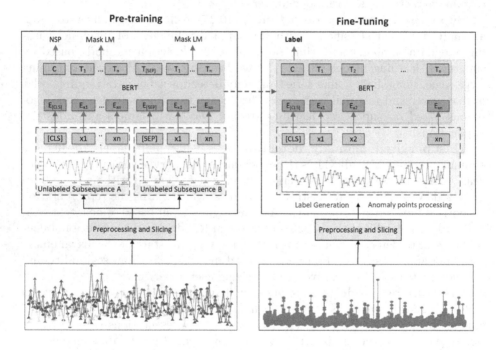

Fig. 1. The framework of TS-Bert.

3.1   Data Preprocessing

In our method, all the input time series are normalized with the maximum and minimum values. In order to map time series to the input format of the Bert model, we expand the normalized time series by *scale* times. The *scale* size is the

Table 1. Notations

Symbol	Description
T	A time series
n	The length of timestamps
Y	The output vector of an anomaly detection algorithm
$scale$	The normalization multiple
$\widetilde{T}$	The time series after preprocessing
p_w	The sequence length in the pre-training phase
h_w	The length of the historical information
s_w	The length of the sliding window when anomaly timestamp is replaced
H	The hidden size of Bert
C	The final hidden vector of Bert
θ	The threshold to generate anomaly detection results using SR
δ	The delay time

normalization multiple which approximately equals to the dictionary size used by the Bert model. The formula of data preprocessing stage is as follows.

$$\widetilde{T} = \frac{T - min(T)}{max(T) - min(T)} \times scale \qquad (1)$$

where $min(T)$ and $max(T)$ are the maximum and minimum value of the time series T respectively.

3.2   Pre-training

The task of the pre-training phase is to learn the behavioral features of time series from massive datasets. Our pre-training task follows the pre-training task of the Bert model, which includes Masked LM and Next Sentence Prediction (NSP). We masks some positions of the input sequence at random, and then predicts the values of those mask positions. Through Masked LM, we obtain a bidirectional pre-trained model. Next Sentence Prediction is to learn the relationships of different time series by judging whether two time series fragments are adjacent. We obtain pre-training data from a large volume of time series, and assuming that there exists no anomaly in the data.

Given k entries of time series $X = (x_1, x_2, ...x_k)$, where x_i is i-th time series with length L_i. We aim to get pre-training data $D = (d_1, d_2, ...d_k)$. The details of pre-training are as follows (also shown in Algorithm 1).

Step 1: For each time series x_i, we preprocess it through the method described in Sect. 3.1. The time series after preprocessing is denoted by $\widetilde{x_i}$.

Step 2: We slice $\widetilde{x_i}$ into subsequences d_i with length p_w. Obviously, $d_i = (s_1, s_2, ...s_m)$ and $m = \lceil \frac{L_i}{p_w} \rceil$. Mapped to the NLP field, d_i is equivalent to a document, and s_i is a sentence in the document.

Algorithm 1. Pre-training the Bert model using the time series data set.

Input: $X = (x_1, x_2, ...x_k)$, where x_i is i-th time series with length L_i; p_w: the subsequence length. scale: the normalization multiple.
Output: Pre-trained Bert model.
1: **for** i in range$(1, k)$ **do**
2:     $\widetilde{x}_i = \frac{x_i - min(x_i)}{max(x_i) - min(x_i)} \times scale$
3:     **for** j in range$(0, \lceil \frac{L_i}{p_w} \rceil)$ **do**
4:         $s_j = x_i[j \times p_w, ..., (j+1) \times p_w]$
5:         $d_i.append(s_j)$
6:     **end for**
7:     $D.append(d_i)$
8: **end for**
9: Build time series dictionary $V = (0, 1, ..., scale, [CLS], [SEP])$.
10: Map D to a format that the Bert model can understand.
11: Pre-train Bert according to the pre-training tasks.
12: Save the parameters of the pre-trained model.

Step 3: Next, we build a dictionary of time series. Time series dictionary $V = (0, 1, ..., scale, [CLS], [SEP])$. [CLS] is used as the aggregate sequence representation of the classification task. Bert learns the relationship between sequences by packing two sequences into a single sequence, and [SEP] is used to separate the two sequences.

Step 4: Then, we map s_i to a format that the Bert model can understand. We round the timestamp values of s_i and map them to the dictionary to get the encoding in the dictionary, named input_ids.

Step 5: Finally, we pre-train Bert according to the pre-training tasks Masked LM and NSP to obtain a pre-trained model.

3.3 Fine-Tuning

The purpose of fine-tuning is to fine-tune the parameters of the pre-trained model according to the target dataset. We plug the inputs of TSAD task and outputs into Bert and fine-tune all the parameters end-to-end. Given a time series T with n timestamps, the detailed process of fine-tuning is shown below (also illustrated in Algorithm 2).

Data Preprocessing. We preprocess T through the method described in Sect. 3.1. The time series after preprocessing is named $\widetilde{T}$.

Label Generation. The fine-tuning of the Bert model requires labels. However, obtaining large-scale, high-quality time series anomaly detection label data requires a lot of manpower and material resources. To alleviate this problem, we use the state-of-the-art anomaly detection method, Spectral Residual (SR) [15] to generate the labels in the training data on fine-tuning stage. Through SR, we get the label vector $Y \in R^n$, where $y_i \in \{0, 1\}$. SR is a light-weighted method which adds little overhead to the entire model.

Algorithm 2. Fine-tune based on time series anomaly detection tasks.

Input: T: time series; n: the length of time series; h_w: the history window size; s_w:
　　the slice window size;
Output: The Bert model after fine tuning.
　1: Build time series dictionary V = (0, 1, ..., scale, [CLS], [SEP]).
　2: **for** i in range(0, n) **do**
　3:　　$\widetilde{T}_i = \frac{T_i - min(T)}{max(T) - min(T)} \times scale$
　4:　　$Y_i = SR(\widetilde{T}_i)$
　5:　　**if** $Y_i == 0$ **then**
　6:　　　$T_i^* = \widetilde{T}_i$
　7:　　**else**
　8:　　　$T_i^* = \frac{\sum_{j=0}^{s_w - 1} \widetilde{T}_{i - s_w + j}}{s_w}$
　9:　　**end if**
　10:　　**if** $i \leq h_w$ **then**
　11:　　　$S_i = [T_{i-h_w+1}^*, T_{i-h_w+2}^*, \cdots, T_i]$
　12:　　**else**
　13:　　　$S_i = [T_0, \cdots, T_1^*, \cdots, T_i]$
　14:　　**end if**
　15:　　Map S_i to a format that the Bert model can understand.
　16: **end for**
　17: Add additional output layer to the Bert.
　18: Fine tune the model according the feature.
　19: **return** The Bert model after fine tuning.

Time Series Slicing and Cleaning. We detect anomaly events at the timestamp t based on S_i, which contains the historical information and the value of the time series at t. The length of the historical information we take is h_w. The prerequisite of high accuracy of the model is that there are no abnormal points in the historical information. However, in practice, historical information is likely to contain abnormal points. Therefore, we need to clean these abnormal points. Specifically, the anomaly timestamps are replaced with the mean value of the sliding window s_w on the left of the sequence. The length of s_w cannot exceed h_w. The calculation of S_i is shown in the following formula.

$$S_i = \begin{cases} [T_{i-h_w+1}^*, T_{i-h_w+2}^*, \cdots, T_i] & i > h\_w \\ [T_0, \cdots, T_1^*, \cdots, T_i] & i \leq h\_w \end{cases} \tag{2}$$

$$T_j^* = \begin{cases} T_j & j \text{ is nomal} \\ \frac{\sum_0^{s_w - 1} T_{j-s_w+i}}{s_w} & j \text{ is anomal} \end{cases} \tag{3}$$

Input Representation Mapping. First, we build a time series dictionary, and then map $S[i]$ to the dictionary, named input_ids. The creation of the dictionary and mapping method to dictionary is the same as the pre-training stage. Then, we get the input representation of Bert according input_ids.

The Addition of a Decoder. To fine-tune Bert on time series anomaly detection dataset, we use the final hidden vector $C \in R^H$(H is the hidden size of

Bert) as the aggregate representation. The new parameters introduced during fine-tuning are classification layer weights $W \in R^{2 \times H}$. We compute a standard classification loss with C and W, i.e., $log(softmax(CW^T))$.

4 Evaluation

In this section, we conduct simulations to verify the effectiveness of TS-Bert. Firstly, the evaluation settings are described. Then the evaluation results on public datasets are provided.

4.1 Settings

Datasets. Two public datasets are adopted in the evaluations, namely KPI dataset [1] and Yahoo dataset [2]. The detailed statistics and settings of these two datasets are shown in Table 2. Figure 2 and Fig. 3 illustrate a sequence in the two datasets, respectively.

Table 2. Statistics of datasets

Dataset	Curves	Training set size	Testing set size	Anomaly Rate
KPI	58	3004066	2918847	134114/2.26%
Yahoo	367	286483	286483	3896/0.68%

KPI dataset collects KPI (Key Performance Indicators) data from many Internet companies after desensitization. These KPI can be roughly divided into two types: service KPI and machine KPI. Service KPI is a performance indicators reflecting the scale and quality of Web services. Machine KPI refers to the performance indicators that can reflect the health status of the machine.

Yahoo dataset is an open dataset for anomaly detection released by Yahoo lab. Part of the time series is synthetic (i.e., simulated); while the other part comes from the real traffic of Yahoo services. The anomaly points in the simulated curves are algorithmically generated and those in the real-traffic curves are

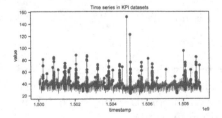
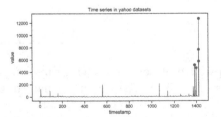

Fig. 2. A user's time series in KPI data set.   **Fig. 3.** A time series in Yahoo data set

labeled by editors manually. A time series from the real traffic of Yahoo services is shown in Fig. 3. Since the Yahoo data set does not divide the training set and the test set. We randomly divide the time series of the Yahoo dataset into two halves, the first half is utilized for fine-tuning the pre-trained model Bert while the second half is leveraged for evaluation.

Benchmarks. We compare TS-Bert with state-of-the-art models for TSAD, including SR [15], SR+CNN [15], FFT [14], DONUT [20]. In view of the fact that the training process of SR+CNN requires normal data provided by Microsoft, which cannot be obtained, we directly quote the result of [15].

Metrics. The usual metrics used in the TSAD problem are precision, recall and F1-score. In this paper, we also use this set of metrics to indicate the performance of our model. For ease of reference, we use Metrics-1 to represent it.

In practice, anomalous observations usually form contiguous segments since they occur in a contiguous manner. In this case, the authors in [1] propose to use a "delayed" manner to calculate TP (true positive), TN(true negative), FP (false positive), FN (false negative). Assume that the delay is δ, the calculation rules of TP, TN, FP and FN are as follows.

- *For a marked continuous abnormal segments:* If there is an anomaly point detected correctly and its timestamp is at most δ steps after the first anomaly of the segment, it is considered that the anomaly detection algorithm has successfully detected the whole continuous abnormal interval, so each abnormal point in the abnormal segment is counted as TP; otherwise, each abnormal point in the continuous abnormal segment is counted as FN.
- *For time points without abnormal markings:* If the anomaly detection algorithm outputs an anomaly, it is counted as FP; otherwise, it is counted as TN.

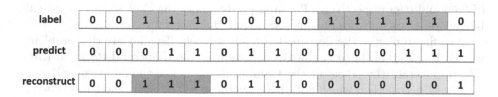

Fig. 4. Illustration of the evaluation protocol. There are 15 contiguous points in the time series, where the first row indicates the labels list of time series; the second row shows the anomaly detection results; and the third row shows adjusted results according to the evaluation protocol.

When $\delta = 2$, the illustration for the evaluation protocol are shown in Fig. 4. There are two abnormal segments in the original time series. When $\delta = 2$, the anomaly detection algorithm successfully detected the first continuous abnormal segment, but failed to detect the second continuous abnormal segment.

This set of metrics is also adopted in this paper to indicate the performance of our model. For ease of reference, we use Metrics-2 to represent it.

Parameter Settings. The frame structure of the Bert model we used is consistent with the uncased L12 H-768 A12($Bert^{Base}$) released by Google. It contains 12-layer, 768-hidden, 12-heads with 110M parameters for the encoder. In our model, we set *scale* as 30000 which is approximately equal to $Bert^{Base}$ dictionary size. We set the sequence length p_w in the pre-training phase as 50. p_w is equal to half of the maximum sequence length in the fine-tuning phase, because the pre-training phase packs them into a sequence when learning the relationship between the two sequences. The history length h_w is set to 100, which limits the size of historical data. The length of the sliding window when anomaly timestamp is replaced s_w is set as 1 in KPI dataset, 5 in Yahoo dataset through a grid search.

We obtain pre-training data from a large volume of time series, and assuming that there exists no anomaly in the data. The pre-training data set is selected from the normal segments of the KPI and Yahoo datasets, and Cluster-trace-v2018 [3] of provided by the Alibaba Open Cluster Trace Program. Cluster-trace-v2018 includes about 4000 machines in a periods of 8 days. There are both online services (aka long running applications, LRA) and batch workloads colocated in every machine in the cluster. Over the past year, the scale of colocation of online services and batch workloads have greatly increased resulting in the improvement of the overall resource utilization.

The pre-training data contains 4530 time series. According to p_w, the time series are slicing into 320838 subsequences with length 50 which contain 16041900 timestamps totally. The pre-training tasks are consistent with the original Bert model, which include Masked LM and Next Sentence Prediction (NSP). The hyperparameter settings for pre-training are shown in Table 2. After pre-training, the pre-training dataset on the Masked LM and NSP tasks has an accuracy of 0.69 and 0.99 respectively. Through pre-training, the Bert model learns the behavioral characteristics of time series from massive data.

The fine-tuning of the Bert model requires labels. To alleviate this problem, we use the state-of-the-art anomaly detection method, Spectral Residual (SR) [15] to generate the labels in the training data on fine-tuning stage. SR is a light-weighted method which adds little overhead to the entire model. Following [15], we set the threshold θ as 3 to generate anomaly detection results using SR. The hyper-parameter max_seq_length in the fine-tuning stage is the same as the pre-training stage, and the other hyper-parameters as $Bert_{Base}$ (Table 3).

Table 3. Hyperameter settings for pre-training

Parameter name	Description	Values
max_seq_length	The maximum total input sequence length	100
max_predictions_per_seq	The maximum number of masked LM per sequence	20
train_batch_size	Total batch size for training	32
eval_batch_size	Total batch size for eval	8
learning_rate	The initial learning rate for Adam	5e−5
num_warmup_steps	The number of warmup steps	10000

4.2 Results

Comparison with SOTAs. For TS-Bert, we set hyper-parameters as Table 3 described. As mentioned, anomalies usually occur in a continuous segment, thus it is also reasonable to use Metric-2 to evaluate the model's performance. We report *Precision, Recall* and F_1-*score* under Metric-2 separately for each dataset. We set the delay time $\delta = 7$ for KPI dataset, $\delta = 3$ for Yahoo dataset. As shown in Table 4, TS-Bert shows excellent generalization capability and achieves the best F1-scores consistently on two public datasets. Specifically, we achieve 21% and 11% improvement over the best state-of-the-art performance on KPI dataset and Yahoo dataset. Besides, we find that the detection time of TS-Bert is milliseconds, which can meet the needs of most online applications.

Table 4. Results on unsupervised model.

Model	KPI			Yahoo		
	Precision	Recall	F1-score	Precision	Recall	F1-score
FFT	0.382	0.534	0.445	0.599	0.301	0.400
DONUT	0.371	0.326	0.347	0.013	0.825	0.026
SR	0.782	0.675	0.725	0.535	0.780	0.634
SR+CNN	0.797	0.747	0.771	0.816	0.542	0.652
TS-Bert	0.897	0.976	**0.935**	0.801	0.664	**0.726**

Intuitively, using the labels generated by the SR [15] method and training the model, the accuracy of the model is at most infinitely close to the SR method. However, before fine-tuning, we used large-scale time series data to learn the semantic information within and between segments of the normal time series. Fine-tuning only uses a small amount of data set to fine-tune the model parameters after pre-training. If the knowledge learned in the pre-training stage is sufficient, the model is sufficient to learn normal time series features. Therefore, fine-tuning on a pre-trained model using large-scale data, the anomaly detection results obtained is better than the label generation method used in the fine-tuning process.

In the previous experiments, we can see that the TS-Bert model shows convincing results in the case of unsupervised anomaly detection. In addition, we can obtain more satisfactory results when the exception label is available. For example, using only half of the training set labels to fine-tune Bert on the KPI data set, the F_1 scores reached 0.938 on the test dataset.

Impact of Indicator Metrics. In practice, anomalies usually occur in a continuous segment thus it is also reasonable to use Metric-2 to evaluate the model's performance. But we require a model to detect anomalies as soon as possible to take quick actions, if the delay of Metric-2 is too long, it will damage the user experience.

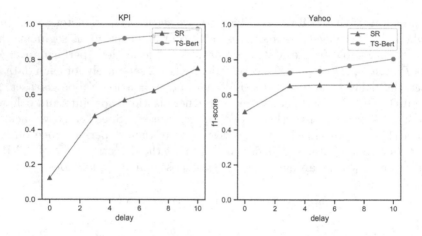

Fig. 5. Comparison with different delay value.

In this subsection, we compare TS-Bert with the state-of-the-art anomaly detection method SR under Metric-2 with different delay values. Figure 5 illustrates the F_1 scores of TS-Bert and SR for different delay value of δ on two datasets. Notice that the F_1 scores becomes larger as the delay δ increases. Overall, our model TS-Bert achieves better performance, even when the acceptable delay is very small.

Impact of s_w. We also analyze the influence of s_w to the performance of TS-Bert. We use the historical information and the current value to detect anomaly events, which is sensitive to irregular and abnormal instance in the history data. To alleviate this problem, we replace those anomaly timestamps in the history data with the mean value of the sliding window s_w on the left of the sequence. The length of s_w cannot exceed h_w.

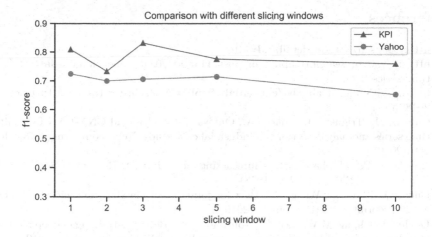

Fig. 6. Comparison with different slicing windows.

In order to exclude the impact of evaluation indicators on the results, we use the Metric-1 as the indicator. Figure 6 shown the F_1 scores of TS-Bert for different s_w metrics on two dataset. We notice that different settings of s_w achieve similar performer on two datasets. That is, when the value of s_w is relatively small, the mean value of the sliding window is closer to the normal value of the point. When s_w is relatively large, the mean value of the sliding window is quite different from the normal value of the point. For example, if s_w is set as 100, the F_1 scores only 0.31 on KPI dataset.

5 Conclusion

In this paper, we introduce the pattern of pre-training and fine-tuning to the field of TSAD and propose a novel framework TS-Bert based on Bert model in NLP to solve the TSAD problem. TS-Bert includes two phases: pre-training and fine-tuning. Through pre-training on massive time series datasets, the TS-Bert model learns the behavioral characteristics of time series very well. We have made some modifications thus to improve the detection accuracy. In addition, to alleviate the problem of missing labeled data, we use the SR method to generate labels on the fine-tuning phase. Evaluation results show that TS-Bert outperforms the state-of-the-art solution on two public datasets. Specifically, we achieve 21% and 11% accuracy increase on KPI dataset and yahoo dataset, respectively.

Acknowledgements. Supported by the National Key Research and Development Program of China 2017YFB1010001.

References

1. http://iops.ai/dataset_detail/?id=10/
2. https://yahooresearch.tumblr.com/post/114590420346/a-benchmark-dataset-for-time-series-anomaly
3. https://github.com/alibaba/clusterdata/blob/v2018/cluster-trace-v2018/trace_2018.md
4. Canizo, M., Triguero, I., Conde, A., Onieva, E.: Multi-head CNN-RNN for multitime series anomaly detection: an industrial case study. Neurocomputing **363**, 246–260 (2019)
5. Chatfield, C.: The holt-winters forecasting procedure. J. Roy. Stat. Soc.: Ser. C (Appl. Stat.) **27**(3), 264–279 (1978)
6. Chen, Q., Zhuo, Z., Wang, W.: Bert for joint intent classification and slot filling. arXiv preprint arXiv:1902.10909 (2019)
7. Devlin, J., Chang, M.W., Lee, K., Toutanova, K.: Bert: pre-training of deep bidirectional transformers for language understanding. arXiv preprint arXiv:1810.04805 (2018)
8. Görnitz, N., Kloft, M., Rieck, K., Brefeld, U.: Toward supervised anomaly detection. J. Artif. Intell. Res. **46**, 235–262 (2013)
9. Laptev, N., Amizadeh, S., Flint, I.: Generic and scalable framework for automated time-series anomaly detection. In: Proceedings of the 21th ACM SIGKDD International Conference on Knowledge Discovery and Data Mining, pp. 1939–1947 (2015)
10. Liu, D., et al.: Opprentice: Towards practical and automatic anomaly detection through machine learning. In: Proceedings of the 2015 Internet Measurement Conference, pp. 211–224 (2015)
11. Lu, W., Ghorbani, A.A.: Network anomaly detection based on wavelet analysis. EURASIP J. Adv. Signal Process. **2009**, 1–16 (2008)
12. Malhotra, P., Ramakrishnan, A., Anand, G., Vig, L., Agarwal, P., Shroff, G.: LSTM-based encoder-decoder for multi-sensor anomaly detection. arXiv preprint arXiv:1607.00148 (2016)
13. Radford, A., Wu, J., Child, R., Luan, D., Amodei, D., Sutskever, I.: Language models are unsupervised multitask learners. OpenAI blog **1**(8), 9 (2019)
14. Rasheed, F., Peng, P., Alhajj, R., Rokne, J.: Fourier transform based spatial outlier mining. In: Corchado, E., Yin, H. (eds.) IDEAL 2009. LNCS, vol. 5788, pp. 317–324. Springer, Heidelberg (2009). https://doi.org/10.1007/978-3-642-04394-9_39
15. Ren, H., et al.: Time-series anomaly detection service at microsoft. In: Proceedings of the 25th ACM SIGKDD International Conference on Knowledge Discovery & Data Mining, pp. 3009–3017 (2019)
16. Said, S.E., Dickey, D.A.: Testing for unit roots in autoregressive-moving average models of unknown order. Biometrika **71**(3), 599–607 (1984)
17. Shipmon, D., Gurevitch, J., Piselli, P.M., Edwards, S.: Time series anomaly detection: detection of anomalous drops with limited features and sparse examples in noisy periodic data (2017)
18. Vig, J., Ramea, K.: Comparison of transfer-learning approaches for response selection in multi-turn conversations. In: Workshop on DSTC7 (2019)
19. Wu, X., Lv, S., Zang, L., Han, J., Hu, S.: Conditional BERT Contextual Augmentation. In: Rodrigues, J.M.F., et al. (eds.) ICCS 2019. LNCS, vol. 11539, pp. 84–95. Springer, Cham (2019). https://doi.org/10.1007/978-3-030-22747-0_7

20. Xu, H., et al.: Unsupervised anomaly detection via variational auto-encoder for seasonal KPIs in web applications. In: Proceedings of the 2018 World Wide Web Conference, pp. 187–196 (2018)
21. Yang, W., Zhang, H., Lin, J.: Simple applications of bert for ad hoc document retrieval. arXiv preprint arXiv:1903.10972 (2019)
22. Zhang, C., et al.: A deep neural network for unsupervised anomaly detection and diagnosis in multivariate time series data. In: Proceedings of the AAAI Conference on Artificial Intelligence, vol. 33, pp. 1409–1416 (2019)
23. Zhang, Y., Ge, Z., Greenberg, A., Roughan, M.: Network anomography. In: Proceedings of the 5th ACM SIGCOMM Conference on Internet Measurement, p. 30 (2005)

Relation Order Histograms as a Network Embedding Tool

Radosław Łazarz$^{(\boxtimes)}$ and Michał Idzik

Institute of Computer Science, AGH University of Science and Technology,
Kraków, Poland
{lazarz,miidzik}@agh.edu.pl

Abstract. In this work, we introduce a novel graph embedding technique called NERO (Network Embedding based on Relation Order histograms). Its performance is assessed using a number of well-known classification problems and a newly introduced benchmark dealing with detailed laminae venation networks. The proposed algorithm achieves results surpassing those attained by other kernel-type methods and comparable with many state-of-the-art GNNs while requiring no GPU support and being able to handle relatively large input data. It is also demonstrated that the produced representation can be easily paired with existing model interpretation techniques to provide an overview of the individual edge and vertex influence on the investigated process.

Keywords: Graph classification · Graph embedding · Representation learning · Complex networks

1 Introduction

Due to their expressiveness and flexibility, complex networks appear in numerous real-world applications, such as sociology, bibliometrics, biochemistry, or telecommunications (to name a few) [36]. However, there are practical issues (i.a. lack of a universal notion of similarity, the fact that there are no bounds on the total vertex number or the number of single vertex neighbours) that make utilisation of such data representation challenging and cumbersome, especially in the context of machine learning. Thus, finding a way to bridge the gap between the graph domain and the vector-driven mainstream science is one of the most important topics in structural pattern processing [4].

1.1 Related Solutions

Classical answers to that problem tend to utilise the concept of graph kernels—either explicit (bivariate functions whose results are equivalent to graph inner

---

The research presented in this paper was financed using the funds assigned by the Polish Ministry of Science and Higher Education to AGH University of Science and Technology.

M. Paszynski et al. (Eds.): ICCS 2021, LNCS 12743, pp. 224–237, 2021.
https://doi.org/10.1007/978-3-030-77964-1_18

products) or implicit (operators mapping the graphs into the vector spaces, where there is a wide range of kernel functions available). Notable examples include counting the number of matching shortest path in input graphs [3], comparing estimated distributions of graphlets (small subgraphs with up to 5 nodes) [28], or relying on pairwise Wasserstein distances between graphs in the dataset [30].

Recently, there was a resurgence of interest in graph representation learning. Deep convolutional neural networks achieved tremendous results in computer vision tasks and, as a consequence, numerous works tried to adapt the ideas behind their success to the graph analysis domain [36]. While differing in implementation details, those approaches are collectively referred to as graph neural networks (GNNs). Multiple directions are being currently pursued in the field, such as imitating the popular U-Net architectures [12], leveraging the transformer self-attention mechanism [22], designing adaptive pooling operators [35], employing Gaussian mixture models [16], or utilising anchor graphs [1].

Nonetheless, despite the considerable achievements of GNN techniques, one should remember that their effectiveness comes with a high demand for memory, computational power, and specialised hardware. Hence, it may still be of interest to seek progress with kernel methods, especially in case of large networks.

2 Basic Notations

Graph $G = (\mathbf{V}, \mathbf{E}, \mathbf{T_V}, \mathbf{T_E})$ is an ordered tuple comprised of a finite ordered set of vertices $\mathbf{V} = \{v_1, \ldots, v_{n_V}\}$, a finite ordered set of edges $\mathbf{E} = \{e_1, \ldots, e_{n_E}\} \mid \mathbf{E} \subseteq \mathbf{V} \times \mathbf{V}$, a finite ordered set of vertex trait functions $\mathbf{T_V} = \{t_{V_1}, \ldots, t_{V_{n_{T_V}}}\}$, and a set of edge trait functions $\mathbf{T_E} = \{t_{E_1}, \ldots, t_{E_{n_{T_E}}}\}$, such that $\forall_{t_V \in \mathbf{T_V}} \operatorname{dom}(t_V) = \mathbf{V}$ and $\forall_{t_E \in \mathbf{T_E}} \operatorname{dom}(t_E) = \mathbf{E}$. We call given trait a label if its codomain is discrete, or an attribute if it is continuous. Throughout this work we assume that all graphs are undirected, i.e. $(v, u) \in \mathbf{E} \implies (u, v) \in \mathbf{E}$.

A k-dimensional array of size m_1-by-$\cdots$-by-m_k is denoted as $\mathbf{A} = S^{m_1 \times \cdots \times m_k}$, where S is the set of possible element values. $\mathbf{A}_{i_1, \ldots, i_k}$ represents its element on position $i_1, \ldots, i_k$. Using $*$ as a position component represents taking all elements along the given dimension—e.g. $\mathbf{A}_{*, j}$ is the jth column of $\mathbf{A}$. The notation $(\mathbf{A}_{i_1, \ldots, i_k})_C$, where C is the set of conditions, is used to describe an array sub-fragment—e.g. $(\mathbf{A}_{i, j})_{1 \leq i \leq 3, 1 \leq j \leq 3}$ is the upper left 3-by-3 sub-array. Additionally, $\mathbf{0}^{m_1 \times \cdots \times m_k}$ is the m_1-by-$\cdots$-by-m_k array filled with zeros.

3 Relation Order Embedding

The main idea behind the proposed embedding method is inspired by previous research by Bagrow et al. [2] and the later improvements suggested by Czech et al. [6–8]. The former introduced the concept of a B-matrix—a way of portraying an arbitrary network that contains information about its structure at multiple levels of abstraction and allows to express between-graph similarity as the euclidean distance. For a given graph, the B-matrix $B_{k,l}$ contains the number of vertices that have a degree of l in the distance-k-graph corresponding to the kth matrix row. A distance-k-graph, in turn, contains all the original nodes and connects with edges those of them that were initially separated by the shortest path distance equal to k. The latter expanded the framework and demonstrated that it is possible to improve its performance by utilising the node-edge distance [6] or employing alternative centrality measures [7]. It also explored the possibilities of obtaining embeddings of large graphs in a distributed fashion [8]. However, all those attempts were focused on purely structural data and assumed the lack of labels or attributes, severely limiting their applicability.

This work aims to further develop the aforementioned techniques and adapt them to support graphs with any type of traits. It is based on the observation, that two crucial components are common for all of them: assembling the embeddings from feature histograms (thus making them invariant to changes in edge and node processing order) and doing so not only for a starting graph but for the whole distance-k family (therefore capturing both local and global relations between elements).

We start by reducing the problem to a vertex-traits-only one by converting the input graph to an edge-node bipartite form (see Algorithm 3.1). This decision not only simplifies the subsequent steps but also potentially boosts the expressive power of the method (as shown in [6]).

Algorithm 3.1. BIPARTITEEDGEVERTEXGRAPH(G)

$(\mathbf{V}, \mathbf{E}, \mathbf{T_V}, \mathbf{T_E}) \leftarrow G$

$\mathbf{V_B} = \mathbf{V} \cup \mathbf{E}$

$\mathbf{E_B} = \{(v, e) \mid v \in \mathbf{V} \land e \in \mathbf{E} \land v \in e\}$

$\mathbf{T_B} = \left\{ \left(\mathbf{V_B} \ni v \mapsto t_B(v) = \begin{cases} t(v) & v \in \mathrm{dom}(t) \\ \varnothing & \text{otherwise} \end{cases} \right) \middle| t \in \mathbf{T_V} \cup \mathbf{T_E} \right\}$

$G_B \leftarrow (\mathbf{V_B}, \mathbf{E_B}, \mathbf{T_B}, \varnothing)$

return G_B

Next, we need to introduce the notion of elements relation order. Two graph elements are in a relation of order $r + 1$ with one another if they are separated by a distance r in the corresponding graph. As a consequence, a node is in order-1 relation with itself, order-2 relation with edges incident to it, and order-3 relation with its direct neighbours.

Now, it is possible to define in the broadest terms the excepted form of the discussed embedding output. NERO (Network Embedding based on Relation Order histograms) should produce a three-dimensional tensor $\mathbf{N}$, where $\mathbf{N}_{i,j,k}$ is the number of order-i relations between vertices for which the first relation element is annotated with trait value j and the second with trait value k. If there is only a single discrete trait present, then $\mathbf{N}_{3,*,*}$ is equivalent to a standard graph attribute matrix, and $\mathbf{N}_{2k+1,*,*}$ to its distance-k graph incarnations.

As we can see, the core architecture of the previously discussed techniques is still upheld. There is a histogram as a basic building block (but this time

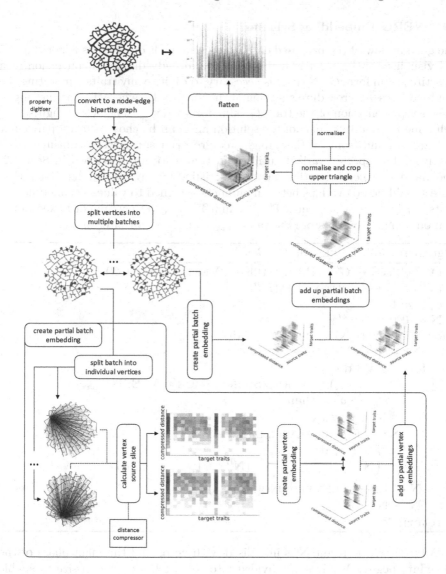

Fig. 1. NERO graph embedding workflow.

of trait connectivity, not of node centrality) and focus enumerating and representing multiple levels of abstraction (generalised as a relation order). However, such general definition necessarily omits important practical details (e.g. sometimes traits are continuous and long-range relations often don't need the same representational resolution as the close ones)—the subsequent sections provide a more comprehensive overview of the embedding algorithm and its intended implementation.

3.1   NERO Embedding Scheme

The general flow of the proposed embedding scheme is outlined as Algorithm 3.2 and visualised in Fig. 1. The first step is the already discussed conversion to a bipartite graph form G'. Next, it is necessary to digitise any traits annotating the graph vertices—a procedure resulting in the matrix $\mathbf{T}'$, where $\mathbf{T}'_{i,j}$ is the single-integer representation of the trait T_j value for vertex v_i. The actual digitisation technique is a modular part of the solution and can be chosen with a particular problem in mind (some of them might require prior supervised or unsupervised training). The specific variants utilised in this work are described in Sect. 3.2. One mandatory condition that has to be satisfied by any applied digitiser is that there should be no overlaps between numbers assigned to values among different traits ($\forall_{j_1,j_2}\ j_1 < j_2 \implies \max \mathbf{T}'_{*,j_1} < \min \mathbf{T}'_{*,j_2}$), so they are not mixed in the final outcome. We also store the value $t_{\max} = \max \mathbf{T}'$ for later usage.

Algorithm 3.2. EMBEDGRAPH(G, n_b)

$(\mathbf{V}, \mathbf{E}, \mathbf{T}, \varnothing) \leftarrow G' \leftarrow$ BIPARTITEEDGEVERTEXGRAPH(G)
$(\mathbf{T}', t_{\max}) \leftarrow$ DIGITISEDTRAITS($\mathbf{T}$)
$\delta_{\max} = 1$
$\mathbf{N} \leftarrow \mathbf{0}^{\delta_{\max} \times t_{\max} \times t_{\max}}$
for $\mathbf{V}^{\mathbf{b}} \in$ VERTEXBATCHES($\mathbf{V}, n_b$) **do**
$\quad \mathbf{N}^{\mathbf{b}} \leftarrow \mathbf{0}^{\delta_{\max} \times t_{\max} \times t_{\max}}$
$\quad$ **for** $v_i \in \mathbf{V}^{\mathbf{b}}$ **do**
$\quad\quad (\mathbf{N}^{\mathbf{b},\mathbf{v}}, d_{\max}) \leftarrow$ VERTEXSOURCESLICE($v, \mathbf{V}, \mathbf{E}, \mathbf{T}', t_{\max}$)
$\quad\quad$ **if** $d_{\max} > \delta_{\max}$ **then**
$\quad\quad\quad \delta_{\max} = d_{\max}$
$\quad\quad \mathbf{N}^{\mathbf{b}} \leftarrow$ ENLARGEEMBEDDING($\mathbf{N}^{\mathbf{b}}, \delta_{\max}$)
$\quad\quad$ **for** $j \in \mathbf{T}'_{i,*}$ **do**
$\quad\quad\quad \left(\mathbf{N}^{\mathbf{b}}_{d,j,*}\right)_{1 \le d \le d_{\max}} \leftarrow \left(\mathbf{N}^{\mathbf{b}}_{d,j,*}\right)_{1 \le d \le d_{\max}} + \mathbf{N}^{\mathbf{b},\mathbf{v}}$
$\quad \mathbf{N} \leftarrow$ ENLARGEEMBEDDING($\mathbf{N}, \delta_{\max}$)
$\quad \mathbf{N} \leftarrow \mathbf{N} + \mathbf{N}^{\mathbf{b}}$
return $\mathbf{N}$

After that, the result $\mathbf{N}$ is initialised with zeros and the main phase of the procedure begins. Vertices are divided into n_b batches of chosen size to enable the possibility of map-reduce style parallel computations (every batch can be processed independently). Then, for each vertex v in a batch, a source slice is calculated, as specified in Algorithm 3.3.

The goal of the slice computation is to create matrix $\mathbf{N}^{\mathbf{b},\mathbf{v}}$ containing information about all the relations in which the current vertex is the first element ($\mathbf{N}^{\mathbf{b},\mathbf{v}}_{i,j}$ is the number of vertices with trait value j in order-i relation with the vertex v). To do so, we first need to calculate all the pairwise distances between the current vertex and all the other ones (e.g. by BFS graph traversal). However, using raw distance values turns out to be impractical unless the graph has a low diameter. Instead, the distances are compressed using one of the options presented in Sect. 3.3.

Algorithm 3.3. VERTEXSOURCESLICE($v, \mathbf{V}, \mathbf{E}, \mathbf{T}', t_{\max}$)

$\quad \mathbf{D} \leftarrow$ SHORTESTDISTANCES($v, \mathbf{V}, \mathbf{E}$)

$\quad (\mathbf{D}', d_{\max}) \leftarrow$ COMPRESSEDDISTANCES($\mathbf{D}$)

$\quad \mathbf{N}^{\mathbf{b},\mathbf{v}} \leftarrow \mathbf{0}^{d_{\max} \times t_{\max}}$

$\quad$ **for** $i \in \{1, 2, \ldots, n_V\}, j \in \{1, 2, \ldots, n_T\}$ **do**

$\qquad \mathbf{N}^{\mathbf{b},\mathbf{v}}{}_{\mathbf{D}'_i, \mathbf{T}'_{i,j}} \leftarrow \mathbf{N}^{\mathbf{b},\mathbf{v}}{}_{\mathbf{D}'_i, \mathbf{T}'_{i,j}} + 1$

$\quad$ **return** ($\mathbf{N}^{\mathbf{b},\mathbf{v}}, d_{\max}$)

The slices are then added to appropriate regions of tensor $\mathbf{N}^{\mathbf{b}}$, iteratively creating a partial embedding. After that, the batch outcomes are summed up position-wise and form a complete result $\mathbf{N}$ (the accumulator is resized whenever necessary based on current $\delta_{\max}$. Finally, the obtained values are normalised (as described in Sect. 3.4) and reshaped into a flat vector to facilitate later usage.

3.2 Trait Digitisation

Discrete traits were digitised in the simplest possible way—by consecutively assigning a natural number to a value whenever a new one is encountered. In the case of continuous traits, two alternatives sharing the same basic scheme were tested. They both utilise a fixed number n_h of non-overlapping bins with boundaries determined by fitting them to the training data. Trait values are then mapped to the number of the bin they fit in, with two special numbers reserved for those belonging outside of the bin scope and for vertices without specified trait value.

The first variant (annotated $\varnothing$ and commonly known as equal-width bins [14]) was used as a baseline. It finds the minimal and maximal trait value in the set and then divides that range into same-size chunks. The second one (annotated A) tries to adapt bin edges to the way the values are distributed. It starts by estimating the said distribution—to do so it uses $n_i \gg n_h$ equal-width bins to create a value histogram for each sample and then calculates their mean over the whole training set. Finally, it picks edge positions so that there would be an equal probability of falling into each one of them.

3.3 Relation Order Compression

Relation order compression is a non-linear mapping between the actual between-vertex distance and the declared relation order. The rationale for employing it is twofold. Firstly—it directly reduces the dimensionality of the resulting embedding, decreasing the memory requirements and speeding up the later operations. Secondly—the experiences from the computer vision field [29] strongly suggest that the higher the relation order, the less important it is to know its exact value (i.e. if a vertex is on the opposite side of the graph it might not be crucial to differentiate whether it lies exactly one step further than another one or not).

We utilised three types of compression: no compression at all ($\varnothing$) as a baseline, and two bin-based compression schemes similar to trait digitising ones discussed in the previous section. Bin edges b_i were determined by the following formulas: $b_i = \textsc{Fibonacci}(i)$ (a standard Fibonacci sequence, denoted by F) and $b_{i-1} = \lfloor a^i i + bi + c \rfloor$ with $a = 1.074$, $b = 1.257$, $c = -3.178$ (denoted by V, a solution inspired by VGG19 [29] CNN layers receptive field sizes). Whichever one was used, the first two boundaries were set to fixed values of $b_0 = 0$ and $b_1 = 1$, while target number of bins was set to a fixed value n_r.

3.4   Normalisation

The relative values of individual NERO embedding components differ in magnitude, as the relation order density is not uniform (there are much more moderately distanced vertices than those in close vicinity or in highly eccentric pairs). Thus, it is practical to normalise them before any further processing. Throughout this work, all values were always subject to the so-called standard scaling—i.e. standardised by removing the mean and scaling to unit variance (baseline $\varnothing$). Additionally, they were sometimes scaled relative to the maximum value per given relation order—$\forall_i \, \mathbf{N}_{i,**} \leftarrow \frac{\mathbf{N}_{i,**}}{\max \mathbf{N}_{i,**}}$ (variant R).

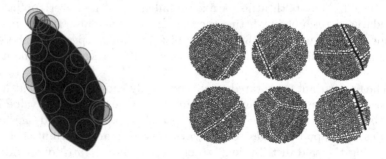

(a) Accepted and discarded circles.   (b) Examples of cropped fragments.

Fig. 2. Obtaining leaf fragment samples.

4   Laminae Venation Graphs

Leaf venation networks of flowering plants form intricate anastomosing structures, that evolved to serve numerous functions, from efficient water transportation to mechanical reinforcement of the lamina [26]. As a result, they are intensively studied by numerous research fields, such as plant paleoecology, plant physiology, developmental biology, or species taxonomy [24].

From the graph embedding point of view, processing of such networks creates an interesting computational challenge - not only because of the hefty numbers of vertices and edges but also due to their large radii making shortest path computations significantly harder [18]. However, popular datasets tend to focus almost solely on data coming from either biochemistry, computer vision, or social sciences [20]. In this work, we introduce an additional set of novel benchmark problems that require correct recognition of organism genus on the basis of small vasculature fragments.

The results obtained by Ronellenfitsch et al. [25] were utilised as the foundation for the subsequent sample generation. They consist of 212 venation graphs belonging to 144 species from 37 genera and were obtained by a multi-stage procedure, involving chemical staining, high-resolution scanning and numerous image transformations. Unfortunately, the samples are strongly unbalanced, with 102 belonging to *Protium* and 25 to *Bursera* genus. Furthermore, it was necessary to remove 44 of them due to various defects caused by unsuccessful extraction or accidental laminae damage. The remaining ones are highly reticulate large planar graphs comprised of up to 300000 elements, with two continuous edge attributes—their lengths and diameters.

Algorithm 4.1. CROPSAMPLES$(G, n_a, \tau_{A_c}, \tau_{\varrho_v}, \tau_{r_c})$

$\quad S_c \leftarrow \varnothing$

$\quad (\mathbf{V}, \mathbf{E}, \{T_x\}, \mathbf{T_E}) \leftarrow G$

$\quad \mathbf{V^u} \leftarrow \mathbf{V}$

$\quad \mathbf{X} \leftarrow \{T_x(v) \mid v \in \mathbf{V}\}$

$\quad \mathbf{H} \leftarrow \text{CONVEXHULL}(\mathbf{X})$

$\quad A \leftarrow \text{SHOELACEAREA}(\mathbf{H})$

$\quad \varrho_v \leftarrow \frac{|\mathbf{V}|}{A}$

$\quad A_c \leftarrow \tau_{A_c} A$

$\quad r_c \leftarrow \sqrt{\frac{A_c}{\pi}}$

$\quad \textbf{while } n_a > 0 \wedge |\mathbf{V^u}| > 0 \textbf{ do}$

$\quad\quad v_c \leftarrow \text{RANDOMCHOICE}(\mathbf{V^u})$

$\quad\quad x_c \leftarrow T_x(v_c)$

$\quad\quad \mathbf{V^c} \leftarrow \{v \mid v \in \mathbf{V} \wedge \|T_x(v) - x_c\| < r_c\}$

$\quad\quad \textbf{if } \frac{|\mathbf{V^c}|}{A_c} > \tau_{\varrho_v} \varrho_v \textbf{ then}$

$\quad\quad\quad \mathbf{V^u} \leftarrow \mathbf{V^u} \setminus \{v \mid v \in \mathbf{V} \wedge \|T_x(v) - x_c\| < \tau_{r_c} r_c\}$

$\quad\quad\quad \mathbf{E^c} \leftarrow \{e = (v_b, v_e) \mid e \in \mathbf{E} \wedge \{v_b, v_e\} \subseteq \mathbf{V_c}\}$

$\quad\quad\quad G_c \leftarrow (\mathbf{V^c}, \mathbf{E^c}, \{T_x\}, \mathbf{T_E})$

$\quad\quad\quad S_c \leftarrow S \cup \{G_c\}$

$\quad\quad \textbf{else}$

$\quad\quad\quad \mathbf{V^u} \leftarrow \mathbf{V^u} \setminus \{v_c\}$

$\quad\quad n_a \leftarrow n_a - 1$

$\quad \textbf{return } S_c$

Algorithm 4.1 was then employed in order to convert those networks into the final datasets by cropping out meaningful circular fragments. It starts by finding the convex hull $\mathbf{H}$ of a given venation graph. Next, the area A of the obtained polygon is calculated using the well-known shoelace formula and used to estimate the vertex density ϱ_v together with the target cropping radius r_c (the latter depending on the target area percentage parameter τ_{A_c}). Then, a central vertex v_c is randomly selected from the available vertices pool $\mathbf{V}^u$ and together with all its neighbours up to the distance r_c forms a crop candidate set $\mathbf{V}^c$.

If the candidate density is higher than a chosen percentage τ_{ϱ_v} of the average one, the crop area is considered to be sufficiently filled with content. In this case, the vertex set is supplemented by all the edges $\mathbf{E}^c$ incident to its members and forms a new sample G_c. After that, vertices closer than τ_{r_c} times the radius r_c from the centre are excluded from the $\mathbf{V}^u$ pool to ensure the lack of overlaps between the produced samples. On the other hand, if the density criterion is not satisfied only vertex v_c leaves the pool and the attempt is counted to be a failed one. The aforementioned loop is repeated until there are n_a failed attempts registered and then proceeds to the next leaf. Figure 2a depicts a possible outcome of the procedure (green circles are accepted and the red ones are omitted due to containing excessive free space) and Fig. 2b shows examples of isolated fragments.

Presented cropping scheme enables the production of venation datasets with chosen fragments radii and distribution, three of which were used to evaluate the proposed embedding technique. Each of them was generated with a different focus in mind: LAMINAEBIG ($\tau_{A_c} = 0.05$, $\tau_{\varrho_v} = 0.7$, $\tau_{r_c} = 2.0$) contains a smaller number of densely packed graphs with large diameter, LAMINAESMALL ($\tau_{A_c} = 0.005$, $\tau_{\varrho_v} = 0.7$, $\tau_{r_c} = 2.0$) increases samples count while decreasing their size, and LAMINAELIGHT ($\tau_{A_c} = 0.01$, $\tau_{\varrho_v} = 0.9$, $\tau_{r_c} = 3.0$) is intended to have lower computational demands due to a moderate number of high quality smaller radius samples. A brief summary of their characteristics can be found in Table 1. The cropping attempts threshold was always set to $n_a = 100$.

5   Experiments and Validation

Multiple configurations of the proposed algorithm are utilised throughout this section. Whenever it happens, the specific version is stated using the NERO$_{D,C,N}$ signature, where D is the chosen trait digitisation method, C specifies the relation order compression scheme, and N is the utilised normalisation. All calculations were performed using a Ryzen 7 3700X 3.6 GHz CPU. GNNs training was supported by a GeForce RTX 2080Ti 11 GB GPU.

Table 1. Benchmark datasets characteristics.

Name	Graphs	Classes	Average nodes	Average edges	Node labels	Edge labels	Node attrib.	Edge attrib.
DD	1178	2	248.32	715.66	+ (1)	–	–	–
MUTAG	188	2	17.93	19.79	+ (1)	+ (1)	–	–
PROTEINS	1113	2	39.06	72.82	+ (1)	–	+ (1)	–
NCI1	4110	2	29.87	32.30	+ (1)	–	–	–
LAMINAEBIG	2044	36	3585.35	4467.58	–	–	–	+ (2)
LAMINAESMALL	17924	36	392.67	473.82	–	–	–	+ (2)
LAMINAELIGHT	4110	36	820.74	1004.31	–	–	–	+ (2)

5.1   Graph Classification

The graph classification task is one of the standard ways of assessing representation robustness. In order to evaluate our proposed embedding scheme, we performed a series of tests on a diverse selection of benchmark problems, including the popular options from TUDataset [20] and three new ones introduced in this work. The first group contains macromolecular graphs representing the spatial arrangement of amino acid structures (PROTEINS [11], DD [27]) or small molecules described as individual atoms and chemical bonds between them (MUTAG [9], NCI1 [32]) and the second focuses on large-diameter laminae venation networks (Sect. 4 presents a more detailed overview of its origins). A brief comparison of their key characteristics is collected in Table 1.

The testing procedure was arranged following the guidelines presented in [20]. Accuracy scores were obtained as a result of stratified 10-fold cross-validation. For each split the training set was further subdivided into the actual training set and an additional validation set, the latter consisting of 10% of the initial samples and keeping the class frequencies as close as possible to the original ones. The optimal parameters of each utilised algorithm were estimated using a grid search over sets of available values with validation accuracy used as a ranking basis. The winning configuration was then retrained once more on the larger group of training samples and evaluated on the test batch previously set aside. Finally, the result was calculated as a mean of the 10 partial ones acquired in an aforementioned way.

In case of the classic benchmarks (Table 2), there were 3 repetitions of the said experiment—the reported value is the average outcome of the whole cross-validation procedure and the ± notation is employed to show the standard deviation of the mean accuracy. Obtained scores are compared with those achieved by a wide range of well-established solutions, representing both the graph kernel family (WL [27], GR [28], SP [3], WL-OA [17], WWL [30]) and the GNN approach (GIN-ϵ [33], GIN-ϵ-JK [34], U2GNN [22], g-U-Net [12], DDGK [1], GIC [16], PPGN [19], HGP-SL [35], GraphSAGE [15], GAT [31], EdgePool [10], PSCN [23]). The sources of the provided values are referenced in corresponding rows. Due to differences in applied measurement methodologies, the declared errors should not be directly compared with one another and should instead be treated only as an approximate credibility gauge. The scores are declared as not available (N/A) if they were not reported in any publications known to the authors (in case of algorithms our solution is being compared to)[1], would make no sense in the context (different trait digitisation techniques when there are no continuous attributes present)[2], or were not computationally feasible (e.g. no relation order compression for large graphs)[3].

NERO runs used an Extremely Randomised Trees [13] classifier with number of decision sub-trees equal to $n_X = 5000$ as its final step. Number of digitiser bins was chosen from $n_h \in \{5, 10, 20\}$ (with $n_i = 500$ where applicable), similarly for the relation order compression $n_r \in \{5, 10, 20\}$.

For the laminae benchmarks (Table 3) only a single experiment run was performed and the ± symbol denotes the standard deviation of the accu-

Table 2. Classification accuracies on popular TUDatasets benchmarks.

	DD	MUTAG	PROTEINS	NCI1
WL [30]	78.29% ± 0.30%	85.78% ± 0.83%	74.99% ± 0.28%	85.83% ± 0.09%
GR [20,22]	78.45% ± 0.26%	81.58% ± 2.11%	71.67% ± 0.55%	66.0% ± 0.4%
SP [20,21,35]	78.72% ± 3.89%	85.8% ± 0.2%	75.6% ± 0.7%	74.4% ± 0.1%
WL-OA [30]	79.15% ± 0.33%	87.15% ± 1.82%	**76.37% ± 0.30%**	**86.08% ± 0.27%**
WWL [30]	**79.69% ± 0.50%**	**87.27% ± 1.50%**	74.28% ± 0.56%	85.75% ± 0.25%
GIN-ϵ [20]	71.90% ± 0.70%	84.36% ± 1.01%	72.2% ± 0.6%	77.7% ± 0.8%
GIN-ϵ-JK [20]	73.82% ± 0.97%	83.36% ± 1.81%	72.2% ± 0.7%	78.3% ± 0.3%
U2GNN [22]	80.23% ± 1.48%	89.97% ± 3.65%	78.53% ± 4.07%	N/A[1]
g-U-Net [12]	82.43%	N/A[1]	77.68%	N/A[1]
DDGK [1]	**83.1% ± 12.7%**	91.58% ± 6.74%	N/A[1]	68.10% ± 2.30%
GIC [16]	N/A[1]	**94.44% ± 4.30%**	77,65% ± 3.21%	**84.08% ± 1.77%**
PPGN [19]	N/A[1]	90.55% ± 8.70%	77,20% ± 4,73%	83.19% ± 1.11%
HGP-SL [35]	80.96% ± 1.26%	N/A[1]	**84.91% ± 1.62%**	78.45% ± 0.77%
SAGE [22,35]	75.78% ± 3.91%	79.80% ± 13,9%	74.01% ± 4,27%	74.73% ± 1.34%
GAT [22,35]	77.30% ± 3.68%	89.40% ± 6,10%	74.72% ± 4,01%	74.90% ± 1.72%
EdgePool [35]	79.20% ± 2.61%	N/A[1]	82.38% ± 0.82%	76.56% ± 1.01%
PSCN [16]	N/A[1]	92.63% ± 4.21%	75.89% ± 2.76%	78.59% ± 1.89%
NERO$_{A,V,R}$	79.40% ± 0.49%	86.49% ± 0.89%	**77.89% ± 0.60%**	80.40% ± 0.25%
NERO$_{A,F,R}$	**80.45% ± 0.18%**	86.55% ± 0.25%	77.30% ± 0.30%	80.01% ± 0.44%
NERO$_{A,V,\varnothing}$	79.15% ± 0.28%	**88.68% ± 0.25%**	76.82% ± 0.44%	**81.63% ± 0.14%**
NERO$_{A,F,\varnothing}$	79.40% ± 0.26%	88.10% ± 1.32%	76.88% ± 0.19%	80.89% ± 0.29%
NERO$_{A,\varnothing,R}$	N/A[3]	85.64% ± 0.41%	76.55% ± 0.19%	79.80% ± 0.20%
NERO$_{\varnothing,V,R}$	N/A[2]	N/A[2]	75.68% ± 0.29%	N/A[2]

Table 3. Classification accuracies on laminae fragments benchmark datasets.

	LAMINAEBIG	LAMINAESMALL	LAMINAELIGHT
WL	71.6% ± 2.0%	**66.9% ± 1.5%**	66.0% ± 2.7%
GR	**73.5% ± 2.7%**	65.9% ± 1.5%	64.7% ± 3.4%
SP	N/A[4]	N/A[4]	**67.5% ± 3.3%**
GINE-ϵ	80.7% ± 5.0%	**83.26% ± 0.91%**	76.0% ± 2.2%
GINE-ϵ-JK	**84.2% ± 3.0%**	82.71% ± 0.93%	**79.1% ± 2.2%**
NERO$_{A,V,R}$	**94.2% ± 4.2%**	**85.2% ± 2.8%**	**86.6% ± 4.1%**
NERO$_{A,V,\varnothing}$	88.5% ± 6.9%	83.7% ± 4.4%	85.1% ± 3.4%
NERO$_{\varnothing,V,R}$	66.3% ± 5.9%	65.4% ± 5.1%	70.5% ± 8.2%

racy itself (estimated over the 10 cross-validation splits). As a consequence, the potential error of such estimation is much higher that the one demonstrated for the previous benchmark group. The NERO parameters were set to $n_h = 10$ and $n_r = 20$ and the RBF kernel SVM [5] with L2 penalty parameter $C \in \{0.1, 1.0, 10.0, 100.0\}$ was the classifier of choice. Some scores were declared as not available (N/A) due to exceeding per-algorithm computation time limit

$(12\,\mathrm{h})^4$. Methods supporting only discrete edge labels had them provided after being digitised by scheme A.

5.2 Results Interpretability

An additional feature of the proposed NERO embedding is the fact, that the final tensor $\mathbf{N}$ is always the combination of individual node and edge contributions $\mathbf{N}^{b,v}$. This, in turn, enables few useful routines. First, as already mentioned, it is possible to parallelise the computation in accordance with the map-reduce paradigm. Second, when completion time is still an issue, one can sample a smaller vertex subset and calculate only the associated embedding fragments, resulting in an approximate solution. Finally, if there is any information associated with individual embedding elements available (e.g. feature importance scores), it can be traced back to the network components that contributed to that particular cell. Figure 3b shows a proof of concept of that application—a venation graph is painted using a heatmap obtained from the impurity values provided by an extra-trees classifier.

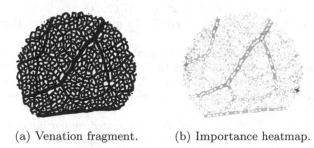

(a) Venation fragment.   (b) Importance heatmap.

Fig. 3. Visualisation of element importances based on extra-trees impurity scores.

6   Conclusions and Future Works

The presented embedding technique performed similarly or better than all the other graph kernels it was compared to on all but one of the utilised benchmark datasets. Obtained accuracies were often on-par with general-purpose GNN methods, without the need for costly GPU-based learning computations. This property was especially useful in case of large laminae venation graphs, where many of those models stopped being applicable due to memory constraints. The technique itself is highly modular and could be further adapted to the better suit a given problem. Lastly, its outcomes can be directly associated with the original nodes and edges, facilitating any attempts at results explanation.

Nevertheless, there are still topics requiring additional studies. The embedding may be further enriched by incorporating supplementary information about the structure itself, such as node centrality measures. Advanced trait digitisation schemes can be designed utilising tiling algorithms, clustering procedures, and fuzzy boundaries. Finally, the real potential of the method interpretability should be assessed in a separate investigation.

Acknowledgements. The authors would like to express their sincerest gratitude to Jana Lasser for providing a unique venation dataset and for her useful tips about its usage.

References

1. Al-Rfou, R., Perozzi, B., Zelle, D.: DDGK: learning graph representations for deep divergence graph kernels. In: The World Wide Web Conference, pp. 37–48 (2019)
2. Bagrow, J.P., Bollt, E.M., Skufca, J.D., Ben-Avraham, D.: Portraits of complex networks. EPL (Europhys. Lett.) **81**(6), 68004 (2008)
3. Borgwardt, K.M., Kriegel, H.P.: Shortest-path kernels on graphs. In: Fifth IEEE International Conference on Data Mining (ICDM 2005), p. 8-pp. IEEE (2005)
4. Bunke, H., Riesen, K.: Towards the unification of structural and statistical pattern recognition. Pattern Recogn. Lett. **33**(7), 811–825 (2012)
5. Chang, C.C., Lin, C.J.: LIBSVM: a library for support vector machines. ACM Trans. Intell. Syst. Technol. (TIST) **2**(3), 1–27 (2011)
6. Czech, W.: Invariants of distance k-graphs for graph embedding. Pattern Recogn. Lett. **33**(15), 1968–1979 (2012)
7. Czech, W., Łazarz, R.: A method of analysis and visualization of structured datasets based on centrality information. In: Rutkowski, L., Korytkowski, M., Scherer, R., Tadeusiewicz, R., Zadeh, L.A., Zurada, J.M. (eds.) ICAISC 2016. LNCS (LNAI), vol. 9693, pp. 429–441. Springer, Cham (2016). https://doi.org/10.1007/978-3-319-39384-1_37
8. Czech, W., Mielczarek, W., Dzwinel, W.: Distributed computing of distance-based graph invariants for analysis and visualization of complex networks. Concurr. Comput.: Pract. Exp. **29**(9), e4054 (2017)
9. Debnath, A.K., Lopez de Compadre, R.L., Debnath, G., Shusterman, A.J., Hansch, C.: Structure-activity relationship of mutagenic aromatic and heteroaromatic nitro compounds. Correlation with molecular orbital energies and hydrophobicity. J. Med. Chem. **34**(2), 786–797 (1991)
10. Diehl, F.: Edge contraction pooling for graph neural networks. CoRR (2019)
11. Dobson, P.D., Doig, A.J.: Distinguishing enzyme structures from non-enzymes without alignments. J. Mol. Biol. **330**(4), 771–783 (2003)
12. Gao, H., Ji, S.: Graph U-nets. In: International Conference on Machine Learning, pp. 2083–2092 (2019)
13. Geurts, P., Ernst, D., Wehenkel, L.: Extremely randomized trees. Mach. Learn. **63**(1), 3–42 (2006)
14. Hacibeyoglu, M., Ibrahim, M.H.: EF_unique: an improved version of unsupervised equal frequency discretization method. Arab. J. Sci. Eng. **43**(12), 7695–7704 (2018)
15. Hamilton, W., Ying, Z., Leskovec, J.: Inductive representation learning on large graphs. In: Advances in Neural Information Processing Systems, pp. 1024–1034 (2017)
16. Jiang, J., Cui, Z., Xu, C., Yang, J.: Gaussian-induced convolution for graphs. In: Proceedings of the AAAI Conference on Artificial Intelligence, vol. 33, pp. 4007–4014 (2019)

17. Kriege, N.M., Giscard, P.L., Wilson, R.: On valid optimal assignment kernels and applications to graph classification. Adv. Neural. Inf. Process. Syst. **29**, 1623–1631 (2016)
18. Kuhn, F., Schneider, P.: Computing shortest paths and diameter in the hybrid network model. In: Proceedings of the 39th Symposium on Principles of Distributed Computing, pp. 109–118 (2020)
19. Maron, H., Ben-Hamu, H., Serviansky, H., Lipman, Y.: Provably powerful graph networks. In: Advances in Neural Information Processing Systems, pp. 2156–2167 (2019)
20. Morris, C., Kriege, N.M., Bause, F., Kersting, K., Mutzel, P., Neumann, M.: TUDataset: a collection of benchmark datasets for learning with graphs. arXiv preprint arXiv:2007.08663 (2020)
21. Neumann, M., Garnett, R., Bauckhage, C., Kersting, K.: Propagation kernels: efficient graph kernels from propagated information. Mach. Learn. **102**(2), 209–245 (2015). https://doi.org/10.1007/s10994-015-5517-9
22. Nguyen, D., Nguyen, T., Phung, D.: Universal self-attention network for graph classification. arXiv preprint arXiv:1909.11855 (2019)
23. Niepert, M., Ahmed, M., Kutzkov, K.: Learning convolutional neural networks for graphs. In: International Conference on Machine Learning, pp. 2014–2023 (2016)
24. Rapacz, M., Łazarz, R.: Automatic extraction of leaf venation complex networks. In: ECAI 2020, vol. 325, pp. 1914–1921. IOS Press (2020)
25. Ronellenfitsch, H., Lasser, J., Daly, D.C., Katifori, E.: Topological phenotypes constitute a new dimension in the phenotypic space of leaf venation networks. PLoS Computat. Biol. **11**(12), e1004680 (2015)
26. Roth-Nebelsick, A., Uhl, D., Mosbrugger, V., Kerp, H.: Evolution and function of leaf venation architecture: a review. Ann. Bot. **87**(5), 553–566 (2001)
27. Shervashidze, N., Schweitzer, P., Van Leeuwen, E.J., Mehlhorn, K., Borgwardt, K.M.: Weisfeiler-lehman graph kernels. J. Mach. Learn. Res. **12**(9) (2011)
28. Shervashidze, N., Vishwanathan, S., Petri, T., Mehlhorn, K., Borgwardt, K.: Efficient graphlet kernels for large graph comparison. In: Artificial Intelligence and Statistics, pp. 488–495 (2009)
29. Simonyan, K., Zisserman, A.: Very deep convolutional networks for large-scale image recognition. arXiv preprint arXiv:1409.1556 (2014)
30. Togninalli, M., Ghisu, E., Llinares-López, F., Rieck, B., Borgwardt, K.: Wasserstein weisfeiler-lehman graph kernels. In: Advances in Neural Information Processing Systems, pp. 6439–6449 (2019)
31. Veličković, P., Cucurull, G., Casanova, A., Romero, A., Liò, P., Bengio, Y.: Graph attention networks. In: International Conference on Learning Representations (2018)
32. Wale, N., Watson, I.A., Karypis, G.: Comparison of descriptor spaces for chemical compound retrieval and classification. Knowl. Inf. Syst. **14**(3), 347–375 (2008)
33. Xu, K., Hu, W., Leskovec, J., Jegelka, S.: How powerful are graph neural networks? In: International Conference on Learning Representations (2018)
34. Xu, K., Li, C., Tian, Y., Sonobe, T., Kawarabayashi, K.i., Jegelka, S.: Representation learning on graphs with jumping knowledge networks. In: International Conference on Machine Learning, pp. 5453–5462 (2018)
35. Zhang, Z., et al.: Hierarchical graph pooling with structure learning. arXiv preprint arXiv:1911.05954 (2019)
36. Zhang, Z., Cui, P., Zhu, W.: Deep learning on graphs: a survey. IEEE Trans. Knowl. Data Eng. (2020)

Desensitization Due to Overstimulation: A Second-Order Adaptive Network Model

Alex Korthouwer, David Noordberg, and Jan Treur[✉]

Social AI Group, Vrije Universiteit Amsterdam, Amsterdam, The Netherlands
{a.korthouwer,d.c.r.noordberg}@student.vu.nl, j.treur@vu.nl

Abstract. In this paper, a second-order adaptive network model is presented for the effects of supernormal stimuli. The model describes via the underlying mechanisms in the brain how responses on stimuli occur in absence of a supernormal stimulus. Moreover, it describes how these normal responses are lost after a supernormal stimulus occurs and due to that adaptive desensitization takes place. By simulated example scenarios, it was evaluated that the model describes the expected dynamics. By stationary point analysis correctness of the implemented model with respect to its design was verified.

Keywords: Overstimulation · Desensitization · Adaptive network model

1 Introduction

Supernormal stimuli are everywhere. They can be seen on billboards, you see them on TV, and you can even find them when you take a walk in the forest. Supernormal stimuli differ from normal stimuli. A stimulus is a change in the external or internal environment to which an organism reacts. This reaction is also referred to as a response. A supernormal stimulus is an extreme version of a particular stimulus that people already have a tendency to respond to. This extreme version of an already existing stimulus creates a stronger response than the regular stimulus does. These supernormal stimuli are found both in the animal kingdom and in the human world. Cuckoo birds are a good example of this [1]. Cuckoo birds lay their eggs in other birds' nests so that these other birds feed their offspring. But why do these other birds feed the cuckoo chicks and not their own? This is due to the supernormal stimuli of the cuckoo chicks, which have a larger and redder beak than the other chicks. These supernormal stimuli of the cuckoo chick produce a stronger response of the parent birds than the stimuli of their own chicks do.

Such effects of supernormal stimuli are explored computationally in this paper. First, in Sect. 2 some related background knowledge is briefly discussed and the research question is formulated. Next, in Sect. 3 the modeling approach used is discussed. In Sect. 4 the introduced computational model is described in some detail. Section 5 illustrates the model by simulation experiments performed. In Sect. 6 verification of the implemented model with respect to its design description based on analysis of stationary points is discussed. Finally, Sect. 7 is a discussion.

© Springer Nature Switzerland AG 2021
M. Paszynski et al. (Eds.): ICCS 2021, LNCS 12743, pp. 238–249, 2021.
https://doi.org/10.1007/978-3-030-77964-1_19

2   Background: Related Work and Research Question

Superstimuli often exist in the human world and companies such as advertising agencies make grateful use of this phenomenon. People are triggered and seduced by commercials filled with astonishing landscapes, beautiful people and delicious-looking food, which are often an exaggeration of reality. Because these stimuli are often over-exaggerated, turning into supernormal stimuli, people react more strongly to this, but they will never be completely satisfied with the purchased product, and they will look for new products to satisfy their buying drive [5]. This phenomenon is known as *desensitization*.

As a general principle, after prolonged or multiple exposure to a particular stimulus or stimuli, the brain will adapt and the brain becomes less sensitive to this stimulus; for example, homestatic excitability regulation [7, 11] refers to this principle. This principle is also reflected in the brain's dopamine release. Dopamine is a substance that is part of the brain's reward system, as a neurotransmitter. These are chemicals in the brain that support transfer of information from one nerve cell to another. Dopamine makes one feel satisfied and rewarded. Dopamine is not constantly produced, but is released during certain actions or situations such as eating, exercising, sex, or drugs. Dopamine is also released when people make purchases. In line with the principle described here, when the brain is exposed to a particular stimulus in a prolonged way, or when the brain is multiple times exposed to a particular stimulus, less dopamine is released for experiencing the same amount of stimuli. In other words, the brain desensitizes due to overstimulation. To experience the same effect, people will have to be exposed to these stimuli for longer, or they have to be exposed more often to these stimuli. Our research question derives from this:

Can mechanisms in the brain indeed explain desensitization as a result of exposure to supernormal stimuli?

3   The Network-Oriented Modeling Approach Used

In this part of the paper the modeling approach used is explained. Traditionally, modelling the world's processes often is based on isolation and separation assumptions. These generally serve as a means to reduce the complexity of the problems. For example, in physics concerning gravity, usually the force from a planet on an object is taken into account, but not all other forces in the universe acting upon this object. Sometimes these assumptions do not hold. For example, Aristotle considered that some internal processes are separated from the body [3]. Later Kim [6] and others from philosophy of mind disputed that the mind and the body could be separated.

However, the problem with modeling such processes is not a particular type of isolation or separation assumption, but with the notion of separation itself. Interaction between states in a model often need cyclic dependencies which make separation assumptions difficult to apply. In philosophy of the mind, the idea is that mental states are caused by input states and other mental states, and some mental states influence outgoing states [6]. This takes place according to a cyclic causal network of mental states within which each mental state has a causal, or functional role; e.g., see:

'Mental events are conceived as nodes in a complex causal network that engages in causal transactions with the outside world by receiving sensory inputs and emitting behavioral outputs. (…) to explain what a given mental state is, we need to refer to other mental states, and explaining these can only be expected to require reference to further mental states, on so on – a process that can go on in an unending regress, or loop back in a circle.' [6], pp. 104–105.

The network-oriented modeling approach described in [9] follows this dynamic perspective on causality and has been applied to obtain the network model introduced here. Within this approach, at each given time t each node (also called state) Y of such a network has a state value $Y(t)$, a real number usually in the interval [0,1]. The approach uses the following *network characteristics* to describe network models (called temporal-causal networks):

- **Connectivity characteristics**
  Each connection from a state X to another state Y has a connection weight $\omega_{X,Y}$ which represents the strength of the connection
- **Aggregation characteristics**
  Each state Y has a combination function $c_Y(\ldots)$ which is used to combine the single casual impacts $\omega_{X,Y}X(t)$ from different states X on state Y; for selection of combination functions from the available library, weights $\gamma_{i,Y}$ are used and the combination functions usually have parameters that can be indicated in general by $\pi_{i,j,Y}$
- **Timing characteristics**
  Each state Y has a speed factor η_Y for the speed of change of Y defining how fast the state changes upon its causal impact.

The choice of combination functions $c_Y(\ldots)$ can be problem-dependent. Within the available dedicated software environment, they can be chosen from a library for them, but can also be easily added. The combination function library, which has over 40 combination functions, also includes a facility to compose new combination functions from the available ones by (mathematical) function composition. The combination functions that have been used in this paper are basic; they are shown in Table 1.

Using these network characteristics, a network model can be simulated in a manner described in [9] as follows:

- For each time point t the aggregated impact is calculated using the combination function $c_Y(\ldots)$ by

$$\mathbf{aggimpact}_Y(t) = c_Y(\omega_{X_1,Y}X_1(t), \ldots, \omega_{X_k,Y}X_k(t)) \tag{1}$$

- Then for the time step from t to $t + \Delta t$ for each state Y the value is adjusted using the aggregated impact and the speed factor

$$Y(t + \Delta t) = Y(t) + \eta_Y[\mathbf{aggimpact}_Y(t) - Y(t)\Delta]t \tag{2}$$

- Hence the following differential equation is obtained:

$$\frac{dY(t)}{dt} = \eta_Y[\mathbf{aggimpact}_Y(t) - Y(t)]$$

Table 1. Basic combination functions from the library used in the presented model

	Notation	Formula	Parameters
Advanced logistic sum	$\mathbf{alogistic}_{\sigma,\tau}(V_1, ...,V_k)$	$\left[\frac{1}{1+e^{-\sigma(V_1+\cdots+V_k-\tau)}} - \frac{1}{1+e^{\sigma\tau}}\right](1 + e^{-\sigma\tau})$	Steepness $\sigma > 0$ Excitability threshold τ
Hebbian learning	$\mathbf{hebb}_\mu(V_1, V_2, W)$	$V_1 V_2(1 - W) + \mu W$	Persistence factor $\mu > 0$
Stepmod	$\mathbf{stepmod}_{\rho,\delta}(V_1, ...,V_k)$	0 if $t \bmod \rho < \delta$, else 1	Repetition ρ Duration of 0 δ

To model adaptive networks, the notion of *self-modeling network* (also called *reified network*) is applied, where the adaptive network characteristics are represented within the network by additional network states, called *self-model* states; see [10]. Shortly, adding a self-model for a temporal-causal network is done in the way that for some of the states Y of the base network and some of its related network structure characteristics for connectivity, aggregation and timing (in particular, some from $\omega_{X,Y}$, $\gamma_{i,Y}$, $\pi_{i,j,Y}$, η_Y), additional network states $\mathbf{W}_{X,Y}$, $\mathbf{C}_{i,Y}$, $\mathbf{P}_{i,j,Y}$, $\mathbf{H}_Y$ (self-model states) are introduced:

(a) **Connectivity self-model**

- Self-model states $\mathbf{W}_{X_i,Y}$ are added representing connectivity characteristics, in particular connection weights $\omega_{X_i,Y}$

(b) **Aggregation self-model**

- Self-model states $\mathbf{C}_{j,Y}$ are added representing aggregation characteristics, in particular combination function weights $\gamma_{i,Y}$
- Self-model states $\mathbf{P}_{i,j,Y}$ are added representing aggregation characteristics, in particular combination function parameters $\pi_{i,j,Y}$

(c) **Timing self-model**

- Self-model states $\mathbf{H}_Y$ are added representing timing characteristics, in particular speed factors η_Y

Here $\mathbf{W}$ refers to ω, $\mathbf{C}$ refers to γ, $\mathbf{P}$ refers to π, and $\mathbf{H}$ refers to η, respectively. For the processing, these self-model states define the dynamics of state Y in a canonical manner according to Eq. (2) whereby $\omega_{X,Y}$, $\gamma_{i,Y}$, $\pi_{i,j,Y}$, η_Y are replaced by the state values of $\mathbf{W}_{X,Y}$, $\mathbf{C}_{i,Y}$, $\mathbf{P}_{i,j,Y}$, $\mathbf{H}_Y$ at time t, respectively. These special effects are defined by outgoing connections of these self-model states, which give them their specific roles. Moreover, as for any other state in the network, the incoming connections and other network characteristics relating to the self-model states (including their combination functions) give them their dynamics.

As the outcome of the addition of a self-model is also a temporal-causal network model itself, as has been shown in detail in [10], Ch 10, this construction can easily be applied iteratively to obtain multiple levels or orders of self-models.

4 The Adaptive Network Model for Desensitization Due to Overstimulation

The proposed network model is based on a couple of assumptions. Firstly, dopamine is used for the reward system for humans, but when exposed too much the human brain becomes less receptive to the stimuli [5]. The second assumption which was used is that advertisements often do try to trigger emotions such that people tend to get a positive emotion associated to the stimulus [5]. Within the introduced network model, three adaptation principles are applied to the network, two first-order adaptation principles (Hebbian learning and Excitability modulation) and one second-order adaptation principle (Exposure accelerates adaptation). They are as follows.

- **Hebbian learning**
  When a cell is repetitive and persistent in activating another cell, there will be a metabolic process making that this cell's influence on the other cell increases [4]. For example, this can increase the preparation for an action when the stimulus representation affects the preparation of an action.
- **Excitability modulation**
  Depending on activation, excitability thresholds are adapted as a form of regulation of neuronal excitability, which to some extent is similar to the adaptation of neurons' internal properties according to 'homeostatic regulation' to guarantee a prefered level of activation; e.g., [7, 11].
- **Exposure accelerates adaptation**
  When the exposure to a stimulus becomes stronger, the adaptation speed will increase. When applied to Hebbian learning, this will make the connection change faster when exposed for some time [8].

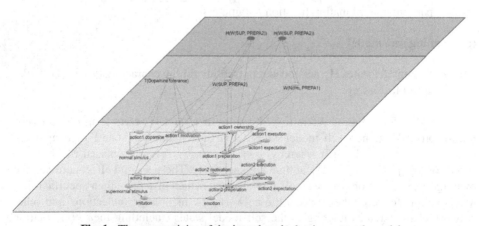

Fig. 1. The connectivity of the introduced adaptive network model

As another principle, ownership of an action makes that the action activation will be performed more frequently [2]. Using these principles as building blocks, the network model has been designed with connectivity as visualized in Fig. 1; see also the explanations of the states in Table 2. The exact dynamics of the model have been defined in the role matrices specifications shown in Figs. 2 and 3.

Table 2. Nomenclature: explanation of the states

State		Explanation
X_1	srs$_{normal}$	Sensory representation state for normal stimulus
X_2	dopamine$_{a_1}$	Dopamine level related to action a$_1$
X_3	mot$_{a_1}$	Motivation level related to action a$_1$
X_4	prep$_{a_1}$	Preparation state for action a$_1$
X_5	exe$_{a_1}$	Execution state for action a$_1$
X_6	exp$_{a_1}$	Expectation (expected effect) for action a$_1$
X_7	own$_{a_1}$	Ownership state for action a$_1$
X_8	srs$_{supernormal}$	Sensory representation state for supernormal stimulus
X_9	dopamine$_{a_2}$	Dopamine level related to action a$_2$
X_{10}	imitation	Imitation for supernormal stimulus
X_{11}	emotion	Emotion for supernormal stimulus
X_{12}	mot$_{a_2}$	Motivation level related to action a$_2$
X_{13}	prep$_{a_2}$	Preparation state for action a$_2$
X_{14}	exe$_{a_2}$	Execution state for action a$_2$
X_{15}	exp$_{a_2}$	Expectation (expected effect) for action a$_2$
X_{16}	own$_{a_2}$	Ownership state for action a$_2$
X_{17}	$\mathbf{T}_{dopetolerance}$	First-order self-model state for tolerance for dopamine
X_{18}	$\mathbf{W}_{srs_{supernormal},\,prep_{a_2}}$	First-order self-model state for the weight of the connection from srs$_{supernormal}$ to prep$_{a_2}$
X_{19}	$\mathbf{W}_{srs_{normal},\,prep_{a_1}}$	First-order self-model state for the weight of the connection from srs$_{normal}$ to prep$_{a_1}$
X_{20}	$\mathbf{H W}_{srs_{supernormal},\,prep_{a_2}}$	Second-order self-model state for the adaptation speed of the weight of the connection from srs$_{supernormal}$ to prep$_{a_2}$
X_{21}	$\mathbf{H W}_{srs_{normal},\,prep_{a_1}}$	Second-order self-model state for the adaptation speed of the weight of the connection from srs$_{normal}$ to prep$_{a_1}$

In role matrix **mb**, each state has its own row in which the other states that have effect on it are listed. In role matrix **mcw**, in the corresponding row the connections weights for these connections are specified. Note that for an adaptive connection weight, this is indicated by the name of the self-model state representing this connection weight (here X_{18} and X_{19} in the rows for the preparation states prep$_{a_1}$ and prep$_{a_2}$).

mb	base connectivity	1	2	3	4	mcw	connection weights	1	2	3	4
X_1	srs_{normal}	X_1				X_1	srs_{normal}	1			
X_2	$dopamine_{a1}$	X_1				X_2	$dopamine_{a1}$	1			
X_3	mot_{a1}	X_2				X_3	mot_{a1}	1			
X_4	$prep_{a1}$	X_1	X_3	X_6		X_4	$prep_{a1}$	X_{18}	0.5	0.5	
X_5	exe_{a1}	X_4	X_7			X_5	exe_{a1}	0.5	0.5		
X_6	exp_{a1}	X_4				X_6	exp_{a1}	1			
X_7	own_{a1}	X_1	X_3	X_4		X_7	own_{a1}	0.33	0.33	0.33	
X_8	$srs_{supernormal}$	X_8				X_8	$srs_{supernormal}$	1			
X_9	$dopamine_{a2}$	X_8				X_9	$dopamine_{a2}$	1			
X_{10}	imitation	X_8				X_{10}	imitation	1			
X_{11}	emotion	X_{10}				X_{11}	emotion	1			
X_{12}	mot_{a2}	X_9	X_{11}			X_{12}	mot_{a2}	1	1		
X_{13}	$prep_{a2}$	X_8	X_{12}	X_{15}		X_{13}	$prep_{a2}$	X_{19}	1	1	
X_{14}	exe_{a2}	X_{13}	X_{16}			X_{14}	exe_{a2}	1	1		
X_{15}	exp_{a2}	X_{13}				X_{15}	exp_{a2}	1			
X_{16}	own_{a2}	X_8	X_{11}	X_{12}	X_{13}	X_{16}	own_{a2}	1	1	1	1
X_{17}	$T_{dopetolerance}$	X_2	X_9	X_{17}		X_{17}	$T_{dopetolerance}$	1	1	1	
X_{18}	$W_{srs_{supernormal},prep_{a2}}$	X_8	X_{13}	X_{18}		X_{18}	$W_{srs_{supernormal},prep_{a2}}$	1	1	1	
X_{19}	$W_{srs_{normal},prep_{a1}}$	X_1	X_4	X_{19}		X_{19}	$W_{srs_{normal},prep_{a1}}$	1	1	1	
X_{20}	$Hw_{srs_{supernormal},prep_{a2}}$	X_8	X_{13}	X_{18}	X_{20}	X_{20}	$Hw_{srs_{supernormal},prep_{a2}}$	1	1	1	1
X_{21}	$Hw_{srs_{normal},prep_{a1}}$	X_1	X_4	X_{19}	X_{21}	X_{21}	$Hw_{srs_{normal},prep_{a1}}$	1	1	1	1

Fig. 2. Role matrices **mb** and **mcw** specifying the connectivity characteristics of the network model

In this model two actions occur, one of which is triggered by a supernormal stimulus and one by a normal stimulus. For both actions dopamine release increases the motivation of the participant to perform the action. It also increases the preparation for the respective action. The supernormal stimuli also create emotional imitation, since supernormal stimuli often play to the emotion by mimicking feeling. The emotion and the dopamine determine the motivation for performing the action relating to the supernormal stimulus. For the normal stimulus, the motivation is solely dependent on the dopamine released. At the moment the total amount of dopamine released from all actions is higher then the threshold for dopamine, the motivation will start to have a higher tolerance for dopamine. Each ownership state will be determined by the stimulus, the motivation, the preparation, and for the supernormal stimulus action it is also dependent on the emotion of the respective action. Moreover, the expectation (expected effect) of an action also influences the preparation of an action and the other way around.

mcfw	combination function weights	1 alogistic	2 hebb	3 stepmod
X_1	srs$_{normal}$	1		
X_2	dopamine$_{a1}$	1		
X_3	mot$_{a1}$	1		
X_4	prep$_{a1}$	1		
X_5	exe$_{a1}$	1		
X_6	exp$_{a1}$	1		
X_7	own$_{a1}$	1		
X_8	srs$_{supernormal}$			1
X_9	dopamine$_{a2}$	1		
X_{10}	imitation	1		
X_{11}	emotion	1		
X_{12}	mot$_{a2}$	1		
X_{13}	prep$_{a2}$	1		
X_{14}	exe$_{a2}$	1		
X_{15}	exp$_{a2}$	1		
X_{16}	own$_{a2}$	1		
X_{17}	$\mathbf{T}$dopetolerance	1		
X_{18}	$\mathbf{W}$srs$_{supernormal}$,prep$_{a2}$		1	
X_{19}	$\mathbf{W}$srs$_{normal}$,prep$_{a1}$		1	
X_{20}	$\mathbf{Hw}$srs$_{supernormal}$,prep$_{a2}$	1		
X_{21}	$\mathbf{Hw}$srs$_{normal}$,prep$_{a1}$	1		

ms	speed factors	1
X_1	srs$_{normal}$	0
X_2	dopamine$_{a1}$	0.2
X_3	mot$_{a1}$	0.2
X_4	prep$_{a1}$	0.2
X_5	exe$_{a1}$	0.2
X_6	exp$_{a1}$	0.2
X_7	own$_{a1}$	0.2
X_8	srs$_{supernormal}$	2
X_9	dopamine$_{a2}$	0.2
X_{10}	imitation	0.2
X_{11}	emotion	0.2
X_{12}	mot$_{a2}$	0.2
X_{13}	prep$_{a2}$	0.2
X_{14}	exe$_{a2}$	0.2
X_{15}	exp$_{a2}$	0.2
X_{16}	own$_{a2}$	0.2
X_{17}	$\mathbf{T}$dopetolerance	0.2
X_{18}	$\mathbf{W}$srs$_{supernormal}$,prep$_{a2}$	X_{20}
X_{19}	$\mathbf{W}$srs$_{normal}$,prep$_{a1}$	X_{21}
X_{20}	$\mathbf{Hw}$srs$_{supernormal}$,prep$_{a2}$	0.1
X_{21}	$\mathbf{Hw}$srs$_{normal}$,prep$_{a1}$	0.1

mcfp	combination function	1 alogistic		2 hebb		3 stepmod	
	parameter	1 σ	2 τ	1 μ	2	1 ρ	2 δ
X_1	srs$_{normal}$	5	0.2				
X_2	dopamine$_{a1}$	5	0.2				
X_3	mot$_{a1}$	5	0.2				
X_4	prep$_{a1}$	5	X_{17}				
X_5	exe$_{a1}$	5	0.2				
X_6	exp$_{a1}$	5	0.2				
X_7	own$_{a1}$	5	0.2				
X_8	srs$_{supernormal}$					100	40
X_9	dopamine$_{a2}$	5	0.2				
X_{10}	imitation	5	0.2				
X_{11}	emotion	5	0.2				
X_{12}	mot$_{a2}$	5	0.2				
X_{13}	prep$_{a2}$	5	X_{17}				
X_{14}	exe$_{a2}$	5	0.2				
X_{15}	exp$_{a2}$	5	0.2				
X_{16}	own$_{a2}$	5	0.2				
X_{17}	$\mathbf{T}$dopetolerance	5	0.6				
X_{18}	$\mathbf{W}$srs$_{supernormal}$,prep$_{a2}$			0.95			
X_{19}	$\mathbf{W}$srs$_{normal}$,prep$_{a1}$			0.95			
X_{20}	$\mathbf{Hw}$srs$_{supernormal}$,prep$_{a2}$	5	0.8				
X_{21}	$\mathbf{Hw}$srs$_{normal}$,prep$_{a1}$	5	0.8				

Fig. 3. Role matrices **mcfw**, **mcfp** and **ms** specifying the aggregation and timing characteristics of the network model

5 Simulation Results

From the simulated scenarios, the following one is presented. First (from time 0 to time 40), only a normal stimulus occurs. As a result the states related to action a_1 are activated (see Fig. 4). After time 40, a supernormal stimulus occurs. As a result the base states related to action a_2 are activated. But what also can be seen in Fig. 4, at the same time the activations of the states related to action a_1 that are triggered by the normal stimulus (which is still present) drop.

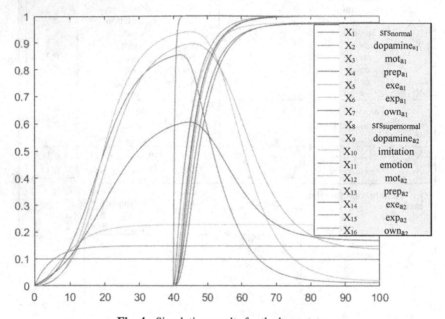

Fig. 4. Simulation results for the base states

The explanation for this drop can be found in Fig. 5, where the values of the adaptation states (the first- and second-order self-model states) have been visualized. When the supernormal stimulus arrives (in addition to the normal stimulus), the normal stimulus does not seem to have much effect anymore because the **T**-state for the dopamine tolerance (the blue line) goes up between time 40 and 50. This indicates desensitisation. Hence the action does not happen anymore when the stronger supernormal stimulus emerges which prevents the normal action from getting much effect.

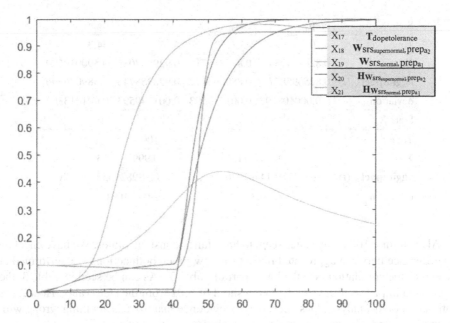

Fig. 5. Simulation results for the first- and second-order self-model states

6 Verification of the Network Model by Stationary Point Analysis

The dynamics of the implemented model were mathematically verified against its design specification by inspecting stationary points, i.e., points for some state Y where $dY(t)/dt = 0$. Based on difference Eq. (2), the following criterion is obtained for a state Y to have a stationary point at t (e.g., [9], Ch 12):

$$\eta_Y = 0 \text{ or}$$
$$\mathbf{aggimpact}_Y(t) = Y(t) \tag{3}$$

where according to (1)

$$\mathbf{aggimpact}_Y(t) = \mathbf{c}_Y(\omega_{X_1,Y}X_1(t), \ldots, \omega_{X_k,Y}X_k(t)) \tag{}$$

with $X_1, \ldots, X_k$ the states from which the state Y has incoming connections. In the example simulation, it can be observed that shortly after the individual starts being exposed to the supernormal stimulus, all of the states regarding the normal stimulus seem to reach a maximum shortly after arrival at time 40 of the supernormal stimulus. Hence the addition of the supernormal stimuli does make the action a_1 be performed less. The numerical data of the simulation do confirm this since the difference between the left hand side and right hand side of criterion (3) is less then 0.01 for all indicated states related to action a_1 within a few time units after the supernormal stimulus is added (see the upper part of Table 3).

Table 3. Mathematical analysis of stationary points and equilibrium values

State X_i	$prep_{a_1}$	exe_{a_1}	exp_{a_1}	own_{a_1}
Time t	42.0	45.3	44.7	44.3
$X_i(t)$	0.826204054	0.88181571	0.929487085	0.590019599
aggimpact$_{X_i}(t)$	0.825695857	0.880066117	0.928358746	0.589006296
deviation	−0.000508197	0.001749593	0.001749593	0.001013302

State X_i	exe_{a_1}	exe_{a_2}
Time t	100	100
$X_i(t)$	0.147271358	0.999819353
aggimpact$_{X_i}(t)$	0.144670364	0.999829352
deviation	0.002600994	−0.001001

Also, at time 100 many states seem to have fairly constant values. We have checked the difference between aggregated impact for how much both actions are performed at the end of the simulation (see the lower part of Table 3). As can be seen in Table 3, the aggregated impact is always within a threshold of 0.01 from the state value. Hence, the stationary points analysis does not show any evidence that there is anything wrong with the implemented model in comparison to its design specification.

7   Discussion

In this paper an adaptive temporal-causal network model has been introduced by which the effect of overstimulation on the performance of actions is modeled. Simulation experiments have been performed in with which a supernormal stimulus and/or normal stimulus was present and led to a desensitization adaptation. The model has been mathematically verified by analysis of stationary points for one of the simulation scenarios.

The proposed model has been built according to the adaptive network modeling approach based on self-modeling networks (or reified networks) described in [10]. This method of modeling makes it possible to easily design any adaptive network model, which in this case describes how by adaptation supernormal stimuli distract people from the day to day activities.

In future work, simuli with different gradations can be explored and further validation of the model may be performed when suitable data sets are available.

References

1. Barrett, D.: Supernormal Stimuli: How Primal Urges Overran Their Evolutionary Purpose. W.W. Norton & Company, New York (2010)
2. Haggard, P.: Human volition: towards a neuroscience of will. Nat. Rev. Neurosci. **8**, 934–946 (2008)

3. Hardie, R.: Aristotle's Physics (1933)
4. Hebb, D.O.: The Organization of Behavior. A Neuropsychological Theory. Wiley, New York (1949)
5. Hendlin, Y.H.: I am a fake loop: the effects of advertising-based artificial selection. Biosemiotics **12**, 131–156 (2019)
6. Kim, J.: Philosophy of Mind. Westview Press, Boulder (1996)
7. Misonou, H.: Homeostatic regulation of neuronal excitability by K(_) channels in normal and diseased brains. Neuroscientist **16**, 51–64 (2010)
8. Robinson, B.L., Harper, N.S., McAlpine, D.: Meta-adaptation in the auditory midbrain under cortical influence. Nat. Commun. **7**, 13442 (2016)
9. Treur, J.: Network-Oriented Modeling: Addressing Complexity of Cognitive, Affective and Social Interactions. Springer, Cham (2016). https://doi.org/10.1007/978-3-319-45213-5
10. Treur, J.: Network-Oriented Modeling for Adaptive Networks: Designing Higher-Order Adaptive Biological, Mental and Social Network Models. Springer, Cham (2020). https://doi.org/10.1007/978-3-030-31445-3
11. Williams, A.H., O'Leary, T., Marder, E.: Homeostatic regulation of neuronal excitability. Scholarpedia **8**, 1656 (2013)

A Modified Deep Q-Network Algorithm Applied to the Evacuation Problem

Marcin Skulimowski[✉] [iD]

Faculty of Physics and Applied Informatics, University of Lodz,
Pomorska 149/153, 90-236 Lodz, Poland
marcin.skulimowski@uni.lodz.pl

Abstract. In this paper, we consider reinforcement learning applied to the evacuation problem. Namely, we propose a modification of the deep Q-network (DQN) algorithm, enabling more than one action to be taken before each update of the Q function. To test the algorithm, we have created a simple environment that enables evacuation modelling. We present and discuss the results of preliminary tests. In particular, we compare the performance of the modified DQN algorithm with its regular version.

Keywords: Reinforcement Learning · Deep Q-Networks · Evacuation

1 Introduction

Reinforcement Learning (RL) is a subdomain of machine learning that allows the agent to gain experience and achieve goals only by interacting with its environment (without any "teacher"). There has been considerable interest in RL and its applications in many domains in recent years, e.g. games, robotics, finance and optimization (see [1,2]). In this paper, we consider RL applied to the *evacuation problem*, i.e. to model humans in evacuation scenarios. The problem has been widely studied, mainly using cellular automata (CA) [3,4]. Developments in RL have led to approaches based on various RL algorithms [5–8] also in conjunction with CA [9]. In the most common RL approach to the evacuation problem, computationally expensive multi-agent methods are applied (see, e.g. [5,8]). It means that each person is considered as a single agent. The approach discussed in this paper is different. Namely, we consider one agent that must evacuate all people from the room as quickly as possible through any available exit. The approach is similar to the one proposed by Sharma et al. [6]. The difference is that we consider evacuation from one room and Sharma et al. consider evacuation from many rooms modelled as a graph. Moreover, they use a deep Q-network algorithm and their agent acts by moving one person at a time step. This paper presents some modification of the deep Q-network (DQN) algorithm, enabling more than one action to be taken before each Q function update. Namely, our agent takes as many actions as is the number of persons in the room. Only then the Q function is updated. To test the proposed algorithm, we have created

© Springer Nature Switzerland AG 2021
M. Paszynski et al. (Eds.): ICCS 2021, LNCS 12743, pp. 250–256, 2021.
https://doi.org/10.1007/978-3-030-77964-1_20

a grid-based evacuation environment that enables modelling evacuation from rooms with many exits and additional interior walls. The paper is organized as follows. Main RL concepts and deep Q-networks are shortly presented in the next section. Section 3 describes, in short, the created evacuation environment. The proposed modification of the deep Q-network algorithm is presented in Sect. 4. Some conclusions are drawn in the final section.

2 Q-Learning and Deep Q-Networks

In RL the sequential decision problem is modelled as a *Markov decision process* (MDP). Under this framework, an *agent* interacts with its *environment*, and at each time step t, it receives information about the environment's *state* $S_t \in \mathcal{S}$. Using this state, the agent selects an *action* $A_t \in \mathcal{A}$ and then receives a *reward* $R_{t+1} \in \mathbb{R}$. The action changes the environment's state to $S_{t+1} \in \mathcal{S}$. The agent behaves according to a policy $\pi(a|s)$, which is a probability distribution over the set $\mathcal{S} \times \mathcal{A}$. The goal of the agent is to learn the *optimal policy* that *maximizes the expected total discounted reward*. The *value* of taking action a in state s under a policy π is defined by:

$$Q_\pi(s,a) \equiv \mathbb{E}[R_1 + \gamma R_2 + \dots | S_o = s, A_o = a, \pi]$$

where $\gamma \in [0,1]$ is a *discount factor*. We call $Q_\pi(s,a)$ the *action-value function for policy* π. The *optimal value* is defined as $Q_*(s,a) = max_\pi Q_\pi(s,a)$. An *optimal policy* is derived by selecting the highest valued action in each state. *Q-learning* algorithm allows obtaining estimates for the optimal action values [10]. In cases where the number of state-action pairs is so large that it is not feasible to learn $Q(s,a)$ for each pair (s,a) separately, the agent can learn a parameterized value function $Q(s,a;\boldsymbol{\theta}_t)$. After taking action A_t in state S_t, obtaining the reward R_{t+1} and observing the next state S_{t+1} parameters are updated as follows [11]:

$$\boldsymbol{\theta}_{t+1} = \boldsymbol{\theta}_t + \alpha(Y_t^Q - Q((S_t, A_t; \boldsymbol{\theta}_t)) \nabla_{\boldsymbol{\theta}_t} Q(S_t, A_t; \boldsymbol{\theta}_t)) \tag{1}$$

where α is a *learning rate* and the *target value* Y_t^Q is defined as:

$$Y_t^Q \equiv R_{t+1} + \gamma \max_a Q(S_{t+1}, a; \boldsymbol{\theta}_t) \tag{2}$$

During the learning, the agent selects actions according to the so-called ε-greedy policy: with probability $1 - \varepsilon$, it selects the highest valued action, and with probability ε, it selects a random action. Learning takes place in the following loop (we omit subscript t for simplicity):

Initialize the value function Q;
foreach *episode* **do**
    Initialize S;
    while S *is not the terminal state* **do**
        Select action A according to an ε-greedy policy derived from Q;
        Take action A, observe R and the next state S';
        Update the value function $Q(S, A; \boldsymbol{\theta})$ towards the target value Y^Q;
        $S \to S'$
    end
end

A *deep Q network (DQN)* is a multi-layered neural network that for a given input state s returns $Q(s; \cdot; \boldsymbol{\theta})$, where $\boldsymbol{\theta}$ are network parameters [11]. Using the neural network as an approximator can cause learning instability. To prevent this Mnih et al. [2], proposed using the so-called *target network* (with parameters $\boldsymbol{\theta}^-$), which is the same as the *online network* (with parameters $\boldsymbol{\theta}$). Parameters $\boldsymbol{\theta}^-$ are copied periodically (each τ time steps) from the online network, thus reducing undesirable correlations with the target:

$$Y_t^{DQN} \equiv R_{t+1} + \gamma \max_a (S_{t+1}, a; \boldsymbol{\theta}_t^-) \tag{3}$$

The second modification to improve the learning stability is the so-called *memory replay* [2], in which the agent's experiences $(s_t, a_t, r_{t+1}, s_{t+1})$ at each time-step are stored. A mini-batch of experiences is sampled uniformly from the memory to update the online network during the learning. Finally, one more improvement of DQN called *Double DQN* (DDQN), which eliminates overestimation of Q values (see [11]). The solution is to use two different function approximators (networks): one (with parameters $\boldsymbol{\theta}$) for selecting the best action and the other (with parameters $\boldsymbol{\theta}^-$) for estimating the value of this action. Thus the target in Double DQN has the following form:

$$Y_t^{DoubleDQN} \equiv R_{t+1} + \gamma Q(S_{t+1}, argmax_a Q(S_{t+1}, a; \boldsymbol{\theta}); \boldsymbol{\theta}_t^-) \tag{4}$$

3 The Room Evacuation Environment

In order to test RL algorithms we have created the following evacuation environment.

A **state** refers to a room with walls (W), exits (E) and people (P) (see Fig. 1). A room of size $x \times y$ can be represented as a $3 \times x \times y$ tensor consisting of 3 matrices $x \times y$. Each matrix is an $x \times y$ grid of 1s and 0s, where digit 1 indicates the position of W, E or P (see Fig. 1). *The terminal state* is an empty room.

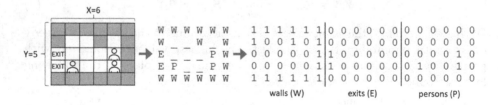

Fig. 1. A tensor representation of a room with 3 persons and 2 exits.

An agent takes **action** by selecting one field from $x \times y$ fields for a given situation in a room. Thus, the number of possible actions at a time step is equal to $x \times y$. It is worth noting that selecting a field is not equivalent to moving to that field because the move is not always possible. If there is one person near the field selected by the agent then this person moves to this field. If there are two or more persons near the selected field, the move is made by a person who comes closer to the exit after the move or by random person.

After each action, the agent receives a **reward**. The reward depends on whether the move on a selected field is possible or not. There are three cases possible:

1. The move to a selected field is possible because the field is empty, and there is a person on an adjacent field who can move. The agent receives $R = +1$ when the selected field is an exit (Fig. 2A) and $R = -1$ in other cases (Fig. 2B).
2. The move to a selected field cannot be made because on an adjacent field there is no person that can move (Fig. 2C) - the agent receives $R = -1$.
3. The move to a selected field is not possible because it is occupied by another person or a wall (Fig. 2D–E) - the agent receives $R = -1$.

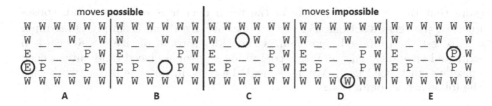

Fig. 2. Possible moves (left side) and impossible moves (right side).

We used the above values of rewards in our tests, but they can be easily changed. The environment can be *static* or *dynamic*. In the static case, the initial number and positions of persons in the room are the same in each learning episode. The dynamic case is more challenging than static - the initial persons' positions in the room are random. During our tests, the dynamic version of the environment was used. The agent aims to evacuate all the people from the room quickly. What is important, the agent has no initial knowledge about the positions of exits and walls in the room.

4   Modified Deep Q-Network and Its Tests

In standard DQN algorithm, the agent, at each time step, selects one action using an ε-greedy policy derived from Q. Next, the agent takes this action, observes a reward and the next state. After that, the update of Q is performed. However, in the evacuation scenario, people in the room can move simultaneously. It means that more than one action can be taken at each time step. The crucial point is that $x \times y$-dimensional vector $Q(S, \cdot; \boldsymbol{\theta})$ contains values of all $x \times y$ (possible and impossible) actions in the room. Consequently, instead of one action, we can select p actions $A_1, A_2, ..., A_p$ from $Q(S, \cdot; \boldsymbol{\theta})$ for a given state S, where p is the number of persons in the room. Namely, we can apply an ε-p-greedy policy during the learning, i.e. with probability $1-\varepsilon$ we select p highest valued actions, and with probability ε we select p random actions. Note that actions $A_1, A_2, ..., A_p$ are in some sense *mutually independent* because they correspond to different fields of room. After taking all actions $A_1, A_2, ..., A_p$ the state S changes to S'. Then, we can update Q. The while loop in the proposed algorithm has the following form:

while S *is not the terminal state* **do**

> Select actions $A_1, A_2, ..., A_p$ according to an ε-n-greedy policy derived from Q;
>
> Take actions $A_1, A_2, ..., A_p$, where p is the number of people in the room, observe rewards $R_1, R_2, ..., R_p$ and the next state S';
>
> Find the targets:
> $$Y_i = R_i + \gamma \underset{a}{max} Q(S', a; \boldsymbol{\theta})$$
> where $i = 1, 2, ..., p$;
>
> Perform a gradient descent step on $(Y_i - Q(S, A_i, \boldsymbol{\theta}))^2$;

end

Note that the number of people in the room decreases over time and, consequently, the agent's actions. For $p = 1$ the above algorithm is simply Q-learning.

To test the proposed algorithm, we consider a room of size 8×8 with two exits and an additional wall inside containing 18 people placed randomly at each episode's start (see Fig. 3). We implemented a convolutional neural network with the following configuration:

```
Conv2d(3,280,kernel_size=(5,5),stride=(1,1),padding=(2,2))
Conv2d(280,144,kernel_size=(3,3),stride=(1,1),padding=(1,1))
Conv2d(144,64,kernel_size=(3,3),stride=(1,1),padding=(1,1))
Conv2d(64,64,kernel_size=(3,3),stride=(1,1),padding=(1,1))
Conv2d(64,32,kernel_size=(3,3),stride=(1,1),padding=(1,1))
Flatten(start_dim=1,end_dim=-1)
Linear(in_features=2048,out_features=64)
```

The ReLU function is used for all layers, except the output layer, where a linear activation is used. The Adagrad optimizer with default parameters and a learning rate equal to 0.01 is used for training. We tested two agents. The first agent (DQN agent) uses the double DQN algorithm with the target network and memory replay. The second agent (MDQN agent) uses the *modified* double DQN algorithm with the target network. Both agents have the same values of the parameters: $\varepsilon = 0.1$, $\gamma = 0.9$, $\tau = 40$ (target network update frequency). Moreover, we set *memory replay* $= 500$ and *batch size* $= 100$ for the DQN agent. Both agents were trained for 1300 episodes. Figure 3 (left side) shows changes in total reward per episode. We can see that the DQN agent learns faster than the MDQN agent. The reason for this seems rather obvious. The DQN agent learns to take only one action at each time step. The MDQN agent has to learn to take more actions; consequently, it learns slower. On the right side of Fig. 3, we can see that the number of time steps required to evacuate all persons from the room is much smaller in the case of MDQN. Figure 4 indicates a significant difference between both agents. The DQN agent evacuates persons one by one.

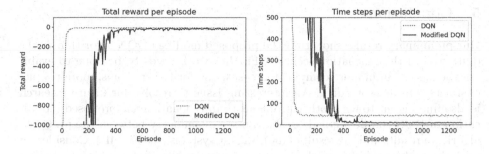

Fig. 3. The evacuation of 18 persons through 2 exits: total reward (left side), time steps (right side) - averages over 5 tests.

It takes one person and moves it from its initial position to the exit. Then the next person is evacuated. We can see in Fig. 4A that three persons marked with frames stay in place, waiting for their turns. In the MDQN agent's case (Fig. 4B), more than one person is moved to exits each time step. Figure 4B also shows that the MDQN agent evacuates persons via the nearest exit even if there are two exits located on the opposite walls.

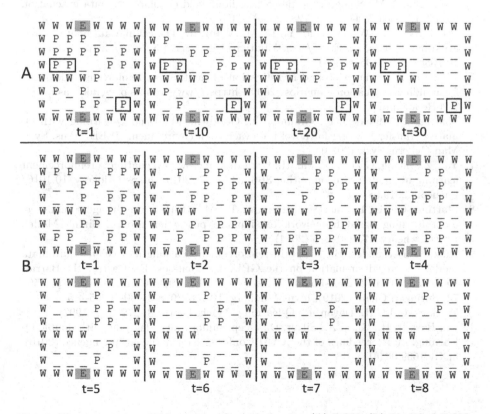

Fig. 4. The evacuation of 18 persons through 2 exits: (A) DQN, (B) Modified DQN.

5 Conclusions

Our preliminary results show that the proposed modified DQN algorithm works quite well in the evacuation scenario. Further work needs to be done to evaluate the algorithm in more complicated cases, e.g. for larger rooms, more people, obstacles and fire spreading. An interesting issue to resolve for future research is also finding out to what other problems than evacuation the proposed modification of DQN can be applied to. We can see a similarity between the proposed algorithm results and the result of multi-agent systems (see Fig. 4B). Considering that the multi-agent RL methods are computationally expensive, the question arises whether the proposed algorithm can be an alternative to the multi-agent approach in some cases.

References

1. Barto, A.G., Thomas, P.S., Sutton, R.S.: Some recent applications of reinforcement learning. In: Proceedings of the Eighteenth Yale Workshop on Adaptive and Learning Systems (2017)
2. Mnih, V., Kavukcuoglu, K., Silver, D., et al.: Human-level control through deep reinforcement learning. Nature **518**(7540), 529–533 (2015)
3. Gwizdałła, T.M.: Some properties of the floor field cellular automata evacuation model. Physica A **419**, 718–728 (2015)
4. Hu, J., Gao, X., Wei, J., Guo, Y., Li, M., Wang, J.: The cellular automata evacuation model based on Er/M/1 distribution. Phys. Scr. **95** (2019). https://doi.org/10.1088/1402-4896/ab4061
5. Wharton, A.: Simulation and investigation of multi-agent reinforcement learning for building evacuation scenarios (2009). https://www.robots.ox.ac.uk/~ash/4YP%20Report.pdf
6. Sharma, J., Andersen, P., Granmo, O., Goodwin, M.: Deep Q-learning with Q-matrix transfer learning for novel fire evacuation environment. IEEE Trans. Syst. Man Cybern. Syst. (2020)
7. Yao, Z., Zhang, G., Lu, D., Liu, H.: Data-driven crowd evacuation: a reinforcement learning method. Neurocomputing **366**, 314–327 (2019). https://doi.org/10.1016/j.neucom.2019.08.021
8. Martinez-Gil, F., Lozano, M., Fernandez, F.: MARL-Ped: a multi-agent reinforcement learning based framework to simulate pedestrian groups. Simul. Model. Practi. Theory **47**, 259–275 (2014)
9. Ruiz, S., Hernández, B.: A hybrid reinforcement learning and cellular automata model for crowd simulation on the GPU. In: Meneses, E., Castro, H., Barrios Hernández, C.J., Ramos-Pollan, R. (eds.) CARLA 2018. CCIS, vol. 979, pp. 59–74. Springer, Cham (2019). https://doi.org/10.1007/978-3-030-16205-4_5
10. Watkins, C.J.C.H., Dayan, P.: Q-learning. Mach. Learn. **8**, 279–292 (1992)
11. van Hasselt, H., Guez, A., Silver, D.: Deep reinforcement learning with double Q-learning. In: Proceedings of the AAAI Conference on Artificial Intelligence (2016). arXiv:1509.06461

Human-Like Storyteller: A Hierarchical Network with Gated Memory for Visual Storytelling

Lu Zhang[1,2], Yawei Kong[1,2], Fang Fang[2], Cong Cao[2(✉)], Yanan Cao[2],
Yanbing Liu[2], and Can Ma[2]

[1] Institute of Information Engineering, Chinese Academy of Sciences, Beijing, China
{zhanglu0101,kongyawei}@iie.ac.cn
[2] School of Cyber Security, University of Chinese Academy of Sciences,
Beijing, China
{fangfang0703,caocong,caoyanan,liuyanbing,macan}@iie.ac.cn

Abstract. Different from the visual captioning that describes an image concretely, the visual storytelling aims at generating an imaginative paragraph with a deep understanding of the given image stream. It is more challenging for the requirements of inferring contextual relationships among images. Intuitively, humans tend to tell the story around a central idea that is constantly expressed with the continuation of the storytelling. Therefore, we propose the Human-Like StoryTeller (HLST), a hierarchical neural network with a gated memory module, which imitates the storytelling process of human beings. First, we utilize the hierarchical decoder to integrate the context information effectively. Second, we introduce the memory module as the story's central idea to enhance the coherence of generated stories. And the multi-head attention mechanism with a self adjust query is employed to initialize the memory module, which distils the salient information of the visual semantic features. Finally, we equip the memory module with a gated mechanism to guide the story generation dynamically. During the generation process, the expressed information contained in memory is erased with the control of the read and write gate. The experimental results indicate that our approach significantly outperforms all state-of-the-art (SOTA) methods.

Keywords: Visual storytelling · Central idea · Gated memory

1 Introduction

Recently, the tasks of combining vision and text have made a great stride, such as the high-profile visual captioning [3], whose purpose is to generate literal descriptions based on the images or the videos. To further investigate the model's capabilities in generating structured paragraphs under more complicated scenarios,

L. Zhang and Y. Kong—Equal contribution.

© Springer Nature Switzerland AG 2021
M. Paszynski et al. (Eds.): ICCS 2021, LNCS 12743, pp. 257–270, 2021.
https://doi.org/10.1007/978-3-030-77964-1_21

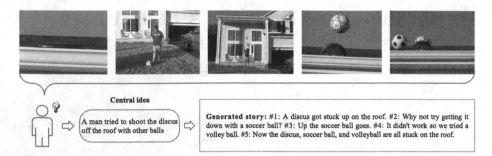

Fig. 1. An example of visual storytelling. "#i" indicates that this is the i-th sentence.

visual storytelling has been proposed by [9]. This task aims to generate a coherent and expressive story with a given temporal image stream, which not only expects an intuitive understanding of image ground content but also requires plentiful emotion as well as imagination. It is more challenging since the model must have the capabilities of inferring contextual relationships that are not explicitly depicted in the images.

Encouraged by the success of the Seq2Seq model in visual captioning, most visual storytelling methods usually employ this typical framework that consists of a visual encoder and a sentence decoder. The visual encoder transforms the image stream to feature vectors and then the sentence decoder generates every storyline. Based on the Seq2Seq framework, Kim et al. [11] and Jung et al. [10] optimize the specific architecture with maximum likelihood estimation (MLE) method; Wang et al. [22] and Hu et al. [7] employ reinforcement learning or adversarial training strategies to improve performance; Yang et al. [23] and Li and Li [12] focus on generating more expressive results by drawing into external knowledge or performing additional processing on the dataset. Although progresses have been made, the generated story still lacks centrality and have lots of semantic repetition, which significantly reduces the coherence and readability.

Intuitively, the contextual description of a story will revolve around a central idea. As shown in Fig. 1, all the contents of the five storylines are related to the central idea - "A man tried to shoot the discus off the roof with other balls". If there is no guidance of it, the second sentence may be about "playing football" instead of "getting it down with a soccer ball". The former just depicts the intuitive content of the image, resulting in the incoherence of the context. Therefore, it is critical to model the central idea during the storytelling process.

Towards filling these gaps, we propose the Human-Like StoryTeller (HLST), a hierarchical neural network with the gated memory module. First, considering the importance of the context in generating coherent stories, we introduce a hierarchical decoder that includes the narration decoder and the sentence decoder. The narration decoder constructs a semantic concept for guiding the sentence decoder to generate every storyline, which makes the generating process in chronological order rather than parallel. Second, we utilize the multi-head attention mechanism with a self adjust query to obtain global memory as our central idea. The self adjust query questions the model "What is the story about?" and the attention mechanism solves it by focusing on different pieces of salient information within the visual semantic features. Then, the model grasps the central idea to generate more coherent and relevant story. Finally, we equip the memory module with a gated mechanism to guide the story generation dynamically. The memory information is gradually erased under the control of the read and write gate, which improves the story's diversity and informativeness. We conduct the experiments on the VIST dataset and the results show that HLST significantly outperforms all baseline models in terms of automatic evaluations and human judgments. Further qualitative analysis indicates that the generated stories by HLST are highly coherent with human understanding.

Our main contributions are summarized as follows:

- To our knowledge, we are the first one to introduce the concept of central idea to benefit the task of visual storytelling.
- We propose the memory module as the central idea to enhance the coherence of stories. It guides the generation process of the hierarchical decoder that integrates the contextual information effectively.
- We equip the memory module with a gated mechanism to dynamically express the central idea. The gated mechanism is conducive to generate more informative stories by removing redundant information.
- Our approach achieves state-of-the-art (SOTA) results, in terms of automatic metrics and human judgments. By introducing the central idea, the generated stories are more coherent and diverse.

2   Related Work

In early visual to language tasks, visual captioning task achieves impressive results [20]. Generally, most visual captioning models utilize the CNN to extract the features of the image or video and send them to a decoder for generating a sentence caption. Take one step further, the visual storytelling is expected to generate an expressive and coherent paragraph with a temporal image stream instead of a single image. Notably, this task is more difficult because it not only focuses on the objective descriptions of the visual objects but also requires to consider the contextual coherence with a deeper understanding of the inputs.

Park and Kim [16] has made pioneering research to explore the visual storytelling task, which retrieves a sequence of natural sentences for an image stream. For the better development of this field, Huang et al. [9] releases a more compatible and sophisticated dataset, named VIST. The VIST is composed of visual-story pairs, in which each item contains five images and the corresponding sentences. In addition, they first employ the Seq2Seq framework to generate stories, which naturally extends the single-image captioning technique of [3] to multiple images. Hence, the subsequent endeavors are concentrating on improving the specific architectures. Kim et al. [11] and Gonzalez-Rico [4] aim at incorporating the contextual information. To alleviate the repetitiveness, Hsu [6] proposes

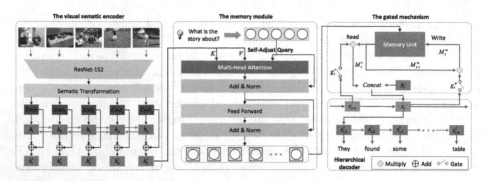

Fig. 2. Model overview. The detailed HLST includes three important modules: a visual semantic encoder, a gated memory module and a hierarchical decoder.

the inter-sentence diverse beam search algorithm. Furthermore, some researchers [2,7,21] strive to incorporate reinforcement learning with rewards or adversarial training strategies for generating more relevant stories. And other studies [8,12,23] are based on drawing into external knowledge or preprocessing data to improve performance. More specifically, Wang et al. [22] first implements a baseline model (XE-ss) as a policy model, which employs a Seq2Seq framework to generate storylines in parallel. They further propose the adversarial reward learning (AREL) to learn an implicit reward for optimizing the policy model.

Intuitively, people usually grasp global information as a central idea and tell the story around it. Hence, we propose HLST, which consists of a gated memory module and a hierarchical decoder. We utilize the multi-head attention [18] with a self adjust query to initialize the memory as the central idea. Inspired by [15,17], we update the memory unit with the gated mechanism to dynamically express the central idea as human beings.

3 Our Approach

In this section, we introduce our Human-Like StoryTeller (HLST) model detailly. As shown in Fig. 2, HLST is composed of three modules: a visual semantic encoder, a gated memory module and a hierarchical decoder. Given five images $V = (v_1, \cdots, v_5)$ in order, the visual semantic encoder obtains the semantic vectors $H' = (h'_1, \cdots, h'_5)$ by integrating the individual and contextual features. Then, we utilize a hierarchical decoder to strengthen the contextual relevance between sentences. It is composed of a narration decoder and a sentence decoder. Moreover, the narration decoder constructs the high-level sentence representations for $S = (s_1, \cdots, s_5)$. And the sentence decoder generates a word sequence $W = (w_{t,1}, w_{t,2}, \cdots, w_{t,n}), w_{t,j} \in \mathbb{V}$ in chronological order based on the corresponding sentence representation s_t. Here, $\mathbb{V}$ is the vocabulary of all output tokens. Moreover, we first introduce the memory module M in this field, which acts as the central idea to enhance coherence in story generation. Specifically, we employ the multi-head attention with a self adjust query to distil the salient information within $H^{'}$. To further dynamically express the central idea, we equip the memory module with the gated mechanism that consists of a read gate and a write gate. With the continuation of the generation process, the information contained in M is eliminated gradually. In the following, we will describe these three parts detailly. For simplicity, we omit all bias in formulations.

3.1   Visual Semantic Encoder

The visual semantic encoder consists of a pre-trained CNN layer, a semantic transformation layer and a Bidirectional Gated Recurrent Units (BiGRU) layer. Given an image stream $V = (v_1, \cdots, v_5), v_i \in \mathbb{R}^{d_v}$, the CNN layer is responsible for extracting the visual features. Then, we map the visual representations into the semantic space to obtain semantic features $f_i \in \mathbb{R}^{d_f}, i \in [1, 5]$ with a linear transformation layer. Here, d_v and d_f is the dimension of the visual and semantic features, respectively. The BiGRU layer further encodes the semantic features as the context vectors $h_i = [\overleftarrow{h_i}; \overrightarrow{h_i}]$, $h_i \in \mathbb{R}^{d_h}$, which integrates the results of the forward and backward calculations. Here, d_h is the number of hidden units. Furthermore, since each sentence in the generated stories corresponds to the specific image, we strengthen the influence of the corresponding image features through a skip connection. The final semantic features $h_i^{'}$ at time-step i is computed as follows:

$$
\begin{aligned}
f_i &= W_1 \cdot \mathrm{CNN}(v_i) \\
h_i &= \mathrm{BiGRU}(h_{i-1}, f_i) \\
h_i^{'} &= W_3(h_i \oplus W_2 f_i)
\end{aligned}
\tag{1}
$$

where $\oplus$ represents the vector concatenation and $\cdot$ denotes the matrix multiplication. $W_1 \in \mathbb{R}^{d_f \times d_v}$, $W_2 \in \mathbb{R}^{d_h \times d_f}$ and $W_3 \in \mathbb{R}^{d_h \times 2d_h}$ are the learnable linear transformation matrices.

3.2   Hierarchical Decoder

Different from many existing works that only use a sentence decoder, we employ the hierarchical decoder to integrate the contextual information effectively. It is composed of a narration decoder and a sentence decoder. The narration decoder is an unidirectional GRU that constructs the sentence representations $s_t \in \mathbb{R}^{d_s}, t \in [1, 5]$. Here, d_s is the number of hidden state units. At each time-step t, the corresponding encoder output $h_t^{'}$ and previous sentence representation s_{t-1} are fed into the narration decoder for calculating s_t. Notably, s_t integrates the information of the generated sentences effectively for taking s_{t-1} as input. Meanwhile, the sentence decoder predicts the next word $w_{t,j}$ based on the s_t and the previous generated word $w_{t,j-1} \in \mathbb{R}^{d_e}$, where d_e is the dimension of the word embedding. The whole generation process can be described as follows:

$$
\begin{aligned}
s_t &= \mathrm{GRU}_n(s_{t-1}, h_i^{'}) \\
s_{t,j}^{'} &= \mathrm{GRU}_s(s_{t,j-1}^{'}, W_4(s_t \oplus w_{t,j-1}))
\end{aligned}
\tag{2}
$$

where $W_4 \in \mathbb{R}^{d_s \times (d_s + d_e)}$ is a learnable projection matrix. Note that GRU_n and GRU_s represent the narration decoder and sentence decoder, respectively. Besides, the sequential structure GRU_n makes the whole generation process in chronological order. And the $s'_{t,j} \in \mathbb{R}^{d_s}$ represents the jth hidden state of the sentence decoder at t^{th} sentence, which is utilized to compute the word probability distribution over the whole vocabulary:

$$p_\theta(w_{t,j}|w_{t,j-1}, s_t, \mathbf{v}) = Softmax(\text{MLP}(s'_{t,j})) \tag{3}$$

where $\text{MLP}(\cdot)$ represents the multi-layer perception that projects the $s'_{t,j}$ to the vocabulary size.

3.3 Multi-head Attention with Self Adjust Query

The multi-head attention mechanism distils the central idea by attending to the visual semantic features $H' = \{h'_1, \ldots, h'_n\}$. However, different from the traditional method, we introduce a self adjust query that is a learnable variable trained with the model. As illustrated in Fig. 2, the self adjust query is equivalent to put a question to the model - "*What is the story about?*". Through the interaction between the self adjust question and visual semantic features, we get the question-aware memory $M_{init} \in \mathbb{R}^{d_h}$ as the central idea, which contains several pieces of salient semantic information. The process is formulated as follows:

$$M_{init} = \text{MHA}(Q^A, H', H')$$
$$\text{MHA}(Q^A, H', H') = \text{Concat}(head_1, \ldots, head_N)W^O \tag{4}$$
$$head_j = \text{Softmax}(\frac{Q^A W_j^Q (H' W_j^K)^T}{\sqrt{d_k}})H' W_j^V$$

where $W_j^Q, W_j^K \in \mathbb{R}^{d_h \times d_k}, W_j^V \in \mathbb{R}^{d_h \times d_v}$ and $W^O \in \mathbb{R}^{Nd_v \times d_h}$ are learnable linear transformation matrices. N is the head number and $d_k = d_v = d_h/N$ in this paper. And $Q^A \in \mathbb{R}^{d_h}$ is the self adjust query vector, which is initialized randomly. Obviously, the attention mechanism grasps the global semantic information, which is utilized to guide the generation process and improves the coherent of generated stories.

3.4 Gated Memory Mechanism

To dynamically control the expression of the story's central idea, we further equip the memory module with gated mechanism that includes a read and a write gate. At different decoding steps, each sentence requires various information to be expressed. Therefore, the read gate is responsible for reading the currently needed content from the memory unit and feeds it to the narration decoder. Then, to avoid elaborating the duplicated information, we employ the write gate to update the memory unit with the current hidden state of the narration decoder, which leads to a decline of the information stored in memory as the decoding processes. We further conduct the ablation experiments to verify the effectiveness of each gate.

At the high-level decoding time t, the read gate $g_t^r \in \mathbb{R}^{d_h}$ is computed with the previous state s_{t-1} of the narration decoder, while the write gate $g_t^w \in \mathbb{R}^{d_h}$ is calculated with the current state s_t. The two gates can be formulated as follows:

$$g_t^r = \sigma(W_r s_{t-1}), \quad g_t^w = \sigma(W_w s_t) \tag{5}$$

where $W_r \in \mathbb{R}^{d_h \times d_s}$ and $W_w \in \mathbb{R}^{d_h \times d_s}$ are learnable parameters. The $\sigma(\cdot)$ is the sigmoid nonlinear activation function and the output value ranges from 0 to 1. Next, the read and write gates are used to update the memory unit as follows:

$$M_t^r = g_t^r \odot M_{t-1}^w, \quad M_t^w = g_t^w \odot M_{t-1}^w \tag{6}$$

Here, $\odot$ denotes the element-wise multiplication and all $M \in \mathbb{R}^{d_h}$. Besides, M_{t-1}^w is the memory contents written back at the previous time-step, which is updated to M_t^w by the write gate g_t^w. And M_t^r is read from M_{t-1}^w with the read gate g_t^r and then is fed into the narration decoder with the encoder output h_t' to compute the current hidden state s_t. Finally, we modify the Eq. (2) as follows:

$$s_t = \text{GRU}(s_{t-1}, W_m(h_t' \oplus M_t^r)) \tag{7}$$

where $W_m \in \mathbb{R}^{d_h \times 2d_h}$ is a linear transformation matrix. It is worth noting that the Eq. (6) is executed several times, which is equivalent to continuously multiplying a matrix between $[0, 1]$. Therefore, the expressed information contained in memory vectors M is gradually decreasing in the decoding process, which is similar to the central idea expressed completely as human beings.

4   Experiments

4.1   Dataset

We conduct experiments on the VIST dataset, which is the most popular dataset for the visual storytelling task [9]. In detail, the dataset includes 10,117 Flickr albums with 210,819 unique images. Each story consists of five images and their corresponding descriptions. For the fair comparison, we follow the same experimental settings as described in [10,22]. Finally, we obtain 40,098 training, 4,988 validation and 5,050 testing samples after filtering the broken images.

4.2   Evaluation Metrics

To evaluate our model comprehensively, we adopt both automatic evaluations and human judgments. Four different automatic metrics are utilized to measure our results, including BLEU [14], ROUGE [13], METEOR [1] and CIDEr [19]. For a fair comparison, we employ the open-source evaluation code[1] as

---

[1] https://github.com/lichengunc/vist_eval.

[22,24]. Since automatic evaluation metrics may not be completely consistent with human judgments, we also invite 6 annotators with the corresponding linguistic background to conduct human evaluations as in [23]. We randomly sample 200 stories from the test dataset, of which each example consists of the image stream and the generated story of different models. We also evaluate the human-generated stories for comparison. Then, all annotators score the results from 1 to 5 in the four aspects: fluency, coherence, relevance and informativeness. In detail, **fluency** mainly evaluates whether the output is grammatically fluent, while **coherence** measures the semantic similarity between sentences. **Relevance** represents the correlation between the generated story and the images. **Informativeness** measures the diversity and richness of outputs.

4.3 Baseline Models

We mainly compare our model with the following representative and competitive frameworks.

Seq2Seq [9] introduces the first dataset for the visual storytelling task and proposes a simple Seq2Seq model. **HARAS** [24] first selects the most representative photos and then composes a story for the album. **GLAC Net** [11] aims to combine global-local (glocal) attention and context cascading mechanisms. **SRT** [21] utilizes reinforcement learning and adversarial training to train the model better. **XE-ss** [22] is a typical encoder-decoder framework that generates story sentences parallel. **GAN** [22] incorporates generative adversarial training based on the XE-ss model. **AREL** [22] learns an implicit reward function and then optimizes policy search with the learned reward function. **XE-TG** [12] exploits textual evidence from similar images to generate coherent and meaningful stories. **HSRL** [8] employs the hierarchical framework trained with the reinforcement learning. **Knowledge** [23] extracts a set of candidate knowledge graphs from the knowledge base and integrates in the attention model. **ReCo-RL** [7] introduces three assessment criteria as the reward function. **INet** [10] proposes a hide-and-tell model that can learn non-local relations across the photo streams.

4.4 Training Details

We extract the image features with a pre-trained ResNet-152 Network proposed by [5], which is widely used in visual storytelling. The feature vector of each image is obtained from the fully-connected (fc) layer that has 2,048 dimensions. Besides, the vocabulary contains the words that occur more than three times in the training set, of which the final size is 9,837. The head number of multi-head attention is set to 4. During training, we set the batch size to 64, and the dimensions of hidden units are all set to 512. We use the Adam optimizer with initial learning rate 10^{-4} to optimize the model. At test time, our stories are produced using the beam search algorithm with the beam size 4. We implement and run all models on a Tesla P4 GPU card with PyTorch[2].

---

[2] https://pytorch.org/.

5   Results and Discussion

5.1   Automatic Evaluation

Table 1. Automatic evaluation results on the VIST dataset. B: BLEU

Models	B1	B2	B3	B4	ROUGE	METEOR	CIDEr
Seq2Seq (Huang et al. 2016)†	52.2	28.4	14.5	8.1	28.5	31.1	6.4
HARAS (Yu et al. 2017)†	56.3	31.2	16.4	9.7	29.1	34.2	7.7
GLAC Net (Kim et al. 2018)†	52.3	28.4	14.8	8.1	28.4	32.4	8.4
SRT (Wang et al. 2018)†	60.5	36.7	20.8	12.5	28.9	33.1	8.5
XE-ss (Wang et al. 2018)	62.3	38.2	22.5	13.7	29.7	34.8	8.7
GAN (Wang et al. 2018)	62.8	38.8	23.0	14.0	29.5	35.0	9.0
AREL(Wang et al. 2018)	63.8	39.1	23.2	14.1	29.5	35.0	9.4
XE-TG (Li and Li 2019)	–	–	–	–	30.0	35.5	8.7
HSRL (Huang et al. 2019)	–	–	–	12.3	30.8	35.2	10.7
Knowledge (Yang et al. 2019)	66.4	39.2	23.1	12.8	29.9	35.2	**12.1**
ReCo-RL (Hu et al.2020)	–	–	–	14.4	30.1	35.2	6.7
INet (Jung et al. 2020)	64.4	40.1	23.9	14.7	29.7	35.6	10.0
Our Model (HLST)	**67.7**	**42.6**	**25.3**	**15.2**	**30.8**	**36.4**	11.3

Table 1 gives the automatic evaluation results on the VIST dataset. The results of all baselines are taken from the corresponding papers and the token "†" means that the results are achieved by [23]. The best performance is highlighted in bold and the results show that our approach significantly outperforms other methods. As shown in Table 1, HLST achieves the best performance on almost all automatic metrics. In detail, all BLEU scores have been improved significantly. Moreover, BLEU2 and BLEU3 exceed the highest score of baseline models by 2.5 and 1.4, respectively. The BLEU metric calculates the n-gram matching degree similarity of the reference and the candidate text and the higher-order BLEU can measure the sentence similarity to some extent. Therefore, the improvements in BLEU scores suggest that HLST is able to generate more coherent and informative stories as human beings. Besides, ROUGE-L and METEOR metrics both achieve state-of-the-art results. The ROUGE-L prefers to measure the recall rate between the ground truth and generated stories. And Huang et al. [9] turn out that the METEOR metric correlates best with human judgments on this task. Hence, the improvements indicate that our HLST can generate high-quality stories. In addition, HLST performs slightly worse than [23] on CIDEr since they equip their model with external knowledge base. And Wang et al. [22] empirically find that the references to the same image sequence are photostream different from each other while CIDEr measures the similarity of a sentence to the majority of the references. Hence, CIDEr may not be suitable for this task. In a word, without any external strategies, such as data preprocessing, knowledge graphs and reinforcement learning, our HLST achieves the best results just by integrating the hierarchical decoder and the gated memory module.

5.2   Human Evaluation

Table 2. The human evaluation results.

Models	Fluency	Relevance	Coherence	Informativeness
XE-ss	4.19	3.72	2.88	2.94
AREL	4.23	4.19	3.25	3.18
HLST	**4.46**	**4.58**	**4.17**	**4.24**
Human	4.65	4.72	4.41	4.59

The human evaluation results are shown in Table 2, which are calculated by averaging all scores from 6 annotators. Obviously, our model significantly exceeds all baselines, especially in coherence and informativeness. For example, compared to the AREL, the coherence score increases from 3.25 to 4.17, and the informativeness score increases from 3.18 to 4.24. The coherence measures whether the output is semantically coherent, while the informativeness evaluates the diversity of the generated stories. Therefore, the high scores of coherence and informativeness suggest that the memory module promotes contextual coherence and the gated mechanism enhances the diversity by reducing semantic duplication. Furthermore, the results are very close to the human's, which suggests that HLST can generate more informative and coherent stories as humans.

5.3   Ablation Study

Table 3. The automatic evaluation results of the ablation study. B: BLEU

Models	B1	B2	B3	B4	ROUGE	METEOR	CIDEr
Basic model	62.1	38.2	22.5	13.7	29.9	35.3	8.1
+ hierarchical decoder	63.4	39.3	23.3	14.2	30.1	35.5	8.7
+ memory	66.3	41.1	24.2	14.5	30	36	9.8
+ read gate	67.0	41.7	24.5	14.4	29.9	35.7	10
+ write gate (HLST)	**67.7**	**42.6**	**25.3**	**15.2**	**30.8**	**36.4**	**11.3**

To investigate the correctness and effectiveness of different modules, we conduct the ablation study and the results are listed in Table 3. The token "+" indicates that we add the corresponding module to the model. Note that our basic model is XE-ss, on which we add the hierarchical decoder, the memory module and the gated mechanism sequentially. We first verify the correctness of the hierarchical decoder. As shown in Table 3, the model with the hierarchical decoder achieves better results. Taking the BLEU1 for an example, the score exceeds the basic model by 1.3 points. Therefore, the model is able to generate more fluent and relevant stories by introducing the hierarchical decoder. We further analyze the effectiveness of the memory module by employing the multi-head attention with self adjust query. And the model with the memory module outperforms the former significantly. The score of BLEU1 is improved to 66.3, which is 2.9 points higher than the model only with a hierarchical decoder. And the results of METEOR and CIDEr are both greatly improved. It suggests that the memory module is important to generate fluent, relevant and coherent stories for integrating the central idea. Furthermore, we conduct experiments to explore the influence of the gated mechanism by sequentially adding the read gate and write gate. The read gate enables the model to read the currently needed information by removing the expressed message. And the improvement of automatic evaluations verifies its effectiveness. Finally, we equip the memory module with both the read and write gate to further filter out the expressed information. Compared with the Basic model, all BLEU scores have increased by about 10%, and CIDEr has increased by 39.5%. This further demonstrates that the central idea is crucial for visual storytelling, which can do a great favour to generating coherent and diverse stories.

5.4   Qualitative Analysis

Fig. 3. An example of qualitative comparison. We mainly compare our model with the competitive baseline AREL and the human annotations.

We take out an image stream from the test dataset and compare the story's qualities from AREL, HLST and the ground truth for qualitative analysis. As shown in Fig. 3, the image stream depicts a story about the wedding. Clearly, the story of HLST is more diverse and expressive than the AREL. In this example, the story of AREL contains a lot of duplicate phrases and sentences, e.g., "It was a beautiful day" and "The bride and groom were very/so happy to be married".

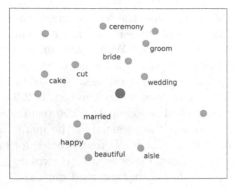

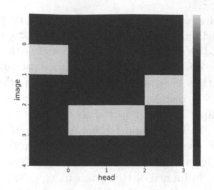

Fig. 4. Memory visualization. (Color figure online)

Fig. 5. Attention visualization.

The semantic repetition significantly reduces the readability and informativeness of the whole story. However, by introducing the multi-head attention with self adjust query, our model grasps the central idea "wedding" and generates the story that contains corresponding keywords, i.e., "wedding cake" and "ceremony". And the semantic repetition is also improved for equipping the memory module with the gated mechanism. Furthermore, HLST has the capability of capturing the chronological relationship with the hierarchical decoder, such as the phrase "After the ceremony".

To verify whether the performance improvements are owing to the central idea, we further conduct data analysis, including memory-aware words distribution and attention visualization. Specifically, we compute the cosine similarity between the memory and the generated word embeddings. As shown in Fig. 4, the red dot represents the memory, and the blue dots are generated words in the story. For simplicity, we remove the common words, such as "a", "the" and "for" etc. The distance between the blue dot and red dot measures the correlation between the corresponding words and the central idea. Hence, the words "bride", "groom" and "wedding" are more relevant than the words "aisle". It indicates that the memory module grasps the central idea and improves the informative of the generated story. We also visualize the attention of this example in Fig. 5. The x-coordinate indicates the number of head and the y-coordinate indicates the input image stream. The lighter the color is, the more important the image is. We can see that the HLST pays more attention to the salient images, i.e. image2, image3 and image4, which is coherent with human understanding. The image3 and image4 explicitly demonstrate that "wedding" is the central idea of the story. Besides, although the scenes in the first two pictures are similar, the bride in image2 is more prominent than the groom in image1. Therefore, the image2 is more relevant with the central idea "wedding" than image1. That is why HLST generates the coherent and diverse stories as human beings.

6   Conclusion

In this paper, we propose a hierarchical neural network with the gated memory module to imitate the process of human storytelling. We utilize a hierarchical decoder to integrate the contextual information, including a narration decoder and a sentence decoder. Besides, the multi-head attention with self adjust query is employed to capture the salient information to initialize the memory unit as the central idea. Furthermore, we equip the memory module with the gated mechanism that includes a read gate and a write gate. The automatic evaluation results and human judgments show that our HLST outperforms all state-of-the-art methods significantly.

Acknowledgement. This research is supported by the National Key R&D Program of China (No.2017YFC0820700, No.2018YFB1004700).

References

1. Banerjee, S., Lavie, A.: Meteor: an automatic metric for MT evaluation with improved correlation with human judgments. In: Proceedings of the ACL Workshop on Intrinsic and Extrinsic Evaluation Measures for Machine Translation and/or Summarization, pp. 65–72 (2005)
2. Chen, Z., Zhang, X., Boedihardjo, A.P., Dai, J., Lu, C.T.: Multimodal storytelling via generative adversarial imitation learning. In: Proceedings of the Twenty-Sixth International Joint Conference on Artificial Intelligence, IJCAI 2017, pp. 3967–3973 (2017)
3. Devlin, J., et al.: Language models for image captioning: the quirks and what works. arXiv preprint arXiv:1505.01809 (2015)
4. Gonzalez-Rico, D., Pineda, G.F.: Contextualize, show and tell: a neural visual storyteller. CoRR abs/1806.00738 (2018)
5. He, K., Zhang, X., Ren, S., Sun, J.: Deep residual learning for image recognition. In: Proceedings of the IEEE Conference on Computer Vision and Pattern Recognition, pp. 770–778 (2016)
6. Hsu, C., Chen, S., Hsieh, M., Ku, L.: Using inter-sentence diverse beam search to reduce redundancy in visual storytelling. CoRR abs/1805.11867 (2018)
7. Hu, J., Cheng, Y., Gan, Z., Liu, J., Gao, J., Neubig, G.: What makes a good story? Designing composite rewards for visual storytelling. In: Proceedings of the AAAI Conference on Artificial Intelligence, vol. 34, no. 05, pp. 7969–7976, April 2020
8. Huang, Q., Gan, Z., Celikyilmaz, A., Wu, D., Wang, J., He, X.: Hierarchically structured reinforcement learning for topically coherent visual story generation. In: Proceedings of the AAAI Conference on Artificial Intelligence, vol. 33, pp. 8465–8472 (2019)
9. Huang, T.H.K., et al.: Visual storytelling. In: Proceedings of the 2016 Conference of the North American Chapter of the Association for Computational Linguistics: Human Language Technologies, pp. 1233–1239 (2016)
10. Jung, Y., Kim, D., Woo, S., Kim, K., Kim, S., Kweon, I.S.: Hide-and-tell: learning to bridge photo streams for visual storytelling. In: Proceedings of the AAAI Conference on Artificial Intelligence, vol. 34, pp. 11213–11220 (2020)

11. Kim, T., Heo, M., Son, S., Park, K., Zhang, B.: GLAC net: GLocal attention cascading networks for multi-image cued story generation. CoRR abs/1805.10973 (2018)
12. Li, T., Li, S.: Incorporating textual evidence in visual storytelling. In: Proceedings of the 1st Workshop on Discourse Structure in Neural NLG, Tokyo, Japan, pp. 13–17. Association for Computational Linguistics, November 2019
13. Lin, C.Y.: ROUGE: a package for automatic evaluation of summaries. In: Text Summarization Branches Out, Barcelona, Spain, pp. 74–81. Association for Computational Linguistics, July 2004
14. Lin, C.Y., Och, F.J.: Automatic evaluation of machine translation quality using longest common subsequence and skip-bigram statistics. In: Proceedings of the 42nd Annual Meeting of the Association for Computational Linguistics (ACL 2004), Barcelona, Spain, pp. 605–612, July 2004
15. Liu, F., Perez, J.: Gated end-to-end memory networks. In: Proceedings of the 15th Conference of the European Chapter of the Association for Computational Linguistics: Volume 1, Long Papers, pp. 1–10 (2017)
16. Park, C.C., Kim, G.: Expressing an image stream with a sequence of natural sentences. In: Cortes, C., Lawrence, N., Lee, D., Sugiyama, M., Garnett, R. (eds.) Advances in Neural Information Processing Systems, vol. 28, pp. 73–81. Curran Associates, Inc. (2015)
17. Sukhbaatar, S., Szlam, A., Weston, J., Fergus, R.: End-to-end memory networks. In: Cortes, C., Lawrence, N., Lee, D., Sugiyama, M., Garnett, R. (eds.) Advances in Neural Information Processing Systems, vol. 28, pp. 2440–2448. Curran Associates, Inc. (2015)
18. Vaswani, A., et al.: Attention is all you need. In: Proceedings of the 31st International Conference on Neural Information Processing Systems, pp. 6000–6010 (2017)
19. Vedantam, R., Zitnick, C.L., Parikh, D.: Cider: consensus-based image description evaluation. In: 2015 IEEE Conference on Computer Vision and Pattern Recognition (CVPR), pp. 4566–4575 (2015)
20. Vinyals, O., Toshev, A., Bengio, S., Erhan, D.: Show and tell: a neural image caption generator. In: 2015 IEEE Conference on Computer Vision and Pattern Recognition (CVPR), pp. 3156–3164 (2015)
21. Wang, J., Fu, J., Tang, J., Li, Z., Mei, T.: Show, reward and tell: automatic generation of narrative paragraph from photo stream by adversarial training. In: Thirty-Second AAAI Conference on Artificial Intelligence (2018)
22. Wang, X., Chen, W., Wang, Y.F., Wang, W.Y.: No metrics are perfect: adversarial reward learning for visual storytelling. In: Proceedings of the 56th Annual Meeting of the Association for Computational Linguistics (vol. 1: Long Papers), pp. 899–909 (2018)
23. Yang, P., et al.: Knowledgeable storyteller: a commonsense-driven generative model for visual storytelling. In: Proceedings of the 28th International Joint Conference on Artificial Intelligence, pp. 5356–5362. AAAI Press (2019)
24. Yu, L., Bansal, M., Berg, T.: Hierarchically-attentive RNN for album summarization and storytelling. In: Proceedings of the 2017 Conference on Empirical Methods in Natural Language Processing, Copenhagen, Denmark, pp. 966–971. Association for Computational Linguistics, September 2017

Discriminative Bayesian Filtering for the Semi-supervised Augmentation of Sequential Observation Data

Michael C. Burkhart[✉][iD]

Adobe Inc., San José, USA
mburkhar@adobe.com

Abstract. We aim to construct a probabilistic classifier to predict a latent, time-dependent boolean label given an observed vector of measurements. Our training data consists of sequences of observations paired with a label for precisely one of the observations in each sequence. As an initial approach, we learn a baseline supervised classifier by training on the labeled observations alone, ignoring the unlabeled observations in each sequence. We then leverage this first classifier and the sequential structure of our data to build a second training set as follows: (1) we apply the first classifier to each unlabeled observation and then (2) we filter the resulting estimates to incorporate information from the labeled observations and create a much larger training set. We describe a Bayesian filtering framework that can be used to perform step 2 and show how a second classifier built using the latter, filtered training set can outperform the initial classifier.

At Adobe, our motivating application entails predicting customer segment membership from readily available proprietary features. We administer surveys to collect label data for our subscribers and then generate feature data for these customers at regular intervals around the survey time. While we can train a supervised classifier using paired feature and label data from the survey time alone, the availability of nearby feature data and the relative expensive of polling drive this semi-supervised approach. We perform an ablation study comparing both a baseline classifier and a likelihood-based augmentation approach to our proposed method and show how our method best improves predictive performance for an in-house classifier.

Keywords: Bayesian filtering · Discriminative modeling · Data augmentation · Semi-supervised learning · Machine learning · Learning from survey data

1 Problem Description and Notation

We aim to predict a binary-valued label of interest $Z_t \in \{0,1\}$ from a vector $X_t \in \mathbb{R}^m$ of measurable features. We are provided a supervised dataset

$$\mathcal{D}_0 = \{(x_\tau^1, z_\tau^1), (x_\tau^2, z_\tau^2), \ldots, (x_\tau^n, z_\tau^n)\}$$

© Springer Nature Switzerland AG 2021
M. Paszynski et al. (Eds.): ICCS 2021, LNCS 12743, pp. 271–283, 2021.
https://doi.org/10.1007/978-3-030-77964-1_22

of n labeled training pairs and an unsupervised dataset

$$\mathcal{D}_1 = \{x^1_{1:\tau-1}, x^1_{\tau+1:T}; x^2_{1:\tau-1}, x^2_{\tau+1:T}; \cdots ; x^n_{1:\tau-1}, x^n_{\tau+1:T}\}$$

of time-indexed feature data for each training instance in a contiguous period $1 \leq t \leq T$ surrounding τ. We adopt the notation $x^i_{1:T} = (x^i_1, x^i_2, \ldots, x^i_T)$ for indexed sequences of data and let $\tau \in \{1, 2, \ldots, T\}$ denote the time for which each sequence $x^i_{1:T}$ of features has an associated label z^i_τ. Strictly speaking, we allow this time τ to be different for each sequence, i.e. τ may depend on i. We suppress the superscript i when describing calculations for a single, generic instance.

If we have reason to believe that the relationship between the observed features and latent labels is stationary, i.e. that $p(z_t|x_t)$ does not depend on the time t, then a natural first approach to solving this problem entails training a supervised classifier on the dataset $\mathcal{D}_0$. *Can the unlabeled sequences of observations in $\mathcal{D}_1$ help us to build a better classifier?* That question, central to the field of semi-supervised learning, motivates our work. We intend to incorporate information from $\mathcal{D}_1$ through a process of data augmentation. In this paper, we develop and validate a novel method for estimating labels for the augmentation process. We develop a discriminative Bayesian filtering framework that provides a principled way to incorporate information from the unlabeled observations with information from the known label provided for a different observation in the same sequence.

Our work focuses on creating new training pairs $(x^i_t, \hat{z}^i_t)$ where x^i_t belongs to one of the sequences in $\mathcal{D}_1$ and $\hat{z}^i_t$ denotes an estimated label for that observation at that time. We combine two sources of knowledge to form our estimate: (1) the snapshot x^i_t of feature data, to which we may apply our original model for $p(z_t|x_t)$ and (2) the ground-truth label z^i_τ that fixes instance i's label at a nearby point in time, to which we may iteratively apply a latent state model. For each point in $\mathcal{D}_1$, we calculate the posterior probability $p(z_t|x_{\tau+1}, \ldots, x_t, z_\tau)$ when $t > \tau$ and $p(z_t|x_t, \ldots, x_{\tau-1}, z_\tau)$ when $t < \tau$. We then take estimates for the posterior that are almost certain (very near to zero or one), threshold them, and use them to form an augmented training set, paired with their corresponding feature-values. We use this larger set to train a second classifier and argue that it tends to have better predictive ability than both the first classifier and a classifier trained using only the first source of knowledge. See Fig. 1 for a visual comparison of these approaches.

Outline. The paper is organized as follows. In the next section, we introduce our filtering framework and describe how to form filtered estimates for a given sequence. In Sect. 3, we show how to use these filtered estimates to create an augmented training dataset. Then in Sect. 4, we compare our classifier trained using filtered data to both the baseline classifier and to an augmented classifier trained using pseudo-labeling [20], a common self-learning approach that augments the training set using estimates for the likelihood alone. We survey related work in Sect. 5 before concluding in Sect. 6.

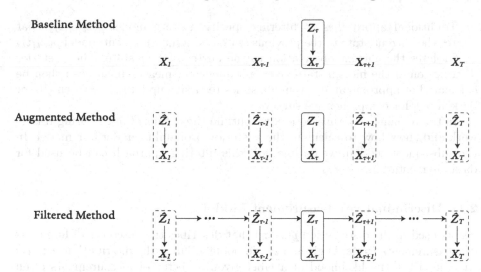

Fig. 1. *Schematic comparing the baseline, augmented, and filtered methods.* The baseline method uses the supervised set $\mathcal{D}_0$ alone, ignoring $\mathcal{D}_1$. The augmented method assigns a probabilistic estimate to each feature-point in $\mathcal{D}_1$ (a type of pseudo-labeling), increasing the available data by a factor of T prior to thresholding. The filtered method additionally incorporates information from the ground truth label to improve its probabilistic estimates.

2 Filtering Methodology

In this section we focus on a single instance i and describe a filtering process that produces predictions for each unlabeled member of the sequence $x_{1:T} = x^i_{1:T}$ of observations using a provided model $p(z_t | x_t)$ and the sequence's corresponding binary label $z_\tau = z^i_\tau$ for some time $\tau = \tau(i)$.

We view the labels and corresponding observations as belonging to a latent state space model. Letting $Z_{1:T} := Z_1, \ldots, Z_T$ denote random variables corresponding to the latent labels and $X_{1:T} := X_1, \ldots, X_T$ denote random variables corresponding to the observations, we model the relationship between these variables according to the Bayesian network:

$$Z_1 \longrightarrow \cdots \longrightarrow Z_{t-1} \longrightarrow Z_t \longrightarrow \cdots \longrightarrow Z_T$$
$$\Big\downarrow \qquad\qquad\qquad \Big\downarrow \qquad\quad \Big\downarrow \qquad\qquad\qquad \Big\downarrow \qquad (1)$$
$$X_1 \qquad\qquad\qquad X_{t-1} \qquad X_t \qquad\qquad\qquad X_T$$

Using this framework, we aim to infer the predictive posterior distribution $p(z_t | x_{\tau+1:t}, z_\tau)$ for times $t > \tau$ and $p(z_t | x_{t:\tau-1}, z_\tau)$ for times $t < \tau$, where Z_t is uncertain. To motivate this exercise, we hypothesize that augmented pairs $(x_t, \hat{z}_t)$ produced using the posterior will better assist in training than those produced using the likelihood, $p(z_t | x_t)$.

Traditional approaches to filtering specify a state model $p(z_t|z_{t-1})$ that relates the current state to the previous state and a measurement model $p(x_t|z_t)$ that relates the current observation to the current latent state. The posterior distribution of the hidden state given a sequence of measurements can then be calculated or approximated through a series recursive updates. See Chen [10] or Särkkä [30] for comprehensive surveys.

In the remainder of this section, we outline the details of our filtering framework and show how to calculate the posterior probability under our model. In the subsequent section, we outline how this filtering approach can be used for data augmentation.

2.1   Discriminative Measurement Model

As opposed to the distribution $p(x_t|z_t)$ that describes the outcomes of hundreds of measurements ($m \gg 1$) given a single boolean label, the distribution $p(z_t|x_t)$ that describes the likelihood of a label given a vector of measurements often proves much more tractable to learn effectively using off-the-shelf classifiers. *Consequently, we approximate the measurement model using* $p(z_t|x_t)$. We apply Bayes' rule and note that, up to a constant depending only on x_t, we may replace $p(x_t|z_t)$ with $p(z_t|x_t)/p(z_t)$. We further assume that this model is stationary, so that $p(z_t|x_t)$ is independent of t. This approach mirrors that of McCallum et al.'s Maximum Entropy Markov Model [22] and the more recent Discriminative Kalman Filter [5,6], as it relies on a discriminative approximation to the measurement model. (In particular, the model is no longer generative, following the terminology of Ng and Jordan [25].)

2.2   Reversible State Model

We specify the state model as as stationary Markov chain

$$\mathbb{P}(Z_t = 1|Z_{t-1} = 0) = \alpha_0, \tag{2a}$$
$$\mathbb{P}(Z_t = 1|Z_{t-1} = 1) = \alpha_1, \tag{2b}$$

for some $0 < \alpha_0, \alpha_1 < 1$ and $2 \le t \le T$. If we let

$$\beta = \alpha_0/(1 + \alpha_0 - \alpha_1), \tag{3}$$

it follows that if

$$\mathbb{P}(Z_{t_0} = 1) = \beta \tag{4}$$

at some time t_0, then $\mathbb{P}(Z_t = 1) = \beta$ for all t. Furthermore, this chain is reversible [11, sec. 6.5], and

$$\mathbb{P}(Z_t = 1|Z_{t+1} = 0) = (1 - \alpha_1)\beta/(1 - \beta) = \alpha_0, \tag{5a}$$
$$\mathbb{P}(Z_t = 1|Z_{t+1} = 1) = \alpha_1, \tag{5b}$$

for all t.

2.3 Filtering Forward in Time

We first describe how to filter forward in time. Starting with the ground-truth label at time τ, we iteratively combine information from our estimate of the previous label (using the state model) with our estimate of the current label given the measurements at that time (using the measurement model). For $t > \tau$, we recursively calculate the posterior

$$\rho_{\tau:t} := \mathbb{P}(Z_t = 1 | X_{\tau+1:t} = x_{\tau+1:t}, Z_\tau = z_\tau) \tag{6}$$

in terms of the likelihood $p_t = \mathbb{P}(Z_t = 1 | X_t = x_t)$ and the previous posterior $\rho_{\tau:t-1}$ as

$$\rho_{\tau:t} \propto \frac{\mathbb{P}(Z_t = 1 | X_t = x_t)}{\mathbb{P}(Z_t = 1)} \mathbb{P}(Z_t = 1 | X_{\tau+1:t-1} = x_{\tau+1:t-1}, Z_\tau = z_\tau)$$

$$\propto \frac{p_t}{\mathbb{P}(Z_t = 1)} \big(\mathbb{P}(Z_t = 1 | Z_{t-1} = 1)\rho_{\tau:t-1} + \mathbb{P}(Z_t = 1 | Z_{t-1} = 0)(1 - \rho_{\tau:t-1}) \big)$$

where we initialize $\rho_{\tau:\tau}$ as the observed point mass $p(Z_\tau = 1)$. It follows from our state model (2) and initialization (4) that

$$\rho_{\tau:t} \propto \frac{p_t}{\beta} \big(\alpha_1 \rho_{\tau:t-1} + \alpha_0 (1 - \rho_{\tau:t-1}) \big) \tag{7}$$

up to a constant depending on the observations alone. To relieve ourselves of that constant, we note that

$$1 - \rho_{\tau:t} \propto \frac{\mathbb{P}(Z_t = 0 | X_t = x_t)}{\mathbb{P}(Z_t = 0)} \mathbb{P}(Z_t = 0 | X_{\tau+1:t-1} = x_{\tau+1:t-1}, Z_\tau = z_\tau)$$

$$\propto \frac{1 - p_t}{\mathbb{P}(Z_t = 0)} \big(\mathbb{P}(Z_t = 0 | Z_{t-1} = 1)\rho_{\tau:t-1} + \mathbb{P}(Z_t = 0 | Z_{t-1} = 0)(1 - \rho_{\tau:t-1}) \big)$$

which itself simplifies to

$$1 - \rho_{\tau:t} \propto \frac{1 - p_t}{1 - \beta} \big((1 - \alpha_1)\rho_{\tau:t-1} + (1 - \alpha_0)(1 - \rho_{\tau:t-1}) \big). \tag{8}$$

Dividing the right-hand side of (7) by the sum of (7) and (8) cancels the constant of proportionality and yields $\rho_{\tau:t}$.

2.4 Filtering Backward in Time

Now, we describe how to filter backward in time. As before, we proceed recursively to calculate

$$\tilde{\rho}_{t:\tau} := \mathbb{P}(Z_t = 1 | X_{t:\tau-1} = x_{t:\tau-1}, Z_\tau = z_\tau) \tag{9}$$

for $t = \tau - 1, \tau - 2, \ldots, 1$ in terms of $\tilde{\rho}_{t+1:\tau}$ and the likelihood p_t:

$$\tilde{\rho}_{t:\tau} \propto \frac{\mathbb{P}(Z_t = 1 | X_t = x_t)}{\mathbb{P}(Z_t = 1)} \mathbb{P}(Z_t = 1 | X_{t+1:\tau-1} = x_{t+1:\tau-1}, Z_\tau = z_\tau)$$

$$\propto \frac{p_t}{\mathbb{P}(Z_t = 1)} \big(\mathbb{P}(Z_t = 1 | Z_{t+1} = 1)\tilde{\rho}_{t+1:\tau} + \mathbb{P}(Z_t = 1 | Z_{t+1} = 0)(1 - \tilde{\rho}_{t+1:\tau}) \big)$$

where, analogously, $\tilde{\rho}_{\tau:\tau}$ is taken to be the observed point mass $p(Z_\tau = 1)$. Upon substituting the state model (2) and initialization (4), we have

$$\tilde{\rho}_{t:\tau} \propto \frac{p_t}{\beta}\left(\alpha_1 \tilde{\rho}_{t+1:\tau} + \alpha_0(1 - \tilde{\rho}_{t+1:\tau})\right). \tag{10}$$

To remove the constant of proportionality, we also calculate:

$$1 - \tilde{\rho}_{t:\tau} \propto \frac{\mathbb{P}(Z_t = 0 | X_t = x_t)}{\mathbb{P}(Z_t = 0)}\mathbb{P}(Z_t = 0 | X_{t+1:\tau-1} = x_{t+1:\tau-1}, Z_\tau = z_\tau)$$

$$\propto \frac{1 - p_t}{\mathbb{P}(Z_t = 0)}\left(\mathbb{P}(Z_t = 0 | Z_{t+1} = 1)\tilde{\rho}_{t+1:\tau} + \mathbb{P}(Z_t = 0 | Z_{t+1} = 0)(1 - \tilde{\rho}_{t+1:\tau})\right)$$

which simplifies to

$$1 - \tilde{\rho}_{t:\tau} \propto \frac{1 - p_t}{1 - \beta}\left((1 - \alpha_1)\tilde{\rho}_{t+1:\tau} + (1 - \alpha_0)(1 - \tilde{\rho}_{t+1:\tau})\right). \tag{11}$$

Dividing the right-hand side of (10) by the sum of (10) and (11) then yields our objective, $\tilde{\rho}_{t:\tau}$.

3   Augmentation Methodology

In this section, we describe our method for data augmentation, relying on the filtering framework developed in the previous section. Given n sequences of measurements $x^i_{1:T}$ with corresponding labels $z^i_\tau \in \{0,1\}$ for $\tau = \tau(i) \in \{1, \ldots, T\}$ and $1 \leq i \leq n$, we define the supervised dataset

$$\mathcal{D}_0 = \{(x^1_\tau, z^1_\tau), (x^2_\tau, z^2_\tau), \ldots, (x^n_\tau, z^n_\tau)\}$$

of features $x^i_\tau \in \mathbb{R}^m$ paired with their corresponding labels $z^i_\tau \in \{0,1\}$, along with the unsupervised dataset containing the unlabeled portions of each sequence,

$$\mathcal{D}_1 = \{x^1_{1:\tau-1}, x^1_{\tau+1:T}; x^2_{1:\tau-1}, x^2_{\tau+1:T}; \ldots ; x^n_{1:\tau-1}, x^n_{\tau+1:T}\}.$$

We augment our supervised dataset $\mathcal{D}_0$ with information from $\mathcal{D}_1$ as follows:

1. We first learn a supervised model $f : \mathbb{R}^m \to [0,1]$ on $\mathcal{D}_0$ such that for inputs $x \in \mathbb{R}^m$,

$$f(x) \approx \mathbb{P}(Z_\tau = 1 | X_\tau = x). \tag{12}$$

   We refer to this probabilistic classifier as the baseline model.
2. For each instance i, we apply the baseline model to each feature-point in $\mathcal{D}_1$ to form

$$\tilde{\mathcal{D}}_1 = \{(x^i_t, f(x^i_t))\}_{x^i_t \in \mathcal{D}_1},$$

   as our stationarity assumption implies $f(x) \approx \mathbb{P}(Z_t = 1 | X_t = x)$ for all x, t.

3. For each i, we apply the filtering equations from the previous section, starting at the time point τ, and filtering forward in time to calculate the posterior estimates $\rho^i_{\tau:t}$ for $t = \tau + 1, \tau + 2, \ldots, T$, and backward in time to determine $\tilde{\rho}^i_{t:\tau}$ for $t = \tau - 1, \tau - 2, \ldots, 1$. We form

$$\check{\mathcal{D}}_1 = \{(x^i_t, \rho^i_{\tau:t})\}_{1 \le i \le n, t \ge \tau} \cup \{(x^i_t, \tilde{\rho}^i_{t:\tau})\}_{1 \le i \le n, t < \tau}$$

and threshold a subset of these points in the manner we will describe in Sect. 3.1.

We use a held-out validation dataset to select parameters α_0, α_1 from (2) for the underlying state model.* These control the propensity for an instance to maintain a label from one time step to the next. As $\mathbb{E}[Z_t] = \beta = \mathbb{E}[Z_\tau]$ can be approximated from the training data, once an optimal value for α_0 has been selected, this value along with an empirical approximation for $\mathbb{E}[Z_\tau]$ can be used with (3) to select a good value for α_1.

3.1 Thresholding to Create Binary Labels

As many classifiers (including the lightGBM and XGBoost models) expect binary (non-probabilistic) labels for training data, we threshold the filtered labels and form a binary-valued set $\check{\mathcal{D}}'_1$ from the filtered set $\check{\mathcal{D}}_1$ and lower and upper bounds $0 < b_L < b_U < 1$ as follows. For each point $(x, \rho) \in \check{\mathcal{D}}_1$, if $\rho < b_L$, we add $(x, 0)$ to $\check{\mathcal{D}}'_1$; if $b_L < \rho < b_U$, we discard the point; and if $b_U < \rho$, we add $(x, 1)$ to $\check{\mathcal{D}}'_1$. This yields a thresholded, filtered training set $\check{\mathcal{D}}'_1$ with binary-valued labels that contains $\mathcal{D}_0$ as a subset. The parameters $0 < b_L < b_U < 1$ can also be selected using validation (or cross-validation).

We summarize our approach using pseudo-code in Algorithm 1.

4   Ablation Study and Results

In this section, we describe how we applied these methods to real-life customer survey data at Adobe.

4.1   Data Provenance

We surveyed approximately ten thousand Adobe subscribers in October 2019 and a distinct set of approximately equal size in February 2020. Based on survey responses alone, we assigned classifications to each subscriber. For the purposes

---

* Given a set $\{\alpha^k_0\}_{k \in K}$ of candidate values for α_0 and a set $\{\alpha^\ell_1\}_{\ell \in L}$ for α_1, we select parameters via an exhaustive grid search as follows. For each $(k, \ell) \in K \times L$, we apply Algorithm 1 with α^k_0 and α^ℓ_1 to the training set, train a classifier on the resulting filtered dataset, and then evaluate this classifier's predictive performance on the validation set (using AUC). Upon completion, we select the parameter values α^k_0 and α^ℓ_1 that yield the most performant classifier.

Data: labeled dataset $\mathcal{D}_0$ and unlabeled dataset $\mathcal{D}_1$; parameters $0 < \alpha_0, \alpha_1 < 1$
and $0 < b_L < b_U < 1$ obtained from validation
Result: labeled binary-valued dataset $\breve{\mathcal{D}}'_1$ extending $\mathcal{D}_0$ that can be used for
supervised learning
Train a supervised model $f : \mathbb{R}^m \to [0,1]$ on $\mathcal{D}_0$ such that $f(x)$ approximates
$\mathbb{P}(Z_t = 1 | X_t = x)$ for $x \in \mathbb{R}^m$;
Initialize $\breve{\mathcal{D}}'_1 = \mathcal{D}_0$;
for $i = 1, \ldots, n$ **do** #*iterate over instances*
    #*filter forward from* τ;
    let $\rho_{\tau:\tau} = z_\tau^i$ from $\mathcal{D}_0$;
    for $t = \tau + 1, \tau + 2, \ldots, T - 1, T$ **do**
        #*determine predictive posterior*;
        let $\rho_{\tau:t} = \pi_1/(\pi_0 + \pi_1)$ where
        $\pi_0 = (1 - f(x_t^i))\big((1 - \alpha_1)\rho_{\tau:t-1} + (1 - \alpha_0)(1 - \rho_{\tau:t-1})\big)/(1 - \beta)$ and
        $\pi_1 = f(x_t^i)\big(\alpha_1\rho_{\tau:t-1} + \alpha_0(1 - \rho_{\tau:t-1})\big)/\beta$;
        #*threshold*;
        if $\rho_{\tau:t} < b_L$ **then** add $(x_t^i, 0)$ to $\breve{\mathcal{D}}'_1$;
        if $\rho_{\tau:t} > b_U$ **then** add $(x_t^i, 1)$ to $\breve{\mathcal{D}}'_1$;
    end
    #*filter backward from* τ;
    let $\tilde{\rho}_{\tau:\tau} = z_\tau^i$ from $\mathcal{D}_0$;
    for $t = \tau - 1, \tau - 2, \ldots, 2, 1$ **do**
        #*determine predictive posterior*;
        let $\tilde{\rho}_{t:\tau} = \pi_1/(\pi_0 + \pi_1)$ where
        $\pi_0 = (1 - f(x_t^i))\big((1 - \alpha_1)\tilde{\rho}_{t+1:\tau} + (1 - \alpha_0)(1 - \tilde{\rho}_{t+1:\tau})\big)/(1 - \beta)$ and
        $\pi_1 = f(x_t^i)\big(\alpha_1\tilde{\rho}_{t+1:\tau} + \alpha_0(1 - \tilde{\rho}_{t+1:\tau})\big)/\beta$;
        #*threshold*;
        if $\tilde{\rho}_{t:\tau} < b_L$ **then** add $(x_t^i, 0)$ to $\breve{\mathcal{D}}'_1$;
        if $\tilde{\rho}_{t:\tau} > b_U$ **then** add $(x_t^i, 1)$ to $\breve{\mathcal{D}}'_1$;
    end
end

Algorithm 1: Discriminative Bayesian Filtering for Data Augmentation

of testing this algorithm, we considered only a single survey category. We engineered hundreds of proprietary features for each user. We then trained a supervised classifier using the ground-truth survey data to predict segment membership given feature data. We generated features for the surveyed subscribers for each of the twelve months between April 2019 and March 2020, inclusive.

We split the surveyed subscribers randomly into seven partitions of approximately equal size. Each partition was further split into a training, validation, and test set at a 70%–15%–15% ratio, respectively. For each partition, we took the supervised training set $\mathcal{D}_0$ to be the training-set subscribers with paired features and ground-truth survey results at their respective survey times and the unsupervised set $\mathcal{D}_1$ to be the features of the training-set subscribers calculated during the year-long period surrounding their survey dates.

4.2   Ablation with Different Supervised Classification Methods

As our method works with any supervised classifier, we learn f in (12) using a variety of different methods: a LightGBM classifier, an XGBoost classifier, and a feedforward neural network classifier. LightGBM [15] attempted to improve upon other gradient boosting machines [12] by performing gradient-based sampling for the estimation of information gain and by bundling infrequently co-occurring features. Our implementation used the default parameters, to avoid over-tuning. XGBoost [9], a predecessor to LightGBM, introduced a novel algorithm for handling sparse data and allowed for unprecedented scalability. Our implementation relied on the default parameters. The neural network we used comprised of a single hidden layer of 30 neurons with rectified linear activation [24] and L^2-regularization for network weights [17], trained with L-BFGS [21].

For each of the seven data partitions, we compare the predictive performance (on the held-out test set) of three different approaches to building a supervised classifier:

- For the *baseline* method, we take our classifier to be the original f trained on the supervised dataset $\mathcal{D}_0$.
- For the *augmented* method, we apply the baseline model to $\mathcal{D}_1$ and threshold the results as described in Sect. 3.1 using thresholds selected for performance on the validation dataset; we then train a classifier on the subsequent dataset.
- For the *filtered* method, we train a classifier using the dataset obtained by applying Algorithm 1. Parameters $0 < \alpha_0, \alpha_1 < 1$ and $0 < b_L < b_U < 1$ were chosen to maximize predictive performance on the validation set.

We aim to isolate the effects of augmenting the training dataset with filtered, posterior-based estimates from the benefits obtained by simply employing likelihood-based estimates.

To illustrate this difference, consider the following example. Suppose we administer a survey in June to Alice, Bob, and ten thousand other subscribers. We train a classifier using the data from June and apply it to Alice and Bob's feature data for July to predict that Alice has a 90% chance of belonging to the segment of interest in July while Bob has a 1% chance. If our lower-bound threshold is $u_L = 2\%$ and our upper threshold is $u_B = 95\%$, then for the *augmented* training set, we would include only Bob's features for July, with a negative label. Suppose further that our ground truth labels in June indicate that Alice and Bob both belonged to the segment of interest that month. If subscribers have an 80% chance of maintaining a positive label from one month to the next ($\alpha_1 = 0.8$ in Eq. 2b), and the segment of interest includes 30% of the population ($\beta = 0.3$ in Eq. 3), then Alice has a 99% chance of belonging to the segment given both her June and July data, while Bob has a 9% chance (see eqs. (7) and (8)). Thus, with the same thresholds, our *filtered* training set would include only Alice's features for July, with a positive label.

We performed all numerical tests on a 15-in. 2018 MacBook Pro (2.6 GHz Intel Core i7 Processor; 16 GB 2400 MHz DDR4 Memory). We used Python 3.8.6 with the following packages (versioning in parentheses): lightgbm (3.0.0), numpy (1.18.5), pandas (1.1.4), scikit-learn (0.23.20), & xgboost (1.2.1).

4.3 Results

We measure model performance using AUC and report our results in Table 1. Mean performance increases with each successive approach (baseline to augmented to filtered methods) for each of the three types of classification methods used. For the LightGBM classifier, we have reason to prefer the filtered training method over the baseline (avg. $+7.0\%$ improvement in AUC; $p = 0.00787$; paired sample t-test with 6 degrees of freedom against a one-sided alternative) and over the augmented training method (avg. $+5.7\%$ improvement in AUC; $p = 0.0333$; same test).

Table 1. *Classification AUC (Area Under the receiver operating characteristic Curve) for each of the methods tested.* The mean column reports the average AUC over the 7 independent trials for each method.

Classifier	Method	Trial 1	Trial 2	Trial 3	Trial 4	Trial 5	Trial 6	Trial 7	Mean
LightGBM	Baseline	0.611	0.564	0.698	0.564	0.645	0.636	0.588	0.615
	Augmented	0.611	0.536	0.689	0.525	0.720	0.611	0.687	0.626
	Filtered	0.630	0.582	0.704	0.620	0.705	0.668	0.690	0.657
XGBoost	Baseline	0.568	0.553	0.669	0.590	0.657	0.688	0.621	0.621
	Augmented	0.555	0.564	0.677	0.601	0.654	0.663	0.652	0.624
	Filtered	0.664	0.580	0.619	0.600	0.696	0.662	0.645	0.638
Feedforward NN	Baseline	0.556	0.520	0.517	0.485	0.621	0.581	0.514	0.542
	Augmented	0.591	0.551	0.511	0.458	0.634	0.558	0.570	0.553
	Filtered	0.589	0.554	0.539	0.460	0.640	0.565	0.595	0.563

5   Related Work

We now describe how our proposed algorithm for data augmentation relates to both filtering and semi-supervised learning.

5.1   Discriminative Bayesian Filtering

State-space models having the graphical form (1) relate an unobserved Markovian sequence $(Z_1, Z_2, \dots)$ of interest to a series of observed measurements $(X_1, X_2, \dots)$, where each measurement depends only on the current hidden state. Also known as hidden Markov models, they have been extensively studied due to their wide range of applications. Discriminative variants, including Maximum Entropy Markov Models [22] and Conditional Random Fields [18], allow one to use $p(z_t|x_t)$ instead of $p(x_t|z_t)$ for inference. Moving from a generative to a discriminative approach allows one to more directly consider the distribution of states given observations, at the cost of no longer learning the joint distribution of the hidden and observed variables. (In particular, after training such a model, one cannot sample observations.) Applications include human motion tracking [16,33] and neural modeling [2–4].

5.2 Data Augmentation and Semi-supervised Learning

Given a small set of labeled training data and an additional set of unlabeled points, semi-supervised methods attempt to leverage the location of the unlabeled training points to learn a better classifier than could be obtained from the labeled training set alone [8,38]. For example, graph-based approaches use pairwise similarities between labeled and unlabeled feature-points to construct a graph, and then let labeled points pass their labels to their unlabeled neighbors. In the sense that our filtering process passes information along the Bayesian graph (1), our method is similar in spirit to graph-based methods—like Label Propagation [39] and Label Spreading [37]—though our graph derives from the natural temporal relationship of our measurements and not from the locations of the feature-points.

Self-learning refers to the practice of using the predictions from one classifier in order to train a second classifier [31,34,35]. For example, Pseudo-labeling [20] assigns predicted labels to unlabeled features and adds them to a supervised training set, in an approach equivalent to minimum entropy regularization [13]. Recent work in deep learning elaborates heavily on this idea, where latent representations gained from intermediate network layers play a crucial role [7,14,19,28,29,32,40]. A common failure mode for self-learners in general entails inadvertently misclassifying points and then propagating these erroneous labels to their unlabeled neighbors. Recent proposals to reduce this so-called confirmation bias mostly focus on deep neural networks [1,26,36].

6 Conclusions and Directions for Future Research

In this paper, we considered a semi-supervised learning problem involving sequential observation data and showed how filtering could be employed for data augmentation. We described a method that leverages the predictive posterior to augment the training dataset and compared it to a standard pseudo-labeling approach that performs augmentation using likelihood-based estimates. We then showed how classifiers trained on the filtered dataset can outperform those trained using pseudo-labeling.

The particular problem we considered consisted of time-delineated sequences of features, each coupled with a single time-dependent label. This problem corresponds to a relatively basic case, where each subscriber is surveyed at a single point in time. One can imagine having ground truth labels for a subscriber or instance at multiple points in time. In such a case, Bayesian smoothing may be applied to determine the predictive posterior for the points in time between the two labels. For more complex relationships between observations, belief propagation [27] or expectation propagation [23] may be worth exploring.

Acknowledgements. The author would like to thank his manager Binjie Lai, her manager Xiang Wu, and his coworkers at Adobe, especially Eunyee Koh for performing an internal review. The author is also grateful to the anonymous reviewers for their thoughtful feedback and to his former advisor Matthew T. Harrison for inspiring this discriminative filtering approach.

References

1. Arazo, E., Ortego, D., Albert, P., O'Connor, N.E., McGuinness, K.: Pseudo-labeling and confirmation bias in deep semi-supervised learning. In: International Joint Conference on Neural Networks, vol. 3, pp. 189–194 (2020)
2. Batty, E., et al.: Behavenet: nonlinear embedding and Bayesian neural decoding of behavioral videos. In: Advances in Neural Information Processing Systems, pp. 15706–15717 (2019)
3. Brandman, D.M., et al.: Rapid calibration of an intracortical brain-computer interface for people with tetraplegia. J. Neural Eng. **15**(2), 026007 (2018)
4. Brandman, D.M., Burkhart, M.C., Kelemen, J., Franco, B., Harrison, M.T., Hochberg, L.R.: Robust closed-loop control of a cursor in a person with tetraplegia using Gaussian process regression. Neural Comput. **30**(11), 2986–3008 (2018)
5. Burkhart, M.C.: A discriminative approach to bayesian filtering with applications to human neural decoding. Ph.D. thesis, Brown University, Division of Applied Mathematics, Providence, U.S.A. (2019)
6. Burkhart, M.C., Brandman, D.M., Franco, B., Hochberg, L.R., Harrison, M.T.: The discriminative Kalman filter for Bayesian filtering with nonlinear and non-gaussian observation models. Neural Comput. **32**(5), 969–1017 (2020)
7. Burkhart, M.C., Shan, K.: Deep low-density separation for semi-supervised classification. In: Krzhizhanovskaya, V.V., et al. (eds.) ICCS 2020. LNCS, vol. 12139, pp. 297–311. Springer, Cham (2020). https://doi.org/10.1007/978-3-030-50420-5_22
8. Chapelle, O., Schölkopf, B., Zien, A. (eds.): Semi-Supervised Learning. MIT Press, Cambridge (2006)
9. Chen, T., Guestrin, C.: XGBoost: a scalable tree boosting system. In: International Conference on Knowledge Discovery and Data Mining, pp. 785–794 (2016)
10. Chen, Z.: Bayesian filtering: from Kalman filters to particle filters, and beyond. Technical report, McMaster U (2003)
11. Durrett, R.: Probability: Theory and Examples. Cambridge University Press, Cambridge (2010)
12. Friedman, J.H.: Greedy function approximation: a gradient boosting machine. Ann. Statist. **29**(5), 1189–1232 (2001)
13. Grandvalet, Y., Bengio, Y.: Semi-supervised learning by entropy minimization. In: Advances in Neural Information Processing Systems, pp. 529–536 (2004)
14. Iscen, A., Tolias, G., Avrithis, Y., Chum, O.: Label propagation for deep semi-supervised learning. In: Conference on Computer Vision and Pattern Recognition (2019)
15. Ke, G., et al.: LightGBM: a highly efficient gradient boosting decision tree. In: Advances in Neural Information Processing Systems, pp. 3146–3154 (2017)
16. Kim, M., Pavlovic, V.: Discriminative learning for dynamic state prediction. IEEE Trans. Pattern Anal. Mach. Intell. **31**(10), 1847–1861 (2009)
17. Krogh, A., Hertz, J.A.: A simple weight decay can improve generalization. In: Advances in Neural Information Processing Systems, pp. 950–957 (1991)
18. Lafferty, J., McCallum, A., Pereira, F.: Conditional random fields: probabilistic models for segmenting and labeling sequence data. In: International Conference on Machine Learning (2001)
19. Laine, S., Aila, T.: Temporal ensembling for semi-supervised learning. In: International Conference on Learning Representations (2017)
20. Lee, D.H.: Pseudo-label: the simple and efficient semi-supervised learning method for deep neural networks. In: ICML Workshop on Challenges in Representation Learning (2013)

21. Liu, D.C., Nocedal, J.: On the limited memory method for large scale optimization. Math. Program. **45**(3), 503–528 (1989)
22. McCallum, A., Freitag, D., Pereira, F.: Maximum entropy Markov models for information extraction and segmentation. In: International Conference on Machine Learning, pp. 591–598 (2000)
23. Minka, T.P.: Expectation propagation for approximate Bayesian inference. In: Uncertainty in Artificial Intelligence (2001)
24. Nair, V., Hinton, G.: Rectified linear units improve restricted Boltzmann machines (2010)
25. Ng, A., Jordan, M.: On discriminative vs. generative classifiers: a comparison of logistic regression and Naive Bayes. In: Advances in Neural Information Processing Systems, vol. 14, pp. 841–848 (2002)
26. Oliver, A., Odena, A., Raffel, C.A., Cubuk, E.D., Goodfellow, I.: Realistic evaluation of deep semi-supervised learning algorithms. In: Advances in Neural Information Processing Systems, pp. 3235–3246 (2018)
27. Pearl, J.: Reverend Bayes on inference engines: a distributed hierarchical approach. In: Proceedings of Association for the Advancement of Artificial Intelligence, pp. 133–136 (1982)
28. Rasmus, A., Berglund, M., Honkala, M., Valpola, H., Raiko, T.: Semi-supervised learning with ladder networks. In: Advances in Neural Information Processing Systems, pp. 3546–3554 (2015)
29. Sajjadi, M., Javanmardi, M., Tasdizen, T.: Regularization with stochastic transformations and perturbations for deep semi-supervised learning. In: Advances in Neural Information Processing Systems, pp. 1163–1171 (2016)
30. Särkkä, S.: Bayesian Filtering and Smoothing. Cambridge University Press, Cambridge (2013)
31. Scudder III, H.J.: Probability of error for some adaptive pattern-recognition machines. IEEE Trans. Inf. Theory **11**(3), 363–371 (1965)
32. Tarvainen, A., Valpola, H.: Mean teachers are better role models: Weight-averaged consistency targets improve semi-supervised deep learning results. In: Advances in Neural Information Processing Systems, pp. 1195–1204 (2017)
33. Taycher, L., Shakhnarovich, G., Demirdjian, D., Darrell, T.: Conditional random people: tracking humans with CRFs and grid filters. In: Computer Vision and Pattern Recognition (2006)
34. Whitney, M., Sarkar, A.: Bootstrapping via graph propagation. In: Proceedings of Association for Computational Linguistics, vol. 1, pp. 620–628 (2012)
35. Yarkowsky, D.: Unsupervised word sense disambiguation rivaling supervised methods. In: Proceedings of Association for Computational Linguistics, pp. 189–196 (1995)
36. Zhang, H., Cisse, M., Dauphin, Y.N., Lopez-Paz, D.: mixup: beyond empirical risk minimization. In: International Conference on Learning Representations (2018)
37. Zhou, D., Bousquet, O., Lal, T.N., Weston, J., Schölkopf, B.: Learning with local and global consistency. In: Advances in Neural Information Processing Systems, pp. 321–328 (2004)
38. Zhu, X.: Semi-supervised learning literature survey. Technical report, TR 1530, U. Wisconsin-Madison (2005)
39. Zhu, X., Ghahramani, Z.: Learning from labeled and unlabeled data with label propagation. Technical report, CMU-CALD-02-107, Carnegie Mellon University (2002)
40. Zhuang, C., Ding, X., Murli, D., Yamins, D.: Local label propagation for large-scale semi-supervised learning (2019). arXiv:1905.11581

TSAX is Trending

Muhammad Marwan Muhammad Fuad[✉]

Coventry University, Coventry CV1 5FB, UK
ad0263@coventry.ac.uk

Abstract. Time series mining is an important branch of data mining, as time series data is ubiquitous and has many applications in several domains. The main task in time series mining is classification. Time series representation methods play an important role in time series classification and other time series mining tasks. One of the most popular representation methods of time series data is the Symbolic Aggregate approXimation (SAX). The secret behind its popularity is its simplicity and efficiency. SAX has however one major drawback, which is its inability to represent trend information. Several methods have been proposed to enable SAX to capture trend information, but this comes at the expense of complex processing, preprocessing, or post-processing procedures. In this paper we present a new modification of SAX that we call Trending SAX (TSAX), which only adds minimal complexity to SAX, but substantially improves its performance in time series classification. This is validated experimentally on 50 datasets. The results show the superior performance of our method, as it gives a smaller classification error on 39 datasets compared with SAX.

Keywords: Symbolic Aggregate Approximation (SAX) · Time series classification · Trending SAX (TSAX)

1 Introduction

Time series mining has witnessed substantial interest in the last two decades because of the popularity of time series data, as much of the data in the world is in the form of time series [15].

Time series mining deals with several tasks such as classification, clustering, segmentation, query-by-content, anomaly detection, prediction, and others.

Time series classification (TSC) is the most important time series mining task. TSC has a wide variety of applications in medicine, industry, meteorology, finance, and many other domains. This variety of applications is the reason why TSC has gained increasing attention over the last two decades [2, 4, 6, 8, 22].

The most common TSC method is the *k-nearest-neighbor* (kNN), which applies the similarity measures to the object to be classified to determine its best classification based on the existing data that has already been classified. The performance of a classification algorithm is measured by the percentage of objects identified as the correct class [15]. kNN can be applied to raw time series or to lower-dimension representations of the time series. These representation methods are widely used in time series mining as

© Springer Nature Switzerland AG 2021
M. Paszynski et al. (Eds.): ICCS 2021, LNCS 12743, pp. 284–296, 2021.
https://doi.org/10.1007/978-3-030-77964-1_23

time series may contain noise or outliers. Besides, performing TSC on lower-dimension representations of time series is more efficient than applying it directly to raw data. The most widely used kNN method is 1NN.

There have been a multitude of time series representation methods in the literature, to name a few; the Piecewise Aggregate Approximation (PAA) [9, 23], Adaptive Piecewise Constant Approximation (APCA) [10], and the Clipping Technique [21].

The Symbolic Aggregate Approximation (SAX) [11, 12] is one of the most popular time series representation methods. The secret behind this popularity is its efficiency and simplicity. SAX is in fact based on PAA, but it applies further steps so that the time series is converted into a sequence of symbols.

Researchers have, however, pointed to a certain drawback in SAX, which is its inability to capture trend information. This is an important feature in many TSC applications.

In this work we present a new modification of SAX that captures the trend information with very small additional storage and computational costs compared with the original SAX. The extensive experiments we conducted on a wide variety of time series datasets in a TSC task show a substantial boost in performance.

The rest of the paper is organized as follows: Sect. 2 discusses background material on time series data, mainly SAX. Section 3 introduces our new method TSAX. In Sect. 4 we conduct extensive experiments that compare TSAX against SAX. We conclude with Sect. 5.

2 Background

A time series $T = (t_1, t_2, \ldots, t_n)$ is an ordered collection of n measurements at timestamps t_n. Time series data appear in a wide variety of applications.

Given a time series dataset D of s time series, each of n dimensions. Each time series $T_i, i \in \{1, 2, \ldots, s\}$ is associated with a class label $L(T_i)$; $L(T_i) \in \{1, 2, \ldots, c\}$. Given a set of unlabeled time series U, the purpose of the TSC task is to map each time series in U to one of the classes in $\{1, 2, \ldots, c\}$. TSC algorithms involve some processing or filtering of the time series values prior or during constructing the classifier. The particularity of TSC, which makes it different from traditional classification tasks, is the natural temporal ordering of the attributes [1].

$1NN$ with the Euclidean distance or the Dynamic Time Warping (DTW), applied to raw time series, has been widely used in TSC, but the Euclidean distance is weak in terms of accuracy [20], and DWT, although much more accurate, is computationally expensive.

Instead of applying different tasks directly to raw time series data, time series representation methods project the data on lower-dimension spaces and perform the different time series mining tasks in those low-dimension spaces.

A large number of time series representation methods have been proposed in the literature. The Piecewise Aggregate Approximation (PAA) [9, 23] is one of the first and most simple, yet effective, time series representation methods in the literature.

PAA divides a time series T of n-dimensions into m equal-sized segments and maps each segment to a point of a lower m-dimensional space, where each point in this space is the mean of values of the data points falling within this segment as shown in Fig. 1.

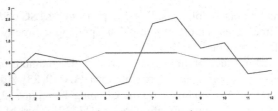

Fig. 1. PAA representation

PAA is the predecessor of another very popular time series representation method, which is the Symbolic Aggregate Approximation – SAX [11, 12]. SAX performs the discretization by dividing a time series T into m equal-sized segments (words). For each segment, the mean value of the points within that segment is computed. Aggregating these m coefficients forms the PAA representation of T. Each coefficient is then mapped to a symbol according to a set of breakpoints that divide the distribution space into α equiprobable regions, where α is the *alphabet size* specified by the user. The locations of the breakpoints are determined using a statistical lookup table for each value of α. These lookup tables are based on the assumption that normalized time series subsequences have a highly Gaussian distribution [12].

It is worth mentioning that some researchers applied optimization, using genetic algorithms and differential evolution, to obtain the locations of the breakpoints [16, 18]. This gave better results than the original lookup tables based on the Gaussian assumption.

To summarize, SAX is applied to normalized time series in three steps as follows:

1. The dimensionality of the time series is reduced using PAA.
2. The resulting PAA representation is discretized by determining the number and locations of the breakpoints. The number of the breakpoints *nrBreakPoints* is: *nrBreakPoints* $= \alpha - 1$. As for their locations, they are determined, as mentioned above, by using Gaussian lookup tables. The interval between two successive breakpoints is assigned to a symbol of the alphabet, and each segment of PAA that lies within that interval is discretized by that symbol.
3. The last step of SAX is using the following distance:

$$MINDIST\left(\hat{S}, \hat{T}\right) = \sqrt{\frac{n}{m}} \sqrt{\sum_{i=1}^{m} (dist(\hat{s}_i, \hat{t}_i))^2} \tag{1}$$

Where n is the length of the original time series, m is the number of segments, $\hat{S}$ and $\hat{T}$ are the symbolic representations of the two time series S and T, respectively, and where the function $dist()$ is implemented by using the appropriate lookup table. For instance, the lookup table for an alphabet size of 4 is the one shown in Table 1. Figure 2 shows an example of SAX for $\alpha = 4$.

Table 1. The lookup table of *MINDIST* for alphabet size = 4.

	a	b	c	d
a	0	0	0.67	1.34
b	0	0	0	0.67
c	0.67	0	0	0
d	1.34	0.67	0	0

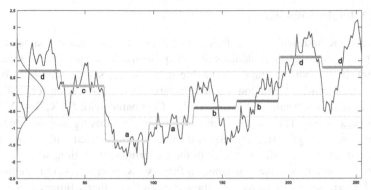

Fig. 2. Example of SAX for $\alpha = 4$, $m = 8$. In the first step the time series, whose length is 256, is discretized using PAA, and then each segment is mapped to the corresponding symbol. This results in the final SAX representation for this time series, which is *dcaabbdd*

3 Trending SAX (TSAX)

3.1 Motivation and Related Work

The efficiency and simplicity of SAX made of it one of the most popular time series representation methods. But SAX has a main drawback, which is its lack of a mechanism to capture the trends of segments. For example, the two segments: $S_1 = [-6, -1, +7, +8]$ and $S_2 = [+9, +3, +1, -5]$ have the same PAA coefficient which is +2, so their SAX representation is the same, although, as we can clearly see, their trends are completely different.

Several researchers have reported this flaw in SAX and attempted to propose different solutions to handle it. In [14] the authors present 1d-SAX which incorporates trend information by applying linear regression to each segment of the time series and then discretizing it using a symbol that incorporates both trend and mean of the segment at the same time. The method applies a complex representation and training scheme. In [24] another extension of SAX that takes into account both trend and value is proposed. This method first transforms each real value into a binary bit. This representation is used in a first pruning step. In the second step the trend is represented as a binary bit. This method requires two passes through the data which makes it inefficient for large datasets. [7] proposes to consider both value and trend information by combining PAA with an extended version of the clipping technique [21]. However, obtaining good results requires prior knowledge of the data. In [17] and [19] two methods to add trend information to

SAX are proposed. The two methods are quite simple, but the improvement is rather small.

In addition to all these drawbacks we mentioned above of each of these methods, which attempt to enable SAX to capture trend information, the main disadvantage of these methods (except [17] and [19]) is that they apply a complex representation and training method, whereas the main merit of SAX is its simplicity with no need for any training stage.

3.2 TSAX Implementation

Our proposed method, which we call *Trending SAX* (TSAX), has the same simplicity as the original SAX. It has an additional, but quite small, storage and computational requirement. It has extra features that enable it to capture segment trend information and it uses this trend information when comparing two time series for TSC, which enables it to obtain substantially better results. Compared with SAX, TSAX has two additional parameters. This might seem as a downside to using our new method, as parameters usually require training. But our method can be used without any training of the parameters, as we will show later in the experimental section, so as such, these parameters are very much like constants. In fact, we wanted to test our method under strictly unbiased conditions, so we gave these parameters rather arbitrary values, yet the performance was superior to that of SAX, as we will see in the experiments.

Before we introduce our method, we have to say that although in this paper we apply it to TSC only, the method can be applied to other time series mining tasks.

In the following we present the different steps of TSAX: Let T be a time series in a n-dimensional space to be transformed by TSAX into a $2m$-dimensional space. After normalization, T is segmented into m equal-sized segments (words) w_i, $i = 1, \ldots, m$, where each word w_i is the mean of the data points lying within that segment (the same steps of PAA described above). In the next step each segment is represented by the corresponding alphabet symbol exactly as in SAX, where α, the alphabet size, is determined by the user.

In the following step, for each word w_i, the trend tr_i of the corresponding segment of time series raw data is calculated by first computing the linear regression of the time series raw data of w_i, then tr_i is represented as $\nearrow$ if it is an upward trending line or by $\searrow$ if it is a downward trending line, so after executing all the above steps, each time series T will be represented in TSAX as:

$$T \rightarrow [w_1 w_2 w_3 \ldots w_m tr_1 tr_2 tr_3 \ldots tr_m] \tag{2}$$

Where tr_i takes one of the values $\nearrow$ or $\searrow$ depending on whether the corresponding linear regression line of the raw time series corresponding to w_i is an upward or downward trend, respectively. We call the first m components of the above representation $[w_1 w_2 w_3 \ldots w_m]$ the *symbolic part* of the TSAX representation, whereas $[tr_1 tr_2 tr_3 \ldots tr_m]$ is called the *trend part* of the TSAX representation.

Notice that the values of the end points of the linear regression line of each segment are real numbers, so the probability that they are equal is zero, so the trend is either $\nearrow$ or $\searrow$. However, because of the way real number are represented on a computer, the two end points could theoretically have the same value, so, for completion, we can choose to represent this (very) special case, where the two values are equal, as either $\nearrow$ or $\searrow$. But this does not have any practical importance.

Also notice, which is important as we will show later, that trend has two values only, so it can be represented by a Boolean variable.

Let T, T' be two time series whose TSAX representations are:

$[w_1 w_2 w_3 \ldots w_m tr_1 tr_2 tr_3 \ldots tr_m]$ and $\left[w'_1 w'_2 w'_3 \ldots w'_m tr'_1 tr'_2 tr'_3 \ldots tr'_m \right]$, respectively. The symbolic parts of the two representations are compared the same way as in SAX, as we will show later. The trend parts are compared as follows: each of tr_i is compared to its counterpart tr'_i, $i = 1, \ldots, m$ according to the following truth table:

tr_i	tr'_i	trendMatch
$\nearrow$	$\nearrow$	$==$
$\searrow$	$\nearrow$	$\neq\neq$
$\nearrow$	$\searrow$	$\neq\neq$
$\searrow$	$\searrow$	$==$

Where *trendMatch* is also represented by a Boolean variable. The symbol ($==$) means that the two trends tr_i and tr'_i are the same, whereas the symbol ($\neq\neq$) means that the two trends tr_i and tr'_i are opposite.

Finally, the TSAX distance between the time series T, T', whose TSAX representations are denoted $\hat{T}$, $\hat{T}'$, respectively, is:

$$TSAX\_DIST\left(\hat{T}, \hat{T}'\right) = \sqrt{\frac{n}{m}} \sqrt{\sum_{i=1}^{m} \left(dist\left(\hat{t}_i, \hat{t}'_i\right)\right)^2 + rew \times k_1 + pen \times k_2} \qquad (3)$$

Where k_1 is the number of times the variable *trendMatch* takes the value ($==$), i.e. the number of times the two time series have segments with matching trends (to their counterparts in the other time series), and k_2 is the number of times the variable *trendMatch* takes the value ($\neq\neq$), i.e. the number of times the two time series have segments with opposite trends. *rew* (stands for "reward") and *pen* (stands for "penalty") are two parameters of our method. We will discuss how they are selected in the experimental section of this paper.

Notice that Eq. (3) is a distance, so when two time series have more segments with matching trends it means the two time series should be viewed as "closer" to each other, i.e. *rew* should take a negative value, whereas *pen* takes a positive value.

3.3 Illustrating Example

In order to better explain our method, we give an example to show how it is applied. Let T, T' be the following two time series of length 16, i.e. $n = 16$ in Eq. (3):

$$T = [-7.1 - 1.1 - 1.3 - 1.5 - 1.4 - 1.3 - 1.0 \ 4.5 \ 9.2 \ 1.0 \ 1.2 \ 9.6 \ 6.1 \ 1.4 - 6.4 - 2.6]$$

$$T' = [-9.9 - 1.4 - 1.5 - 1.6 - 1.6 - 1.3 - 1.0 - 3.5 \ 7.1 \ 1.2 \ 1.1 \ 1.0 \ 7.9 \ 4.6 \ 4.8 \ 5.6]$$

After normalization (values are rounded up because of space limitations):

$$T \rightarrow [-1.6 - 0.4 - 0.4 - 0.4 - 0.4 - 0.4 \ 0.3 \ 0.8 \ 1.8 \ 0.1 \ 0.1 \ 1.9 \ 1.2 \ 0.2 - 1.5 - 0.7]$$

$$T' \rightarrow [-2.4 - 0.5 - 0.5 - 0.5 - 0.5 \ 0.4 - 0.4 - 1.0 \ 1.4 \ 0.1 \ 0.1 \ 0.1 \ 1.6 \ 0.9 \ 0.9 \ 1.1]$$

If we choose $\frac{n}{m} = 4$, which is the value used in the original SAX, and the one used in the experimental section of this paper, then $m = 4$. The PAA representations, where each component is the mean of four successive values of T, T', are the following:

$$T\_PAA \rightarrow [-0.700 \ 0.075 \ 0.975 - 0.200]$$

$$T'\_PAA \rightarrow [-0.975 - 0.375 \ 0.425 \ 1.125]$$

For alphabet size $\alpha = 4$ (the one used in the experiments), the symbolic parts of the TSAX representations are the following:

$$T\_TSAX\_SymbolicPart = [acdb]$$

$$T'\_TSAX\_SymbolicPart = [abcd]$$

As for the trend part, we need to compute the linear regression of each segment of the raw data. For instance, the linear regression of segment $[-1.6 - 0.4 - 0.4 - 0.4]$ is $[-1.2206 - 0.8725 - 0.5245 - 0.1764]$, so the trend for this segment of T is $\nearrow$. Continuing in the same manner for the other segments we get:

$$T\_TSAX\_TrendPart = [\nearrow \ \nearrow \ \nearrow \ \searrow]$$

$$T'\_TSAX\_TrendPart = [\nearrow \ \searrow \ \searrow \ \searrow]$$

When comparing the trend match between T and T' we get: $[== \neq\neq \neq\neq ==]$, so $k_1 = 2$ and $k_2 = 2$ (for this illustrating example using short time series, and rounding up values to a one-digit mantissa, it happened that k_1 and k_2 are equal. Real time series are usually much longer).

3.4 Complexity Considerations

The TSAX representations for all the time series in the dataset are to be computed and stored offline. The length of a TSAX representation is $2m$, whereas that of SAX is m. However, the actual additional storage requirement is much less than that because the additional storage requirement of TSAX, which is the trend part, uses Boolean variables, which require one bit each, whereas the symbols used to represent SAX (and the symbolic part of TSAX), use characters, which require 8 bits each. What is more, in many cases the trend of several successive segments of a time series is the same. This gives the possibility of using the common lossless compression technique *Run Length Encoding* (RLE) [5].

As for the computational complexity when applying Eq. (3). This is also a tiny additional cost compared with SAX, because as we can see from Eq. (3), the two additional computations are $rew \times k_1$ and $pen \times k_2$. Each of them requires one multiplication operation, and m logical comparison operations, which have small computational latency, especially compared with the other very costly operation (sqrt) required to compute Eq. (1) in SAX.

So as we can see, the additional storage and computational costs that TSAX adds are quite small. During experiments, we did not notice any difference in computational time.

4 Experiments

We tested TSAX against SAX in a TSC task using 1NN, which we presented in Sect. 1. We avoided any choice that could bias our method, so we chose the time series archive which is suggested by the authors of SAX to test time series mining methods: *UCR Time Series Classification Archive* [3]. This archive contains 128 time series datasets. To avoid "cherry picking", the archive managers suggest testing any new method on all the datasets, which is not possible to present here for space limitations. However, we decided to do the second best thing to avoid favoring our method; because the datasets are arranged in alphabetical order of their names, we simply decided to test both TSAX and SAX on the first 50 datasets of the archive. Because the name of a dataset is by no means related to its nature, this choice is random enough to show that we did not choose to experiment on datasets that might bias TSAX over SAX. Besides, in other experiments, which we do not report here, on other randomly chosen datasets from the archive, we got similar results.

The datasets in the archive are divided into test and train, but neither SAX nor TSAX need training (we will show later why TSAX does not need training), so the two methods were applied directly to the test datasets, which is the same protocol used to validate SAX in the original paper.

As for the value of the alphabet size α for both SAX and TSAX, SAX can be applied to any value between 3–20, but in [13] Lin et al., the authors of SAX say concerning the choice of the alphabet size "a value of 3 or 4 works well for most time series datasets. In most experiments, we choose $\alpha = 4$ for simplicity", so we too chose $\alpha = 4$ (We conducted other experiments, not shown here, for $\alpha = 3$, and we got similar results). We also chose the same compression ratio (i.e. n/m in Eqs. (1) and (3)).

TSAX has two parameters that SAX does not have, these are *rew* and *pen*. In order to obtain the best performance of TSAX, these two parameters need to be trained on each dataset. However, because our intention was to propose a method that has the same simplicity as SAX, we chose the same value for *rew* and *pen* for all the datasets. And although even in this case, we could at least test different values on different datasets to decide which ones work best, we decided to push our method to the limit, so we chose, rather arbitrarily, $rew = -1$ and $pen = 1$. The rationale behind this choice is that we thought when two segments have the same trend this is equivalent to the distance

Table 2. The datasets, the number of time series in each dataset, the number of classes, the length of the time series, the classification errors of SAX, and the classification errors of TSAX. The best result for each dataset is shown in boldface printing.

Dataset	Nr. of TS	Nr. of Classes	Length	SAX	TSAX
Adiac	391	37	176	0.974	**0.852**
ArrowHead	175	3	251	0.571	**0.309**
Beef	30	5	470	0.667	**0.167**
BeetleFly	20	2	512	**0.250**	0.300
BirdChicken	20	2	512	**0.350**	0.500
Car	60	4	577	0.550	**0.350**
CBF	900	3	128	**0.236**	0.579
ChlorineConcentration	3840	3	166	0.742	**0.495**
CinCECGTorso	1380	4	1639	0.304	**0.223**
Coffee	28	2	286	0.464	**0.250**
Computers	250	2	720	0.564	**0.444**
CricketX	390	12	300	**0.513**	0.759
CricketY	390	12	300	**0.531**	0.741
CricketZ	390	12	300	**0.485**	0.769
DiatomSizeReduction	306	4	345	0.696	**0.160**
DistalPhalanxOutlineAgeGroup	139	3	80	0.717	**0.240**
DistalPhalanxOutlineCorrect	276	2	80	0.412	**0.317**
DistalPhalanxTW	139	6	80	0.722	**0.320**
Earthquakes	139	2	512	0.317	**0.180**
ECG200	100	2	96	0.220	**0.190**
ECG5000	4500	5	140	0.195	**0.077**
ECGFiveDays	861	2	136	0.475	**0.236**
ElectricDevices	7711	7	96	0.879	**0.584**
FaceAll	1690	14	131	0.571	**0.463**
FaceFour	88	4	350	**0.205**	0.330
FacesUCR	2050	14	131	0.476	**0.447**
FiftyWords	455	50	270	**0.415**	0.499
Fish	175	7	463	0.851	**0.280**
FordA	1320	2	500	0.377	**0.332**
FordB	810	2	500	0.451	**0.444**
GunPoint	150	2	150	0.260	**0.167**
Ham	105	2	431	0.486	**0.419**
HandOutlines	370	2	2709	0.283	**0.187**
Haptics	308	5	1092	0.718	**0.640**
Herring	64	2	512	**0.406**	0.469
InlineSkate	550	7	1882	0.800	**0.769**
InsectWingbeatSound	1980	11	256	0.494	**0.480**
ItalyPowerDemand	1029	2	24	0.459	**0.070**
LargeKitchenAppliances	375	3	720	0.643	**0.592**
Lightning2	61	2	637	**0.311**	0.492
Lightning7	73	7	319	**0.507**	0.808
Mallat	2345	8	1024	0.668	**0.309**
Meat	60	3	448	0.667	**0.533**
MedicalImages	760	10	99	0.599	**0.539**
MiddlePhalanxOutlineAgeGroup	154	3	80	0.730	**0.292**
MiddlePhalanxOutlineCorrect	291	2	80	0.647	**0.430**
MiddlePhalanxTW	154	6	80	0.601	**0.431**
MoteStrain	1252	2	84	0.277	**0.165**
NonInvasiveFetalECGThorax1	1965	42	750	0.908	**0.789**
NonInvasiveFetalECGThorax2	1965	42	750	0.871	**0.704**
				11/50	39/50

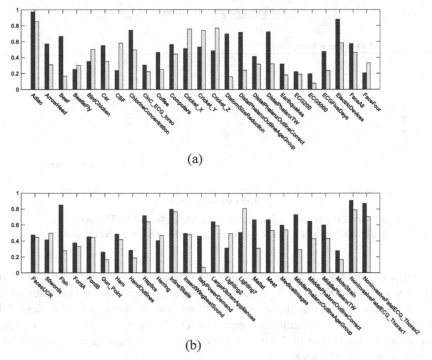

(a)

(b)

Fig. 3. Bar chart comparisons of the classification errors between TSAX (in yellow) and SAX (in blue) on the first 25 datasets (a) and the second 25 datasets. The figure shows the superior performance of TSAX over SAX.

between two symbols (In Table 1: 1.34–0.67, rounded to the closest integer, and taken as a negative value), so we chose $rew = -1$ and we simply chose *pen* to be the opposite of *rew*. We know this choice is rather arbitrary, which is intentional, because we wanted to show that the performance of TSAX is independent of the choice of the parameters *rew* and *pen*, so as such, they are treated as constants.

Table 2 and Fig. 3 show the classification errors of the 50 datasets on which we conducted our experiments for TSAX and SAX. As we can see, TSAX clearly outperforms SAX in TSC as it gave better results, i.e. a lower classification error, in 39 out of the 50 datasets tested, whereas SAX gave a lower classification error in 11 datasets only. In some datasets, like (Beef), (DiatomSizeReduction), (DistalPhalanxOutlineAgeGroup), (ItalyPowerDemand), the gain in applying TSAX is substantial, whereas the only dataset we see the substantial gain in applying SAX is (CBF) (which is a synthetic dataset).

An interesting phenomenon that we notice is that for certain "groups" of similar datasets, such as (CricketX), (CricketY), (CricketZ), and (DiatomSizeReduction), (DistalPhalanxOutlineAgeGroup), (DistalPhalanxOutlineCorrect), (DistalPhalanx-TW), or (MiddlePhalanxOutlineAgeGroup), (MiddlePhalanxOutlineCorrect), (MiddlePhalanxTW), the performance is usually the same. We believe this validates our motivation that for some time series datasets, trend information is quite crucial in performing TSC, whereas for fewer others it is not important.

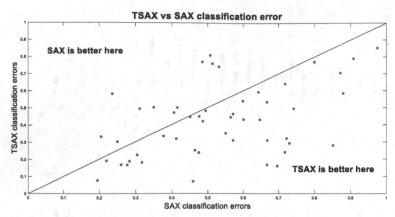

Fig. 4. Comparison of the classification errors of TSAX versus SAX. The bottom-right region is where TSAX performs better, and the top-left region is where SAX performs better.

Finally, in Fig. 4 we show, on one plot, a comparison of the classification errors for TSAX versus SAX on all 50 datasets. This global view of the outcomes of the experiments summarizes our above mentioned findings. The bottom-right region of the figure shows where the classification errors of TSAX are lower than those of SAX, whereas the top-left region of the figure shows where the classification errors of SAX are lower than those of TSAX.

5   Conclusion

In this paper we presented TSAX, which is a modification of SAX - one of the most popular time series representation methods. TSAX enhances SAX by providing it with features that capture segment trending information. TSAX has almost the same simplicity and efficiency of SAX, but it has a better performance in time series classification, which we showed through extensive experiments on 50 datasets, where TSAX gave a lower classification error than SAX on 39 of these datasets. For some datasets the improvement in performance was substantial.

In the experiments we conducted we meant to focus on simplicity. But we can even improve the performance of TSAX further by finding the best values for the parameters *rew* and *pen*.

Although we tested TSAX in a time series classification task only, we believe, which is one direction of future work, that TSAX can give better results than SAX in other time series mining tasks as well.

Like SAX, TSAX compares two time series segment by segment, we are investigating comparing two time series on a several-segment basis at a time. This will be of interest on both the symbolic and the trend parts of the TSAX representation. We are particularly interested in integrating some techniques from string comparisons that are used in the bioinformatics community.

References

1. Bagnall, A., Lines, J., Bostrom, A., Large, J., Keogh, E.: The great time series classification bake off: a review and experimental evaluation of recent algorithmic advances. Data Min. Knowl. Disc. **31**(3), 606–660 (2016). https://doi.org/10.1007/s10618-016-0483-9
2. Baydogan, M., Runger, G., Tuv, E.: A bag-of-features framework to classify time series. IEEE Trans. Pattern Anal. Mach. Intell. **25**(11), 2796–2802 (2013)
3. Dau, H.A., et al.: The UCR Time Series Classification Archive (2019). https://www.cs.ucr.edu/~eamonn/time_series_data_2018/
4. Fawaz, H.I., Forestier, G., Weber, J., Idoumghar, L., Muller, P.A.: Adversarial attacks on deep neural networks for time series classification. In: Proceedings of the 2019 International Joint Conference on Neural Networks (IJCNN), Budapest, Hungary, 14–19 July 2019 (2019)
5. Golomb, S.W.: Run-length encodings. IEEE Trans. Inf. Theory **12**(7), 399–401 (1966)
6. Hatami, N., Gavet, Y., Debayle, J.: Bag of recurrence patterns representation for time-series classification. Pattern Anal. Appl. **22**(3), 877–887 (2018). https://doi.org/10.1007/s10044-018-0703-6
7. Kane, A.: Trend and value based time series representation for similarity search. In: 2017 IEEE Third International Conference on Multimedia Big Data (BigMM), p. 252 (2017)
8. Karim, F., Majumdar, S., Darabi, H.: Insights into LSTM fully convolutional networks for time series classification. IEEE Access **7**, 67718–67725 (2019). https://doi.org/10.1109/ACCESS.2019.2916828
9. Keogh, E., Chakrabarti, K., Pazzani, M., Mehrotra, S.: Dimensionality reduction for fast similarity search in large time series databases. J. Knowl. Inform. Syst. **3**(3), 263–286 (2000)
10. Keogh, E., Chakrabarti, K., Pazzani, M., Mehrotra, S.: Locally adaptive dimensionality reduction for similarity search in large time series databases. In: SIGMOD, pp. 151–162 (2001)
11. Lin, J., Keogh, E., Lonardi, S., Chiu, B.Y.: A symbolic representation of time series, with implications for streaming algorithms. In: DMKD 2003, pp. 2–11 (2003)
12. Lin, J., Keogh, E., Wei, L., Lonardi, S.: Experiencing SAX: a novel symbolic representation of time series. Data Min. Knowl. Discov. **15**(2), 107–144 (2007)
13. Lin, J., Khade, R., Li, Y.: Rotation-invariant similarity in time series using bag-of-patterns representation. J. Intell. Inf. Syst. **39**, 287–315 (2012)
14. Malinowski, S., Guyet, T., Quiniou, R., Tavenard, R.: 1d-SAX: a novel symbolic representation for time series. In: Tucker, A., Höppner, F., Siebes, A., Swift, S. (eds.) IDA 2013. LNCS, vol. 8207, pp. 273–284. Springer, Heidelberg (2013). https://doi.org/10.1007/978-3-642-41398-8_24
15. Maimon, O., Rokach, L. (eds.): Data Mining and Knowledge Discovery Handbook. Springer, New York (2005). https://doi.org/10.1007/b107408
16. Muhammad Fuad, M.M.: Differential evolution versus genetic algorithms: towards symbolic aggregate approximation of non-normalized time series. In: Sixteenth International Database Engineering & Applications Symposium– IDEAS 2012, Prague, Czech Republic, 8–10 August, 2012. BytePress/ACM (2012)
17. Muhammad Fuad, M.M.: Extreme-SAX: extreme points based symbolic representation for time series classification. In: Song, M., Song, I.-Y., Kotsis, G., Tjoa, A.M., Khalil, I. (eds.) DaWaK 2020. LNCS, vol. 12393, pp. 122–130. Springer, Cham (2020). https://doi.org/10.1007/978-3-030-59065-9_10
18. Muhammad Fuad, M.M.: Genetic algorithms-based symbolic aggregate approximation. In: Cuzzocrea, A., Dayal, U. (eds.) DaWaK 2012. LNCS, vol. 7448, pp. 105–116. Springer, Heidelberg (2012). https://doi.org/10.1007/978-3-642-32584-7_9

19. Muhammad Fuad, M.M.: Modifying the symbolic aggregate approximation method to capture segment trend information. In: Torra, V., Narukawa, Y., Nin, J., Agell, N. (eds.) MDAI 2020. LNCS (LNAI), vol. 12256, pp. 230–239. Springer, Cham (2020). https://doi.org/10.1007/978-3-030-57524-3_19
20. Ratanamahatana, C., Keogh, E.: Making time-series classification more accurate using learned constraints. In: Proceedings of SIAM International Conference on Data Mining, pp. 11–22 (2004)
21. Ratanamahatana, C., Keogh, E., Bagnall, A.J., Lonardi, S.: A novel bit level time series representation with implication of similarity search and clustering. In: Ho, T.B., Cheung, D., Liu, H. (eds.) PAKDD 2005. LNCS (LNAI), vol. 3518, pp. 771–777. Springer, Heidelberg (2005). https://doi.org/10.1007/11430919_90
22. Wang, Z., Yan, W., Oates, T.: Time series classification from scratch with deep neural networks: a strong baseline. In: Proceedings of International Joint Conference on Neural Networks (IJCNN), May 2017, pp. 1578–1585 (2017)
23. Yi, B.K., Faloutsos, C.: Fast time sequence indexing for arbitrary LP norms. In: Proceedings of the 26th International Conference on Very Large Databases, Cairo, Egypt (2000)
24. Zhang, T., Yue, D., Gu, Y., Wang, Y., Yu, G.: Adaptive correlation analysis in stream time series with sliding windows. Comput. Math. Appl. **57**(6), 937–948 (2009)

MultiEmo: Multilingual, Multilevel, Multidomain Sentiment Analysis Corpus of Consumer Reviews

Jan Kocoń[✉][ID], Piotr Miłkowski[✉][ID], and Kamil Kanclerz[✉][ID]

Wrocław University of Science and Technology, Wrocław, Poland
{jan.kocon,piotr.milkowski,kamil.kanclerz}@pwr.edu.pl

Abstract. This article presents MultiEmo, a new benchmark data set for the multilingual sentiment analysis task including 11 languages. The collection contains consumer reviews from four domains: medicine, hotels, products and university. The original reviews in Polish contained 8,216 documents consisting of 57,466 sentences. The reviews were manually annotated with sentiment at the level of the whole document and at the level of a sentence (3 annotators per element). We achieved a high Positive Specific Agreement value of 0.91 for texts and 0.88 for sentences. The collection was then translated automatically into English, Chinese, Italian, Japanese, Russian, German, Spanish, French, Dutch and Portuguese. MultiEmo is publicly available under the MIT Licence. We present the results of the evaluation using the latest cross-lingual deep learning models such as XLM-RoBERTa, MultiFiT and LASER+BiLSTM. We have taken into account 3 aspects in the context of comparing the quality of the models: multilingualism, multilevel and multidomain knowledge transfer ability.

1 Introduction

Sentiment analysis has become very popular in recent years in many areas using natural language text processing. These include topics such as prediction of future events including security issues in the world [25]. There is also great interest in the analysis of consumer opinions [6,15,16] especially among product manufacturers who want to know the general reactions of customers to their products and thus improve them. Consumer reviews allow for the recognition of specific customer preferences, which facilitates good marketing decisions. With the increase in the number of reviews, especially for products sold on the global market (for which reviews are available in many languages), it is necessary to develop an effective method of multilingual analysis of the sentiment of a review, which would also be able to evaluate not only the sentiment of the entire opinion, but also its components, e.g. aspects or features of the product, whose sentiment is expressed at the level of sentences [24]. It is important that the method should also work in as many domains as possible [1,17,18].

© Springer Nature Switzerland AG 2021
M. Paszynski et al. (Eds.): ICCS 2021, LNCS 12743, pp. 297–312, 2021.
https://doi.org/10.1007/978-3-030-77964-1_24

In this work we present MultiEmo, a multilanguage benchmark corpus of consumer opinions, developed on the basis of PolEmo 2.0 [19]. The original collection was created to fill the gap in datasets annotated with sentiments for low-resource language, such as Polish. However, the results of this work show that perhaps treating Polish as a low-resource language is no longer correct (Sect. 7). It can certainly be said that the number of corpora annotated with sentiment for the Polish one is very small (low-resource in this domain, Sect. 3). Low-resource languages often provide a wealth of information related to the culture of the people who speak them. This knowledge concerns intangible cultural heritage, which allows a better understanding of the processes that have shaped a given society, its value system and traditions. These factors are important in the process of determining the sentiment of texts written by a person belonging to a particular cultural group.

MultiEmo allows building and evaluating a sentiment recognition model for both high-resource and low-resource languages at the level of the whole text, as well as single sentences, and for different domains. A high level of Positive Specific Agreement (PSA) [13] was achieved for this set, which is 0.91 for annotations at the text level and 0.81 at the sentence level. It turns out that the collection is very well suited for the evaluation of modern deep language models, especially cross-lingual ones. To the best of our knowledge, there is no other such large publicly available dataset annotated with a sentiment, allowing simultaneous evaluation of models in 3 different aspects (3M): multilingual, multilevel and multidomain.

We also present the results of classification using selected recent deep language models: XLM-RoBERTa [4], MultiFiT [8] and the proposed new combination of LASER+BiLSTM, using the Language-Agnostic SEntence Representations (LASER) [2] model to evaluate the quality of cross-lingual sentiment recognition zero-shot transfer learning task.

Table 1. The description of the review sources, with review domain, author type, subject type and domain subcorpus size (number of documents). For two domains potentially neutral texts were added as part of articles related to the domain.

ID	Name	Source	Author	Subject	Size
H	Hotels	tripadvisor.com	Guest	Hotel	3456
M	Medicine	znanylekarz.pl	Patient	Doctor	2772
U	University	polwro.pl	student	Professor	484
P	Products	ceneo.pl	customer	Product	504
H	Hotels	ehotelarstwo.com	Neutral texts		500
M	medicine	naukawpolsce.pap.pl	Neutral texts		500

2   Related Work

In recent years, the development of Transformer-based language models has led to significant improvements in cross-lingual language understanding (XLU). This would not have been possible without an increasing number of benchmark sets, which make it possible to test the quality of new language models and compare them with existing ones. The pre-training and fine-tuning approach allows for state-of-the-art results for a large number of NLP tasks. Among the popular pre-trained models, two groups can be distinguished. The first of them are monolingual models, e.g.: BERT [7] or RoBERTa [21]. The second group are multilingual models, e.g.: LASER [2], XLM-RoBERTa [4], or MultiFiT [8] In this article we will focus mainly on the second group and we compare their effectiveness in aspects not only related to cross-lingual tasks, but also multidomain and multilevel. There are many benchmark data sets on which the above mentioned models are tested. In general, they can also be divided into similar groups. The following datasets can be listed in the monolingual group: GLUE [27], KLEJ [23] or CoLA [28]. In the multilingual group, the examples are: XGLUE [20] or XTREME [14].

Most of the mentioned language models support over 100 languages, e.g. LASER, mBERT, XLM, XLM-RoBERTa, fastText-RCSLS. However, there are models that are pre-trained in a much smaller number of languages, e.g. Unicoder (15 languages) or MultiFiT (7 languages). In the context of multilingual benchmark data sets, the number of supported languages is usually even smaller. The largest number of languages is XTREME (40 languages), XGLUE (19 languages) and XNLI (15 languages). However, the tasks in these datasets are mostly unique to the individual languages, i.e. they are not their translations. Additionally, there are no sets for which different levels of annotation (e.g. document level and sentence level) or other phenomena, e.g. cross-domain knowledge transfer, can be studied at the same time (i.e. on the same texts, translated into many languages). Moreover, low-resource languages are highly underrepresented in most of the sub-tasks of these benchmarks.

An important problem from the perspective of multilingual sentiment analysis is the small number of benchmark sets. None of the previously mentioned sets contain multilingual data for this task. To the best of our knowledge, there is no set for this task, which contains accurate translations of the training and test instances for many languages, additionally taking into account multidomain and multilevel aspects. We found two collections close to the one we need, but both of them did not meet our objectives. One of the existing datasets is a collection of the SemEval-2016-Task-5 [22]. One of its subtask (Out-of-domain Aspect-Based Sentiment Analysis) contains data sets for 8 languages. These are consumer reviews from different sources, but each language contains a different number of them and they are not translations of the same reviews in different languages. The next most conceptually similar set to MultiEmo is Multilanguage Tweets Corpus [9]. This collection contains 2794 tweets in Polish (1397 positive and 1397 negative), 4272 tweets in Slovenian (2312 positive and 1950 negative) and 3554 tweets in Croatian (2129 positive and 1425 negative). Then the Google Translate tool was used to translate these tweets into English. However, this data was not translated into other languages, and there were different texts within the non-English collections. Due to a lack of data, we decided to prepare our own collection.

3   MultiEmo Sentiment Corpus

The motivation to prepare the source corpus for MultiEmo were works devoted to domain-oriented sentiment analysis, where the model is trained on annotated reviews from the source domain and tested on other [10]. A newer work on this subject describes a study on the Amazon Product Data collection [11]. However, this collection contains ratings assigned to the reviews by the authors of the texts. Additionally, annotations are assigned at the level of the entire document. The initial idea was to have a corpus of reviews that would be evaluated by the recipients, not the authors of the content. Annotations should also be assigned not only at the level of the whole document, but also at the level of individual sentences, which makes it easier to assess aspects of the opinion. The last important feature was that the collection would be multidomain in order to be able to study models in the cross-domain knowledge transfer task. Four domains presented in Table 1 were chosen to build the initial corpus. Initial set of annotation tags contained 6 different ratings: 1) Strong Positive (SP), 2) Weak Positive (WP), 3) Neutral (0), 4) Weak Negative (WN), 5) Strong Negative (SN), 6) Ambivalent (AMB). The annotators were asked not to judge the strength of sentiment when distinguishing between strong and weak categories. If the review was entirely positive or entirely negative, then it received a strong category. If the positive aspects outweighed the negative ones, then weak. If the positive and negative aspects were balanced, then the texts were marked as AMB. These rules were applied both to the entire text level and the sentence level. The final Positive Specific Agreement on a part of corpus containing 50 documents was 90% (meta) and 87% (sentence).

Table 2. PSA for WP/WN/AMB tags merged into one tag (AMB) at the (L)evel of (T)ext and (S)entence for the following (D)omains: (H)otels, (M)edicine, (P)roducts, (S)chool and (A)ll. Abbreviations: Strong Positive (SP), Neutral (0), Strong Negative (SN), Ambivalent (AMB).

L	D	SN	0	AMB	SP	A
T	H	91.92	99.42	78.50	91.62	89.39
	M	94.09	99.05	70.25	96.28	93.43
	P	94.06	100.0	77.82	85.95	89.07
	S	87.50	00.00	80.78	92.52	88.32
	A	92.87	99.18	76.87	93.48	90.91
S	H	93.78	88.40	65.64	93.05	89.83
	M	90.43	91.84	59.40	93.43	90.13
	P	91.27	48.42	41.22	90.84	79.12
	S	79.21	26.56	45.48	81.39	65.68
	A	91.92	87.21	56.82	92.12	87.50

After annotating the whole corpus, it turned out that PSA for weak categories (WP, WN, AMB) is low and does not exceed 40%. Distinguishing between the significance of positive and negative aspects was a difficult task. It was decided to merge the WP, WN and AMB categories into one AMB category. Table 2 presents the PSA value after the weak category merging procedure. After this operation, the total PSA value has increased from 83% to 91% for annotations at the text level and from 85% to 88% for annotations at the sentence level.

Table 3. The number of texts/sentences for each evaluation type in train/dev/test sets. Average length (Avg len) of line is calculated from merged set.

Type	Domain	Train	Dev	Test	SUM	Avg len
SDT	Hotels	3165	396	395	3956	773
	Medicine	2618	327	327	3272	782
	Products	387	49	48	484	756
	School	403	50	51	504	427
DOT	!Hotels	3408	427	-	3835	737
	!Medicine	3955	496	-	4451	740
	!Products	6186	774	-	6960	757
	!School	6170	772	-	6942	778
MDT	All	6573	823	820	8216	754
SDS	Hotels	19881	2485	2485	24851	92
	Medicine	18126	2265	2266	22657	111
	Products	5942	743	742	7427	98
	School	2025	253	253	2531	110
DOS	!Hotels	26093	3262	-	29355	108
	!Medicine	27848	3481	-	31329	95
	!Products	40032	5004	-	45036	101
	!School	43949	5494	-	49443	101
MDS	All	45974	5745	5747	57466	101

Table 3 shows the number of texts and sentences annotated by linguists for all evaluation types, with division into the number of elements within training, validation and test sets as well as average line length of each combined set. Finally, the corpus has been translated into 10 languages using the DeepL[1] tool: English, Chinese, Italian, Japanese, Russian, German, Spanish, French, Dutch and Portuguese. Its translations are of better quality than those generated by Microsoft Translator Hub [26]. DeepL achieves the best results when translating German texts into English or French. The semantic correctness of the translations does not guarantee the precise preservation of the sentiment associated with a given

---

[1] https://www.deepl.com/.

text. However, in a situation where we have limited resources and want to use information about the cultural background of authors writing in a low-resource language, machine translation is one of the best solutions. MultiEmo[2] corpus is available under the MIT Licence.

4   Chosen Language Models

We have chosen XLM-RoBERTa [4] and MultiFiT [8] language models to perform analysis of sentiment recognition task and LASER [2] to test cross-lingual zero-shot transfer task capability using MultiEmo. The first model, Unsupervised Cross-lingual Representation Learning at Scale (XLM-RoBERTa), is a large multillingual language model, trained on 2.5TB of filtered CommonCrawl data, using self-supervised training techniques to achieve state-of-the-art performance in cross-lingual understanding. Unfortunately, usage of this model is a very resource-intensive process due to its complexity. The second model, Efficient Multi-lingual Language Model Fine-tuning (MultiFiT), is based on Universal Language Model Fine-Tuning (ULMFiT) [12] with number of improvements: 1) usage of SentencePiece subword tokenization instead of word-based tokenization, significantly reducing vocabulary size for morphologically rich languages, and 2) Quasi-Recurrent Neural Network (QRNN) [3] which are up to 16 times faster at train and test time comparing to long short-term memory (LSTM) neural networks due to increased parallelism. The last approach is our proposal to use LASER embeddings as an input for the neural network based on bidirectional long short-term memory (BiLSTM) architecture. During the literature review we did not find such an application directly. LASER is capable of calculating sentence embeddings for 93 languages, therefore solution prepared on one language can be used on other language without any additional training and allows performing sentiment recognition zero-shot cross-lingual transfer task. The main advantage of this multilingual approach is that a preparation of individual model for each language can be avoided. This significantly reduces the training time and memory usage. The second advantage is that it is not necessary to translate the text into each language separately. This results in a reduction of training time and the computational resources usage.

5   Multidimensional Evaluation

In order to present the multidimensional evaluation possibilities of MultiEmo, we have conducted several types of evaluation. The first three evaluation processes focused on the multilingual aspect of the sentiment corpus. The first one was to check whether models trained on LASER embeddings of texts in one language would be equally effective in sentiment analysis of texts in another language as models trained on LASER embeddings of texts in the same language as the test set. We chose 11 different languages available in MultiEmo

---

[2] https://clarin-pl.eu/dspace/handle/11321/798.

Sentiment Corpus: Chinese, Dutch, English, French, German, Italian, Japanese, Polish, Portuguese, Russian and Spanish. The second type of evaluation aimed to check whether models trained on LASER embeddings of texts in languages other than Polish will be able to effectively analyze sentiment in texts in Polish as well as the model trained only on LASER embeddings of texts in Polish. The third evaluation focused on measuring the effectiveness of classifiers in the task of sentiment analysis in texts written in 10 different languages: Chinese, Dutch, English, French, German, Italian, Japanese, Portuguese, Russian and Spanish. We decided to evaluate 3 different classifiers: bidirectional long short-term memory network trained on language-agnostic sentence embeddings (LASER+BiLSTM), MultiFiT and XLM-RoBERTa. The fourth evaluation focused on the multilevel aspect of the MultiEmo Sentiment Corpus. In the evaluation process we focused on checking the effectiveness of 3 classifiers (LASER+BiLSTM, MultiFiT and XLM-RoBERTa) in the sentiment recognition of single sentences. A single sentence provides far less information than a multisentence opinion. Such a small amount of information makes it difficult to correctly determine the sentiment of a review. Therefore, we decided to test the same 3 classifiers that were used in the evaluation process on text-level annotations to see if they will be equally effective in the classification of sentence-level annotations. The fifth evaluation aims to take advantage of the multidomain aspect of MultiEmo Sentiment Corpus. The sentiment of a given word often depends on the domain of the whole text. Depending on the subject of the text, the word may have positive, neutral or negative sentiment. Moreover, correct recognition of the sentiment of a text regardless of its field is an even more difficult task and requires good quality texts from many domains. During this process we evaluated 3 classifiers (LASER+BiLSTM, MultiFiT and XLM-RoBERTa) in the task of sentiment recognition in texts from a single domain. We conducted the evaluation both when the classifiers were trained on a set containing only texts from the same domain (SD) and when the training set contained texts from multiple domains (MD).

During the evaluation process we trained 30 instances of each model and then conducted evaluation on a given test set. After that we conducted statistical tests to verify the statistical significance of differences between evaluation results of each model. We decided to use independent samples t-test, as the evaluation results concerned different models. Before we conducted the test, we checked its assumptions and if any of the samples did not meet them, we used the non-parametric Mann Whitney U test. The values in bold in each table with the results of a particular evaluation, presented in Sect. 6 mean that a given model performed significantly better than the others. It should be mentioned that monolingual models are in fact multilingual models tuned using a single language set. In our five experiments we counted how many "cases" the model was better than the others by counting the number of occurrences of the best result in all variants in a single experiment.

Table 4 presents the average F1-score values for each of the labels as well as global F1-score, micro-AUC and macro-AUC for the MultiEmo evaluation of bidirectional long short-term memory network models trained on language-agnostic sentence embeddings. Significant differences between performance of the models trained on texts in Polish and the models trained in the same language as the test set were observed in 26 out of 70 cases (37%). The models achieved different results mainly in case of neutral and ambivalent texts, which are much more diverse than texts characterized by strong and uniform emotions, e.g. strongly positive and strongly negative.

Table 4. Average F1-scores for the MultiEmo evaluation of LASER+BiLSTM models trained on texts in Polish and the ones trained on texts in the same language as the test set. The values in bold refer to model that achieved significantly better results than the other one. Abbreviations: Strong Positive (SP), Neutral (0), Strong Negative (SN), Ambivalent (AMB).

Test lang	Train lang	SP	0	SN	AMB	F1	micro	macro
Chinese	Polish	5.62	**9.60**	2.47	**14.89**	4.17	45.05	49.23
	Chinese	**16.45**	0.72	**18.70**	0.66	**12.64**	**62.19**	52.45
Dutch	Polish	65.70	65.46	78.51	40.01	69.48	79.65	75.19
	Dutch	67.62	**73.71**	78.66	39.59	70.48	80.32	76.94
English	Polish	67.33	67.49	79.17	**42.49**	70.29	80.20	76.37
	English	69.89	71.21	77.45	35.53	70.07	80.04	76.08
French	Polish	66.02	**66.18**	78.61	**39.96**	69.54	79.74	75.45
	French	62.47	59.48	76.78	30.81	66.92	77.99	72.52
German	Polish	65.39	63.34	78.20	**38.52**	68.83	79.22	74.62
	German	**70.37**	65.07	78.76	34.81	70.43	80.29	75.48
Italian	Polish	66.18	62.80	79.08	**40.97**	69.45	79.63	75.22
	Italian	70.00	**69.77**	80.07	35.30	71.86	81.24	76.73
Japanese	Polish	**36.86**	**17.84**	**34.15**	**11.00**	**19.33**	**75.55**	**69.29**
	Japanese	3.05	0.75	21.35	0.00	12.10	60.99	50.57
Portuguese	Polish	66.59	65.13	79.53	**41.17**	70.06	80.04	75.76
	Portuguese	67.42	66.57	77.29	32.61	69.00	79.33	74.61
Russian	Polish	65.47	**64.18**	79.02	**39.22**	69.38	79.59	**75.01**
	Russian	65.46	43.54	75.43	31.19	65.43	76.95	70.56
Spanish	Polish	66.67	**65.50**	79.44	40.91	70.07	80.05	75.72
	Spanish	65.02	56.33	75.41	38.23	66.68	77.79	73.77

Table 5 shows average F1-scores for the MultiEmo evaluation of long short-term memory neural network models trained on language-agnostic sentence embeddings on the test set containing only texts in Polish. The results of models trained on texts in languages different than Polish were compared with the results of the model trained only on texts in Polish. On the basis of statistical tests described in Sect. 5, significant differences in model results were observed in 3 out of 70 cases (4.3%). The worst results were observed for models trained on Chinese and Japanese texts.

The MultiEmo multilingual evaluation results of different classifiers are presented in Table 6. We decided to choose three classifiers: LASER+BiLSTM, MultiFiT and XLM-RoBERTa. MultiFiT achieved the best results in 32 out of 49 cases (65%). XLM-RoBERTa outperformed other models in 38 out of 70 cases (54%). Both MultiFiT and XLM-RoBERTa obtained better results than LASER+BiLSTM in every case. MultiFiT performed better than XLM-RoBERTa in 4 out of 7 languages (57%).

6   Results

The results of the evaluation on the MultiEmo sentence-based multidomain dataset are described in Table 7. MultiFiT outperformed other models in 28 out of 28 cases (100%). XLM-RoBERTa achieved the best results in 13 out of 42 cases (31%).

Table 5. Average F1-scores for the MultiEmo evaluation of LASER+BiLSTM models on the test set containing only texts in Polish. The values in bold refer to models that achieved significantly better results than the model trained on texts in Polish. Abbreviations: Strong Positive (SP), Neutral (0), Strong Negative (SN), Ambivalent (AMB).

Train lang.	SP	0	SN	AMB	F1	micro	macro
Polish	66.49	63.20	78.44	36.74	69.04	79.36	74.49
Chinese	27.20	1.70	21.48	1.32	30.22	53.48	47.84
Dutch	66.36	**70.85**	77.24	37.30	68.97	79.31	75.67
English	**71.44**	67.54	77.52	36.18	70.50	80.33	75.99
French	62.46	53.98	75.89	25.68	65.79	77.20	71.12
German	**71.17**	64.40	78.25	31.21	70.09	80.06	74.74
Italian	70.46	70.33	79.64	33.92	71.50	81.00	76.60
Japanese	2.83	2.46	58.83	0.00	42.06	61.37	50.61
Portuguese	63.40	57.57	74.29	29.08	65.59	77.06	71.94
Russian	63.29	37.47	74.60	26.37	63.85	75.90	68.76
Spanish	63.84	50.81	73.57	35.38	64.80	76.54	72.04

Table 6. Average F1-scores for the MultiEmo evaluation of three different classifiers: LASER+BiLSTM, MultiFiT and XLM-RoBERTa. For languages not supported by MultiFiT, an evaluation was carried out for the LASER+BiLSTM and XLM-RoBERTa classifiers. The values in bold refer to model that achieved significantly better results than the other ones. Abbreviations: Strong Positive (SP), Neutral (0), Strong Negative (SN), Ambivalent (AMB).

Language	Classifier	SP	0	SN	AMB	F1	micro	macro
Chinese	LASER+BiLSTM	16.45	0.72	18.70	0.66	12.64	62.19	52.45
	MultiFiT	85.81	95.02	86.78	**59.91**	83.19	88.79	87.64
	XLM-RoBERTa	**86.34**	**95.69**	**87.99**	57.13	**84.05**	**89.37**	**87.92**
Dutch	LASER+BiLSTM	67.62	73.71	78.66	39.59	70.48	80.32	76.94
	XLM-RoBERTa	**84.00**	**96.39**	**86.31**	**53.20**	**82.45**	**88.30**	**86.76**
English	LASER+BiLSTM	69.89	71.21	77.45	35.53	70.07	80.04	76.08
	XLM-RoBERTa	**85.96**	**93.76**	**88.67**	**60.47**	**84.87**	**89.91**	**88.48**
French	LASER+BiLSTM	62.47	59.48	76.78	30.81	66.92	77.99	72.52
	MultiFiT	**86.48**	**96.04**	**87.49**	**57.42**	**83.63**	**89.09**	**87.76**
	XLM-RoBERTa	83.88	95.60	86.18	51.81	81.93	87.96	86.43
German	LASER+BiLSTM	70.37	65.07	78.76	34.81	70.43	80.29	75.48
	MultiFiT	**85.85**	**96.52**	**88.21**	**60.35**	**84.22**	**89.48**	**88.28**
	XLM-RoBERTa	82.16	89.83	86.86	59.06	82.74	88.49	86.85
Italian	LASER+BiLSTM	70.00	69.77	80.07	35.30	71.86	81.24	76.73
	MultiFiT	**86.18**	**96.04**	**87.87**	57.91	83.70	89.13	87.82
	XLM-RoBERTa	85.36	93.75	87.65	**59.06**	**84.06**	**89.37**	**87.87**
Japanese	LASER+BiLSTM	3.05	0.75	21.35	0.00	12.10	60.99	50.57
	MultiFiT	83.39	**95.77**	**87.63**	58.09	82.61	88.41	87.35
	XLM-RoBERTa	**84.54**	93.60	87.41	**58.80**	**83.67**	**89.11**	**87.54**
Portuguese	LASER+BiLSTM	67.42	66.57	77.29	32.61	69.00	79.33	74.61
	XLM-RoBERTa	**85.85**	**96.87**	**86.88**	**55.69**	**83.40**	**88.93**	**87.62**
Russian	LASER+BiLSTM	65.46	43.54	75.43	31.19	65.43	76.95	70.56
	MultiFiT	**85.54**	**96.40**	86.95	**59.72**	**83.43**	**88.96**	**87.87**
	XLM-RoBERTa	82.95	90.93	**86.96**	58.94	83.22	88.81	87.10
Spanish	LASER+BiLSTM	65.02	56.33	75.41	38.23	66.68	77.79	73.77
	MultiFiT	**86.67**	95.98	**87.36**	**59.45**	**83.81**	**89.21**	**88.05**
	XLM-RoBERTa	86.28	**96.64**	87.05	56.59	83.56	89.04	87.83

Table 8 shows the evaluation results on MultiEmo single-domain and multidomain datasets. We decided to evaluate three classifiers: LASER+BiLSTM, MultiFiT and XLM-RoBERTa. In case of single domain datasets MultiFiT obtained the best results in 8 out of 16 cases (50%). XLM-RoBERTa outperformed other

Table 7. Average F1-scores for the evaluation on the MultiEmo sentence-based multidomain dataset. Classifiers: LASER+BiLSTM, MultiFiT, XLM-RoBERTa. For languages not supported by MultiFiT, an evaluation was carried out for the LASER + BiLSTM and XLM-RoBERTa classifiers. The values in bold refer to model that achieved significantly better results than the other ones. Abbreviations: Strong Positive (SP), Neutral (0), Strong Negative (SN), Ambivalent (AMB).

Language	Classifier	SP	0	SN	AMB	F1	micro	macro
English	LASER+BiLSTM	**50.36**	21.97	64.33	21.65	50.95	67.30	62.17
	XLM-RoBERTa	41.34	**39.44**	**68.39**	**28.44**	**58.07**	**72.05**	**66.74**
German	LASER+BiLSTM	41.40	14.79	62.45	26.70	47.50	65.00	60.88
	MultiFiT	**76.19**	**73.17**	**77.66**	**47.25**	**72.91**	**81.94**	**79.20**
	XLM-RoBERTa	25.06	23.93	62.73	16.78	49.75	66.50	60.14
Italian	LASER+BiLSTM	44.50	27.59	64.84	24.87	49.74	66.49	62.66
	MultiFiT	**76.47**	**74.00**	**77.51**	**47.85**	**73.10**	**82.07**	**79.51**
	XLM-RoBERTa	36.20	36.26	66.76	24.32	55.71	70.47	64.87
Japanese	LASER+BiLSTM	2.46	5.30	8.82	1.56	5.14	55.24	50.93
	MultiFiT	**73.36**	**70.73**	**75.22**	**45.40**	**70.63**	**80.42**	**77.40**
	XLM-RoBERTa	45.53	44.05	69.6	30.41	59.90	73.27	68.26
Polish	LASER+BiLSTM	45.82	40.02	66.53	28.51	53.44	68.96	64.78
	XLM-RoBERTa	**52.07**	**49.81**	**73.41**	**36.59**	**64.54**	**76.36**	**71.78**
Russian	LASER+BiLSTM	46.87	4.93	61.36	18.28	46.18	64.12	59.48
	MultiFiT	**76.87**	**73.89**	**77.68**	**47.64**	**73.33**	**82.22**	**79.55**
	XLM-RoBERTa	44.82	42.67	69.74	30.07	59.89	73.26	68.17

models in 10 out of 24 cases (42%). LASER+BiLSTM turned out to be the best in 6 out of 24 cases (25%). It outperformed other models in the review domain, achieving the best results in 5 out of 6 cases (83%). In case of multi domain evaluation XLM-RoBERTa outperformed other models in 18 out of 24 cases (75%). MultiFiT achieved the best results in 2 out of 16 cases (12.50%). The only case where LASER+BiLSTM achieved the best results were texts about products written in Japanese.

Table 8. Average F1-scores for the evaluation on the MultiEmo single-domain (**SD**) and multidomain (**MD**) datasets. The languages of the individual datasets: **DE** – German, **EN** – English, **IT** – Italian, **JP** – Japanese, **PL** – Polish, **RU** – Russian. Classifiers: LASER+BiLSTM, MultiFiT, XLM-RoBERTa. For languages not supported by MultiFiT, an evaluation was carried out for the LASER + BiLSTM and XLM-RoBERTa classifiers. The values in bold refer to model that achieved significantly better results than the other ones. Abbreviations: Strong Positive (SP), Neutral (0), Strong Negative (SN), Ambivalent (AMB).

Type	Test set domain	Classifier	DE	EN	IT	JP	PL	RU
SD	Hotels	LASER+BiLSTM	75.19	74.18	77.47	7.59	68.35	61.27
		MultiFiT	**83.29**	–	**83.03**	**80.00**	–	82.03
		XLM-RoBERTa	79.49	**83.04**	73.92	77.97	**82.03**	**83.29**
	Medicine	LASER+BiLSTM	63.61	50.76	75.84	10.09	81.96	80.12
		MultiFiT	**87.77**	–	**85.32**	**85.93**	–	**87.46**
		XLM-RoBERTa	81.65	**84.71**	82.26	83.18	**82.57**	83.18
	Products	LASER+BiLSTM	58.33	39.58	**75.00**	20.83	66.67	68.75
		MultiFiT	**76.60**	–	65.96	65.96	–	72.34
		XLM-RoBERTa	72.92	**72.92**	72.92	**72.92**	**72.92**	72.92
	University	LASER+BiLSTM	**70.00**	74.00	**72.00**	38.00	**64.00**	**68.00**
		MultiFiT	56.00	–	66.00	52.00	–	66.00
		XLM-RoBERTa	56.00	56.00	56.00	**56.00**	56.00	56.00
MD	Hotels	LASER+BiLSTM	64.30	62.78	63.29	1.77	75.44	58.99
		MultiFiT	82.53	–	83.54	**84.56**	–	81.52
		XLM-RoBERTa	**86.84**	**85.82**	**86.08**	83.80	**87.85**	**84.81**
	Medicine	LASER+BiLSTM	75.23	78.90	75.23	2.75	83.49	51.07
		MultiFiT	87.16	–	86.07	84.10	–	85.02
		XLM-RoBERTa	**89.60**	**88.07**	**89.07**	**89.30**	**87.46**	**88.69**
	Products	LASER+BiLSTM	62.50	72.92	64.58	**95.83**	64.58	75.00
		MultiFiT	78.72	–	**87.23**	74.47	–	80.85
		XLM-RoBERTa	79.17	72.92	68.75	83.33	**77.08**	79.17
	University	LASER+BiLSTM	76.00	58.00	36.00	16.00	70.00	48.00
		MultiFiT	76.00	–	76.00	80.00	–	80.00
		XLM-RoBERTa	**78.00**	**80.00**	**92.00**	**88.00**	**80.00**	**84.00**

7   Conclusions and Future Work

MultiEmo service[3] with all models is available through the CLARIN-PL Language Technology Centre[4]. The source code is available on the MutliEmo GitHub page[5]. In the case of LASER+BiLSTM model evaluation few differences were found between the model trained on texts in Polish and the model trained on

---

[3] http://ws.clarin-pl.eu/multiemo.

[4] http://clarin-pl.eu/.

[5] https://github.com/CLARIN-PL/multiemo.

texts in the same language as the test set. Similarly, statistical tests showed few differences in the effectiveness of models trained on texts in different languages in the task of sentiment recognition of texts in Polish. Low values of the model average F1-scores in the case of texts in Chinese and Japanese may be related to a significantly worse quality of text translations compared to translations into languages more similar to Polish, such as English or German. On the other hand, similar values of the average F1-scores for the multilingual model in the case of Polish and translated texts may be related to the high similarity of the model used for machine translation and the multilingual model. The authors of DeepL do not provide information regarding this subject.

In Table 4 presenting a comparison of LASER+BiLSTM models tested on pairs of texts in Polish and the language of training data, the biggest differences are observed for the classes with the smallest number representation in the set. Analyzing F1, micro and macro results, significant differences are only for Asian languages. The results are significantly worse for these two languages than for the others. This may be due to a much smaller number of data for the LASER model for these languages, because in Table 6 the results obtained for these languages on XLM-RoBERTa and MultiFiT models are much better. Unfortunately, we do not have access to the training resources of the source models to make this clear. The results for the other languages indicate that regardless of the configuration choice, the results within a pair of two languages do not differ significantly from each other. There is a possibility that the source models (DeepL and LASER) were trained on similar data for these language pairs. On the other hand, LASER supports 93 languages and DeepL only 12. We are not able to evaluate the other languages supported by LASER, but it can be assumed that if the data representation in the source model was at a similar level as for the examined languages with a high score, we can expect equally high results for such languages. Another experiment was to compare models trained on different languages and tested only on Polish (Table 5). Aggregate results for the LASER+BiLSTM model show that models created on translations of the original set are of comparable or worse quality than the model trained on Polish. Results for some single classes turn out to be even better for models built on translations than on the model built on the original corpus. Such cases are observed for Dutch, English and German. It is possible that in the data to create source models (LASER and DeepL) for these languages there is significantly larger number of translation examples. Further work should examine the quality of the translations for individual language pairs and check the correlation between the quality of the translation and the results of models based on these translations. Table 6 shows the results of different deep multi-language models built on different MultiEmo language versions for whole texts. Similar results are available in Table 7 for models built on single sentences. The aggregate results (F1, macro, micro) show a clear superiority of XLM-RoBERTa and MultiFiT models over the zero-shot transfer learning approach. The probable cause of these differences is the use of much more texts to create DeepL, XLM-RoBERTa and MultiFiT models, compared to the LASER model. On the other hand, in the absence of a good machine translation tool, the LASER+BiLSTM model for most languages still achieves results that are at least in some business applications already acceptable. The results also show that translating a text into another language using a good quality translator allows to obtain a model with results comparable to those obtained for a model built for the source language. Moreover, it has been shown that the Polish language achieves more and more satisfactory support in known SOTA tools and models, and perhaps assigning this language to the low-resource category [5] is no longer justified. Otherwise, the conclusion is that we can also get very good quality models for high-resource languages from rare resources in low-resource languages.

Table 8 shows the results of models trained on the selected domain (SD) and on all domains simultaneously (MD). The results show that in the context of domain adaptation it is not possible to clearly indicate the best model to represent a single domain (SD variants). Differences were also found in different languages within the same domain. In case one model was trained on all domains, the most domain-agnostic sentiment representation has the XLM-RoBERTa.

MultiFiT achieved the best results in the greatest number of cases. The disadvantage of this model is the small number of supported languages (only 7). XLM-RoBERTa most often achieved the second best results, except the multidomain evaluation, where it outperformed other classifiers. LASER+BiLSTM as the only zero-shot classifier obtained worse results in almost every case. In our further research, we would like to address the detailed analysis of the impact of translations on sentiment analysis. Apart from the quality of the translations as such, a relatively interesting issue seems to be a direct change in the sentiment of the text during the translation.

Acknowledgements. Funded by the Polish Ministry of Education and Science, CLARIN-PL Project.

References

1. Al-Moslmi, T., Omar, N., Abdullah, S., Albared, M.: Approaches to cross-domain sentiment analysis: a systematic literature review. IEEE Access **5**, 16173–16192 (2017)
2. Artetxe, M., Schwenk, H.: Massively multilingual sentence embeddings for zero-shot cross-lingual transfer and beyond. Trans. Assoc. Comput. Linguist. **7**, 597–610 (2019)
3. Bradbury, J., Merity, S., Xiong, C., Socher, R.: Quasi-recurrent neural networks. arXiv preprint arXiv:1611.01576 (2016)
4. Conneau, A., et al.: Unsupervised cross-lingual representation learning at scale. In: Proceedings of the 58th Annual Meeting of the Association for Computational Linguistics, pp. 8440–8451. Association for Computational Linguistics, July 2020. https://doi.org/10.18653/v1/2020.acl-main.747
5. Dadas, S., Perełkiewicz, M., Poświata, R.: Evaluation of sentence representations in polish. arXiv preprint arXiv:1910.11834 (2019)
6. Day, M.Y., Lin, Y.D.: Deep learning for sentiment analysis on google play consumer review. In: 2017 IEEE international conference on information reuse and integration (IRI), pp. 382–388. IEEE (2017)

7. Devlin, J., Chang, M.W., Lee, K., Toutanova, K.: Bert: Pre-training of deep bidirectional transformers for language understanding. In: Proceedings of the 2019 Conference of the North American Chapter of the Association for Computational Linguistics: Human Language Technologies, Volume 1 (Long and Short Papers), pp. 4171–4186 (2019)
8. Eisenschlos, J., Ruder, S., Czapla, P., Kadras, M., Gugger, S., Howard, J.: Multifit: efficient multi-lingual language model fine-tuning. In: Proceedings of the 2019 Conference on Empirical Methods in Natural Language Processing and the 9th International Joint Conference on Natural Language Processing (EMNLP-IJCNLP), pp. 5706–5711 (2019)
9. Galeshchuk, S., Qiu, J., Jourdan, J.: Sentiment analysis for multilingual corpora. In: Proceedings of the 7th Workshop on Balto-Slavic Natural Language Processing, Florence, Italy, pp. 120–125. Association for Computational Linguistics, August 2019. https://doi.org/10.18653/v1/W19-3717
10. Glorot, X., Bordes, A., Bengio, Y.: Domain adaptation for large-scale sentiment classification: a deep learning approach. In: Proceedings of the 28th International Conference on Machine Learning (ICML 2011), pp. 513–520 (2011)
11. He, R., McAuley, J.: Ups and downs: Modeling the visual evolution of fashion trends with one-class collaborative filtering. In: proceedings of the 25th International Conference on World Wide Web, pp. 507–517. International World Wide Web Conferences Steering Committee (2016)
12. Howard, J., Ruder, S.: Universal language model fine-tuning for text classification. arXiv preprint arXiv:1801.06146 (2018)
13. Hripcsak, G., Rothschild, A.S.: Technical brief: agreement, the F-measure, and reliability in information retrieval. JAMIA **12**(3), 296–298 (2005). https://doi.org/10.1197/jamia.M1733
14. Hu, J., Ruder, S., Siddhant, A., Neubig, G., Firat, O., Johnson, M.: Xtreme: a massively multilingual multi-task benchmark for evaluating cross-lingual generalization. arXiv preprint arXiv:2003.11080 (2020)
15. Kanclerz, K., Miłkowski, P., Kocoń, J.: Cross-lingual deep neural transfer learning in sentiment analysis. Procedia Comput. Sci. **176**, 128–137 (2020)
16. Kocoń, J., Zaśko-Zielińska, M., Miłkowski, P.: Multi-level analysis and recognition of the text sentiment on the example of consumer opinions. In: Proceedings of the International Conference on Recent Advances in Natural Language Processing (RANLP 2019), pp. 559–567 (2019)
17. Kocoń, J., et al.: Recognition of emotions, valence and arousal in large-scale multi-domain text reviews. In: Human Language Technologies as a Challenge for Computer Science and Linguistics, pp. 274-280 (2019). ISBN 978-83-65988-31-7
18. Kocoń, J., et al.: Propagation of emotions, arousal and polarity in WordNet using Heterogeneous Structured Synset Embeddings. In: Proceedings of the 10th International Global Wordnet Conference (GWC'19), (2019)
19. Kocoń, J., Miłkowski, P., Zaśko-Zielińska, M.: Multi-level sentiment analysis of PolEmo 2.0: Extended corpus of multi-domain consumer reviews. In: Proceedings of the 23rd Conference on Computational Natural Language Learning (CoNLL), pp. 980–991 (2019)
20. Liang, Y., et al.: Xglue: a new benchmark dataset for cross-lingual pre-training, understanding and generation. arXiv preprint arXiv:2004.01401 (2020)
21. Liu, Y., et al.: Roberta: a robustly optimized Bert pretraining approach. arXiv preprint arXiv:1907.11692 (2019)

22. Pontiki, M., et al.: SemEval-2016 task 5: aspect based sentiment analysis. In: Proceedings of the 10th International Workshop on Semantic Evaluation (SemEval-2016), San Diego, California, pp. 19–30. Association for Computational Linguistics, June 2016. https://doi.org/10.18653/v1/S16-1002
23. Rybak, P., Mroczkowski, R., Tracz, J., Gawlik, I.: KLEJ: comprehensive benchmark for polish language understanding. In: Proceedings of the 58th Annual Meeting of the Association for Computational Linguistics, pp. 1191–1201. Association for Computational Linguistics, July 2020
24. Shoukry, A., Rafea, A.: Sentence-level Arabic sentiment analysis. In: 2012 International Conference on Collaboration Technologies and Systems (CTS), pp. 546–550. IEEE (2012)
25. Subramaniyaswamy, V., Logesh, R., Abejith, M., Umasankar, S., Umamakeswari, A.: Sentiment analysis of tweets for estimating criticality and security of events. J. Organ. End User Comput. (JOEUC) 29(4), 51–71 (2017)
26. Volkart, L., Bouillon, P., Girletti, S.: Statistical vs. neural machine translation: a comparison of MTH and DeepL at swiss post's language service. In: Proceedings of the 40th Conference Translating and the Computer, AsLing, pp. 145–150 (2018) iD: unige:111777
27. Wang, A., Singh, A., Michael, J., Hill, F., Levy, O., Bowman, S.: Glue: a multi-task benchmark and analysis platform for natural language understanding. In: Proceedings of the 2018 EMNLP Workshop BlackboxNLP: Analyzing and Interpreting Neural Networks for NLP, pp. 353–355 (2018)
28. Warstadt, A., Singh, A., Bowman, S.R.: Neural network acceptability judgments. Trans. Assoc. Comput. Linguist. 7, 625–641 (2019)

Artificial Intelligence
and High-Performance Computing
for Advanced Simulations

Outlier Removal for Isogeometric Spectral Approximation with the Optimally-Blended Quadratures

Quanling Deng[1](✉)[ID] and Victor M. Calo[2][ID]

[1] Department of Mathematics, University of Wisconsin–Madison,
Madison, WI 53706, USA
Quanling.Deng@math.wisc.edu
[2] Curtin Institute for Computation and School of Electrical Engineering,
Computing and Mathematical Sciences, Curtin University, P.O. Box U1987,
Perth, WA 6845, Australia
Victor.Calo@curtin.edu.au

Abstract. It is well-known that outliers appear in the high-frequency region in the approximate spectrum of isogeometric analysis of the second-order elliptic operator. Recently, the outliers have been eliminated by a boundary penalty technique. The essential idea is to impose extra conditions arising from the differential equation at the domain boundary. In this paper, we extend the idea to remove outliers in the supercon-vergent approximate spectrum of isogeometric analysis with optimally-blended quadrature rules. We show numerically that the eigenvalue errors are of superconvergence rate h^{2p+2} and the overall spectrum is outlier-free. The condition number and stiffness of the resulting algebraic system are reduced significantly. Various numerical examples demonstrate the performance of the proposed method.

Keywords: Isogeometric analysis · Boundary penalty · Spectrum · Eigenvalue · Superconvergence · Optimally-blended quadrature

1 Introduction

Isogeometric analysis (IGA) is a widely-used analysis tool that combines the classical finite element analysis with computer-aided design and analysis tools. It was introduced in 2005 [6,16]. There is a rich literature since its first development; see an overview paper [20] and the references therein. In particular, a

This work and visit of Quanling Deng in Krakow was partially supported by National Science Centre, Poland grant no. 017/26/M/ST1/00281. This publication was made possible in part by the CSIRO Professorial Chair in Computational Geoscience at Curtin University and the Deep Earth Imaging Enterprise Future Science Platforms of the Commonwealth Scientific Industrial Research Organisation, CSIRO, of Australia. Additional support was provided by the European Union's Horizon 2020 Research and Innovation Program of the Marie Sklodowska-Curie grant agreement No. 777778 and the Curtin Institute for Computation.

© Springer Nature Switzerland AG 2021
M. Paszynski et al. (Eds.): ICCS 2021, LNCS 12743, pp. 315–328, 2021.
https://doi.org/10.1007/978-3-030-77964-1_25

rich literature on IGA has been shown that the method outperforms the classical finite element method (FEM) on the spectral approximations of the second-order elliptic operators. The initial work [7] showed that the spectral errors of IGA were significantly smaller when compared with FEM approximations. In [17], the authors further explored the advantages of IGA on the spectral approximations.

To further reduce the spectral errors on IGA spectral approximations, on one hand, the recent work [5,23] introduced Gauss-Legendre and Gauss-Lobatto optimally-blended quadrature rules. By invoking the dispersion analysis which was unified with the spectral analysis in [18], the spectral errors were shown to be superconvergent with two extra orders. The work [10] generalized the blended rules to arbitrary p-th order IGA with maximal continuity. Along the line, the work [9] further studied the computational efficiency and the work [1,4,13,14,22] studied its applications.

On the other hand, the spectral errors in the highest-frequency regions are much larger than in the lower-frequency region. There is a thin layer in the highest-frequency region which is referred to as "outliers". The outliers in isogeometric spectral approximations were first observed in [7] in 2006. The question of how to efficiently remove the outliers remained open until recently. In [11], the authors removed the outliers by a boundary penalty technique. The main idea is to impose higher-order consistency conditions on the boundaries to the isogeometric spectral approximations. The work proposed to impose these conditions weakly. Outliers are eliminated and the condition numbers of the systems are reduced significantly.

In this paper, we propose to further reduce spectral errors by combining the optimally-blended quadrature rules and the boundary penalty technique. To illustrate the idea, we focus on tensor-product meshes on rectangular domains. We first develop the method in 1D and obtain the generalized matrix eigenvalue problems. We then apply the tensor-product structure to generate the matrix problems in multiple dimensions. By using the optimally-blended quadrature rules, we retain the eigenvalue superconvergence rate. By applying the boundary penalty technique, we remove the outliers in the superconvergent spectrum. The method also reduces the condition numbers.

The rest of this paper is organized as follows. Section 2 presents the problem and its discretization by the standard isogeometric analysis. Section 3 concerns the Gauss-Legendre and Gauss-Lobatto quadrature rules. We then present the optimally-blended rules. In Sect. 4, we apply the boundary penalty technique developed in [11] to the IGA setting with blending rules. Section 5 collects numerical results that demonstrate the performance of the proposed method. In particular, we perform the numerical study on the condition numbers. Concluding remarks are presented in Sect. 6.

2   Problem Setting

Let $\Omega = [0,1]^d \subset \mathbb{R}^d, d = 1, 2, 3$ be a bounded open domain with Lipschitz boundary $\partial\Omega$. We use the standard notation for the Hilbert and Sobolev spaces.

For a measurable subset $S \subseteq \Omega$, we denote by $(\cdot, \cdot)_S$ and $\| \cdot \|_S$ the L^2-inner product and its norm, respectively. We omit the subscripts when it is clear in the context. For any integer $m \geq 1$, we denote the H^m-norm and H^m-seminorm as $\| \cdot \|_{H^m(S)}$ and $| \cdot |_{H^m(S)}$, respectively. In particular, we denote by $H_0^1(\Omega)$ the Sobolev space with functions in $H^1(\Omega)$ that are vanishing at the boundary. We consider the classical second-order elliptic eigenvalue problem: Find the eigenpairs $(\lambda, u) \in \mathbb{R}^+ \times H_0^1(\Omega)$ with $\|u\|_\Omega = 1$ such that

$$
\begin{aligned}
-\Delta u &= \lambda u \quad &&\text{in} \quad \Omega, \\
u &= 0 \quad &&\text{on} \quad \partial\Omega,
\end{aligned}
\tag{1}
$$

where $\Delta = \nabla^2$ is the Laplacian. The variational formulation of (1) is to find $\lambda \in \mathbb{R}^+$ and $u \in H_0^1(\Omega)$ with $\|u\|_\Omega = 1$ such that

$$
a(w, u) = \lambda b(w, u), \quad \forall\, w \in H_0^1(\Omega),
\tag{2}
$$

where the bilinear forms

$$
a(v, w) := (\nabla v, \nabla w)_\Omega, \qquad b(v, w) := (v, w)_\Omega.
\tag{3}
$$

It is well-known that the eigenvalue problem (2) has a countable set of positive eigenvalues (see, for example, [2, Sec. 9.8])

$$
0 < \lambda_1 < \lambda_2 \leq \lambda_3 \leq \cdots
$$

and an associated set of orthonormal eigenfunctions $\{u_j\}_{j=1}^\infty$, that is, $(u_j, u_k) = \delta_{jk}$, where $\delta_{jk} = 1$ is the Kronecker delta. Consequently, the eigenfunctions are also orthogonal in the energy inner product since there holds $a(u_j, u_k) = \lambda_j b(u_j, u_k) = \lambda_j \delta_{jk}$.

At the discretize level, we first discretize the domain Ω with a uniform tensor-product mesh. We denote a general element as τ and its collection as $\mathcal{T}_h$ such that $\overline{\Omega} = \cup_{\tau \in \mathcal{T}_h} \tau$. Let $h = \max_{\tau \in \mathcal{T}_h} \mathrm{diameter}(\tau)$. In the IGA setting, for simplicity, we use the B-splines. The B-spline basis functions in 1D are defined by using the Cox-de Boor recursion formula; we refer to [8,21] for details. Let $X = \{x_0, x_1, \cdots, x_m\}$ be a knot vector with knots x_j, that is, a nondecreasing sequence of real numbers. The j-th B-spline basis function of degree p, denoted as $\phi_p^j(x)$, is defined recursively as

$$
\begin{aligned}
\phi_0^j(x) &= \begin{cases} 1, & \text{if } x_j \leq x < x_{j+1}, \\ 0, & \text{otherwise,} \end{cases} \\
\phi_p^j(x) &= \frac{x - x_j}{x_{j+p} - x_j} \phi_{p-1}^j(x) + \frac{x_{j+p+1} - x}{x_{j+p+1} - x_{j+1}} \phi_{p-1}^{j+1}(x).
\end{aligned}
\tag{4}
$$

A tensor-product of these 1D B-splines produces the B-spline basis functions in multiple dimensions. We define the multi-dimensional approximation space as $V_p^h \subset H_0^1(\Omega)$ with (see [3,15] for details):

$$
V_p^h = \mathrm{span}\{\phi_j^p\}_{j=1}^{N_h} = \begin{cases} \mathrm{span}\{\phi_{p_x}^{j_x}(x)\}_{j_x=1}^{N_x}, & \text{in 1D}, \\ \mathrm{span}\{\phi_{p_x}^{j_x}(x)\phi_{p_y}^{j_y}(y)\}_{j_x,j_y=1}^{N_x,N_y}, & \text{in 2D}, \\ \mathrm{span}\{\phi_{p_x}^{j_x}(x)\phi_{p_y}^{j_y}(y)\phi_{p_z}^{j_z}(z)\}_{j_x,j_y,j_z=1}^{N_x,N_y,N_z}, & \text{in 3D}, \end{cases}
$$

where p_x, p_y, p_z specify the approximation order in each dimension. N_x, N_y, N_z is the total number of basis functions in each dimension and N_h is the total number of degrees of freedom. The isogeometric analysis of (1) in variational formulation seeks $\lambda^h \in \mathbb{R}$ and $u^h \in V_p^h$ with $\|u^h\|_\Omega = 1$ such that

$$a(w^h, u^h) = \lambda^h b(w^h, u^h), \quad \forall \, w^h \in V_p^h. \tag{5}$$

At the algebraic level, we approximate the eigenfunctions as a linear combination of the B-spline basis functions and substitute all the B-spline basis functions for w^h in (5). This leads to the generalized matrix eigenvalue problem

$$\mathbf{KU} = \lambda^h \mathbf{MU}, \tag{6}$$

where $\mathbf{K}_{kl} = a(\phi_p^k, \phi_p^l), \mathbf{M}_{kl} = b(\phi_p^k, \phi_p^l)$, and $\mathbf{U}$ is the corresponding representation of the eigenvector as the coefficients of the B-spline basis functions. In practice, we evaluate the integrals involved in the bilinear forms $a(\cdot, \cdot)$ and $b(\cdot, \cdot)$ numerically using quadrature rules. In the next section, we present the Gauss–Legendre and Gauss–Lobatto quadrature rules, followed by their optimally-blended rules.

3   Quadrature Rules and Optimal Blending

In this section, we first present the classic Gauss-type quadrature rules and then present the optimally-blended rules developed recently in [5]. While the optimally-blended rules have been developed in [10] for arbitrary order isogeometric elements, we focus on the lower-order cases for simplicity.

3.1   Gaussian Quadrature Rules

Gaussian quadrature rules are well-known and we present these rules by following the book [19]. On a reference interval $\hat{\tau}$, a quadrature rule is of the form

$$\int_{\hat{\tau}} \hat{f}(\hat{x}) \, d\hat{x} \approx \sum_{l=1}^{m} \hat{\varpi}_l \hat{f}(\hat{n}_l), \tag{7}$$

where $\hat{\varpi}_l$ are the weights, $\hat{n}_l$ are the nodes, and m is the number of quadrature points. We list the lower-order Gauss-Legendre and Gauss-Lobatto quadrature rules in 1D below; we refer to [19] for rules with more points. The Gauss-Legendre quadrature rules for $m = 1, 2, 3, 4$ in the reference interval $[-1, 1]$ are as follows:

$$
\begin{aligned}
m = 1: \quad & \hat{n}_1 = 0, \quad \hat{\varpi}_1 = 2; \\[2mm]
m = 2: \quad & \hat{n}_1 = \pm\frac{\sqrt{3}}{3}, \quad \hat{\varpi}_1 = 1; \\[2mm]
m = 3: \quad & \hat{n}_1 = 0, \quad \hat{n}_{2,3} = \pm\sqrt{\frac{3}{5}}, \quad \hat{\varpi}_1 = \frac{8}{9}, \quad \hat{\varpi}_{2,3} = \frac{5}{9}; \\[2mm]
m = 4: \quad & \hat{n}_{1,2,3,4} = \pm\sqrt{\frac{3}{7} \mp \frac{2}{7}\sqrt{\frac{6}{5}}}, \quad \hat{\varpi}_{1,2,3,4} = \frac{18 \pm \sqrt{30}}{36}.
\end{aligned}
\tag{8}
$$

A Gauss-Legendre quadrature rule with m points integrates exactly a polynomial of degree $2m - 1$ or less. The Gauss-Lobatto quadrature rules with $m = 2, 3, 4, 5$ in the reference interval $[-1, 1]$ are as follows:

$$
\begin{aligned}
m = 2: \quad & \hat{n}_{1,2} = \pm 1, \quad \hat{\varpi}_1 = 1; \\[4pt]
m = 3: \quad & \hat{n}_1 = 0, \ \hat{n}_{2,3} = \pm 1, \quad \hat{\varpi}_1 = \frac{4}{3}, \ \hat{\varpi}_{2,3} = \frac{1}{3}; \\[4pt]
m = 4: \quad & \hat{n}_{1,2} = \pm\sqrt{\frac{1}{5}}, \ \hat{n}_{3,4} = \pm 1, \quad \hat{\varpi}_{1,2} = \frac{5}{6}, \ \hat{\varpi}_{3,4} = \frac{1}{6}; \\[4pt]
m = 5: \quad & \hat{n}_1 = 0, \ \hat{n}_{2,3} = \pm\sqrt{\frac{3}{7}}, \ \hat{n}_{4,5} = \pm 1, \ \hat{\varpi}_1 = \frac{32}{45}, \ \hat{\varpi}_{2,3} = \frac{49}{90}, \hat{\varpi}_{4,5} = \frac{1}{10}.
\end{aligned}
\tag{9}
$$

A Gauss-Lobatto quadrature rule with m points integrates exactly a polynomial of degree $2m - 3$ or less.

For each element τ, there is a one-to-one and onto mapping σ such that $\tau = \sigma(\hat{\tau})$, which leads to the correspondence between the functions on τ and $\hat{\tau}$. Let J_τ be the corresponding Jacobian of the mapping. Using the mapping, (7) induces a quadrature rule over the element τ given by

$$
\int_\tau f(\boldsymbol{x}) \ \mathrm{d}\boldsymbol{x} \approx \sum_{l=1}^{N_q} \varpi_{l,\tau} f(n_{l,\tau}),
\tag{10}
$$

where $\varpi_{l,\tau} = \det(J_\tau)\hat{\varpi}_l$ and $n_{l,\tau} = \sigma(\hat{n}_l)$. For blending rules, we denote by Q a general quadrature rule, by G_m the $m-$point Gauss-Legendre quadrature rule, by L_m the $m-$point Gauss-Lobatto quadrature rule, by Q_η a blended rule, and by O_p the optimally-blended rule for the p-th order isogeometric analysis.

3.2 Optimal Blending Quadrature Rules

Let $Q_1 = \{\varpi_{l,\tau}^{(1)}, n_{l,\tau}^{(1)}\}_{l=1}^{m_1}$ and $Q_2 = \{\varpi_{l,\tau}^{(2)}, n_{l,\tau}^{(2)}\}_{l=1}^{m_2}$ be two quadrature rules. We define the blended quadrature rule as

$$
Q_\eta = \eta Q_1 + (1 - \eta) Q_2,
\tag{11}
$$

where $\eta \in \mathbb{R}$ is a blending parameter. We note that the blending parameter can be both positive and negative. The blended rule Q_η for the integration of a function f is understood as

$$
\int_\tau f(\boldsymbol{x}) \ \mathrm{d}\boldsymbol{x} \approx \tau \sum_{l=1}^{m_1} \varpi_{l,\tau}^{(1)} f(n_{l,\tau}^{(1)}) + (1 - \tau) \sum_{l=1}^{m_2} \varpi_{l,\tau}^{(2)} f(n_{l,\tau}^{(2)}).
\tag{12}
$$

The blended rules for isogeometric analysis of the eigenvalue problem can reduce significantly the spectral errors. In particular, the optimally-blended rules deliver two extra orders of convergence for the eigenvalue errors; see [5, 10] for the developments. We present the optimally-blended rule for the p-th order isogeometric elements

$$O_p = Q_\eta = \eta G_{p+1} + (1 - \eta)L_{p+1}, \tag{13}$$

where the optimal blending parameters are given in the Table 1 below.

Table 1. Optimal blending parameters for isogeometric elements.

p	1	2	3	4	5	6	7
η	$\frac{1}{2}$	$\frac{1}{3}$	$-\frac{3}{2}$	$-\frac{79}{5}$	-174	$-\frac{91177}{35}$	$-\frac{105013}{2}$

We remark that there are other optimally-blended quadrature rules for isogeometric elements developed in [5, 10]. Non-standard quadrature rules are developed in [9] and they are shown to be equivalent with the optimally-blended rules. They all lead to the same stiffness and mass matrices. Herein, for simplicity, we only adopt the blended rules in the form of (13).

4   The Boundary Penalty Technique

Outliers appear in the isogeometric spectral approximations when using C^2 cubic B-splines and higher-order approximations. These outliers can be removed by a boundary penalty technique. In this section, we first recall the boundary penalty technique introduced recently in [11]. We present the idea for 1D problem with $\Omega = [0, 1]$ and $\partial\Omega = \{0, 1\}$. We then generalize it at the algebraic level using the tensor-product structure for the problem in multiple dimensions. We denote

$$\alpha = \lfloor \frac{p-1}{2} \rfloor = \begin{cases} \frac{p-1}{2}, & p \text{ is odd,} \\ \frac{p-2}{2}, & p \text{ is even.} \end{cases} \tag{14}$$

The isogeometric analysis of (1) in 1D with the boundary penalty technique is: Find $\tilde{\lambda}^h \in \mathbb{R}$ and $\tilde{u}^h \in V_p^h$ such that

$$\tilde{a}(w^h, \tilde{u}^h) = \tilde{\lambda}^h \tilde{b}(w^h, \tilde{u}^h), \quad \forall\, w^h \in V_p^h. \tag{15}$$

Herein, for $w, v \in V_p^h$

$$\tilde{a}(w, v) = \int_0^1 w'v'\, dx + \sum_{\ell=1}^{\alpha} \eta_{a,\ell}\pi^2 h^{6\ell-3}\Big(w^{(2\ell)}(0)v^{(2\ell)}(0) + w^{(2\ell)}(1)v^{(2\ell)}(1)\Big),$$
$$\tag{16a}$$

$$\tilde{b}(w, v) = \int_0^1 wv\, dx + \sum_{\ell=1}^{\alpha} \eta_{b,\ell} h^{6\ell-1}\Big(w^{(2\ell)}(0)v^{(2\ell)}(0) + w^{(2\ell)}(1)v^{(2\ell)}(1)\Big), \tag{16b}$$

where $\eta_{a,\ell}, \eta_{b,\ell}$ are penalty parameters set to $\eta_{a,\ell} = \eta_{b,\ell} = 1$ in default. The superscript (2ℓ) denotes the 2ℓ-th derivative. We further approximate the inner-products by the quadrature rules discussed above. For p-th order element and $\tau \in \mathcal{T}_h$, we denote $G_{p+1} = \{\varpi_{l,\tau}^G, n_{l,\tau}^G\}_{l=1}^{p+1}$ and $L_{p+1} = \{\varpi_{l,\tau}^L, n_{l,\tau}^L\}_{l=1}^{p+1}$. Applying the optimally-blended quadrature rules in the form of (13) to (15), we obtain the approximated form

$$\tilde{a}_h(w^h, \tilde{u}^h) = \tilde{\lambda}^h \tilde{b}_h(w^h, \tilde{u}^h), \quad \forall\, w^h \in V_p^h, \tag{17}$$

where for $w, v \in V_p^h$

$$\tilde{a}_h(w,v) = \sum_{\tau \in \mathcal{T}_h} \sum_{l=1}^{p+1} \left(\eta \varpi_{l,\tau}^G \nabla w(n_{l,\tau}^G) \cdot \nabla v(n_{l,\tau}^G) + (1-\eta)\varpi_{l,\tau}^L \nabla w(n_{l,\tau}^L) \cdot \nabla v(n_{l,\tau}^L) \right)$$
$$+ \sum_{\ell=1}^{\alpha} \eta_{a,\ell} \pi^2 h^{6\ell-3} \left(w^{(2\ell)}(0)v^{(2\ell)}(0) + w^{(2\ell)}(1)v^{(2\ell)}(1) \right), \tag{18}$$

and

$$\tilde{b}_h(w,v) = \sum_{\tau \in \mathcal{T}_h} \sum_{l=1}^{p+1} \left(\eta \varpi_{l,\tau}^G w(n_{l,\tau}^G) \cdot v(n_{l,\tau}^G) + (1-\eta)\varpi_{l,\tau}^L w(n_{l,\tau}^L) \cdot v(n_{l,\tau}^L) \right)$$
$$+ \sum_{\ell=1}^{\alpha} \eta_{b,\ell} h^{6\ell-1} \left(w^{(2\ell)}(0)v^{(2\ell)}(0) + w^{(2\ell)}(1)v^{(2\ell)}(1) \right). \tag{19}$$

With this in mind, we arrive at the matrix eigenvalue problem

$$\tilde{\mathbf{K}}\mathbf{U} = \tilde{\lambda}^h \tilde{\mathbf{M}}\mathbf{U}, \tag{20}$$

where $\tilde{\mathbf{K}}_{kl} = \tilde{\mathbf{K}}_{kl}^{1D} = \tilde{a}_h(\phi_p^k, \phi_p^l), \tilde{\mathbf{M}}_{kl} = \tilde{\mathbf{M}}_{kl}^{1D} = \tilde{b}_h(\phi_p^k, \phi_p^l)$ in 1D. Using the tensor-product structure and introducing the outer-product $\otimes$ (also known as the Kronecker product), the corresponding 2D matrices (see [5] for details) are

$$\tilde{\mathbf{K}} = \tilde{\mathbf{K}}^{2D} = \tilde{\mathbf{K}}_x^{1D} \otimes \tilde{\mathbf{M}}_y^{1D} + \tilde{\mathbf{M}}_x^{1D} \otimes \tilde{\mathbf{K}}_y^{1D},$$
$$\tilde{\mathbf{M}} = \tilde{\mathbf{M}}^{2D} = \tilde{\mathbf{M}}_x^{1D} \otimes \tilde{\mathbf{M}}_y^{1D}, \tag{21}$$

and the 3D matrices are

$$\tilde{\mathbf{K}} = \tilde{\mathbf{K}}^{3D} = \tilde{\mathbf{K}}_x^{1D} \otimes \tilde{\mathbf{M}}_y^{1D} \otimes \tilde{\mathbf{M}}_z^{1D} + \tilde{\mathbf{M}}_x^{1D} \otimes \tilde{\mathbf{K}}_y^{1D} \otimes \tilde{\mathbf{M}}_z^{1D} + \tilde{\mathbf{M}}_x^{1D} \otimes \tilde{\mathbf{M}}_y^{1D} \otimes \tilde{\mathbf{K}}_z^{1D},$$
$$\tilde{\mathbf{M}} = \tilde{\mathbf{M}}^{3D} = \tilde{\mathbf{M}}_x^{1D} \otimes \tilde{\mathbf{M}}_y^{1D} \otimes \tilde{\mathbf{M}}_z^{1D}, \tag{22}$$

where $\tilde{\mathbf{K}}_q^{1D}, \tilde{\mathbf{M}}_q^{1D}, q = x, y, z$ are 1D matrices generated from the modified bilinear forms in (17).

5   Numerical Examples

In this section, we present numerical tests to demonstrate the performance of the method. We consider the problem (1) with $d = 1, 2, 3$. The 1D problem has true eigenpairs $\left(\lambda_j = j^2\pi^2, u_j = \sin(j\pi x)\right), j = 1, 2, \cdots$, the 2D problem has true eigenpairs $\left(\lambda_{jk} = (j^2 + k^2)\pi^2, u_{jk} = \sin(j\pi x)\sin(k\pi y)\right), j, k = 1, 2, \cdots$, and the 3D problem has true eigenpairs $\left(\lambda_{jkl} = (j^2 + k^2 + l^2)\pi^2, u_{jkl} = \sin(j\pi x)\sin(k\pi y)\sin(l\pi z)\right), j, k, l = 1, 2, \cdots$. We sort both the exact and approximate eigenvalues in ascending order. Since outliers appear in the spectrum for cubic $(p = 3)$ and higher order isogeometric elements, we focus on p-th order elements with $p \geq 3$.

5.1   Numerical Study on Error Convergence Rates and Outliers

To study the errors, we consider both the H^1-seminorm and L^2-norm for the eigenfunctions. The optimal convergence rates in H^1-seminorm and L^2-norm are h^p and h^{p+1} for p-th order elements, respectively. For the eigenvalues, we consider the relative eigenvalue errors defined as $\frac{|\tilde{\lambda}_j^h - \lambda_j|}{\lambda_j}$. The optimal convergence rate is h^{2p} for p-th order elements. We denote by ρ_p the convergence rate of p-th order isogeometric elements.

Table 2 shows the eigenvalue and eigenfunction errors for $p = 3, 4, 5$ in 1D while Table 3 shows the eigenvalue errors for the problems in 2D and 3D. The first eigenvalue error approaches the machine precision fast. Thus, we calculate the convergence rates with coarser meshes. In all these scenarios, the eigenfunction errors are convergent optimally, while the eigenvalue errors are superconvergent with two extra orders, i.e., h^{2p+2}. These results are in good agreement with the theoretical predictions.

Table 2. Errors and convergence rates for the first and sixth eigenpairs in 1D when using IGA with optimally-blended quadratures and the boundary penalty technique.

p	N	$\frac{|\tilde{\lambda}_1^h - \lambda_1|}{\lambda_1}$	$|u_1 - \tilde{u}_1^h|_{H^1}$	$\|u_1 - \tilde{u}_1^h\|_{L^2}$	$\frac{|\tilde{\lambda}_6^h - \lambda_6|}{\lambda_6}$	$|u_6 - \tilde{u}_6^h|_{H^1}$	$\|u_6 - \tilde{u}_6^h\|_{L^2}$
3	5	3.52e−7	4.95e−3	1.67e−4	3.05e−1	1.83e1	9.71e−1
	10	1.32e−9	5.75e−4	9.25e−6	3.21e−3	1.50	3.38e−2
	20	5.09e−12	7.05e−5	5.60e−7	9.57e−6	1.12e−1	1.01e−3
	40	1.12e−13	8.77e−6	3.47e−8	3.45e−8	1.20e−2	4.94e−5
	ρ_3	8.04	3.05	4.08	7.76	3.54	4.79
4	5	6.90e−9	5.26e−4	1.80e−5	3.05e−1	1.87e1	9.91e−1
	10	6.31e−12	2.91e−5	4.72e−7	7.59e−4	6.38e−1	1.45e−2
	20	5.75e−14	1.76e−6	1.41e−8	4.42e−7	1.91e−2	1.75e−4
	ρ_4	10.09	4.11	5.16	9.70	4.97	6.23
	5	1.13e−10	5.65e−5	1.97e−6	3.06e−1	1.89e1	1.00
	10	1.15e−14	1.48e−6	2.43e−8	1.41e−4	2.63e−1	6.17e−3
	20	1.27e−14	4.42e−8	3.54e−10	1.72e−8	3.30e−3	3.06e−5
	ρ_5	13.26	5.16	6.22	12.04	6.24	7.50

Table 3. Errors for the first and sixth eigenvalues in 2D and 3D when using IGA with optimally-blended quadratures and the boundary penalty technique.

d	N	p 3 $\frac{\|\tilde{\lambda}_1^h-\lambda_1\|}{\lambda_1}$	$\frac{\|\tilde{\lambda}_6^h-\lambda_6\|}{\lambda_6}$	4 $\frac{\|\tilde{\lambda}_1^h-\lambda_1\|}{\lambda_1}$	$\frac{\|\tilde{\lambda}_6^h-\lambda_6\|}{\lambda_6}$	5 $\frac{\|\tilde{\lambda}_1^h-\lambda_1\|}{\lambda_1}$	$\frac{\|\tilde{\lambda}_6^h-\lambda_6\|}{\lambda_6}$
2	3	2.28e−5	1.63e−4	1.32e−6	8.52e−3	6.55e−8	1.30e−7
	6	8.05e−8	6.10e−4	1.09e−9	9.03e−5	1.29e−11	1.15e−5
	12	3.07e−10	1.94e−6	1.07e−12	6.06e−8	5.21e−14	1.61e−9
	24	1.31e−12	7.15e−9	1.35e−14	5.37e−11	9.45e−14	2.86e−13
	ρ_p	8.02	8.19	10.12	10.54	7.95	12.8
3	2	6.78e−4	3.33e−1	1.00e−4	3.33e−1	1.28e−5	3.33e−1
	4	2.15e−6	6.03e−4	6.74e−8	8.92e−5	1.79e−9	1.14e−5
	8	7.94e−9	1.92e−6	5.96e−11	5.99e−8	5.30e−13	1.59e−9
	16	3.06e−11	7.07e−9	3.00e−15	5.30e−11	4.66e−15	3.66e−13
	ρ_p	8.13	8.48	11.5	10.82	12.26	13.19

Figures 1, 2 and 3 show the overall spectral errors when using the standard IGA and IGA with optimally-blended rules and the boundary penalty technique. The polynomial degrees are $p \in \{3,4,5\}$. There are 100 elements in 1D, 20 × 20 elements in 2D, and 20 × 20 × 20 elements in 3D, respectively. In all the scenarios, we observe that there are outliers in the IGA spectra. These outliers are eliminated by the proposed method. Moreover, we observe that these spectral errors are reduced significantly, especially in the high-frequency regions.

5.2   Numerical Study on Condition Numbers

We study the condition numbers to show further advantages of the method. Since the stiffness and mass matrices are symmetric, the condition numbers of the generalized matrix eigenvalue problems (6) and (20) are given by

$$\gamma := \frac{\lambda_{\max}^h}{\lambda_{\min}^h}, \qquad \tilde{\gamma} := \frac{\tilde{\lambda}_{\max}^h}{\tilde{\lambda}_{\min}^h}, \tag{23}$$

where $\lambda_{\max}^h, \tilde{\lambda}_{\max}^h$ are the largest eigenvalues and $\lambda_{\min}^h, \tilde{\lambda}_{\min}^h$ are the smallest eigenvalues of IGA and the proposed method, respectively. The condition number characterizes the stiffness of the system. We follow the recent work of soft-finite element method (SoftFEM) [12] and define the *condition number reduction ratio* of the method with respect to IGA as

$$\rho := \frac{\gamma}{\tilde{\gamma}} = \frac{\lambda_{\max}^h}{\tilde{\lambda}_{\max}^h} \cdot \frac{\tilde{\lambda}_{\min}^h}{\lambda_{\min}^h}. \tag{24}$$

In general, one has $\lambda_{\min}^h \approx \tilde{\lambda}_{\min}^h$ for IGA and the proposed method with sufficient number of elements (in practice, these methods with only a few elements

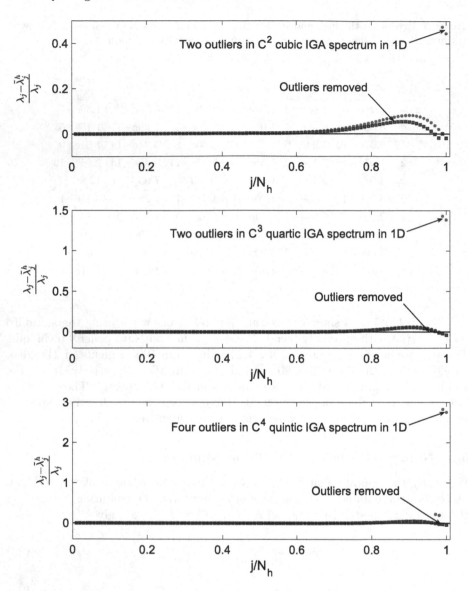

Fig. 1. Outliers in IGA spectra and their eliminations when using the IGA with optimally-blended rules and the boundary penalty technique. There are 100 elements in 1D with polynomial degrees $p \in \{3, 4, 5\}$.

already lead to good approximations to the smallest eigenvalues). Thus, the condition number reduction ratio is mainly characterized by the ratio of the largest eigenvalues. Finally, we define the *condition number reduction percentage* as

$$\varrho = 100\frac{\gamma - \tilde{\gamma}}{\gamma}\,\% = 100(1 - \rho^{-1})\,\%. \tag{25}$$

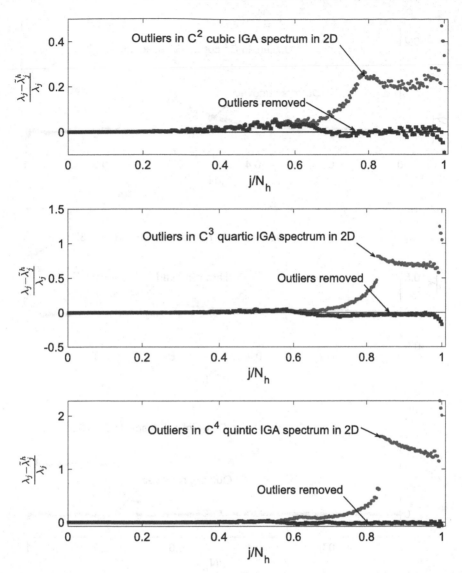

Fig. 2. Outliers in IGA spectra and their eliminations when using the IGA with optimally-blended rules and the boundary penalty technique. There are 20×20 elements in 2D with polynomial degrees $p \in \{3, 4, 5\}$.

Table 4 shows the smallest and largest eigenvalues, condition numbers and their reduction ratios and percentages for 1D, 2D, and 3D problems. We observe that the condition numbers of the proposed method are significantly smaller. For higher-order isogeometric elements, there are more outliers and these outliers pollute a larger high-frequency region. Consequently, they lead to larger errors in the high-frequency region. The proposed method reduces more errors for

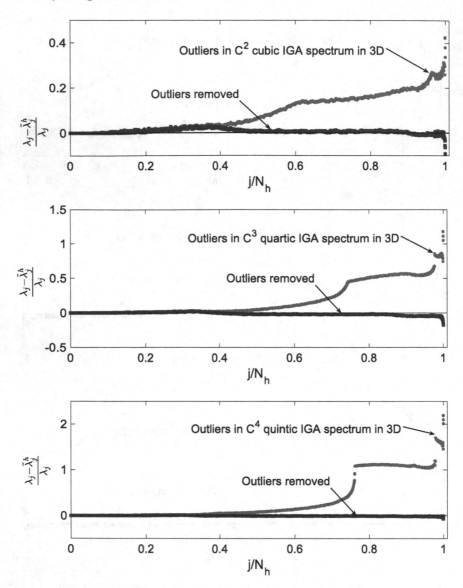

Fig. 3. Outliers in IGA spectra and their eliminations when using the IGA with optimally-blended rules and the boundary penalty technique. There are $20 \times 20 \times 20$ elements in 3D with polynomial degrees $p \in \{3, 4, 5\}$.

higher-order elements. This leads to small errors in the high-frequency region also for high-order elements. The condition number of the proposed method reduces by about 32% for C^2 cubic, 60% for C^3 quartic, and 75% for C^4 quintic elements. This holds valid for both 2D and 3D problems.

Table 4. Minimal and maximal eigenvalues, condition numbers, reduction ratios and percentages when using IGA and IGA with optimally-blended quadratures and the boundary penalty technique. The polynomial degrees are $p \in \{3, 4, 5\}$. There are 100, 48×48 elements, and $16 \times 16 \times 16$ elements in 1D, 2D, and 3D, respectively.

d	p	$\lambda_{\min}^h$	$\lambda_{\max}^h$	$\tilde{\lambda}_{\max}^h$	γ	$\tilde{\gamma}$	ρ	ϱ
	3	9.87	1.46e5	9.87e4	1.47e4	1.00e4	1.47	32.17%
1	4	9.87	2.45e5	9.87e4	2.48e4	1.00e4	2.48	59.69%
	5	9.87	3.93e5	1.00e5	3.98e4	1.02e4	3.92	74.47%
	3	1.97e1	6.71e4	4.55e4	3.40e3	2.30e3	1.47	32.17%
2	4	1.97e1	1.13e5	4.55e4	5.72e3	2.30e3	2.48	59.69%
	5	1.97e1	1.81e5	4.57e4	9.17e3	2.31e3	3.96	74.77%
	3	2.96e1	1.12e4	7.58e3	3.78e2	2.56e2	1.48	32.23%
3	4	2.96e1	1.88e4	7.58e3	6.36e2	2.56e2	2.48	59.72%
	5	2.96e1	3.02e4	7.59e3	1.02e3	2.56e2	3.98	74.89%

6   Concluding Remarks

We improve the isogeometric spectral approximations by combining the two ideas: optimally-blended quadratures and a boundary penalty technique. As a result, we obtained a superconvergence of rate h^{2p+2} for the eigenvalue errors and eliminated the outliers in the spectra. The technique can be also used to improve the spectral approximations of the Neumann eigenvalue problems. These improvements lead to a better spatial discretization for the time-dependent partial differential equations, which in return, improve the overall performance of numerical methods. As future work, it would be interesting to study the method for higher-order differential operators and nonlinear application problems.

References

1. Bartoň, M., Calo, V., Deng, Q., Puzyrev, V.: Generalization of the Pythagorean eigenvalue error theorem and its application to Isogeometric analysis. In: Di Pietro, D.A., Ern, A., Formaggia, L. (eds.) Numerical Methods for PDEs. SSSS, vol. 15, pp. 147–170. Springer, Cham (2018). https://doi.org/10.1007/978-3-319-94676-4_6
2. Brezis, H.: Functional analysis. Sobolev spaces and partial differential equations. Universitext, Springer, New York (2011)
3. Buffa, A., De Falco, C., Sangalli, G.: Isogeometric analysis: new stable elements for the Stokes equation. Int. J. Numer. Methods Fluids (2010)
4. Calo, V., Deng, Q., Puzyrev, V.: Quadrature blending for isogeometric analysis. Procedia Comput. Sci. **108**, 798–807 (2017)
5. Calo, V., Deng, Q., Puzyrev, V.: Dispersion optimized quadratures for isogeometric analysis. J. Comput. Appl. Math. **355**, 283–300 (2019)
6. Cottrell, J.A., Hughes, T.J.R., Bazilevs, Y.: Isogeometric Analysis: Toward Integration of CAD and FEA. Wiley, New York (2009)

7. Cottrell, J.A., Reali, A., Bazilevs, Y., Hughes, T.J.R.: Isogeometric analysis of structural vibrations. Comput. Methods Appl. Mech. Eng. **195**(41), 5257–5296 (2006)
8. De Boor, C.: A Practical Guide to Splines, vol. 27. Springer, New York (1978)
9. Deng, Q., Bartoň, M., Puzyrev, V., Calo, V.: Dispersion-minimizing quadrature rules for C1 quadratic isogeometric analysis. Comput. Methods Appl. Mech. Eng. **328**, 554–564 (2018)
10. Deng, Q., Calo, V.: Dispersion-minimized mass for isogeometric analysis. Comput. Methods Appl. Mech. Eng. **341**, 71–92 (2018)
11. Deng, Q., Calo, V.M.: A boundary penalization technique to remove outliers from isogeometric analysis on tensor-product meshes. Comput. Methods Appl. Mech. Eng. **383**, 113907 (2021)
12. Deng, Q., Ern, A.: SoftFEM: revisiting the spectral finite element approximation of elliptic operators. arXiv preprint arXiv:2011.06953 (2020)
13. Deng, Q., Puzyrev, V., Calo, V.: Isogeometric spectral approximation for elliptic differential operators. J. Comput. Sci. (2018)
14. Deng, Q., Puzyrev, V., Calo, V.: Optimal spectral approximation of 2n-order differential operators by mixed isogeometric analysis. Comput. Methods Appl. Mech. Eng. **343**, 297–313 (2019)
15. Evans, J.A., Hughes, T.J.: Isogeometric divergence-conforming B-splines for the Darcy-Stokes-Brinkman equations. Math. Models Methods Appl. Sci. **23**(04), 671–741 (2013)
16. Hughes, T.J.R., Cottrell, J.A., Bazilevs, Y.: Isogeometric analysis: CAD, finite elements, NURBS, exact geometry and mesh refinement. Comput. Methods Appl. Mech. Eng. **194**(39), 4135–4195 (2005)
17. Hughes, T.J.R., Evans, J.A., Reali, A.: Finite element and NURBS approximations of eigenvalue, boundary-value, and initial-value problems. Comput. Methods Appl. Mech. Eng. **272**, 290–320 (2014)
18. Hughes, T.J.R., Reali, A., Sangalli, G.: Duality and unified analysis of discrete approximations in structural dynamics and wave propagation: comparison of p-method finite elements with k-method NURBS. Comput. Methods Appl. Mech. Eng. **197**(49), 4104–4124 (2008)
19. Kythe, P.K., Schäferkotter, M.R.: Handbook of Computational Methods for Integration. CRC Press, Boca Raton (2004)
20. Nguyen, V.P., Anitescu, C., Bordas, S.P., Rabczuk, T.: Isogeometric analysis: an overview and computer implementation aspects. Math. Comput. Simul. **117**, 89–116 (2015)
21. Piegl, L., Tiller, W.: The NURBS book. Springer, Heidelberg (1997). https://doi.org/10.1007/978-3-642-59223-2
22. Puzyrev, V., Deng, Q., Calo, V.: Spectral approximation properties of isogeometric analysis with variable continuity. Comput. Methods Appl. Mech. Eng. **334**, 22–39 (2018)
23. Puzyrev, V., Deng, Q., Calo, V.M.: Dispersion-optimized quadrature rules for isogeometric analysis: modified inner products, their dispersion properties, and optimally blended schemes. Comput. Methods Appl. Mech. Eng. **320**, 421–443 (2017)

Socio-cognitive Evolution Strategies

Aleksandra Urbańczyk[1], Bartosz Nowak[1], Patryk Orzechowski[1,2],
Jason H. Moore[2], Marek Kisiel-Dorohinicki[1], and Aleksander Byrski[1(✉)]

[1] Institute of Computer Science, AGH University of Science and Technology,
Kraków, Poland
{aurbanczyk,doroh,olekb}@agh.edu.pl, bnowak@student.agh.edu.pl
[2] Penn Institute for Biomedical Informatics, University of Pennsylvania,
Philadelphia, PA, USA
jhmoore@upenn.edu

Abstract. Socio-cognitive computing is a paradigm developed for the last several years, it consists in introducing into metaheuristics mechanisms inspired by inter-individual learning and cognition. It was successfully applied in hybridizing ACO and PSO metaheuristics. In this paper we have followed our previous experiences in order to hybridize the acclaimed evolution strategies. The newly constructed hybrids were applied to popular benchmarks and compared with their referential versions.

Keywords: Metaheuristics · Socio-cognitive computing · Hybrid algorithms · Evolution strategies

1 Introduction

Tackling difficult optimization problems requires using metaheuristics [21], very often it is needed to create new ones [34], e.g. hybridizing the existing algorithms [30]. It is a well-known fact, that metaheuritics are very often inspired by nature, therefore their hybridizations often put together different phenomena observed in the real-world.

An interesting theory, which already has become a basis for efficient hybrid algorithms, is Social Cognitive Theory introduced by Bandura [2]. This theory is used in psychology, education, and communication and assumes that portions of an individual's acquisition of knowledge can be directly related to observing others in the course of their social interactions, their experiences, and outside media influences [3]. Thus, the individuals use this gathered information to guide their behaviors, not solely learning them by themselves (e.g., during the course of trials and errors). They can replicate others' deeds (trial and error) and predict

The research presented in this paper has been financially supported by: Polish National Science Center Grant no. 2019/35/O/ST6/00570 "Socio-cognitive inspirations in classic metaheuristics."; Polish Ministry of Science and Higher Education funds assigned to AGH University of Science and Technology.

M. Paszynski et al. (Eds.): ICCS 2021, LNCS 12743, pp. 329–342, 2021.
https://doi.org/10.1007/978-3-030-77964-1_26

the consequences based on observations, thus possibly reaching their goals sooner (cf. Bandura's Bobo Doll experiment [4]).

Of course many social metaheuristics, processing a number of individuals, especially in the case when the individuals can be perceived as somewhat autonomous (e.g. EMAS [10,23]) already use certain socio-cognitive inspirations, however we already introduced dedicated mechanisms rooted in Social-Cognitive Theory to selected metaheuristics (socio-cognitive ACO [11] and socio-cognitive PSO [8]), obtaining good results comparing to the reference algorithms.

This paper is devoted to hybridization of the socio-cognitive ideas with classic evolution strategies by Rechenberg and Schwefel [29]. Up-to-now we have researched hybridizing popular algorithms with strong social component (ACO [13], PSO [17]). In this and subsequent papers, we would like to try hybridizing well-known classic metaheuristics, such as genetic algorithm, evolution strategies, clonal selection algorithm, differential evolution and many others [30]. Our work is motivated (among others) by the famous "no free lunch" theorem by Wolpert and Macready [34] – its main conclusion is that we have to tune our metaheuristics, but it is to note that sometimes seeking new metaheuristics is also necessary (especially when no theoretical work has shown that the constructed metaheuristic actually is able to find anything (e.g. Michael Vose has proven that Simple Genetic Algorithm is a feasible computing method [33], Gunter Rudolph researched theoretically Evolution Strategies (see, e.g. [5]) and we have also applied methods similar to the ones used by Michael Vose to prove feasibility of agent-based computing methods [9]).

The main contribution of this paper is a novel hybrid of Evolution Strategies utilizing socio-cognitively inspired mechanism which makes possible to exchange the information among the individuals. The efficiency and efficacy of the novel algorithms are tested using well-known high dimensional, multi-modal benchmark functions.

2   Classic and Hybrid Evolution Strategies

Evolution strategies (ES) are the classic approach in biologically-inspired computing discipline. Devised in the Technical University of Berlin by Rechenberg and Schwefel [25,28] has been developed and applied for more than fifty years. Due to its universality in solving difficult problems, evolution strategy is considered as one of classic metaheuristics. The classic ES incorporate two nature-inspired mechanisms: mutation and selection. Even though there are notions of parents (μ) and offspring (λ), the offspring can be a result of crossover between selected pair of parents but also just a mutation of a selected parent. After the mutation, parents can be either excluded from the pool for next-generation selection (referred ','/comma strategies) or can be saved to be part of a new generation ('+'/plus strategies). The simplest version is 1+1 strategy, were two individuals are compared and the one with better fitness becomes a parent for the next generation. This is an instance of $\mu+\lambda$, where both μ and $\lambda = 1$, but they can equal any greater number on the condition that μ is equal or less than

λ. In the case of μ,λ ES, offspring population ought to be greater then parent in order to have a sufficient number of individuals to select from.

During its long history ES have been modified and hybridized in many ways in order to improve performance in hard problem's solutions. Based on Talbi's flat taxonomy for hybrid metaheuristics [31] we can differentiate between homogeneous and heterogeneous hybrids. The latter ones join two or more types of metaheuristics and in such hybrids ES can serve either as major metaheuristic that is enriched by other one or as a algorithm [6]. Very often heterogeneous hybrid metaheuristics are designed to solve certain problem (e.g., [16,18,24], and thus are hard to be generalised.

On the other hand, homogeneous hybrids use only one type of metaheuristic and emerge from combining parts of algorithm with different parameters. In the case of evolution strategies, such control parameters as mutation strength or population size (step-size) can be manipulated. One of such versions is CMA-ES [15], but many other can be found in the literature, e.g., [1,7,32]).

The existing hybrids Evolution Strategies are often focused on particular application, e.g. vehicle routing problem [26] or optimization of engineering and construction problems [19,27]. At the same time, general-purpose hybrids of Evolution Strategy exists, e.g. CMA-ES hybridization is proposed in [35], a hybrid ES for solving mixed (continuous and discrete) problems was proposed in [22]. Apparently the number of hybrid metaheuristics based on Evolution Strategies is not high, thus it seems that exploring the possibilities of creative hybridization of those algorithms might be interesting and advantageous. Therefore, based on our previous experiences in hybridizing ACO [11] and PSO [8], we propose to introduce socio-cognitive mechanisms into Evolution Strategies.

3   Socio-cognitive Hybridization of the Evolution Strategies

Striving towards better exploration of socio-cognitive inspired hybridization of metaheuristics, we would like to present a first step towards verification of such possibilities, focusing on evolution strategies.

The classic self-adaptive version algorithm for evolution strategies can be described as follows.

1. Initialize parent population $P_\mu = \{i_1, \ldots, i_\mu\}$. Each of the individuals can be described as follows: $I \ni i_k = \{g_{k,1}, \ldots, g_{k,d}, s_{k_1}, \ldots, s_{k,d}\}, k, d \in \mathbb{N}$ stands for an individual containing a genotype ($g_{k,l}$ is $l-th$ gene of $k-th$ genotype). The dimensionality of the considered problem is d. The $s_{k_1}, \ldots, s_{k,d}$ are mutation strategy parameters that will be adapted in order to guide the search.
2. Generate λ offspring individuals forming the offspring population $P_\lambda = \{i_1, \ldots, i_\lambda\}$ in the following procedure:
   - Randomly select ϱ parents from P_μ (if $\varrho = \mu$ take all of them of course).
   - Recombine the ϱ selected parents (traditionally a pair) to form a recombinant individual i_r, using any possible recombination means (traditionally averaging crossover operator was used).

- Mutate the strategy parameter set $s_{r,1}, \ldots, s_{r,d}$ of the recombinant i_r (adapting e.g. the mutation diversities for the next mutation). Traditionally mutation is realized by applying a distortion based on e.g. uniform or Gaussian random distribution, adding or substracting a certain value to (from) a selected gene.
- Mutate the objective parameter set $g_{r_1}, \ldots, g_{r_d}$ of the recombinant i_r using the mutated strategy parameter set to control the statistical properties of the object parameter mutation.

3. Select new parent population (using deterministic truncation selection) from either the offspring population P_λ (this is referred to as comma-selection, usually denoted as "(μ, λ)-selection"), or the offspring P_λ and parent $P\mu$ population (this is referred to as plus-selection, usually denoted as $P(\mu + \lambda)$.
4. Goto 2. until termination criterion fulfilled.

We have decided to introduce the socio-cognitive mechanisms to all basic versions of evolution strategies, namely: $(1 + 1)$, (μ, λ), $(\mu + \lambda)$.

In the novel, socio-cognitive version of Evolution Strategy, we try to increase the exchange of the knowledge among the individuals, so they can get information not only during the mutation and adaptation of their mutation parameters, but also observe others. So inside the second step of the algorithm depicted above, we introduce the following changes:

1. The algorithm stores historically α_{best} and α_{worst} individuals.
2. During the mutation, one of the following three different mutations are realized:
   - Classic mutation realized with γ probability.
   - Modification of the individual towards the historically α_{best} best individuals with probability γ_{good}. This mutation sets the current gene copying the gene from one of the historically best individuals: Assume, $i_b = \{g_{b,1}, \ldots, g_{b,d}, s_{b_1}, \ldots, s_{b,d}\}$ is a randomly picked individual from the historically best ones. The individual about to be mutated is $i_m = \{g_{m,1}, \ldots, g_{m,d}, s_{m_1}, \ldots, s_{m,d}\}$. Let us randomly pick one of the genes of i_m. Assume $1 \leq p \leq d$ is this random value, so the picked gene is $g_{m,p}$. Now we will simply assign the value of this gene to the value of a correspondent gene in i_b, that is $g_{m,p} \leftarrow g_{b,p}$.
   - Modification of the individual trying to avoid the historically α_{bad} worst individuals with probability γ_{bad}. This mutation computes the difference between the current gene and one gene of the historically worst individuals, computes a fraction of this value, multiplying it by β, and adds it to the current gene. The procedure of randomly choosing one of the historically worst individual is similar as it was described above. Let us go to the final step, we have the individual to be mutated i_m, randomly chosen individual belonging to the historically worst i_w. $1 \leq p \leq d$ stands for the random index of the gene. The following assignment is realized: $i_{m,p} \leftarrow \beta \cdot (i_{m,p} - i_{w,p})$.
3. When better or worse (historically) individuals are generated, the lists of α_{best} and α_{worst} individuals are updated.

Three basic versions of Evolution Strategies were modified using this mechanism. Thus the mutation is realized not only in a fully stochastic way, but also reuses the knowledge about the previous generations. Of course such mechanism is not complex, but as it will be shown in the next section, it already produced interesting results.

4   Study Design

The main aim of the experiments was verification of efficacy of global optimization (minimization) of the novel algorithms for the selected benchmark functions (Ackley, De Jong, Rastrigin, and Griewank [12] visualized in Fig. 1) in $d = 10$, 50, 100, 500 and 1000 dimensions. Both the value obtained in the last iteration, and the trajectory of the fitness functions improvements were considered – in certain situations it is desirable to have a relatively fast convergence earlier, in other situations the focus is put on the final result. The equations used are as follows:

- Ackley: $f(x) = -ae^{-b\sqrt{1/n \sum_{i=1}^{n}(x_i^2)}} - e^{1/n \sum_{i=1}^{n} cos(cx_i)} + a + e; a = 20; b = 0.2; c = 2\pi; i \in [1:n]; -32.768 \le x(i) \le 32.768. f(x^{opt}) = 0, x_i^{opt} = 0.$
- De Jong: $f(x) = \sum_{i=1}^{n} x_i^2, i \in [1,n]; -5.12 \le x_i \le 5.12. f(x^{opt}) = 0, x_i^{opt} = 0.$
- Rastrigin: $f(x) = 10n + \sum_{i=1}^{n}(x_i^2 - 10cos(2\pi x_i)), i \in [1,n]; -5.12 \le x_i \le 5.12. f(x^{opt}) = 0, x_i^{opt} = 0.$
- Griewank: $f(x) = \sum_{x=1}^{n} x_i^2/4000 - \prod cos(x_i/\sqrt{i}) + 1, i \in [1,n]; -600 \le x_i \le 600, f(x^{opt}) = 0, x_i^{opt} = 0.$

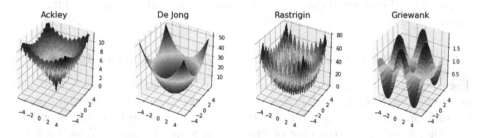

Fig. 1. 3-dimensional visualization of the benchmarks used in the study.

The following algorithms were benchmarked:

- original Evolution Strategy $(1+1)$, (μ, λ) or $(\mu + \lambda)$,
- hybrid Evolution Strategy – with the mechanism of getting closer to the historically best results,
- hybrid Evolution Strategy – with the mechanism of going farther from the historically worst results,
- hybrid Evolution Strategy – with both the above-mentioned mechanism.

The stopping criteria was reaching maximum number of evaluations of fitness function (set as 25,000 for all the experiments). Number of the individuals in the population was μ (1 in the case of (1+1) Evolution Strategy, 20 in the case of other Strategies).

The following settings were used for the algorithms:

- $\mu = 1$, $\lambda = 1$ in the case of $(1 + 1)$ Evolution Strategy.
- $\mu = 20$, $\lambda = 140$ in other cases.
- $\gamma_{good} = 0.4$, $\gamma_{bad} = 0.04$, $\beta = 0.01$.
- $\gamma = 1/($number of dimensions$)$,
- number of the historically best or worst individuals: 5.
- polynomial mutation [20].

Each experiment was replicated 20 times and the mean value of fitness function was taken for the reference. In order to check whether the observed sample had normal distribution we have applied Shapiro-Wilk test with significance threshold of 0.05. Kruskal-Wallis test was used in order to check whether their cumulative distribution functions differed, and finally Dunn's test in order to check which ones were significantly different.

5   Experimental Results

The algorithms were implemented using jMetalPy[1] computing framework. The source code is available on request. The computations were conduced on a Microsoft Windows 10 machine with AMD Ryzen 9 5900X (3.7 GHz) CPU, NVIDIA GeForce RTX 3080 GPU, and 2x8 GB RAM. We have analyzed the speed of improvement of the fitnesses depending on the number of evaluations and compared statistical differences between the algorithms after 25,000 iterations.

5.1   (1 + 1) Evolution Strategy and Its Socio-cognitive Hybrids

The $(1 + 1)$ Evolution Strategy was compared with its hybrids for Rastrigin benchmark in 1000 dimensions (see Fig. 2). A metaheuristic algorithm operating only on one individual (so-called trajectory method) as expected is not too versatile, especially in the case of high-dimensional benchmarks. So, even the original algorithm does not reach the vicinity of the global optimum, stopping around 4000. Its hybrid version with the mechanism of getting closer to the best results is very similar to the original one. The two remaining algorithms perform significantly worse, apparently getting stuck in a local sub-optimum. It seems that in the case of $(1 + 1)$ Evolution Strategy, adding the mechanism of avoiding worse individuals does not help or requires significant improvements.

The more detailed observation of the results presented in Table 1 confirms the findings. The original implementation of Evolution Strategy produces similar results to its hybrid, while the two remaining hybrids are significantly worse.

---

[1] https://github.com/jMetal/jMetalPy.

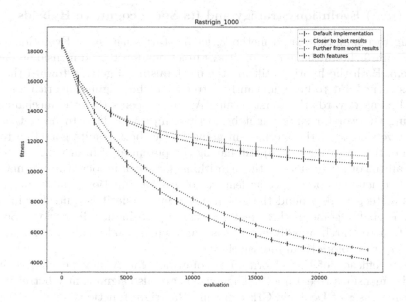

Fig. 2. Trajectory of changes of mean fitness function value (standard deviation whiskers are also shown) for 1000-dimensional Rastrigin problem and $(1 + 1)$ Evolutionary Strategy, depending on the number of fitness function evaluation.

Table 1. Mean and standard deviation of fitness value after 25,000 evaluations of $(1, 1)$ Evolution Strategy and its hybrids for 10, 50, 100, 500 and 1000 dim. problems.

Dimension	10		50		100		500		1000	
	Fit.	Std.	Fit.	Std.	Fit.	Std.	Fit.	Std.	Fit.	Std.
Ackley										
Default	0,27	0,12	2,19	0,28	3,43	0,20	9,89	0,31	14,25	0,21
Closer to best	0,42	0,18	2,63	0,34	4,00	0,24	10,90	0,29	15,15	0,24
Farther from worst	**0,21**	**0,11**	2,88	0,25	5,01	0,33	17,72	1,04	20,67	0,12
Both	0,24	0,09	2,90	0,31	5,23	0,36	19,02	0,92	20,62	0,09
De Jong										
Default	0,00	0,00	0,07	0,02	0,58	0,12	69,97	7,37	490,89	36,76
Closer to best	0,00	0,00	0,11	0,04	0,94	0,18	95,67	6,23	627,27	43,34
Farther from worst	0,00	0,00	0,13	0,04	1,41	0,16	157,25	11,94	1108,31	69,49
Both	0,00	0,00	**0,06**	**0,02**	0,89	0,10	151,14	12,08	1039,64	42,48
Griewank										
Default	0,01	0,01	0,04	0,09	0,08	0,13	0,36	0,15	0,87	0,10
Closer to best	0,01	0,01	**0,02**	**0,04**	**0,03**	**0,04**	0,35	**0,07**	0,97	0,05
Farther from worst	0,01	0,01	0,03	0,05	0,08	0,10	0,53	0,08	1,24	0,03
Both	0,01	0,01	0,03	0,06	0,04	0,11	0,48	0,04	1,21	0,02
Rastrigin										
Default	0,10	0,05	10,96	2,29	50,78	4,66	1178,87	69,39	4027,59	132,27
Closer to best	0,16	0,09	15,02	2,37	64,07	6,08	1394,98	53,55	4661,99	114,84
Farther from worst	0,08	0,05	23,31	2,51	132,28	11,58	3797,29	136,07	10912,88	193,63
Both	0,09	0,05	16,29	3,03	106,35	10,34	3401,20	127,62	10459,14	309,85

5.2 (μ, λ) Evolution Strategy and Its Socio-cognitive Hybrids

Moving to population-based methods, let us start with (μ, λ) Evolution Strategy. Figure 3 shows the curves of fitness functions for 1000-dimensional Ackley problem. Both the hybrids utilizing the mechanisms of getting toward the best results turned out to be significantly better than the original algorithm and its hybrid going towards the worst results. Again, it seems that the mechanism of avoiding the worst results might be further improved. It is to note, that the hybrid version with the mechanism going to the best results got close to the global optimum (around 18.0) still having the potential of improving the result, while all the other versions of the algorithms got stuck in local sub-optima. One should remember about specific features of Ackley function – in 3 dimensions it has a steep peak around the global optimum, while it is quite flat in other areas. Probably some of these features scale up to higher dimensions. So it is important to search broadly in this case, and when good results are found, it is necessary to explore them intensively.

Good efficacy of this algorithm is confirmed for Ackley benchmark in all tested dimensions (see Table 2). The hybrids are also significantly better in several other cases of De Jong, Griewank and Rastrigin functions.

5.3 $(\mu + \lambda)$ Evolution Strategy and Its Socio-cognitive Hybrids

Finally, $(\mu + \lambda)$ Evolution Strategy is examined along with its hybrids, applied to optimization of 1000-dimensional Griewank problem. In Fig. 4 we can see that both hybrids utilizing the mechanism of getting closer to the best results prevailed. They come very close to the optimum and apparently do not loose the diversity. Other two algorithms also do not seem to be stuck, however they are significantly slower than the winning hybrids. However, this part of the experiment show, that $(\mu + \lambda)$ Evolution Strategy was the best starting point for the hybridization.

This observation is further confirmed when looking at Table 3. The final results produced by the hybrids getting closer to the best results actually prevailed for almost all of the tested instances of the problems.

In order to make sure that we have really different average results, we have used the Dunn's test (see Table 4) for excluding the null hypotheses that cumulative distribution function of the sampled final values (the fitness obtained when the stopping condition is met). As it can be seen, a significant majority of the tests were finished with excluding the null-hypothesis.

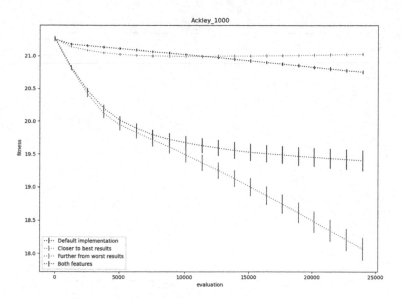

Fig. 3. Trajectory of changes of mean fitness function value (standard deviation whiskers are also shown) for 1000-dimensional Ackley problem and (μ, λ) Evolutionary Strategy, depending on the number of fitness function evaluation.

Table 2. Mean and standard deviation of fitness value after 25,000 evaluations of (μ, λ) Evolution Strategy and its hybrids for 10, 50, 100, 500 and 1000 dim. problems.

Dimension	10		50		100		500		1000	
	Fit.	Std.	Fit.	Std.	Fit.	Std.	Fit.	Std.	Fit.	Std.
Ackley										
Default	0,54	0,19	6,86	0,47	12,52	0,56	20,10	0,07	20,72	0,03
Closer to best	**0,30**	**0,11**	**2,84**	**0,27**	**5,00**	**0,23**	15,01	0,29	17,89	0,19
Farther from worst	0,29	0,13	7,28	0,63	14,40	0,97	20,85	0,05	21,04	0,02
Both	0,13	0,05	2,87	0,25	5,64	0,41	16,87	0,40	19,48	0,21
De Jong										
Default	0,00	0,00	2,77	0,64	33,59	5,17	1937,85	65,87	5797,31	119,30
Closer to best	0,00	0,00	0,15	0,05	1,90	0,33	**315,97**	**26,29**	1646,45	88,98
Farther from worst	0,00	0,00	2,78	0,58	37,02	4,56	2227,48	86,29	6538,94	118,44
Both	0,00	0,00	**0,01**	**0,01**	**1,05**	**0,25**	354,54	14,56	1787,58	109,50
Griewank										
Default	0,01	0,01	0,09	0,08	0,42	0,05	1,49	0,02	2,44	0,02
Closer to best	0,01	0,01	0,01	0,01	0,03	0,01	**0,79**	**0,06**	1,39	0,02
Farther from worst	0,01	0,01	0,10	0,05	0,46	0,08	1,55	0,02	2,63	0,02
Both	0,01	0,01	0,00	0,01	**0,02**	**0,01**	0,84	0,05	1,45	0,02
Rastrigin										
Default	0,34	0,17	62,43	7,90	314,27	19,86	5689,33	113,58	14276,61	141,83
Closer to best	0,15	0,11	16,26	3,02	80,71	6,91	**2188,21**	**74,69**	7190,55	146,55
Farther from worst	0,11	0,23	64,49	6,24	356,50	30,52	6292,99	131,96	15158,68	161,31
Both	**0,03**	**0,02**	**12,22**	**2,80**	**66,64**	**7,91**	2197,37	93,43	7521,77	170,56

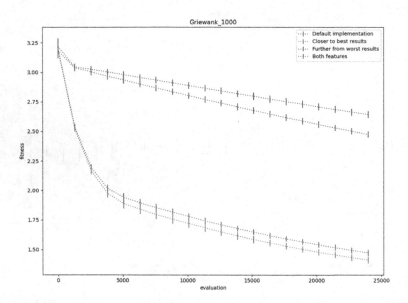

Fig. 4. Trajectory of changes of mean fitness function value (standard deviation whiskers are also shown) for 1000-dimensional Griewank problem and $(\mu + \lambda)$ Evolutionary Strategy, depending on the number of fitness function evaluation.

Table 3. Mean and standard deviation of fitness value after 25,000 evaluations of $(\mu + \lambda)$ Evolution Strategy and its hybrids for 10, 50, 100, 500 and 1000 dim. problems.

Dimension	10		50		100		500		1000	
	Fit.	Std.	Fit.	Std.	Fit.	Std.	Fit.	Std.	Fit.	Std.
Ackley										
Default	0,56	0,17	6,47	0,60	11,92	0,57	20,10	0,07	20,72	0,02
Closer to best	0,40	0,16	**2,87**	**0,19**	**4,83**	**0,26**	**15,02**	**0,27**	**18,03**	**0,16**
Farther from worst	0,27	0,24	6,95	0,71	13,81	0,78	20,63	0,04	20,87	0,02
Both	**0,14**	**0,04**	3,08	0,39	5,94	0,37	17,21	0,52	19,58	0,17
De Jong										
Default	0,00	0,00	2,16	0,45	32,26	3,86	1945,75	63,55	5788,75	120,45
Closer to best	0,00	0,00	0,17	0,05	1,94	0,36	**326,75**	**24,01**	**1647,68**	**80,94**
Farther from worst	0,00	0,00	2,24	0,48	34,10	4,56	2143,40	64,08	6469,60	116,27
Both	0,00	0,00	**0,01**	**0,02**	**1,12**	**0,39**	363,40	23,09	1846,96	95,48
Griewank										
Default	0,02	0,01	0,09	0,10	0,41	0,05	1,48	0,02	2,44	0,02
Closer to best	0,01	0,01	0,03	0,08	**0,04**	**0,04**	**0,80**	**0,05**	1,41	**0,04**
Farther from worst	0,01	0,01	0,08	0,08	0,41	0,08	1,54	0,01	2,62	0,03
Both	0,01	0,01	**0,01**	**0,04**	0,04	0,08	0,83	0,05	1,46	0,02
Rastrigin										
Default	0,19	0,10	54,27	7,80	305,64	17,42	5677,59	93,41	14323,20	148,70
Closer to best	0,12	0,06	17,92	3,42	82,88	7,69	**2235,61**	**93,00**	**7261,88**	**216,26**
Farther from worst	0,07	0,04	49,53	6,56	318,51	22,35	6123,40	118,54	15003,41	221,74
Both	**0,03**	**0,03**	**10,80**	**3,05**	**67,64**	**9,09**	2268,98	108,03	7613,46	234,01

Table 4. P-values generated by Dunn's test for excluding the null hypotheses that cumulative distribution function of the sampled final values (the fitness obtained when the stopping condition is met). The cases when the p-value is lower than 0.05 were enhanced with bold font.

(1+1) Evolution Strategy				
Ackley	Def.	Cl. b.	Far. w.	Both
Def.		**3.90e−02**	**1.40e−12**	**2.03e−09**
Cl. b.			**2.46e−05**	**2.24e−03**
Far. w.				1.00e+00
De Jong	Def.	Cl. b.	Far. w.	Both
Def.		**4.59e−02**	8.92e−14	**2.69e−08**
Cl. b.			**3.08e−06**	**8.30e−03**
Far. w.				4.09e−01
Griewank	Def.	Cl. b.	Far. w.	Both
Def.		4.48e−01	**8.46e−12**	**1.39e−07**
Cl. b.			**6.93e−07**	**8.56e−04**
Far. w.				8.06e−01
Rastrigin	Def.	Cl. b.	Far. w.	Both
Def.		**3.90e−02**	**1.68e−14**	**6.83e−08**
Cl. b.			**1.35e−06**	**1.69e−02**
Far. w.				1.71e−01

(mu, lambda) Evolution Strategy				
Ackley	Def.	Cl. b.	Far. w.	Both
Def.		**3.14e−07**	**3.90e−02**	**3.90e−02**
Cl. b.			**1.93e−15**	**3.90e−02**
Far. w.				**3.14e−07**
De Jong	Def.	Cl. b.	Far. w.	Both
Def.		**3.08e−06**	**3.90e−02**	**1.00e−02**
Cl. b.			**5.82e−14**	3.62e−01
Far. w.				**2.69e−08**
Griewank	Def.	Cl. b.	Far. w.	Both
Def.		**6.20e−07**	**3.90e−02**	**2.67e−02**
Cl. b.			**5.28e−15**	7.96e−02
Far. w.				**1.57e−07**
Rastrigin	Def.	Cl. b.	Far. w.	Both
Def.		**9.33e−07**	**3.90e−02**	**2.11e−02**
Cl. b.			**9.70e−15**	1.20e−01
Far. w.				**1.02e−07**

(mu+lambda) Evolution Strategy				
Ackley	Def.	Cl. b.	Far. w.	Both
Def.		**3.14e−07**	**3.90e−02**	**3.90e−02**
Cl. b.			**1.93e−15**	**3.90e−02**
Far. w.				**3.14e−07**
De Jong	Def.	Cl. b.	Far. w.	Both
Def.		**7.20e−07**	**3.90e−02**	**2.45e−02**
Cl. b.			**6.59e−15**	9.25e−02
Far. w.				**1.34e−07**
Griewank	Def.	Cl. b.	Far. w.	Both
Def.		**1.67e−06**	**3.90e−02**	**1.48e−02**
Cl. b.			**2.33e−14**	2.10e−01
Far. w.				**5.37e−08**
Rastrigin	Def.	Cl. b.	Far. w.	Both
Def.		**1.93e−06**	**4.59e−02**	**1.12e−02**
Cl. b.			**4.45e−14**	2.73e−01
Far. w.				**4.57e−08**

6 Conclusions

In this paper we have presented a novel hybrid algorithms based on classic Evolution Strategies by Schwefel and Rechenberg. The method was tested on a set of well known multi-modal benchmark problems. Three different evolution strategies were analyzed: $(1 + 1)$, (μ, λ) and $(\mu + \lambda)$ and efficacy of the new socio-cognitive hybrids was compared to the baseline model.

The proposed herein socio-cognitive hybrids increase the possibility to exchange information between the individuals by introducing a set of historically best and historically worst solutions, affecting the way in which the individuals are mutated. In the case of the hybrid utilizing the historically worst solutions, an attempt has been made to avoid them during mutation. Conversely, when considering the historically best solutions, during the mutation some of their information was copied.

The $(1 + 1)$ Evolution Strategy turned out not to be an effective solution for hybridization purposes, as the baseline algorithm outperformed hybrids in most of the examined cases. In the case of (μ, λ) Evolution Strategy, the best results were obtained for Ackley benchmark where 'Closer to best' strategy converged much faster and to better solutions. As for $(\mu + \lambda)$ Evolution Strategy was actually better in the case of mechanism going closer to the best up-to-date solutions, for all the considered problem instances.

We conclude, that using information from the best performing individuals may increase the convergence speed of the evolutionary strategies and significantly improve over the baseline model. Both hybrids, that used this feature, namely 'Closer to best' and 'Both', performed usually better than the baseline model. This is actually an expected outcome, as those strategies focus on further local optimizations, thus are more likely to fine-tune better solutions. It is also in line with other research in evolutionary computation (e.g. [14]). Avoiding worst individuals implemented in 'Further from worst' turned out to be somewhat helpful, but only for low dimensional spaces (up to 100 dimensions). When the dimensionality of the problem increased, this hybrid was slowing down convergence and eventually led to less effective solutions.

In future, besides the necessary enhancement, we would like to further explore the social component in the Evolution Strategies hybrids, by testing certain parameters (e.g. the length of the history), but also by introducing new and updated social mechanisms (e.g. different species and relations among them). Other population-based metaheuristics will be also considered.

References

1. Arnold, D.V.: Weighted multirecombination evolution strategies. Theoret. Comput. Sci. **361**(1), 18–37 (2006). foundations of Genetic Algorithms
2. Bandura, A.: Self-efficacy: toward a unifying theory of behavioral change. Psychol. Rev. **84**(2), 191–215 (1977)
3. Bandura, A.: Social Foundations of Thought and Action: A Social Cognitive Theory. Prentice-Hall, Englewood Cliffs (1986)

4. Bandura, A., Ross, D., Ross, S.: Transmission of aggression through the imitation of aggressive models. J. Abnormal Soc. Psychol. **63**(3), 575–582 (1961)
5. Beume, N., Rudolph, G.: Faster s-metric calculation by considering dominated hypervolume as klee's measure problem. In: Kovalerchuk, B. (ed.) Proceedings of the Second IASTED International Conference on Computational Intelligence, San Francisco, California, USA, November 20–22, 2006, pp. 233–238. IASTED/ACTA Press (2006)
6. Blum, C., Puchinger, J., Raidl, G.R., Roli, A.: Hybrid metaheuristics in combinatorial optimization: a survey. Appl. Soft Comput. **11**(6), 4135–4151 (2011). 10.1016/j.asoc.2011.02.032, https://www.sciencedirect.com/science/article/pii/S1568494611000962
7. Brockhoff, D., Auger, A., Hansen, N., Arnold, D.V., Hohm, T.: Mirrored sampling and sequential selection for evolution strategies. In: Schaefer, R., Cotta, C., Kołodziej, J., Rudolph, G. (eds.) PPSN 2010. LNCS, vol. 6238, pp. 11–21. Springer, Heidelberg (2010). https://doi.org/10.1007/978-3-642-15844-5_2
8. Bugajski, I., et al.: Enhancing particle swarm optimization with socio-cognitive inspirations. In: Connolly, M. (ed.) International Conference on Computational Science 2016, ICCS 2016. Procedia Computer Science, vol. 80, pp. 804–813. Elsevier (2016)
9. Byrski, A., Schaefer, R., Smołka, M., Cotta, C.: Asymptotic guarantee of success for multi-agent memetic systems. Bull. Pol. Acad. Sci. Tech. Sci. **61**(1), 257–278 (2013)
10. Byrski, A., Drezewski, R., Siwik, L., Kisiel-Dorohinicki, M.: Evolutionary multi-agent systems. Knowl. Eng. Rev. **30**(2), 171–186 (2015)
11. Byrski, A., Swiderska, E., Lasisz, J., Kisiel-Dorohinicki, M., Lenaerts, T., Samson, D., Indurkhya, B., Nowé, A.: Socio-cognitively inspired ant colony optimization. J. Comput. Sci. **21**, 397–406 (2017)
12. Dieterich, J., Hartke, B.: Empirical review of standard benchmark functions using evolutionary global optimization. Appl. Math. **3**(18A) (2012)
13. Dorigo, M., Stutzle, T.: Ant Colony Optimization. MIT Press, Cambridge (2004)
14. Du, H., Wang, Z., Zhan, W., Guo, J.: Elitism and distance strategy for selection of evolutionary algorithms. IEEE Access **6**, 44531–44541 (2018). https://doi.org/10.1109/ACCESS.2018.2861760
15. Hansen, N., Ostermeier, A.: Completely derandomized self-adaptation in evolution strategies. Evol. Comput. **9**(2), 159–195 (2001)
16. Jamasb, A., Motavalli-Anbaran, S.H., Ghasemi, K.: A novel hybrid algorithm of particle swarm optimization and evolution strategies for geophysical non-linear inverse problems. Pure Appl. Geophys. **176** (2019)
17. Kennedy, J., Eberhart, R.: Particle swarm optimization. In: Proceedings of ICNN 1995 - International Conference on Neural Networks, vol. 4, pp. 1942–1948 (1995)
18. Klose, A.D., Hielscher, A.H.: Hybrid approach for diffuse optical tomography combining evolution strategies and gradient techniques. In: Chance, B., Alfano, R.R., Tromberg, B.J., Tamura, M., Sevick-Muraca, E.M. (eds.) Optical Tomography and Spectroscopy of Tissue IV. International Society for Optics and Photonics, SPIE, vol. 4250, pp. 11–19 (2001)
19. Koulocheris, D., Vrazopoulos, H., Dertimanis, V.: Hybrid evolution strategy for the design of welded beams. In: Proceedings of International Congress on Evolutionary Methods for Design, Optimization and Control with Applications to Industrial Problems EUROGEN 2003. CIMNE Barcelona (2003)

20. Liagkouras, K., Metaxiotis, K.: An elitist polynomial mutation operator for improved performance of moeas in computer networks. In: 2013 22nd International Conference on Computer Communication and Networks (ICCCN). pp. 1–5 (2013). https://doi.org/10.1109/ICCCN.2013.6614105

21. Michalewicz, Z., Fogel, D.: How to Solve It: Modern Heuristics. Springer, Heidelberg (2004). https://doi.org/10.1007/978-3-662-07807-5

22. Moreau-Giraud, L., Lafon, P.: A hybrid evolution strategy for mixed discrete continuous constrained problems. In: Fonlupt, C., Hao, J.-K., Lutton, E., Schoenauer, M., Ronald, E. (eds.) AE 1999. LNCS, vol. 1829, pp. 123–135. Springer, Heidelberg (2000). https://doi.org/10.1007/10721187_9

23. Placzkiewicz, L., et al.: Hybrid swarm and agent-based evolutionary optimization. In: Shi, Y., et al. (eds.) ICCS 2018. LNCS, vol. 10861, pp. 89–102. Springer, Cham (2018). https://doi.org/10.1007/978-3-319-93701-4_7

24. Rabiej, M.: A hybrid immune-evolutionary strategy algorithm for the analysis of the wide-angle X-ray diffraction curves of semicrystalline polymers. J. Appl. Crystallogr. **47**(5), 1502–1511 (2014)

25. Rechenberg, I.: Cybernetic solution path of an experimental problem. Roy. Aircraft Establishment Lib. Transl. 1122 (1965). https://ci.nii.ac.jp/naid/10000137330/en/

26. Repoussis, P., Tarantilis, C., Bräysy, O., Ioannou, G.: A hybrid evolution strategy for the open vehicle routing problem. Comput. Oper. Res. **37**(3), 443–455 (2010). hybrid Metaheuristics

27. dos Santos Coelho, L., Alotto, P.: Electromagnetic device optimization by hybrid evolution strategy approaches. Int. J. Comput. Math. Electr. Electr. Eng **26**(2), 269–279 (2007)

28. Schwefel, H.P.: Numerische Optimierung von Computer-Modellen mittels der Evolutionsstrategie: mit einer vergleichenden Einführung in die Hill-Climbing-und Zufallsstrategie, vol. 1. Springer, Heidelberg (1977). https://doi.org/10.1007/978-3-0348-5927-1

29. Schwefel, H.P.: Evolution and Optimum Seeking. Wiley, New York (1995)

30. Talbi, E.G.: Metaheuristics: From Design to Implementation. Wiley, New York (2009)

31. Talbi, E.G.: A taxonomy of hybrid metaheuristics. J. Heurist. **8**, 541–564 (2002). https://doi.org/10.1023/A:1016540724870

32. Huang, T.-Y., Chen, Y.-Y.: Modified evolution strategies with a diversity-based parent-inclusion scheme. In: Proceedings of the 2000. IEEE International Conference on Control Applications. Conference Proceedings (Cat. No.00CH37162), pp. 379–384 (2000)

33. Vose, M.D.: The Simple Genetic Algorithm - Foundations and Theory. Complex Adaptive Systems. MIT Press, Cambridge (1999)

34. Wolpert, D.H., Macready, W.G.: No free lunch theorems for optimization. Trans. Evol. Comput. **1**(1), 67–82 (1997)

35. Zhang, G., Shi, Y.: Hybrid sampling evolution strategy for solving single objective bound constrained problems. In: 2018 IEEE Congress on Evolutionary Computation (CEC), pp. 1–7 (2018). https://doi.org/10.1109/CEC.2018.8477908

Effective Solution of Ill-Posed Inverse Problems with Stabilized Forward Solver

Marcin Łoś[ID], Robert Schaefer[ID], and Maciej Smołka[✉][ID]

Institute of Computer Science, AGH University of Science and Technology,
Kraków, Poland
{los,schaefer,smolka}@agh.edu.pl
http://www.informatyka.agh.edu.pl/

Abstract. We consider inverse parametric problems for elliptic variational PDEs. They are solved through the minimization of misfit functionals. Main difficulties encountered consist in the misfit multimodality and insensitivity as well as in the weak conditioning of the direct (forward) problem, that therefore requires stabilization. A complex multi-population memetic strategy hp-HMS combined with the Petrov-Galerkin method stabilized by the Demkowicz operator is proposed to overcome obstacles mentioned above. This paper delivers the theoretical motivation for the common inverse/forward error scaling, that can reduce significantly the computational cost of the whole strategy. A short illustrative numerical example is attached at the end of the paper.

Keywords: Inverse problem · Memetic algorithm · Stabilized Petrov-Galerkin method

1 Ill-Conditioned Data Inversion: State of the Art

We will focus on Inverse Problems (IPs) which consist in finding a set of parameters $\mathcal{S}$ that minimize a misfit functional f over an admissible set of parameters $\mathcal{D} \subset \mathbb{R}^N$, i.e.:

$$\mathcal{D} \supset \mathcal{S} = \arg \min_{\omega \in \mathcal{D}} f(d_0, u(\omega)), \tag{1}$$

where $u(\omega) \in U$ is a solution of the forward (direct) problem

$$B(\omega; u(\omega)) = l \tag{2}$$

corresponding to the parameter $\omega \in \mathcal{D}$, $d_0 \in \mathcal{O}$ denotes the observed data and $f : \mathcal{O} \times U \to \mathbb{R}_+$. The forward problem operator $B : \mathcal{D} \times U \to V'$ forms a mathematical model of the studied phenomenon, U is a Hilbert space of forward solutions and $l \in V'$ is a functional over a Hilbert space of "test functions" V.

The work was supported by the National Science Centre, Poland, grant No. 2017/26/M/ST1/00281.

M. Paszynski et al. (Eds.): ICCS 2021, LNCS 12743, pp. 343–357, 2021.
https://doi.org/10.1007/978-3-030-77964-1_27

If (1) has more than one solution $(\text{card}(\mathcal{S}) > 1)$ then it becomes *ill-conditioned*. If $\mathcal{S}$ is not connected, then IP is called *multimodal*. If $\mathcal{S}$ contains an open set in $\mathbb{R}^N$, then we have *insensitivity region (plateau)* in the objective (misfit) landscape (see [11] for details). Traditionally, one can handle IP's ill conditioning by supplementing misfit with a regularization term to make it globally or locally convex (see e.g. [4]). Unfortunately, such methods can produce undesirable artifacts and lead to the loss of information regarding the modeled process. In the worst cases this method can deliver outright false solutions, forced by the regularization term.

A more sophisticated way of dealing with such IPs is to use stochastic global optimization methods which are able to identify separate basins of attraction to global minimizers (see e.g. [13]). Papers [2,3] show the application of a hierarchic evolutionary strategy HGS combined with hp-FEM to the identification of ambiguous Lamme coefficients in linear elasticity. The use of the same combined strategy in the inversion of logging data obtained with DC and AC probes is described in [9,16]. The paper [14] shows the method of misfit insensitivity regions approximation applied to MT data inversion. A hierarchic memetic strategy hp-HMS supplemented with an evolutionary algorithm using *multiwinner selection* [8] separates such areas around multiple global minimizers of the misfit.

The above-mentioned papers show an effective method of solving ill-conditioned IPs when the solution space coincides with the test space of the forward problems (i.e., $U = V$) and the forward operator B is generally coercive[1]. In this paper we show how to extend the method to the class of forward problems in which $U \neq V$ and B satisfies the inf-sup condition only. Such forward problems are also difficult to solve, because of the huge difference between the norms of particular components of their governing equations (e.g. diffusion-convection flow problem). Special numerical methods with stabilization such as DPG [6] have to be applied in such cases. There are few examples of inverse solutions in which DPG was applied for solving forward problem. In this number Bramwell in his dissertation [5] discusses the tomography problem of restoring the squared material slowness coefficient, using the measured displacements resulted from the point-wise harmonic stimulation on the domain surface.

2 hp-HMS as an Effective Inverse Solver

2.1   Evolutionary Core

hp-HMS is a complex strategy that consists of a multi-phase multi-population evolutionary algorithm combined with local search, cluster detection and local approximation methods. The multi-population evolutionary core is responsible for the discovery of problem objective local minima, including more interesting global minima and these, in the inverse problem case, are solutions for (1). This global search is performed by a tree of concurrently running single-population

---

[1] This is not the case of AC and MT problems, but there we also have $U = V$.

evolutionary algorithms, called *demes* in the sequel. The whole tree executes subsequent global steps called *metaepochs*. Each metaepoch in turn consists of running a fixed number of evolutionary steps in each active deme. The tree itself evolves according to the following rules:

- it starts with a single root,
- it has a fixed maximal depth,
- after a metaepoch each deme that has spotted a promising solution and is not at the deepest level can try to *sprout* a child deme with a population located in the neighborhood of the spotted solution.

A new deme can be sprouted only if the newfound solution is not too close to other child demes. The main idea is that the search performed in a child deme is more focused and more accurate than the one performed in the parent. The structure of the tree is then determined by the parent-child relation among demes with depth levels related to levels of search precision. The root searches the most chaotically but over the widest domain and the final solutions result from the most focused search in leaves. Any solution found in leaf demes that has a decent objective value is memorized for the further investigation. The evolution in any deme is controlled with *a local stopping condition* that deactivates stagnant searches. The whole strategy evolution is stopped when *a global stopping condition* is satisfied.

2.2 Handling Multimodality and Insensitivity

The sprouting condition forbids the exploration of neighborhoods of already found solutions. That forces the whole strategy to move to other areas of the computational domain and makes it possible to spot multiple different solutions. In some problems this can mean *all solutions*, of course when using an appropriate global stopping condition. This way, the HMS can handle multi-modal problems of a simpler type having many *isolated* solutions. But in some problems the difficulty level is raised higher with solutions forming *plateaus*, i.e., sets with nonempty interior. The evolutionary HMS core is not capable of coping with such difficulties, but the whole strategy has additional modules for this purpose. First, leaf node populations provide a hint of local minima attraction basins: among them the above-mentioned plateaus. Second, these populations are merged pairwise if they cover the same basin. Third, an evolutionary algorithm with special-type multiwinner-voting-based selection operator is run on those merged populations to enhance the coverage of the basins. Finally, on each of the basins we build a local approximation of the objective and appropriate level sets of these approximations serve as estimates of plateau boundaries.

2.3 Computation Speedup with Dynamic Accuracy Adjustment

It is well-known that inverse computations are very time-expensive. The main weight of these is the cost of direct problem solution. HMS inverse computations cannot entirely avoid this problem but in some classes of IPs we can mitigate it significantly: this is the case of solving the direct problem by means of an adaptive solver. Here we provide the idea of taking the advantage of the direct solver hp-adaptivity, which leads to a version of HMS called hp-HMS.

As it was said before, we assume that we compute the objective $\mathcal{C}(u(\omega)) = f(d_o, u(\omega))$ with an hp-FEM forward problem (2) solver. In some important cases (see below) we have the following estimate

$$\left| \mathcal{C}\left(u_{\frac{h}{2},p+1}(\omega)\right) - \mathcal{C}\left(u(\omega^*)\right)\right| \leq$$
$$A_1 err_{rel}(\omega)^\alpha + A_2 \|u_{h,p}(\omega) - u(\omega)\|_U^\beta + A_3 \|\omega - \omega^*\|_D^\gamma. \quad (3)$$

In formula (3) $u(\omega)$ is the solution of the direct (forward) problem (2), $u_{h,p}(\omega)$ is the hp-FEM approximate solution of (2) for mesh size h and polynomial degree p, $err_{rel}(\omega)$ is a measure of the difference between approximations obtained in two subsequent hp-FEM steps

$$e_{rel}(\omega) = u_{\frac{h}{2},p+1}(\omega) - u_{h,p}(\omega), \quad (4)$$

$A_1, A_2, A_3 > 0$ are positive constants and $\alpha \geq 1$, $\beta \geq 1$. In the simplest case $err_{rel}(\omega) = \|e_{rel}(\omega)\|_V$, but quite often err_{rel} contains other components, e.g., coming from the approximation of the dual problem to (2).

Here we include a proof of (3) in a quite simple but at the same time also quite general case.

Proposition 1. *Assume that there exist $C_u > 0$, $C_C > 0$, $r \geq 1$, $s \geq 1$ such that*

$$\|u(\omega) - u(\omega')\|_U \leq C_u \|\omega - \omega'\|_D^r \quad (5)$$

for every $\omega, \omega' \in \mathcal{D}$ and that

$$|\mathcal{C}(u) - \mathcal{C}(u')| \leq C_C \|u - u'\|^s \quad (6)$$

for all u and u'. Then there exist $A_1 > 0$ and $A_3 > 0$ such that

$$\left| \mathcal{C}\left(u_{\frac{h}{2},p+1}(\omega)\right) - \mathcal{C}\left(u(\omega^*)\right)\right|$$
$$\leq A_1 \|u_{\frac{h}{2},p+1}(\omega) - u_{h,p}(\omega)\|^s + A_1 \|u_{h,p}(\omega) - u(\omega)\|^s + A_3 \|\omega - \omega^*\|^{rs}. \quad (7)$$

Proof. Let us start with a simple equality

$$u_{\frac{h}{2},p+1}(\omega) - u(\omega^*) = u_{\frac{h}{2},p+1}(\omega) - u_{h,p}(\omega) + u_{h,p}(\omega) - u(\omega) + u(\omega) - u(\omega^*).$$

Using the triangle inequality we obtain

$$\|u_{\frac{h}{2},p+1}(\omega) - u(\omega^*)\| \leq \|u_{\frac{h}{2},p+1}(\omega) - u_{h,p}(\omega)\| + \|u_{h,p}(\omega) - u(\omega)\| + \|u(\omega) - u(\omega^*)\|.$$

From (6) and (8) we have that

$$\left| \mathcal{C}\left(u_{\frac{h}{2},p+1}(\omega)\right) - \mathcal{C}\left(u(\omega^*)\right)\right|$$
$$\leq C_C \left(\|u_{\frac{h}{2},p+1}(\omega) - u_{h,p}(\omega)\| + \|u_{h,p}(\omega) - u(\omega)\| + \|u(\omega) - u(\omega^*)\| \right)^s. \quad (8)$$

Then, using Jensen's inequality we obtain

$$\left| \mathcal{C}\left(u_{\frac{h}{2},p+1}(\omega)\right) - \mathcal{C}\left(u(\omega^*)\right)\right|$$
$$\leq 3^{s-1}C_{\mathcal{C}}\left(\|u_{\frac{h}{2},p+1}(\omega) - u_{h,p}(\omega)\|^s + \|u_{h,p}(\omega) - u(\omega)\|^s + \|u(\omega) - u(\omega^*)\|^s\right).$$

Now, the application of (5) completes the proof with

$$A_1 = 3^{s-1}C_{\mathcal{C}}, \quad A_3 = 3^{s-1}C_{\mathcal{C}}C_u.$$

$\square$

Constants C_u and r are related to properties of forward equation (2). Here, we show their derivation in an important special case.

Proposition 2. *Assume that in the forward problem (2) $U = V$ and that the forward operator B has the following properties.*

1. *For every $\omega \in \mathcal{D}$ $B(\omega; u)$ is linear and continuous with respect to u.*
2. *$B(\omega; u)$ is uniformly coercive with respect to u, i.e. there exists $C_b > 0$ independent on ω such that*

$$|\langle B(\omega; u), u \rangle| \geq C_b \|u\|_U^2.$$

3. *$B(\omega; u)$ is uniformly Lipschitz-continuous with respect to ω, i.e. there exist $C_B > 0$ independent on u such that*

$$\|B(\omega; u) - B(\omega'; u)\|_{U'} \leq C_B \|u\|_U \|\omega - \omega'\|_{\mathcal{D}}.$$

Then (5) holds with $r = 1$ and

$$C_u = \frac{C_B}{C_b^2}\|l\|_{U'}.$$

Proof. In the sequel we shall use the following notation

$$B(\omega) : U \ni u \longmapsto B(\omega; u) \in U'.$$

Then, we can state that
$$u(\omega) = B(\omega)^{-1}(l). \tag{9}$$

We have the following sequence of equalities.

$$u(\omega) - u(\omega') = B(\omega)^{-1}(l) - B(\omega')^{-1}(l)$$
$$= \left[B(\omega)^{-1} \circ (B(\omega') - B(\omega)) \circ B(\omega')^{-1}\right](l).$$

Therefore, using operator-norm definition we obtain

$$\|u(\omega) - u(\omega')\| \leq \|B(\omega)^{-1}\| \cdot \|B(\omega') - B(\omega)\| \cdot \|B(\omega')^{-1}\| \cdot \|l\|. \tag{10}$$

Thanks to Assumption 2 we have

$$C_b\|u\|^2 \leq |\langle B(\omega; u), u\rangle| \leq \|B(\omega; u)\|\|u\|.$$

Therefore, again from the definition of the operator norm, it follows that

$$\|B(\omega; u(\omega))\| \geq C_b\|u(\omega)\|.$$

Using (9) we can rewrite the above in the following way

$$\|B(\omega)^{-1}(L)\| \leq \frac{1}{C_b}\|l\|.$$

Hence, for every $\omega \in \mathcal{D}$ we have

$$\|B(\omega)^{-1}\| \leq \frac{1}{C_b}.$$

Assumption 3 can be rewritten in the following way

$$\|B(\omega) - B(\omega')\| \leq C_B\|\omega - \omega'\|.$$

Hence, using the above inequalities along with (10) we obtain

$$\|u(\omega) - u(\omega')\| \leq \frac{1}{C_b} \cdot C_B \cdot \|\omega - \omega'\| \cdot \frac{1}{C_b}\|l\|,$$

which concludes the proof.   □

The first right-hand-side component of (3) contains a power of the error loss $err_{rel}(\omega)$ in a single hp-FEM step. The last component is a power of the error of the inverse problem solution, that is related to the assumed search range on a given level of the HMS tree. When using an hp-adaptive direct solver we can trade the precision of the computations for the savings in time and reversely we can spend more time to obtain more accurate solution. The main idea of the hp-HMS is then to dynamically adjust the accuracy of the misfit evaluation to both a particular value of ω and the (inevitable) inverse problem solution error characterizing a given HMS tree level. The adjustment is realized by keeping the balance between the first and the last right-hand term in (3): the middle term can be neglected due to the high rate of the hp-FEM convergence. Therefore, if δ_j is an assumed precision of the inverse problem solution (i.e., an assumed level of the inverse error) at level j we perform the hp-adaptation of the FEM solution of the forward problem until $err_{rel}(\omega)$ drops below $Ratio(j)\delta_j^{\frac{1}{\alpha}}$, where $Ratio(j)$ is a parameter of the strategy related to the constant $(A_3(A_1)^{-1})^{\frac{1}{\alpha}}$.

Remark 1. Such a dynamical accuracy adjustment can result in a notable reduction of the computational cost of hp-HMS: we refer the reader to papers [2,15] for details.

3   Stabilizing Forward Petrov-Galerkin Based Solver by Using Demkowicz Operator

3.1   Sample Exact Forward Problem

Let us study the following variational forward problem:

$$\left\{ \begin{array}{l} \text{Find } u \in U; \\ b(u,v) = l(v) \ \forall v \in V \end{array} \right\} \Leftrightarrow \left\{ \begin{array}{l} Bu = l, \ B : U \to V'; \\ \langle Bu, v \rangle_{V' \times V} = b(u,v) \ \forall v \in V \end{array} \right\} \tag{11}$$

where U, V are two real Hilbert spaces, $b : U \times V \to \mathbb{R}$ is a bilinear (sesquilinear), continuous form, $l : V \to \mathbb{R}$ continuous linear functional so $|b(u,v)| \leq M\|u\|_U \|v\|_V$, $|l(v)| \leq \|l\|_{V'}\|u\|_U$ where M stands for the norm of the form b.

If b satisfies the following inf-sup condition equivalent to the condition, that B is bounded below:

$$\exists \gamma > 0; \ \forall u \in U \ \sup_{0 \neq v \in V} \left\{ \frac{|b(u,v)|}{\|v\|_V} \geq \gamma \|u\|_U \right\} \tag{12}$$

and l satisfies the compatibility condition

$$l(v) = 0 \ \forall v \in V_0 = \ker(B') = \{v \in V; b(u,v) = 0 \ \forall u \in U\}, \tag{13}$$

or the stronger one $\ker(B') = \{v \in V; b(u,v) = 0 \ \forall u \in U\} = \{0\}$, then (11) has a unique solution that satisfies the stability estimate $\|u\|_U \leq \frac{1}{\gamma}\|l\|_{V'}$, where $B' : (V')' = V \ni v \to \langle \cdot, B'v \rangle_{V' \times V} = b(\cdot, v) \in U'$, because V is reflexive as a Hilbert space. Moreover B is the isomorphism (see e.g. Babuška [1] and references inside).

3.2   Petrov-Galerkin Method

By choosing $U_h \subset U, \ V_h \subset V; \ \dim(U_h) = \dim(V_h) < +\infty$ the problem approximate to (11) can be obtained:

$$\text{Find } u_h \in U_h; \ b(u_h, v) = l(v) \ \forall v \in V_h \tag{14}$$

If (14) satisfies the discrete inf-sup condition:

$$\forall u \in U_h \ \sup_{0 \neq v \in V_h} \left\{ \frac{|b(u,v)|}{\|v\|_V} \right\} \geq \gamma_h \|u\|_U, \tag{15}$$

then it has the unique solution u_h which satisfies the discrete stability condition $\|u_h\|_U \leq \frac{1}{\gamma_h}\|l\|_{V'}$, moreover if u is the solution to (11) then

$$\|u - u_h\|_U \leq \frac{M}{\gamma_h} \inf_{w_h \in U_h} \{\|u - w_h\|_U\}. \tag{16}$$

If the stability constants γ_h have a positive lover bound $\inf_h\{\gamma_h\} = \gamma_0 > 0$, then the Petrov-Galerkin method error converges with the same rate as the best approximation error, because:

$$\|u - u_h\|_U \leq \frac{M}{\gamma_0} \inf_{w_h \in U_h} \{\|u - w_h\|_U\}. \tag{17}$$

Unfortunately, the continuous inf-sup condition does not imply uniform discrete inf-sup condition, so (17) does not hold in general for the arbitrary spaces $U_h \subset U$, $V_h \subset V$ (see Babuška [1]).

3.3 Demkowicz Operator

The way to overcome this obstacle was introduced by Demkowicz and his collaborators [6,7].

Let $\mathcal{H}$ be the set of all finite dimensional subspaces of U and $\mathcal{B}$ the set of all finite dimensional subspaces of V. We are looking for any mapping $\Im : \mathcal{H} \to \mathcal{B}$ so that (14) is symmetric and uniformly stable ((15) is satisfied with a uniform constant) for each pair $(U_h, V_h = \Im(U_h))$, $U_h \in \mathcal{H}$.

Let $R_V : V \to V'$ be the Riesz isometry, then $\forall u \in U$ $\|Bu\|_{V'} = \|R_V^{-1}Bu\|_V$. We define now the linear Demkowicz operator

$$T = R_V^{-1}B : U \to V. \tag{18}$$

T is an isomorphism as a composition of isomorphisms, moreover

$$\forall u \in U \quad \|Bu\|_{V'} = \|R_V^{-1}Bu\|_V = \|Tu\|_V. \tag{19}$$

It can be proved, that $\Im(U_h) \equiv TU_h$ satisfies our needs.

First, setting $V_h = TU_h$ we obtain $\dim(V_h) = \dim(U_h) = n$, because T is the isomorphism and next:

$$\forall 0 \neq u \in U \|Bu\|_{V'} = \sup_{0 \neq v \in V} \left\{ \frac{|b(u,v)|}{\|v\|_V} \right\} = \frac{|b(u,Tu)|}{\|Tu\|_V} \tag{20}$$

Let us introduce the energy norm $\| \cdot \|_E$ on the space U so that

$$\|u\|_E = \|Tu\|_V = \|R_V^{-1}Bu\|_V = \|Bu\|_{V'} = \sup_{0 \neq v \in V} \left\{ \frac{|b(u,v)|}{\|v\|_V} \right\}. \tag{21}$$

Both norms $\| \cdot \|_E$ and $\| \cdot \|_U$ are equivalent, because

$$\gamma\|u\|_U \leq \sup_{0 \neq v \in V} \left\{ \frac{|b(u,v)|}{\|v\|_V} \right\} = \|u\|_E = \|Bu\|_{V'} \leq M\|u\|_U, \tag{22}$$

and it is easy to prove:

Lemma 1. *If $V_h = TU_h$, then the inf-sup constant in (15) equals $\gamma_h = 1$ with respect to the norm $\| \cdot \|_E$ independently on the selection of the space $U_h \subset U$.*

3.4   The Abstract Stabilized Forward Problem

Now we are ready to introduce the following form of the stabilized Petrov-Galerkin forward problem:

$$\text{Find } u_h \in U_h; \; d(u_h, w) = r(w) \;\; \forall w \in U_h, \tag{23}$$

where $d : U \times U \ni u, w \to d(u, w) = b(u, Tw) \in \mathbb{R}$ and $r : U \ni w \to r(w) = l(Tw) \in \mathbb{R}$.

Remark 2. The problems (14) and (23) are equivalent, if $V_h = TU_h$.

Theorem 1.

1. *d is symmetric on $U \times U$*
2. $d(u, u) = \|u\|_E^2 \; \forall u \in U$
3. $|d(u, w)| \leq \|u\|_E \|w\|_E \; \forall u, w \in U$
4. $|r(w)| \leq \|l\|_{V'} \|w\|_E \; \forall w \in U$

Theorem 2. *The problem (23) has the unique solution $u_h \in U_h$ so that $\|u_h\|_E \leq \|l\|_{V'}$. Moreover $\|u - u_h\|_E = \inf_{w \in U_h}\{\|u - w\|_E\}$, where $u \in U$ is the solution to (11).*

Proof of Theorem 1: $d(u, w) = b(u, Tw) = \langle Bu, Tw \rangle_{V' \times V} = (R_V^{-1}Bu, Tw)_V = (Tu, Tw)_V$, so d satisfies 1. Next $d(u, u) = (Tu, Tu)_V = \|Tu\|_V^2 = \|u\|_E^2$ proves 2. and $|d(u, w)| = |(Tu, Tw)_V| \leq \|Tu\|_V \|Tw\|_V = \|u\|_E \|w\|_E$ proves 3. Finally, $|r(w)| \leq \|l\|_{V'} \|Tw\|_V = \|l\|_{V'} \|w\|_E$ proves 4.   □

Proof of Theorem 2: The form d is a scalar product on U inducing the energy norm $\| \cdot \|_E$. Moreover d and r preserve their conditions while restricting to the subspace U_h, so the first thesis follows from Riesz theorem for this subspace. The second thesis immediately follows from the Céa lemma which implies $\|u - u_h\|_E \leq \inf_{w \in U_h}\{\|u - w\|_E\}$, but $\|u - u_h\|_E$ must be less or equal to $\inf_{w \in U_h}\{\|u - w\|_E\}$ because both $u_h, w \in U_h$.   □

Remark 3. All results of Theorems 1 and 2 do not depend on the selection of the approximation subspace $U_h \subset U$.

Remark 4. Both, the energy norm $\| \cdot \|_E$ and Demkowicz isomorphism T depend on the variational problem to be solved. Moreover, the conditions of each problem (23) determined by Theorems 1, 2 can be expressed using the norm $\| \cdot \|_U$ and constants M, γ characterizing the exact problem (11). In particular

$$\|u - u_h\|_U \leq \frac{M}{\gamma} \inf_{w_h \in U_h} \{\|u - w_h\|_U\}, \tag{24}$$

moreover:

$$d(u, u) \geq \gamma^2 \|u\|_U^2, \; |d(u, w)| \leq M^2 \|u\|_U \|w\|_U,$$
$$|r(w)| \leq \|l\|_{V'} M \|w\|_U \;\; \forall u, w \in U. \tag{25}$$

4 Including Demkowicz Operator in the hp-HMS Structure

Let us assume, that the HMS objective $\mathcal{C}(u(\omega))$ fits the misfit of the inverse problem (1) and satisfies the inequality:

$$\begin{aligned}
\mathcal{C}(u(\omega)) &= f(d_o, u(\omega)) \ \forall \omega \in \mathcal{D}, \\
|\mathcal{C}(u) - \mathcal{C}(u')| &\leq C_C \|u - u'\|_U^s \ \forall u, u' \in U,
\end{aligned} \tag{26}$$

where s, C_C are some positive constants and $u(\omega) \in U$ is the solution to the following forward problem:

$$\left\{ \begin{array}{c} \text{Find } u(\omega) \in U; \\ b(\omega; u(\omega), v) = l(v) \ \forall v \in V \end{array} \right\}$$
$$\Updownarrow \tag{27}$$
$$\left\{ \begin{array}{c} B(\omega; u(\omega)) = l, \ \ B(\omega; \cdot) : U \to V'; \\ \langle B(\omega; u, v) \rangle_{V' \times V} = b(\omega; u, v) \ \forall u \in U, \forall v \in V \end{array} \right\}$$

where U, V are two real Hilbert spaces, $b(\omega; \cdot, \cdot) : U \times V \to \mathbb{R}$ is a bilinear (sesquilinear), continuous form, $l : V \to \mathbb{R}$ continuous linear functional so $|b(\omega; u, v)| \leq M_\omega \|u\|_U \|v\|_V$, $|l(v)| \leq \|l\|_{V'} \|u\|_U$ where M_ω stands for the norm of the form $b(\omega; \cdot, \cdot)$.

If $b(\omega; \cdot, \cdot)$ satisfies the following inf-sup condition:

$$\exists \gamma_\omega > 0; \ \forall u \in U \ \sup_{0 \neq v \in V} \left\{ \frac{|b(\omega; u, v)|}{\|v\|_V} \geq \gamma_\omega \|u\|_U \right\} \tag{28}$$

and l satisfies the compatibility condition

$$l(v) = 0 \ \forall v \in V_0 = \ker(B'(\omega, \cdot)) = \{v \in V; b(\omega; u, v) = 0 \ \forall u \in U\}, \tag{29}$$

or the stronger one $\ker(B'(\omega; \cdot)) = \{v \in V; b(\omega; u, v) = 0 \ \forall u \in U\} = \{0\}$, then (27) has a unique solution that satisfies the stability estimate $\|u(\omega)\|_U \leq \frac{1}{\gamma_\omega} \|l\|_{V'}$, where $B'(\omega; \cdot) : (V')' = V \ni v \to \langle \cdot, B'(\omega; \cdot) v \rangle_{V' \times V} = b(\omega; \cdot, v) \in U'$, because V is reflexive as a Hilbert space. Moreover $B(\omega; \cdot)$ is the isomorphism (see e.g. Babuška [1] and references inside). Next we assume that

$$\exists \gamma, M; \ +\infty > M = \sup_{\omega \in \mathcal{D}} \{M_\omega\}, \ 0 < \gamma = \inf_{\omega \in \mathcal{D}} \{\gamma_\omega\} \tag{30}$$

Moreover, we assume that $\mathcal{D}$ is bounded and that

$$\exists C_F > 0; \|B(\omega; u) - B(\omega'; u)\|_{V'} \leq C_F \|u\|_U \|\omega - \omega'\|_\mathcal{D} \ \forall u \in U, \ \forall \omega, \omega' \in \mathcal{D}. \tag{31}$$

In the sequel we shall use the following notation

$$\begin{aligned}
T(\omega; \cdot) &= R_V^{-1} B(\omega; \cdot) : U \to V, \\
d(\omega; u, v) &= b(\omega; u, T(\omega; v)), \ \ D(\omega; u) : U \to U'; \\
\langle D(\omega; u), v \rangle_{U' \times U} &= d(\omega; u, v) = b(\omega; u, T(\omega; v)) \ \forall u, v \in U.
\end{aligned} \tag{32}$$

Proposition 3. *The family of operators* $\{D(\omega;\cdot)\}. \omega \in \mathcal{D}$ *satisfies the assumptions of Proposition 2.*

Proof. Because $\langle D(\omega;u),v \rangle_{U' \times U} = b(\omega;u,T(\omega;v))$ than $D(\omega;\cdot)$ is linear and continuous for every $\omega \in \mathcal{D}$, which satisfies the assumption 1.

Following Remark 4 and the assumption (30) we have

$$\langle D(\omega;u),u \rangle_{U' \times U} = d(\omega;u,u) \geq \gamma_\omega^2 \|u\|^2 \geq \gamma^2 \|u\|^2$$

which proves the uniform coercivity of $D(\omega;\cdot)$ postulated in assumption 2.

Let us evaluate for arbitrary $\omega, \omega' \in \mathcal{D}$, $u,v \in U$ and using (30), (31)

$$
\begin{aligned}
|d(\omega;u,v) &- d(\omega';u,v)| = |b(\omega;u,T(\omega,v)) - b(\omega';u,T(\omega';v))| \\
&\leq |b(\omega;u,T(\omega,v)) - b(\omega';u,T(\omega;v))| + |b(\omega';u,T(\omega;v)) - b(\omega';u,T(\omega';v))| \\
&\leq \|B(\omega;u) - B(\omega';u)\|_{V'} \|T(\omega;v)\|_U + |b(\omega';u,T(\omega;v) - T(\omega';v))| \\
&\leq C_F \|\omega - \omega'\|_{\mathcal{D}} \|u\|_U \|T(\omega;v)\|_U + \|B(\omega')\| \|u\|_U \|T(\omega;v) - T(\omega';v)\|_V \\
&\leq C_F \|u\|_U \|\omega - \omega'\|_{\mathcal{D}} \|B(\omega;v)\|_{V'} + \|B(\omega')\| \|u\|_U \|B(\omega;v) - B(\omega';v)\|_{V'} \\
&\leq C_F \|u\|_U \|\omega - \omega'\|_{\mathcal{D}} \|B(\omega)\| \|v\|_V + \|B(\omega')\| \|u\|_U \cdot C_F \|\omega - \omega'\|_{\mathcal{D}} \|v\|_V \\
&\leq C_F (\|B(\omega)\| + \|B(\omega')\|) \|\omega - \omega'\|_{\mathcal{D}} \|u\|_U \|v\|_V
\end{aligned}
$$

(31) can be rewritten as

$$\|B(\omega) - B(\omega')\| \leq C_F \|\omega - \omega'\|_{\mathcal{D}},$$

therefore, since $\mathcal{D}$ is bounded, we have that

$$M_B = \sup_{\omega \in \mathcal{D}} \|B(\omega)\| < +\infty.$$

Hence, we obtain

$$|d(\omega;u,v) - d(\omega';u,v)| \leq 2C_F M_B \|\omega - \omega'\|_{\mathcal{D}} \|u\|_U \|v\|_V. \tag{33}$$

In other words

$$\|D(\omega;u) - D(\omega';u)\|_{V'} \leq 2C_F M_B \|\omega - \omega'\|_{\mathcal{D}} \|u\|_U, \tag{34}$$

which is exactly assumption 3.   □

Finally, taking into account the assumed inequality (26), all assumptions of the Proposition 1 hold, so the evaluation of the misfit relative error (7) is valid with the following constants $A_1 = 3^{s-1} C_C$, $A_3 = A_1 C_u = \frac{3^{s-1} C_C C_B^2}{C_b^2} \|l\|_{U'}$.

Proposition 4. *All the above considerations authorize the conclusion that the common misfit error scaling described in Sect. 2.3 might be applied in twin adaptive solution of the inverse problem (1) where the forward problems (27) are stabilized by Demkowicz operator, using the recommended value of Ratio(j), i.e.* $(A_3(A_1)^{-1})^{\frac{1}{\alpha}}$.

Remark 5. The above Proposition allows us to apply the dynamical accuracy adjustment technique to problems demanding stabilization as well. An important consequence is that also in this case we can expect computational cost savings similar to those mentioned in Remark 1.

5 Numerical Example

In this section we shall describe a simple illustrative computational example. Our intention is to show a hint for a broader class of problems that can be attacked using stabilized FEM coupled with the accuracy-adjusting HMS. However, the problem itself is forward-only, hence it does not use any inverse solver.

Let us consider a simple advection-dominated diffusion-advection equation on a one-dimensional domain $\Omega = (0,1)$:

$$-\varepsilon u'' + u' = 1, \quad \varepsilon \ll 1 \tag{35}$$

with zero Dirichlet boundary conditions. The exact solution

$$u(x) = x - \frac{e^{x/\varepsilon} - 1}{e^{1/\varepsilon} - 1}$$

exhibits a sharp boundary layer near $x = 1$, which necessitates stabilization. As a starting point we use the standard weak formulation with $U = V = H_0^1(\Omega)$: Find $u \in H_0^1(\Omega)$ such that

$$\varepsilon(u', v') + (u', v) = (1, v) \quad \forall v \in H_0^1(\Omega)$$

To discretize the above continuous formulation, we employ the „practical" DPG method described in [10, Definition 31]. Having chosen the discrete trial space U_h and a discrete test space V^s satisfying $\dim U_h < \dim V^s$, we proceed to solve the following equivalent discrete mixed problem [10, Theorem 39]: Find $u_h \in U_h$ and $\psi \in V^s$ such that

$$\begin{aligned}
(\psi', v') + \varepsilon(u', v') + (u', v) &= (1, v) \quad \forall v \in V^s \\
\varepsilon(w', \psi') + (w', \psi) &= 0 \qquad \forall w \in U_h
\end{aligned} \tag{36}$$

In our numerical example we use uniform mesh with 20 elements. As the trial space U_h we use quadratic B-splines space with C^1 global continuity, and as the discrete test space V^s – cubic B-splines with C^0 global continuity. Figure 1 presents the exact solution (5) for $\varepsilon = 1/125$, and approximate solutions using the described stabilization technique and the standard Galerkin method with trial and test spaces equal to U_h. In the vicinity of the boundary layer the Galerkin solution oscillates heavily, while the stabilized solution remains close to the exact solution.

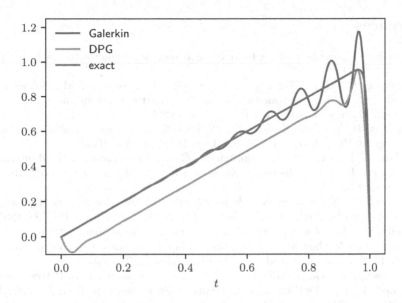

Fig. 1. Exact and approximate solutions of problem (35)

6 Conclusions

The paper refers to the effective stochastic strategy hp–HMS of solving ill conditioned parametric inverse problems. The core of this strategy is a dynamic, common inverse/forward error scaling based on the formula (3). Our earlier papers [2,3,8,9,14,16] verifies this formula and the hp–HMS parameters setting for strongly motivated engineering problems: identification of Lamé coefficients in linear elasticity, inversion of logging data obtained with DC and AC probes and MT data inversion by underground resources investigations. For such cases the associated forward problem (2) have uniformly coercive and Lipschitz continuous operators and can be solved by the Galerkin method with the same test and trial spaces.

Considerations included in Sects. 2–4 allow to extend the above results to more difficult forward problems solved by means of the Petrov-Galerkin method, that needs to be stabilized by the Demkowicz operator. In particular, Proposition 4 specifies new constant values in (3) for the stabilized case and makes it possible to apply the same algorithmic solutions as in non-stabilized cases. The stabilization of the forward problem using the Demkowicz operator can be combined with the standard hp-adaptivity [12], making it appropriate for the use in double-adaptive inverse solvers. Most of the other observations concerning hp-HMS behavior also remain valid.

Our future work shall involve the application of the presented methods in the solution of real-world inverse problems.

References

1. Babuška, I.: Error-bounds for finite element method. Numerische Mathematic **16**, 322–333 (1971)
2. Barabasz, B., Gajda-Zagórska, E., Migórski, S., Paszyński, M., Schaefer, R., Smołka, M.: A hybrid algorithm for solving inverse problems in elasticity. Int. J. Appl. Math. Comput. Sci. **24**(4), 865–886 (2014)
3. Barabasz, B., Migórski, S., Schaefer, R., Paszyński, M.: Multi-deme, twin adaptive strategy hp-HGS. Inverse Probl. Sci. Eng. **19**(1), 3–16 (2011)
4. Beilina, L., Klibanov, M.V.: Approximate Global Convergence and Adaptivity for Coefficient Inverse Problems. Springer, Boston (2012). https://doi.org/10.1007/978-1-4419-7805-9
5. Bramwell, J.: A Discontinuous Petrov-Galerkin Method for Seismic Tomography Problems. Ph.D. thesis, The University of Texas at Austin, Austin, USA (2013)
6. Demkowicz, L., Gopalakrishnan, J.: A class of discontinuous Petrov-Galerkin method. Part I: The transport equation. Comput. Methods Appl. Mech. Eng. **199**, 1558–1572 (2010). https://doi.org/10.1016/j.cma.2010.01.003
7. Demkowicz, L., Gopalakrishnan, J.: A class of discontinuous Petrov-Galerkin method. Part II: Optimal test functions. Numer. Methods Partial Differ. Equ. **27**, 70–105 (2011). https://doi.org/10.1002/num.20640
8. Faliszewski, P., Sawicki, J., Schaefer, R., Smołka, M.: Multiwinner voting in genetic algorithms. IEEE Intell. Syst. **32**(1), 40–48 (2017). https://doi.org/10.1109/MIS.2017.5
9. Gajda-Zagórska, E., Schaefer, R., Smołka, M., Paszyński, M., Pardo, D.: A hybrid method for inversion of 3D DC logging measurements. Natural Comput. **14**(3), 355–374 (2014). https://doi.org/10.1007/s11047-014-9440-y
10. Gopalakrishnan, J.: Five lectures on DPG methods. Master's thesis, Portland State University (2013)
11. Łoś, M., Smołka, M., Schaefer, R., Sawicki, J.: Misfit landforms imposed by ill-conditioned inverse parametric problems. Comput. Sci. **19** (2018). https://doi.org/10.7494/csci.2018.19.2.2781
12. Petrides, S., Demkowicz, L.: An adaptive DPG method for high frequency time-harmonic wave propagation problems. report 16–20. Technical report. The Institute for Computational Engineering and Sciences, The University of Texas at Austin (2016)
13. Preuss, M.: Multimodal Optimization by Means of Evolutionary Algorithms. Natural Computing, Springer, Cham (2015). https://doi.org/10.1007/978-3-319-07407-8
14. Sawicki, J., Łoś, M., Smołka, M., Schaefer, R., Álvarez-Aramberri, J.: Approximating landscape insensitivity regions in solving ill-conditioned inverse problems. Memetic Comput. **10**(3), 279–289 (2018). https://doi.org/10.1007/s12293-018-0258-5
15. Schaefer, R., Kołodziej, J.: Genetic search reinforced by the population hierarchy. In: Foundations of Genetic Algorithms, vol. 7, pp. 383–399. Morgan Kaufman (2003)
16. Smołka, M., Gajda-Zagórska, E., Schaefer, R., Paszyński, M., Pardo, D.: A hybrid method for inversion of 3D AC logging measurements. Appl. Soft Comput. **36**, 422–456 (2015)

Supermodeling - A Meta-procedure for Data Assimilation and Parameters Estimation

Leszek Siwik[✉], Marcin Łoś, and Witold Dzwinel

Department of Computer Science, AGH University of Science and Technology,
Krakow, Poland
siwik@agh.edu.pl

Abstract. The supermodel synchronizes several imperfect instances of a baseline model - e.g., variously parametrized models of a complex system - into a single simulation engine with superior prediction accuracy. In this paper, we present convincing pieces of evidence in support of the hypothesis that supermodeling can be also used as a meta-procedure for fast data assimilation (DA). Thanks ago, the computational time of parameters' estimation in multi-parameter models can be radically shortened. To this end, we compare various supermodeling approaches which employ: (1) three various training schemes, i.e., *nudging*, *weighting* and *assimilation*, (2) three classical data assimilation algorithms, i.e., ABC-SMC, 3DVAR, simplex method, and (3) various coupling schemes between dynamical variables of the ensembled models. We have performed extensive tests on a model of diversified cancer dynamics in the case of tumor growth, recurrence, and remission. We demonstrated that in all the configurations the supermodels are radically more efficient than single models trained by using classical DA schemes. We showed that the tightly coupled supermodel, trained by using the *nudging* scheme synchronizes the best, producing the efficient and the most accurate prognoses about cancer dynamics. Similarly, in the context of the application of supermodeling as the meta-algorithm for data assimilation, the classical 3DVAR algorithm appeared to be the most efficient baseline DA scheme for both the supermodel training and pre-training of the sub-models.

Keywords: Supermodeling · Data assimilation · Tumor dynamics

1 Introduction – The Concept of Supermodeling

The assimilation of the computer model with a real phenomenon through a set of observations is a complex task and its time complexity increases exponentially with the number of parameters. This makes data assimilation (DA) procedures useless when applied for multiscale models such as models of weather dynamics [26, 28] or biological processes like tumor growth [2]. Usually, such the multiscale models are highly parametrized.

© Springer Nature Switzerland AG 2021
M. Paszynski et al. (Eds.): ICCS 2021, LNCS 12743, pp. 358–372, 2021.
https://doi.org/10.1007/978-3-030-77964-1_28

It is well known from the literature that a multi-model ensemble produces more accurate prognoses than a single-model forecast [25]. Consequently, simultaneous estimation of the model parameters for the ensemble Kalman filter (EnKF) is more efficient and accurate than for single data assimilation (DA) algorithm [3]. On the other hand, averaging trajectories from multiple models without synchronization may lead to undesired smoothing and variance reduction [7]. The alternative approach for taking advantage from the many trajectories followed by distinctive models and discovering many "basins of attraction" but without (premature) loss of the trajectories variety is combining models dynamically. The naive approach of this kind was proposed in [10]. The more mature solution was presented in [28] and consists in *dynamic combination of sub-models by introducing connection terms into the model equations that "nudge" the state of one model to the state of every other model in the ensemble, effectively forming a new dynamical system with the values of the connection coefficients as additional parameters.* In our previous paper [19] we posed a hypothesis that by combining a few imperfect, previously pre-trained, dynamical models and synchronizing them through the most sensitive dynamical variables one can overcome the problem of exponential inflation of the parameter space with their number [20]. Instead of estimating multiple parameters, one can train only a few coupling factors (hidden layer of data assimilation) of the model ensemble - the supermodel. In [19] we demonstrated that this meta-procedure can be more efficient than a single ABC-SMC classical data assimilation algorithm. However, in [19], we presented a case study, which uses only: (1) one supermodeling coupling scheme and (2) one classical DA algorithm. To make our hypothesis more credible we present here more extensive experiments.

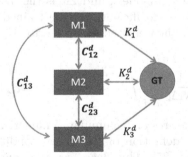

Fig. 1. The supermodeling coupling for 3 sub-models and *ground truth.*

The main contribution of this paper is the comparison of many versions of supermodels employing various training schemes and various classical DA algorithms as the baseline (i.e., for pre-training of the sub-models and training their coupling factors). Additionally, compared to our previous study, we employ a supermodeling scheme for a different and more realistic dynamical model, i.e., the model of tumor dynamics. The results of experiments allow for strengthening considerably the hypothesis about the benefits of using supermodeling as a meta-procedure for data assimilation for improving prediction accuracy.

2   Supermodel Training Schemes

Let us first define the supermodel assumptions [26]. Assuming, that we have $\mu = 1 \ldots M$ sub-models, each described by the $d = 1 \ldots D$ set of dynamical variables, i.e.,

$$\dot{x}_\mu^d = f_\mu^d(x_\mu) \tag{1}$$

the supermodel set of equations looks as follows [28]:

$$\dot{x}_\mu^d = f_\mu^d(x_\mu) + \underbrace{\sum_{v \neq \mu}^{v \neq \mu} C_{\mu v}^d(x_v^d - x_\mu^d)}_{\substack{\text{Synchronization} \\ \text{between sub-models}}} + \underbrace{K_\mu^d(x_{GT}^d - x_\mu^d)}_{\substack{\text{Synchronization} \\ \text{sub-model and the } \textit{ground truth}}} \tag{2}$$

where C_{ij} are the coupling (synchronization) factors between sub-models and K^d are the *nudging* coefficients attracting the models to the *ground truth* (GT) data (see Fig. 1). We have assumed that $K^d = K$. The supermodel behavior is calculated as the average of the sub-models:

$$x_s^d(C, t) \equiv \frac{1}{M} \sum_\mu x_\mu^d(C, t). \tag{3}$$

Additionally, we define the synchronization quality as follows:

$$e^d(t) = \frac{1}{lp} \sum_{(\mu, v)} \frac{1}{N} \sum_{n=0}^{N-1} [x_\mu^d(n\Delta t) - x_v^d(\Delta t)]^2, \tag{4}$$

where N is the number of samples that discretize the trajectory, lp is the number of couplings between the sub-models, and Δt is the discretization interval. As shown in [27], the C tensor coefficients can be trained by using *ground truth* vector x_{GT} by minimizing the weighted squared error $E(C)$ in N following time-steps Δt; i.e.,

$$E(C) = \frac{1}{N\Delta t} \sum_{n=1}^{N} \int_{t_n}^{t_n + \Delta t} |x_s(C, t) - x_{GT}(t)|^2 \gamma^t dt. \tag{5}$$

Error function $E(C)$ measures an accumulated numerical error, which includes the imperfections in the definition of the initial conditions. Discount value γ is from the (0,1) interval [27]. This decreases the contributing factors of the increases in the internal errors.

To develop a fully functional supermodel we have to (1) devise training algorithms for meta-parameters (C_{ij}^d, K_i^d), (2) select a proper set of the sub-models, (3) decide about the connections between them. In this section, we start with the training algorithms.

There exist a few different training algorithms for developing efficient and accurate supermodels. In the forthcoming sections we briefly discuss nudged, assimilated and weighted training schemes described in the previously published papers [19, 26, 28]. Many interesting novel concepts of the supermodels training, in the context of climate/weather forecast, are collected in [17].

2.1 Nudging Supermodel

In this training scheme, during the first stage, the values of C_{ij}^d factors are updated alongside the sub-models according to the following formula [28]:

$$\dot{C}_{\mu v}^d = \alpha(x_\mu^d - x_v^d)(x_{GT}^d - x_\mu^d), \qquad (6)$$

where α is a training constant. The second stage consists in running the coupled sub-models once again but this time C_{ij} are fixed to the values resulting from the first stage while the supermodel trajectory is *nudged* to GT data by correcting the values of K_i^d. The details of this procedure can be found in [27]. As the output of the *nudging* supermodel, we take the average of outputs of the M coupled sub-models obtained during the second stage of training.

2.2 Assimilated Supermodel

This alternative approach, (which is referred to as *assimilated* supermodel) consists in pre-training of the sub-models and estimating coupling factors using well known data assimilation algorithms such as: Kalman filters [3], 3DVAR [23], Blue or Ensemble Blue [1], ABC-SMC [24] or optimization algorithms such as: Evolutionary Algorithms, Differential Evolution [15], Tabu Search [5,6] or classical derivative free optimization techniques [13,14] such as Simplex method [9]. Proposed idea consists in the following steps [19]: (1) Apply a classical data-assimilation algorithm for a short pre-training of a few instances of the baseline model. (2) Create a supermodel from these imperfect sub-models coupled by only the most sensitive dynamical variables. Thus the number of coupling factors k will be small compared to the number of model parameters (e.g., for three sub-models coupled by only one variable $k = 3$). (3) Train these coupling factors using *ground truth* data by applying classical DA algorithm. The nudging coefficients are not required then. So now, the supermodel is described by the following equation:

$$\dot{x}_\mu^i = \underbrace{f_\mu^i(x_\mu)}_{\substack{\text{Pretrained} \\ \text{submodels}}} + \underbrace{\sum_{}^{v \neq \mu} C_{\mu v}^i(x_\mu^i - x_v^i)}_{\text{Submodel coupling}} + \underbrace{K^i(x_{GT} - x_\mu^i)}_{\substack{\text{Synchronization submodels} \\ \text{and ground-truth}}} \qquad (7)$$

What is important, in this procedure the most complex and time-consuming part, i.e., matching all the parameters of the baseline models, is significantly reduced due to a short pre-training only. Moreover, the instances (sub-models) can be created during only one pre-training, selecting the parameters corresponding to the best and the most diverse local minimums of the cost function. The submodels can be also pre-trained parallelly reducing the computational time. In our experiments, shown in the Results section we have assumed that they are calculated in parallel.

The main part of the method is focused then on estimating (only a few) coupling factors. Since now, *nudging* to the *ground truth* is performed by the baseline data assimilation algorithm, so the part of Eq. 2, responsible for nudging the model towards the *ground truth* data, has been eliminated (see formula 7).

2.3   Weighted Supermodel

The weighted sum of sub-models is the next and the most popular model ensembling approach. This time, the procedure is as follows: (1) Apply a classical data assimilation algorithm for short pre-training of a few instances of the over-parametrized baseline models. (2) Ensemble the supermodel as the weighted sum of tentatively pre-trained imperfect sub-models with the weights w_μ matched to GT data by some DA algorithm. Thus the supermodel value x_s^d of the d_{th} dynamic variable is calculated now as:

$$x_s^d = \sum_{\mu=1}^{M} w_\mu x_\mu^d \tag{8}$$

The sum of weights w_μ is normalized to 1. As shown in [28] and [26] the weighted average scheme is equivalent to the *nudging* training scheme for large coupling factors.

3   Model of Tumor Dynamics

For the test case studies we use the model of the tumor (glioma brain cancer) dynamics: its growth and post anti-cancer treatment phases: remission and recurrence [16]. Unlike the *Handy* model considered in [19], the cancer model is more realistic and supported by real data. As shown by Ribba et al. [16] the model can be used in predictive oncology.

Proliferative cancer cells (P) represent fully functional cells that have the ability to multiply. It is also assumed that those cancer cells that are in unfavorable conditions (such as hypoxia or high mechanical pressure) transform into the quiescent cells (Q). Additionally, all of the proliferative and most of the quiescent cells die due to the anti-cancer drug activity (C). The surviving Q cells transform into mutated quiescent cancer cells (QP), which in turn convert to the proliferative (P) cancer cells (in reality even more aggressive ones). This model, though highly simplified, is a good metaphor for the principal processes influencing cancer dynamics during and after treatment. The model is described by the following set of ODEs [16]:

$$\frac{dC}{dt} = -KDE \times C$$

$$\frac{dP}{dt} = \lambda_P \times P(1 - \frac{P^*}{K}) + k_{Q_pP} \times Q_P - k_{PQ} \times P - \gamma_P \times C \times KDE \times P$$

$$\frac{dQ}{dt} = k_{PQ} \times P - \gamma_Q \times C \times KDE \times Q \tag{9}$$

$$\frac{dQ_P}{dt} = \gamma_Q \times C \times KDE \times Q - k_{Q_pP} \times Q_P - \delta_{QP} \times Q_P$$

$$P^* = P + Q + Q_P.$$

At a given moment of time t, the model state vector is $V(t) = [P(t), Q(t), Qp(t), C(t)]$, where $P(t)$, $Q(t)$, $Qp(t)$ are the numbers of proliferating, quiescent and mutated quiescent cancer cells respectively. Assuming that the cancer cells have similar size, based on known average tumor density one can assume that the number of cells is proportional to the volume they occupy. Though, this will be rather a rough estimation knowing the high variability of tumor density from the necrotic center to the tumor surface. The $C(t)$ dynamical variable represents anti-cancer drug concentration during chemotherapy or radiation dose in the case of radiotherapy. The approximate size of the tumor - calculated in the linear scale as MTD (the mean tumor diameter) - can be estimated from the total number of cancer cells $P^* = P + Q + Q_P$, and vice versa.

In [16], seven parameters were defined: λ_P – cell multiplication P; k_{PQ} – transition of P cells to into Q cells; γ_P – dying cells P; γ_Q – dying cells Q; KDE – therapy intensity; $k_{Q_P P}$ – transition of cells Q_P into P; and δ_{QP} – dying cells Q_P. For the initialization of the model, the initial values: P_0, Q_0, and Q_{P0} are given in [16].

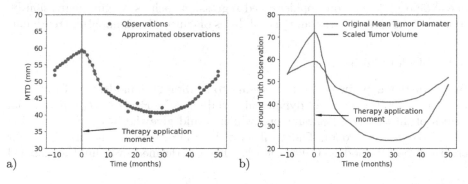

a)   b)

Fig. 2. Data augmentation via approximation of tumor dynamics for a patient with ID = 2 (see [16]) (a), and the comparison of the tumor dynamics measured by MTD and scaled tumor volume (b). (Color figure online)

3.1   *Ground Truth* Data

For the experiments presented and discussed in this paper, the observations of the tumor dynamics for a patient, described in [16] as the patient with ID = 2, have been adopted as the *ground truth* (GT) data. Our choice of the patient was completely random from almost 60 cases described in [16]. As shown in Fig. 2a, data represent the Mean Tumor Diameter (MTD) measurements in time. The Y-axis is scaled in millimeters (MTD) while X-axis (time) in months. The 0 value on X-axis means the start of anticancer therapy. Because glioma tumor has rather a regular shape, MTD can be relatively easy to measure and it reflects well the size of the tumor. In the case of irregular tumors, the cancer volume should be estimated instead. Because, the set of Eqs. 9 describes the tumor time evolution in terms of the number of tumor cells, and assuming that this number is proportional to the volume V, one can easily estimate these volumes from MTD measurements. Therefore, MTD values taken from the Fig. 2a were converted into the volume of the tumor. Assuming that the tumor is a sphere (glioma is rather a cancer of regular spheroidal shape), the conversion has been made according to the Eq. 10:

$$V = \frac{\pi (MTD)^3}{6} \tag{10}$$

The values of V obtained from 10 were scaled by dividing them by 1500 (arbitrary value) so that the volume and corresponding MTD values be of the same order of magnitude. The comparison of the tumor dynamics for the patient with the ID = 2, measured in MTD, and corresponding tumor volumes are shown in Fig. 2b. It is worth mentioning here that the authors of [16] made a mistake assimilating the equations directly to the linear MTD scale. That is why the final results from the paper [16] cannot be credible.

As shown in Fig. 2a, for the purposes of this paper we made a simple data augmentation through the approximation of the measured data (60 points in green in Fig. 2a). In the clinical case of predictive oncology, based on scarce data, one should perform more sophisticated regression such as Kriging regression [8] to predict various scenarios and probabilities of cancer dynamics after treatment.

3.2   Model Calibration

Before further experiments, it is necessary to calibrate the model to GT data and to estimate: initial values of dynamical variables, (C_0, P_0, Q_0, QP_0), and initial values of all the model parameters, which would give the best approximation of the *ground truth* data. These values will be later used as the first guess for creating the sub-models. Calibration has been performed by using the classical

Parameter name	Parameter value
λ_P	0.6801
K	160.14
$K_{Q_P P}$	0.0000001
K_{PQ}	0.41737
γ_P	5.74025
γ_Q	1.343
δ_{QP}	0.6782
KDE	0.09513
C_0	0
P_0	4.7279
Q_0	48.5147
QP_0	0

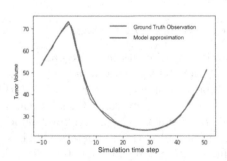

Fig. 3. The cancer model parameters after its calibration to *ground truth* data (left hand side) and the fitting of the calibrated model to *ground truth* for the first 60 time steps (right hand side).

$(\mu + \lambda)$ genetic algorithm (GA) for single-objective optimization. The fitness function has been defined as the root mean square error (RMSE) calculated for the *ground truth data* and tumor dynamics predicted for a given individual. GA has been executed in two phases. In the first one, the genotype of every individual consists of four double-float values representing the initial solutions C_0, P_0, Q_0, QP_0 whereas in the second stage the genotype of every individual consists of eight double-float values representing the parameters of the tumor growth model (i.e., $\lambda_p, K, K_{Q_P P}, K_{PQ}, \gamma_P, \gamma_Q, \delta_{QP}, KDE$). In the experiments, the implementation of GA provided by the pyMOO[1] library has been applied. The calculated values of model parameters alongside the visualization of the tumor growth modeled by the calibrated model during the first 60 time steps are presented in Fig. 3.

4   Experiments

The model of tumor dynamics used as the test case (see Eq. 9) has been selected because: (1) the computational time required for its simulation is reasonably short allowing for numerous repetitions needed in data assimilation and parameters matching procedure; (2) though the model is simple it is non-trivial in terms of the number of parameters; (3) the dynamics of the model can be disturbed anytime by administrating a new treatment what makes its dynamics unpredictable; (4) the model is realistic and can be considerably extended [11,12,21] making it computationally more demanding, what justifies using the supermodeling for its implementation and the use in clinical practice. The experimental verification is focused on the analysis of prediction quality depending on the combination of: (1) the supermodel ensembling technique (*nudging, assimilation, weighted*), (2) the data assimilation and optimization algorithm (ABC-SMC

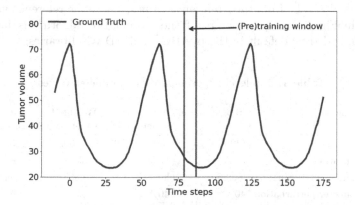

Fig. 4. Tumor dynamics in three drug administration cycles used as the *ground truth* data.

---

[1] https://pymoo.org/algorithms/index.html.

[24], 3DVAR [23], classical derivative free algorithm, i.e., simplex method [9]) and (3) sub-model coupling scheme (i.e., dense or sparse sub-models coupling). For the purposes of this paper, i.e., demonstration of the usefulness and superiority of the supermodeling as a meta-procedure for data assimilation, we have selected a bit unrealistic tumor dynamics case shown in Fig. 4. An oncologist who uses the predictive tool for planning the anti-cancer therapy would like to check all therapy scenarios depending on the tumor behavior. Usually, based on data from the beginning of the therapy, such as in Fig. 2, they simulate further stages of tumor development planning the moments the following drug dose should be administrated to obtain the optimal therapeutic result. For sure, the oncologist would never allow for such an extensive tumor regrowth like that shown in Fig. 4, moreover, he would use all the data from the beginning of therapy administration (not just a fragment which is shown in Fig. 4) for further prediction. However, in our tests, we would like to study what accuracy can we expect for not only the "forward" but also the "backward" predictions. The latter allows for the insight into tumor past what could be very valuable for anti-cancer therapy and for inference of the reasons of the tumor evolution. Taking for our test GT data from the middle of the anti-cancer therapy, and having more complicated "backward" dynamics, we can better estimate the accuracy of the prediction methods we examined. The parameter values presented in Table 1 have been taken arbitrarily based on the preliminary experiments. The *evaluation budget* means the number of evaluations of tumor volume P^* (see Eq. 9) in the subsequent timesteps. In the experiments, the budgets are 30, 50 and 70, respectively. The *pre-training ratio* is the ratio of computational time needed for pretraining of the sub-models (in parallel) to the total training time, i.e., the evaluation budget considered (with training the supermodel coupling coefficients). We have assumed that pre-training is independent and thus can be run in parallel. In the following experiments, this value is set to 40% of the total *evaluation budget*. Initial perturbation means the deviation of the tumor growth model (i.e. λ_p, K, K_{qpp}, K_{pq}, γ_p, γ_q, δ_{qp} and KDE) from their reference values, where the reference values are collected in Fig. 3. The parameters of the assimilation algorithms have been taken as provided by default in the pyABC[2] and ADAO[3] libraries.

Table 1. Top-level experimentation parameter values

Parameter name	Single model	Assimilated super model	Weighted super model	Nudged super model
Evaluation budget	(30, 50, 70)	(30, 50, 70)	(30, 50, 70)	(30, 50, 70)
Pretraining ratio	N/A	40%	40%	40%
(Pre)training window range	[80:86]	[80:86]	[80:86]	[80:86]
Initial parameter perturbation	40%	40%	40%	40%
Number of submodels	N/A	4	4	4

---

[2] https://pyabc.readthedocs.io/en/latest/.
[3] https://docs.salome-platform.org/latest/gui/ADAO/en/index.html#.

4.1   Results

In Tables 2, 3, 4, 5 and 6 the average tumor dynamics percentage prediction errors are presented. The results in individual tables differ with tumor model dynamic variable the sub-models have been coupled by. The best tumor dynamics predictions produced by supermodeling approaches are as follows:

Table 2. Average tumor dynamics prediction errors. Coupling variable: P. (Results are in percentages.)

	Simplex			3DVAR			ABC-SMC		
	30	50	70	30	50	70	30	50	70
Single model	83,64	67,29	58,28	35,40	30,01	22,92	67,29	58,28	46,44
Assimilated supermo	55,72	45,11	42,71	22,40	15,64	13,01	45,11	42,71	34,53
Weighted supermodel	49,51	42,12	39,31	22,23	16,61	16,54	42,12	39,31	29,78
nudged supermodel	30,28	29,66	27,08	22,15	16,96	15,43	30,28	27,08	21,10

Table 3. Average tumor dynamics prediction errors. Coupling variable Q. (Results are in percentages.)

	Simplex			3DVAR			ABC-SMC		
	30	50	70	30	50	70	30	50	70
Single model	78,46	49,57	39,07	40,88	33,22	24,84	57,50	57,09	47,47
Assimilated supermo	50,04	40,81	34,30	24,94	23,72	19,12	42,19	38,45	26,32
Weighted supermodel	34,93	34,33	29,93	22,75	22,10	17,32	51,12	35,91	28,09
nudged supermodel	41,87	24,12	23,84	18,35	17,63	12,75	39,96	31,70	24,58

Table 4. Average tumor dynamics prediction errors. Coupling variable: QP. (Results are in percentages.)

	Simplex			3DVAR			ABC-SMC		
	30	50	70	30	50	70	30	50	70
Single model	56,38	42,96	38,44	43,84	24,10	21,97	72,41	62,04	49,51
Assimilated supermo	37,70	30,51	25,27	21,92	19,54	16,01	44,95	40,47	34,35
Weighted supermodel	34,81	28,81	25,07	21,84	18,72	15,93	43,01	41,94	32,70
nudged supermodel	37,74	30,54	25,32	21,94	19,56	16,03	44,95	40,46	34,35

Table 5. Average tumor dynamics prediction errors. Coupling factor: C. (Results are in percentages.)

	Simplex			3DVAR			ABC-SMC		
	30	50	70	30	50	70	30	50	70
Single model	58,65	51,11	41,59	30,42	23,06	19,34	57,74	51,28	42,03
Assimilated supermo	40,57	33,60	23,03	26,04	20,41	16,01	42,63	36,45	29,01
Weighted supermodel	48,73	32,01	24,24	27,59	22,75	15,71	42,91	37,60	26,38
nudged supermodel	NaN	NaN	NaN	NaN	NaN	NaN	NaN	NaN	NaN

Table 6. Average tumor dynamics prediction errors. Coupling variables: P, Q, QP, C. (Results are in percentages.)

	Simplex			3DVAR			ABC-SMC		
	30	50	70	30	50	70	30	50	70
Single model	58,83	51,91	46,21	31,81	26,03	19,34	65,14	56,14	33,95
Assimilated supermo	35,54	31,73	24,45	25,72	21,16	19,12	49,5	34,19	25,96
Weighted supermodel	30,64	28,24	22,06	19,66	17,12	14,68	41,87	29,67	23,98
nudged supermodel	10,95	9,41	8,69	8,82	8,78	6,22	14,12	9,52	9,35

- when coupled by P – the nudged supermodel with 3DVAR (13,01%),
- when coupled by Q – the nudged supermodel with 3DVAR (12,75%),
- when coupled by QP – the weighted supermodel with 3DVAR (15,93%),
- when coupled by C – the weighted supermodel with 3DVAR (15,71%),
- when coupled by all the dynamic variables – the nudged supermodel with 3DVAR (6,22%).

For comparison, the best single-model predictions have been obtained with 3DVAR as a data assimilation algorithm and the average prediction error was: 21,68%. In Fig. 5 we present the predictions and the accuracy of all compared supermodeling schemes and the fully trained single-model. As one can see, the obtained prediction inaccuracies clearly demonstrates the advantage of the supermodeling approach. Thus, in the case of the single-model (the purple line in Fig. 6) if the DA process of parameters matching is not able to assimilate the model to reality with adequate accuracy, there is no mechanism that would be able to improve the prediction. While for supermodeling the situation is different. Even if the pre-training process assimilates the sub-model parameters very poorly (see the green line in Fig. 6), what gives much worse predictions than the well-trained single-model does, then during the coupling of a sub-model with

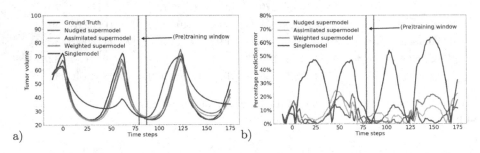

Fig. 5. Tumor dynamics prediction obtained by the single and the supermodels *Left*, and *Right*: the errors comparison

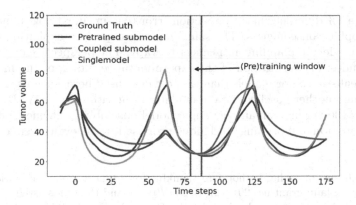

Fig. 6. Tumor dynamics prediction obtained by a pretrained sub-model (green line), the fully trained single-model (purple line) and one of the sub-models from the supermodel (orange line). The supermodel response is the average of the responses of all its member sub-models. (Color figure online)

the other ones, it can be 'corrected' and synchronized with the others. Consequently, the supermodel ensemble, as the average of all the sub-models, is able to produce much better predictions (see the orange line in Fig. 6) than the fully trained single-model.

5   Conclusions and Future Work

In the paper, the dependency of the supermodeling prediction quality on the supermodel ensembling method, the baseline data assimilation algorithm, and the (pre)training budget is analyzed. The combinations of three different supermodeling approaches, i.e., *nudged, assimilated* and *weighted* supermodels with three different data assimilation techniques (ABC-SMC, 3DVAR and SIMPLEX methods) have been analyzed. The tumor dynamics model has been used as the test case and its prediction accuracy has been compared to the generated *ground truth* evolution. The conclusions coming from the presented research are as follows: (1) All the supermodels clearly outperform the single-model accuracy in all the time budgets taken into consideration. The 3DVAR data assimilation algorithm, used as a baseline DA method, gives the highest accuracy. (2) For a single-coupling between the sub-models (only one dynamic variable is used for sub-models coupling) the accuracy of all the supermodel training methods is very similar and does not depend significantly on which specific dynamic variable is used to couple the sub-models. However, for dense connection (the sub-models are interconnected by all the dynamic variables) the average prediction error can be reduced by half and the *nudging* training scheme yields clearly the lowest prediction error. (3) The difference between the best supermodel and single model predictions is high. One should be aware that the more complex model the greater can be the difference. What is important, in many cases even a 1% difference in the prediction error (e.g. in climatology), may result in catastrophic consequences. The same one can expect in exploiting complex 3D tumor models in planning anti-cancer therapy. The perspective of the future works includes: (1) verification of the experimental results applying more complex and realistic tumor models than the test case used here, such as [11,12,21]; (2) applying another baseline data assimilation algorithms as 4DVAR, Kalman filters, etc.; (30 verification another supermodel ensembling schemes such as the "cross-pollination and pruning" technique proposed by Schevenhoven and Selten in [17,18].

Acknowledgement. The Authors are thankful for support from the National Science Centre, Poland grant no. 2016/21/B/ST6/01539 and the funds assigned to AGH University of Science and Technology by the Polish Ministry of Science and Higher Education. This research was supported in part by PLGrid Infrastructure and ACK CYFRONET Krakow. We also thank professor dr Greg Duane, University of Bergen, for his helpful comments and advice.

References

1. Bouttier, B., Courtier, P.: Data assimilation concepts and methods. Meteorological Training Course Lecture Series. European Centre for Medium-Range Weather Forecasts (1999)
2. Dzwinel, W., Kłusek, A., Vasilyev, O.V.: Supermodeling in simulation of melanoma progression. In: International Conference on Computational Science: ICCS 2016, 6–8 June 2016, pp. 999–1010, San Diego, California, USA (2016)
3. Evensen G.: The ensemble Kalman filter: theoretical formulation and practical implementation. In: Seminar on Recent Developments in Data Assimilation for Atmosphere and Ocean, 8–12 September 2003. European Centre for Medium-Range Weather Forecasts (2003)
4. Glover, F.W.: Tabu search. In: Gass, S.I., Harris, C.M. (eds.) Encyclopedia of Operations Research and Management Science, pp. 1537–1544. Springer, Boston (2013). https://doi.org/10.1007/1-4020-0611-X_1034
5. Glover, F.: Tabu search part I. ORSA J. Comput. **1**(3), 190–206 (1989)
6. Glover, F.: Tabu search part II. ORSA J. Comput. **2**(1), 4–32 (1990)
7. Hazeleger, W., van den Hurk, B., Min, E.: Tales of future weather. Nat. Clim. Change **5**, 107–113 (2015)
8. Hengl, T., Heuvelink, G.B., Rossiter, D.G.: About regression-kriging: from equations to case studies. Comput. Geosci. **33**(10), 1301–1315 (2007)
9. Karloff, E.: The simplex algorithm. Linear Program. 23–47 (1991)
10. Kirtman, B.P., Min, D., Schopf, P.S., Schneider, E.K.: A New approach for coupled GCM sensitivity studies. Technical Report. The Center for Ocean-Land-Atmosphere Studies (2003)
11. Klusek, A., Los, M., Paszynski, M., Dzwinel, W.: Efficient model of tumor dynamics simulated in multi-GPU environment. Int. J. High Perform. Comput. Appl. **33**(3), 489–506 (2019)
12. Łoś, M., Paszyński, M., Kłusek, A., Dzwinel, W.: Application of fast isogeometric L2 projection solver for tumor growth simulations. Comput. Methods Appl. Mech. Eng. **306**, 1257–1269 (2017)

13. Powell, M.J.D.: An efficient method for finding the minimum of a function of several variables without calculating derivatives. Comput. J. **7**(2), 155–162 (1964)
14. Powell, M.J.D.: A direct search optimization method that models the objective and constraint functions by linear interpolation. In: In: Gomez, S., Hennart, J.P. (eds.) Advances in Optimization and Numerical Analysis, pp. 51–67. Springer, Netherlands (1994). https://doi.org/10.1007/978-94-015-8330-5_4
15. Qing, A.: Fundamentals of different evolution in differential evolution. In: Fundamentals and Applications in Electrical Engineering, pp. 41–60. Wiley-IEEE Press, September 2009
16. Ribba, B., Kaloshi, G., et al.: A Tumor growth inhibition model for low-grade glioma treated with chemotherapy or radiotherapy. Clin. Cancer Res. **18**(18), 5071–5080 (2012)
17. Schevenhoven, F.: Training of supermodels-in the context of weather and climate forecasting. Ph.D. thesis, University of Bergen, Norway (2021)
18. Schevenhoven, F.J., Selten, F.M.: An efficient training scheme for supermodels. Earth Syst. Dyn. **8**(2), 429–437 (2017)
19. Sendera, M., Duane, G.S., Dzwinel, W.: Supermodeling: the next level of abstraction in the use of data assimilation. In: Krzhizhanovskaya, V.V., et al. (eds.) ICCS 2020. LNCS, vol. 12142, pp. 133–147. Springer, Cham (2020). https://doi.org/10.1007/978-3-030-50433-5_11
20. Siwik, L., Los, M., Klusek, A., Dzwinel, W., Paszyński, M.: Tuning two-dimensional tumor growth simulations. In: Proceedings of 2018 Summer Computer Simulation Conference SCSC 2018, ACM Conference Series, 9–12 July 2018, Bordeoaux, France, p. 8 (2018)
21. Siwik, L., Los, M., Klusek, A., Pingali, K., Dzwinel, W., Paszynski, M.: Supermodeling of tumor dynamics with parallel isogeometric analysis solver. arXiv preprint arXiv:1912.12836 (2019)
22. Suganthan, P.N.: Differential evolution algorithm: recent advances. In: Dediu, A.-H., Martín-Vide, C., Truthe, B. (eds.) TPNC 2012. LNCS, vol. 7505, pp. 30–46. Springer, Heidelberg (2012). https://doi.org/10.1007/978-3-642-33860-1_4
23. Talagrand, O.: Assimilation of observations, an introduction (special issue on data assimilation in meteorology and oceanography: theory and practice). J. Meteorol. Soc. Jpn. **75**(1B), 191–209 (1997)
24. Toni, T., Welch, D., Strelkowa, N., Ipsen, A., Stumpf, M.P.H.: Approximate Bayesian computation scheme for parameter inference and model selection in dynamical systems. J. R. Soc. Interface **6**(31), 187–202
25. Weigel, A.P., Liniger, M.A., Appenzeller, C.: Can multi-model combination really enhance the prediction skill of probabilistic ensemble forecasts? Q. J. R. Meteorol. Soc. **134**(630), 241–260 (2008)
26. Wiegerinck, W., Burgers, W., Selten, F.: On the limit of large couplings and weighted averaged dynamics. In: Kocarev, L. (eds.) Consensus and Synchronization in Complex Networks, pp. 257–275. Springer, Berlin (2013). https://doi.org/10.1007/978-3-642-33359-0_10
27. Wiegerinck, W., Mirchev, M., Burgers, W., Selten, F.: Supermodeling dynamics and learning mechanisms. In: Kocarev, L. (eds.) Consensus and Synchronization in Complex Networks, pp. 227–255. Springer, Heidelberg (2013). https://doi.org/10.1007/978-3-642-33359-0_9
28. Wiegerinck, W., Selten, F.M.: Attractor learning in synchronized chaotic systems in the presence of unresolved scales. Chaos Interdisc. J. Nonlinear Sci. **27**(12), 1269–2001 (2017)

29. van den Berge, L.A., Selten, F.M., Wiegerinck, W., Duane, G.S.: A multi-model ensemble method that combines imperfect models through learning. Earth Syst. Dyn. **2**(1), 161–177 (2011)
30. Yamanaka, A., Maeda, Y., Sasaki, K.: Ensemble Kalman filter-based data assimilation for three-dimensional multi-phase-field model: estimation of anisotropic grain boundary properties. Mater. Des. **165**, 107577 (2019)

AI-Accelerated CFD Simulation Based on OpenFOAM and CPU/GPU Computing

Krzysztof Rojek[1]([⊠]) [iD], Roman Wyrzykowski[1] [iD], and Pawel Gepner[2] [iD]

[1] Czestochowa University of Technology, Czestochowa, Poland
{krojek,roman}@icis.pcz.pl
[2] Warsaw University of Technology, Warszawa, Poland

Abstract. In this paper, we propose a method for accelerating CFD (computational fluid dynamics) simulations by integrating a conventional CFD solver with our AI module. The investigated phenomenon is responsible for chemical mixing. The considered CFD simulations belong to a group of steady-state simulations and utilize the MixIT tool, which is based on the OpenFOAM toolbox. The proposed module is implemented as a CNN (convolutional neural network) supervised learning algorithm. Our method distributes the data by creating a separate AI sub-model for each quantity of the simulated phenomenon. These sub-models can then be pipelined during the inference stage to reduce the execution time or called one-by-one to reduce memory requirements.

We examine the performance of the proposed method depending on the usage of the CPU or GPU platforms. For test experiments with varying quantities conditions, we achieve time-to-solution reductions around a factor of 10. Comparing simulation results based on the histogram comparison method shows the average accuracy for all the quantities around 92%.

Keywords: AI acceleration for CFD · Convolutional neural networks · Chemical mixing · 3D grids · OpenFOAM · MixIT · CPU/GPU computing

1 Introduction

Machine learning and artificial intelligence (AI) methods have become pervasive in recent years due to numerous algorithmic advances and the accessibility of computational power [1,6]. In computational fluid dynamics (CFD), these methods have been used to replace, accelerate or enhance existing solvers [12]. In this work, we focus on the AI-based acceleration of a CFD tool used for chemical mixing simulations.

The authors are grateful to the byteLAKE company for their substantive support. We also thank Valerio Rizzo and Robert Daigle from Lenovo Data Center and Andrzej Jankowski from Intel for their support.

M. Paszynski et al. (Eds.): ICCS 2021, LNCS 12743, pp. 373–385, 2021.
https://doi.org/10.1007/978-3-030-77964-1_29

Chemical mixing is a critical process used in various industries, such as pharmaceutical, cosmetic, food, mineral, and plastic ones. It can include dry blending, emulsification, particle size reduction, paste mixing, and homogenization to achieve your desired custom blend [5].

We propose a collection of domain-specific AI models and a method of integrating them with the stirred tank mixing analysis tool called MixIT. MixIT [10] provides deep insights solutions to solve scale-up and troubleshooting problems. The tool utilizes the OpenFOAM toolbox [13] for meshing, simulation, and data generation. It allows users to design, simulate and visualize phenomena of chemical mixing. More detailed, MixIT provides geometry creation, performs 3-dimensional (3D) CFD flow simulations for stirred reactors, including tracer simulations and heat transfer analysis. Moreover, it allows you to get performance parameters, such as mixing intensity, power per unit volume, blend time, critical suspension speed, gas hold-up, and mass transfer coefficients.

Our goal is to provide an interaction between AI and CFD solvers for much faster analysis and reduced cost of trial & error experiments. The scope of our research includes steady-state simulations, which use an iterative scheme to progress to convergence. Steady-state models perform a mass and energy balance of a process in an equilibrium state, independent of time [2]. In other words, we assume that a solver calculates a set of iterations to achieve the convergence state of the simulated phenomenon. Whence, our method is responsible for predicting the convergence state with the AI models based on a few initial iterations generated by the CFD solver. In this way, we do not need to calculate intermediate iterations to produce the final result, so the time-to-solution is significantly reduced. The proposed AI models make it possible to run many more experiments and better explore the design space before decisions are made.

The contributions of this work are as follows:

- AI-based method that is integrated with a CFD solver and significantly reduces the simulation process by predicting the convergence state of simulation based on initial iterations generated by the CFD solver;
- method of AI integration with the MixIT tool that supports complex simulations with size of $\approx$ 1 million cells based on the OpenFOAM toolbox and high performance computing with both CPUs and graphic processing units (GPUs);
- performance and accuracy analysis of the AI-accelerated simulations.

2 Related Work

Acceleration of CFD simulations is a long-standing problem in many application domains, from industrial applications to fluid effects for computer graphics and animation.

Many papers are focused on the adaptation of CFD codes to hardware architectures exploring modern compute accelerators such as GPU [11,14,15], Intel Xeon Phi [19] or field-programmable gate array (FPGA) [16]. Building a simulator can entail years of engineering effort and often must trade-off generality for accuracy in a narrow range of settings. Among the main disadvantages of such approaches are requirements of in-depth knowledge about complex and extensive CFD codes, expensive and long-term process of portability across new hardware platforms, and, as a result, relatively low-performance improvement compared with the original CFD solver. In many cases, only a small kernel of the solver is optimized.

Recent works have addressed increasing computational performance of CFD simulations by implementing generalized AI models able to simulate various use cases and geometries of simulations [9,12]. It gives the opportunity of achieving lower cost of trial & error experiments, faster prototyping, and parametrization. Current AI frameworks support multiple computing platforms that provide code portability with minimum additional effort.

More recently - and most related to this work - some authors have regarded the fluid simulation process as a supervised regression problem. In [7], the authors present a novel generative model to synthesize fluid simulations from a set of reduced parameters. A convolutional neural network (CNN) is trained on a collection of discrete, parameterizable fluid simulation velocity fields.

In work [21], J. Thompson et al. propose a data-driven approach that leverages the approximation of deep learning to obtain fast and highly realistic simulations. They use a CNN with a highly tailored architecture to solve the linear system. The authors rephrase the learning task as an unsupervised learning problem. The key contribution is to incorporate loss training information from multiple time-steps and perform various forms of data-augmentation.

In paper [7], the authors show that linear functions are less efficient than their non-linear counterparts. In this sense, deep generative models implemented by CNNs show promise for representing data in reduced dimensions due to their capability to tailor non-linear functions to input data.

Work [9] introduces a machine learning framework for the acceleration of Reynolds-averaged Navier-Stokes to predict steady-state turbulent eddy viscosities, given the initial conditions. As a result, they proposed a framework that is hybridized with machine learning.

In [17], the authors present a general framework for learning simulation and give a single model implementation that yields state-of-the-art performance across a variety of challenging physical domains, involving fluids, rigid solids, and deformable materials interacting with one another.

Our method for AI-accelerated CFD simulations is based on utilizing a set of sub-models that are separately trained for each simulated quantity. This approach allows to reduce the memory requirements and operate on large CFD meshes. The proposed data-driven approach provides a low entry barrier for future researchers since the method can be easily tuned when the CFD solver evolves.

3   Simulation of Chemical Mixing with MixIT Tool

3.1   MixIT: Simulation Tool Based on OpenFOAM

MixIT [10] is the next generation collaborative mixing analysis and scale-up tool designed to facilitate comprehensive stirred tank analysis using lab and plant data, empirical correlations, and advanced 3D CFD models. It combines knowledge management techniques and mixing engineering (science) in a unified environment deployable enterprise-wide.

This tool allows users to solve Euler-Lagrange simulations [8] and momentum transfer from the bubbles to the liquid. The liquid flow is described with the incompressible Reynolds-averaged Navier-Stokes equations using the standard k-ε model.

The generation of 3D grids is performed with the OpenFOAM meshing tool *snappyHexMesh* [8]. For Euler-Lagrange simulations, a separate grid for each working volume is created using the preconfigured settings of MixIT. A mesh of the complete domain is generated, and the working volume is defined by the initial condition of the gas volume fraction with the OpenFOAM toolbox.

3.2   Using MixIT Tool for Simulation of Chemical Mixing

The chemical mixing simulation is based on the standard k-ε model. The goal is to compute the converged state of the liquid mixture in a tank equipped with a single impeller and a set of baffles (Fig. 1). Based on different settings of the input parameters, we simulate a set of quantities, including the velocity vector field U, pressure scalar field p, turbulent kinetic energy k of the substance, turbulent dynamic viscosity mut, and turbulent kinetic energy dissipation rate ϵ.

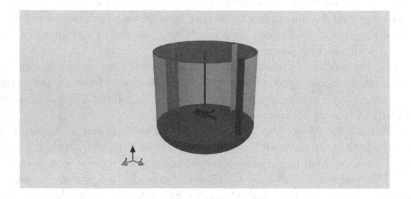

Fig. 1. Scheme of the simulated phenomenon

To simplify the simulation process, we have selected a subset of parameters responsible for the simulation flow. The CFD solver supports many scenarios; however, our research includes three basic case studies:

- assuming the different liquid level of a mixed substance,
- exploring the full range of rotations per minutes (rpm) of the impeller,
- considering different viscosities of the mixed substance.

4   AI-Based Acceleration

4.1   Introduction of AI into Simulation Workflow

Conventional modeling with OpenFOAM involves multiple steps (Fig. 2a). The first step includes pre-processing, where you need to create the geometry and meshing. This step is often carried out with other tools. The next step is the simulation. It is the part that we mainly focus on in this paper by providing the AI-based acceleration. The third step is post-processing (visualization, result analysis).

Our goal is to create solver-specific AI method to ensure the high accuracy of predictions. Our approach belongs to a group of data-driven methods, where we use partial results returned by the CFD solver. The advantage of this method is that it does not require to take into account a complex structure of the simulation, but focus on the data. Such an approach lowers the entry barrier for new CFD adopters compared with other methods, such as a learning-aware approach [4], which is based on the mathematical analysis of solver equations.

Figure 2b presents the general scheme of the AI-accelerated simulation versus the conventional non-AI simulation. It includes (i) the initial results computed by the CFD solver and (ii) the AI-accelerated part executed by the proposed AI module. The CFD solver produces results sequentially iteration by iteration, where each iteration produces intermediate results of the simulation. All intermediate results wrap up into what is called the simulation results. The proposed method takes a set of initial iterations as an input, sends them to our AI module, and generates the final iteration of the simulation. The AI module consists of three stages: (i) data formatting and normalization, (ii) prediction with AI model (inference), and (iii) data export.

Data formatting and normalization translate the data from the OpenFOAM ASCII format to the set of arrays, where each array stores a respective quantity of

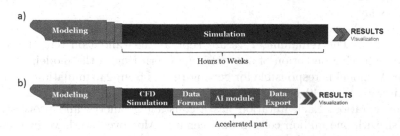

Fig. 2. Scheme of AI-accelerated simulation (b) versus conventional non-AI approach (a)

the simulation (U, p, ϵ, mut, and k). These quantities are normalized depending on a user configuration. The linear normalization is performed based on the following equation:

$$y_i = x_i/max(|maxV|, |minV|) \cdot R, \tag{1}$$

where y_i is the normalized value, x_i is the input value, $maxV$, $minV$ are the maximum and minimum values from all the initial iterations of a given quantity, R is a radius value (in our experiments $R = 1$). When a dataset has a relatively small value of median compared to the maximum value of the set (median value is about 1% of the maximum), then we use a cube normalization with $y_i^* = y_i^3$.

The AI-accelerated simulation is based on supervised learning, where a set of initial iterations is taken as an input and returns the last iteration. For simulating the selected phenomenon with MixIT and conventional non-AI approach, it is required to execute 5000 iterations. At the same time, only the first N_I iterations create the initial iterations that produce input data for the AI module. Moreover, to reduce the memory requirements, the initial dataset is composed of simulation results corresponding to every S_I-th iteration from the N_I initial iterations. The determination of parameters N_I and S_I is the subject of our future work. At this moment, we have empirically selected them by training the AI model with different values of the parameters, up to achieving an acceptable accuracy. For the analyzed phenomenon, we set $N_I = 480$, and $S_I = 20$. As a result, we take the initial data set composed from iterations $20, 40, 60, ..., 480$, so $480/5000 = 0.096$ of the CFD simulation has to be executed to start the inference.

The data export stage includes denormalization of the data and converting them to the OpenFOAM ASCII format. This stage and data formatting one are executed on the CPU, but the prediction with AI model (inference) can be executed on either the CPU or GPU, depending on user preferences.

4.2   Idea of Using AI for Accelerating Simulation

Our neural network is based on the ResNet network [3] organized as residual blocks. In a network with residual blocks, each layer feeds into the next layer and directly into the layers about two hops away. To handle relatively large meshes (about 1 million cells), we have to reduce the original ResNet network to 6 CNN layers.

To train the network, we use 90% of the total data set, referred to as the training data. The remaining 10% are kept as the validation data for model selection, allowing detection of any potential overfitting of the model.

Our AI model is responsible for getting results from 24 simulation iterations (from iterations $20, 40, 60, ..., 480$) as the input, feed the network, and return the final iteration. Each iteration has a constant geometry and processes the 3D mesh with one million cells in our scenario. Moreover, we have five quantities that are taken as the input and returned as the output of the simulation. One of the main challenges here is to store all those data in the memory during the learning. To reduce memory requirements, we create a set of sub-models that independently work on a single quantity. Thanks to this approach, all the sub-models are learned sequentially, which significantly reduces memory requirements. This effect is especially important when the learning process is performed on the GPU.

The proposed strategy also impacts the prediction (inference) part. Since we have a set of sub-models, we can predict the result by calling each sub-model one-by-one to reduce the memory requirements or perform pipeline predictions for each quantity and improve the performance. The created pipelines simultaneously call all the sub-models, where each quantity is predicted independently.

In this way, our method can be executed on the GPU platform (in one-by-one mode), or the CPU platform with a large amount of RAM (in a pipelined mode).

5   Experimental Evaluation

5.1   Hardware and Software Platform

All experiments are performed on the Lenovo platform equipped with two Intel Xeon Gold 6148 CPUs clocked at 2.40 GHz and two NVIDIA V100 GPUs with 16 GB of the global HBM2 memory. The host memory is 400 GB.

For training the models, the CUDA framework v.10.1 with cuDNN v.7.6.5 is used. As a machine learning environment, we utilize TensorFlow v.2.3, the Keras API v.2.4, and python v.3.8. The operating system is Ubuntu 20.04 LTS. For the compilation of codes responsible for data formatting and export from/to OpenFOAM, we use GCC v.9.3. The CFD simulations are executed using the OpenFOAM toolbox v.1906 and MixIT v.4.2.

5.2   Performance and Accuracy Analysis

The first part of the analysis is focused on the accuracy results, while the second one investigates the performance aspects. Since the accuracy evaluation for the regression-based estimation is not so evident as for the classification method, we have selected a set of metrics to validate the achieved results. In the beginning, we compare the contour plots of each quantity created across the XZ cutting plane defined in the center point of the impeller. The results are shown in Figs. 3, 4 and 5, where we can see the converged states computed by the conventional CFD solver (left side) and AI-accelerated approach (right side).

The obtained results show high similarity, especially for the values from the upper bound of the range. Some quantities, such as k, and ϵ have a meager median value (below 1% of the maximum value). As a result, a relatively small error of the prediction impacts the sharp shape of contours for the near-zero values. The higher absolute values, the more smooth shape of the contour plot can be observed.

To estimate the accuracy, we also use a set of statistical metrics. The first two ones are correlation coefficients that measure the extent to which two variables tend to change together. These coefficients describe both the strength and the

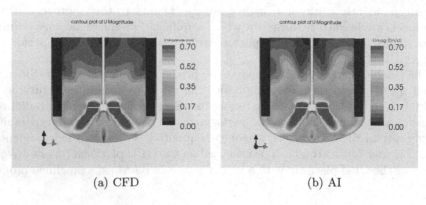

(a) CFD   (b) AI

Fig. 3. Contour plot of the velocity magnitude vector field (U) using either the conventional CFD solver (a) or AI-accelerated approach (b)

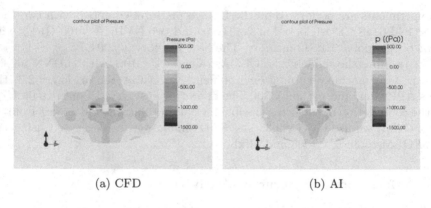

(a) CFD   (b) AI

Fig. 4. Contour plot of the pressure scalar field (p)

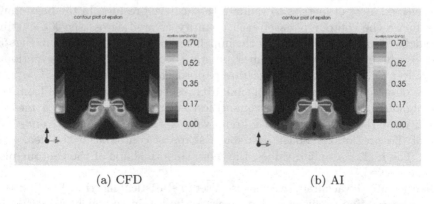

(a) CFD   (b) AI

Fig. 5. Contour plot of turbulent kinetic energy dissipation rate (ϵ)

Table 1. Accuracy results with statistical metrics

Quantity	Pearson's coeff	Spearman's coeff	RMSE	Histogram equaliz. [%]
U	0.99	0.935	0.016	89.1
p	0.993	0.929	0.004	90.1
ϵ	0.983	0.973	0.023	90.3
k	0.943	0.934	0.036	99.4
mut	0.937	0.919	0.147	93.5
Average	0.969	0.938	0.045	92.5

direction of the relationship. Here, we use two coefficients, including the Pearson correlation that estimates the linear relationship between two continuous variables, as well as the Spearman correlation that assesses the monotonic relationship between two continuous or ordinal variables. The Pearson correlation varies from 0.93 for the mut quantity to 0.99 for U and p. The average Pearson correlation for all the quantities is 0.97. It shows a high degree correlation between the CFD (computed by solver) and AI (predicted) values. The Spearman correlation varies from 0.92 to 0.97 with the average value equal to 0.94. It shows a strong monotonic association between the CFD and AI results.

The next statistical metric is the Root Mean Square Error (RMSE). It is the standard deviation of the residuals (differences between the predicted and observed values). Residuals are a measure of how far from the regression line data points are. The implemented data normalization methods ensure that the maximum distance from the X-axis is 1. RMSE varies from 0.004 for the p quantity to 0.15 for the mut quantity. The average RMSE for all the quantities is 0.05. Based on these results, we can conclude that the proposed AI models are well fit.

The last method of accuracy assessment is histogram comparison. In this method, we create histograms for the CFD solver and AI module results and estimate a numerical parameter that expresses how well two histograms match with each other. The histogram comparison is made with the coefficient of determination, which is the percentage of the variance of the dependent variable predicted from the independent variable. The results vary from 89.1% to 99.4%, with an average accuracy of 92.5%. All metrics are included in Table 1.

We have also performed a collective comparison of the results, where we plot a function y(x), where x represents the results obtained from the CFD solver, while y is the prediction. The results are shown in Fig. 6. The black line shows the perfect similarity, while the blue dots reveal the prediction uncertainty.

Now we focus on the performance analysis. We start with comparing the execution time for the AI module executed on the CPU and GPU. In this experiment, the mesh size is one million cells, and the CFD solver is run only on the CPU. The AI-accelerated part is executed in three steps, including data formatting and data export (implemented on CPU), as well as the AI prediction (performed on the CPU or GPU platform). This part includes 90.4% of the sim-

Table 2. Comparison of execution time for CPU and GPU

	CPU	GPU	Ratio GPU/CPU
Data formatting [s]	65.1		
Prediction (inference) [s]	41.6	290.6	7.0
Data export [s]	9.5		
Entire AI module [s]	116.2	365.2	3.1
Full simulation [s]	1483.3	1732.3	1.2

ulation. For the AI-accelerated approach, the full simulation includes 9.6% of all CFD iterations executed by the CFD solver and the AI-accelerated part.

Data formatting takes 65.1 s, while the data export takes 9.5 s. The AI prediction time depends on the selected platform, taking 41.6 s on the CPU and 290.6 s on the GPU platform. So, somewhat surprisingly, the AI-accelerated part (formatting + prediction + export) runs 3.1 times faster on the CPU than in the case when the GPU is engaged. Considering the CFD solver overhead (9.6% of all iterations), we can observe that this is the most time-consuming component of the entire AI-accelerated simulation. So the final speedup of the CPU-based simulation is 1.2 against the case when the GPU is engaged. The performance details are summarized in Table 2, where the execution time for the full simulation (last row) includes executing both the entire AI module and the first 9.6% of the simulation, which takes 1367.2 s.

The reason for the performance loss for the GPU prediction (inference) is a high time of data allocation on the GPU compared with CPU and multiple data transfers between the host and GPU global memory. These transfers are required

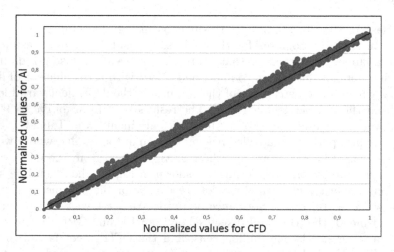

Fig. 6. Comparison of simulation results for the conventional CFD solver and AI-accelerated approach

Table 3. Comparison of execution time for simulation of chemical mixing using either the conventional CFD solver or AI-accelerated approach

	9.6% of sim.	90.4% of sim.	100% of sim.
CFD solver [s]	1367.2	12874.1	14241.2
AI-accelerated (CPU) [s]	1367.2	116.2	1483.3
Speedup	1	110.8	9.6
AI-accelerated (engaging GPU) [s]	1367.2	365.2	1732.3
Speedup	1	35.3	8.2

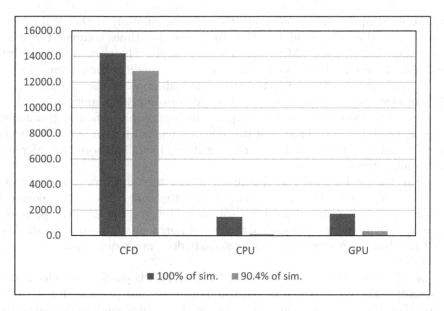

Fig. 7. Advantages in execution time (in seconds) achieved for the AI-accelerated approach over the conventional CFD solver

because each quantity has its sub-model that needs to be sequentially loaded. On the CPU platform, all the predictions of sub-models can be pipelined, and the memory overhead becomes much lower.

The final performance comparison considers the execution time for the conventional CFD solver and AI-accelerated approach. The first 9.6% of the simulation takes 1367.2 s. The remaining part takes 12874.1 s for the conventional CFD solver. Using the AI-accelerated module, this part is computed 110.8 times faster when executed on CPU, and 35.3 times faster when GPU is involved. As result, the entire simulation time is reduced from 14241.2 s to 1483.3 s (9.6× speedup) for the CPU, and to 1732.3 s (8.2× speedup) when the GPU is engaged. These results are summarized in Table 3.

Figure 7 illustrates the performance advantages of the proposed AI-accelerated solution against the conventional CFD solver. Here the blue bars show the execution time for the whole simulation, while the orange ones correspond to executing 90.4% of the simulation. The two bars on the left side correspond to using exclusively the conventional CFD solver, while other bars demonstrate the advantages of using the AI module to reduce the execution time of the simulation. Figure 7 not only illustrates the speedup achieved in this way but also demonstrates that this speedup is finally limited by the time required to perform the initial 9.4% of the simulation using the OpenFOAM CFD solver.

6   Conclusions and Future Works

The proposed approach to accelerate the CFD simulation of chemical mixing allows us to reduce the simulation time by almost ten times compared to using the conventional OpenFOAM CFD solver exclusively. The proposed AI module uses 9.6% of the initial iterations of the solver and predicts the converged state with 92.5% accuracy. It is expected that this reduction in the execution time will translate [20] into decreasing the energy consumption significantly, which means reducing the environmental footprint, including the carbon footprint [18]. However, the reliable evaluation of this effect is the subject of our future work since it requires considering the whole workflow, including both the inference and training stages.

Our method is fully integrated with the MixIT tool and supports 3D meshes with one million cells. Thanks to a data-driven approach, this method does not require a high knowledge of the CFD solvers to integrate it with the proposed solution. Such an integration gives a promising perspective to apply the method for CFD solvers that constantly evolve since it does not require going deep into CFD models.

The AI module is portable across the CPU and GPU platforms, which allows us to utilize the GPU power in the training stage and provides high performance of prediction using the CPU. The proposed pipelined model can separately train each quantity that significantly reduces memory requirements and supports larger meshes on a single node platform.

Our method is still under development. Particular attention will be paid to support more parameters of the CFD simulation of chemical mixing, including shape, position and number of impellers, the shape of the tube, number of baffles, and mesh geometry. Another direction of our research is providing further accuracy improvement and reduce the number of initial iterations required by the AI module.

References

1. Archibald, R., et al.: Integrating deep learning in domain sciences at exascale. arXiv preprint arXiv:2011.11188v1 (2020)
2. Bhatt, D., Zhang, B.W., Zuckerman, D.M.: Steady-state simulations using weighted ensemble path sampling. J. Chem. Phys. **133**(1), 014110 (2010)

3. Chen, D., Hu, F., Nian, G., Yang, T.: Deep residual learning for nonlinear regression. Entropy **22**(2), 193 (2020)
4. Jim, M., Ray, D., Hesthaven, J.S., Rohde, C.: Constraint-aware neural networks for Riemann problems. arXiv preprint arXiv:3327.50678 (2019)
5. Paul, E.L., Atiemo-Obeng, V., Kresta, S.M. (eds.): Handbook of Industrial Mixing: Science and Practice. Wiley, Hoboken (2004)
6. Jouppi, N.P., Young, C., Patil, N., Patterson, D.: A domain-specific architecture for deep neural networks. Commmun. ACM **61**(9), 50–59 (2018)
7. Kim, B., et al.: Deep fluids: a generative network for parameterized fluid simulations. arXiv preprint arXiv:1806.02071v2 (2019)
8. Kreitmayer, D. et al.: CFD-based characterization of the single-use bioreactor XcellerexTM XDR-10 for cell culture process optimization. arXiv preprint arXiv:3461.77983 (2020)
9. Maulik, R., Sharma, H., Patel, S., Lusch, B., Jennings, E.: Accelerating RANS turbulence modeling using potential flow and machine learning. arXiv preprint arXiv:1910.10878 (2019)
10. MixIT: the enterprise mixing analysis tool. https://mixing-solution.com/. Accessed 5 Feb 2021
11. Mostafazadeh, B., Marti, F., Pourghassemi, B., Liu, F., Chandramowlishwaran, A.: Unsteady Navier-Stokes computations on GPU architectures. In: 23rd AIAA Computational Fluid Dynamics Conference (2017). https://doi.org/10.2514/6.2017-4508
12. Obiols-Sales, O., Vishnu, A., Malaya, N., Chandramowlishwaran, A.: CFDNet: a deep learning-based accelerator for fluid simulations. In: Proceedings of the 34th ACM International Conference on Supercomputing (ICS 2020), pp. 1–12. ACM (2020)
13. OpenFOAM: the open source CFD toolbox. https://www.openfoam.com. Accessed 5 Feb 2021
14. Rojek, K., et al.: Adaptation of fluid model EULAG to graphics processing unit architecture. Concurr. Comput. Pract. Exp. **27**(4), 937–957 (2015)
15. Rojek, K., Wyrzykowski, R., Kuczynski, L.: Systematic adaptation of stencil-based 3D MPDATA to GPU architectures. Concurr. Comput. Pract. Exp. **29**(9), e3970 (2017)
16. Rojek, K., Halbiniak, K., Kuczynski, L.: CFD code adaptation to the FPGA architecture. Int. J. High Perform. Comput. Appl. **35**(1), 33–46 (2021)
17. Sanchez-Gonzalez, A., et al.: Learning to simulate complex physics with graph networks. arXiv preprint arXiv:3394.45567 (2020)
18. Schwartz, R., Dodge, J., Smith, N.A., Etzioni, O.: Green AI. Commmun. ACM **63**(12), 54–63 (2020)
19. Szustak, L., Rojek, K., Olas, T., Kuczynski, L., Halbiniak, K., Gepner, P.: Adaptation of MPDATA heterogeneous stencil computation to Intel Xeon Phi coprocessor. Sci. Program., 14 (2015). https://doi.org/10.1155/2015/642705
20. Szustak, L., Wyrzykowski, R., Olas, T., Mele, V.: Correlation of performance optimizations and energy consumption for stencil-based application on Intel Xeon scalable processors. IEEE Trans. Parallel Distrib. Syst. **31**(11), 2582–2593 (2020)
21. Tompson, J., Schlachter, K., Sprechmann, P., Perlin, K.H.: Accelerating Eulerian fluid simulation with convolutional networks. In: ICML2017: Proceedings of the 34th International Conference on Machine Learning, PLMR 70, pp. 3424–3433 (2017)

An Application of a Pseudo-Parabolic Modeling to Texture Image Recognition

Joao B. Florindo$^{(\boxtimes)}$ [iD] and Eduardo Abreu [iD]

Institute of Mathematics, Statistics and Scientific Computing,
University of Campinas, Rua Sérgio Buarque de Holanda, 651, Cidade Universitária
"Zeferino Vaz" - Distr. Barão Geraldo, 13083-859 Campinas, SP, Brazil
{florindo,eabreu}@unicamp.br
http://www.ime.unicamp.br

Abstract. In this work, we present a novel methodology for texture image recognition using a partial differential equation modeling. More specifically, we employ the pseudo-parabolic equation to provide a dynamics to the digital image representation and collect local descriptors from those images evolving in time. For the local descriptors we employ the magnitude and signal binary patterns and a simple histogram of these features was capable of achieving promising results in a classification task. We compare the accuracy over well established benchmark texture databases and the results demonstrate competitiveness, even with the most modern deep learning approaches. The achieved results open space for future investigation on this type of modeling for image analysis, especially when there is no large amount of data for training deep learning models and therefore model-based approaches arise as suitable alternatives.

Keywords: Pseudo-parabolic equation · Texture recognition · Image classification · Computational methods for PDEs

1 Introduction

Texture images (also known as visual textures) can be informally defined as those images in which the most relevant information is not encapsulated within one or a limited set of well-defined objects, but rather all pixels share the same importance in their description. This type of image has found numerous application in material sciences [19], medicine [11], facial recognition [15], remote sensing [33], cybersecurity [27], and agriculture [24], to name but a few fields with increasing research activity.

While deep learning approaches have achieved remarkable success in problems of object recognition and variations of convolutional neural networks have prevailed in the state-of-the-art for this task, texture recognition on the other

Supported by São Paulo Research Foundation (FAPESP), National Council for Scientific and Technological Development, Brazil (CNPq), and PETROBRAS - Brazil.

© Springer Nature Switzerland AG 2021
M. Paszynski et al. (Eds.): ICCS 2021, LNCS 12743, pp. 386–397, 2021.
https://doi.org/10.1007/978-3-030-77964-1_30

hand still remains a challenging problem and the classical paradigm of local image encoders still is competitive with the most modern deep neural networks, presenting some advantages over the last ones, like the fact that they can work well even when there is little data available for training.

In this context, here we present a local texture descriptor based on the action of an operator derived from the Buckley-Leverett partial differential equation (PDE) (see [1, 2] and references cited therein). PDE models have been employed in computer vision at least since the 1980's, especially in image processing. The scale-space theory developed by Witkin [31] and Koenderink [17] are remarkable examples of such applications. The anisotropic diffusion equation of Perona and Malik [23] also represented great advancement in that research front, as it solved the problem of edge smoothing, common in classical diffusion models. Evolutions of this model were later presented and a survey on this topic was developed in [30].

Despite these applications of PDEs in image processing, substantially less research has been devoted to recognition. As illustrated in [4,29], pseudo-parabolic PDEs are promising models for this purpose. An important characteristic of these models is that jump discontinuities in the initial condition are replicated in the solution [10]. This is an important feature in recognition as it allows some control over the smoothing effect and would preserve relevant edges, which are known to be very important in image description.

Based on this context, we propose the use of a pseudo-parabolic equation as an operator acting as a nonlinear filter over the texture image. That image is used as initial condition for the PDE problem and the solution obtained by a numerical scheme developed in [1] is used to compose the image representation. The solution at each time is encoded by a local descriptor. Extending the idea presented in [29], here we propose two local features: the sign and the magnitude binary patterns [13]. The final texture descriptors are provided by simple concatenation of histograms over each time.

The effectiveness of the proposed descriptors is validated on the classification of well established benchmark texture datasets, more exactly, KTH-TIPS-2b [14] and UIUC [18]. The accuracy is compared with the state-of-the-art in texture recognition, including deep learning solutions, and the results demonstrate the potential of our approach, being competitive with the most advanced solutions recently published on this topic.

This paper is structured as follows. In Sect. 2, we present the pseudo-parabolic partial differential equation we are considering to texture image recognition, along with the key aspects of the discretization method being used. In Sect. 3, we introduce the proposed methodology for the application of a pseudo-parabolic modeling to texture datasets recognition. We also highlight the effectiveness of the proposed descriptors and validate them on the classification of well established benchmark and state-of-the-art databases in texture recognition. In Sect. 4, we present some numerical experiments to show the efficiency and accuracy of the computed solutions verifying that the results demonstrate the potential of our approach, being competitive with the most advanced solutions recently available in the specialized literature on this subject matter. In Sect. 5, we concrete to present some results and discussion for the proposed descriptors in the classification of KTH-TIPS-2b and UIUC. We discuss in particular the accuracy of the proposed descriptors compared with other texture descriptors in the literature, confirming its potential as a texture image model. Finally, in the last Sect. 6, we present our concluding remarks.

2 Partial Differential Equation and Numerical Modeling

We consider an advanced simulation approach for the pseudo-parabolic PDE

$$\frac{\partial u}{\partial t} = \nabla \cdot \mathbf{w}, \qquad \text{where} \qquad \mathbf{w} = g(x,y,t) \nabla \left(u + \tau \frac{\partial u}{\partial t} \right), \tag{1}$$

and let $\Omega \subset \mathbb{R}^2$ denote a rectangular domain and $u(\cdot, \cdot, t) : \Omega \to \mathbb{R}$ be a sequence of images that satisfies the pseudo-parabolic equation (1), in which the original image at $t = 0$ corresponds to the initial condition, along with zero flux condition across the domain boundary $\partial\Omega$, $\mathbf{w} \cdot \mathbf{n} = 0$, $(x, y) \in \partial\Omega$. We consider the discretization modeling of the PDE (1) in a uniform partition of Ω into rectangular subdomains $\Omega_{i,j}$, for $i = 1, \ldots, m$ and $j = 1, \ldots, l$, with dimensions $\Delta x \times \Delta y$. The center of each subdomain $\Omega_{i,j}$ is denoted by (x_i, y_j). Given a final time of simulation T, consider a uniform partition of the interval $[0, T]$ into N subintervals, where the time step $\Delta t = T/N$ is subject to a stability condition (see [28,29] for details). We denote the texture configuration frames in the time levels $t_n = n\Delta t$, for $n = 0, \ldots, N$. Let $U_{i,j}^n$ and $W_{i,j}^{n+1}$ be a finite difference approximations for $u(x_i, y_j, t_n)$ and $\mathbf{w}$, respectively, and both related to the pseudo-parabolic PDE modeling of (1). In the following, we employ a stable cell-centered finite difference discretization in space after applying the backward Euler method in time to (1), yielding

$$\frac{U_{i,j}^{n+1} - U_{i,j}^n}{\Delta t} = \frac{W_{i+\frac{1}{2},j}^{n+1} - W_{i-\frac{1}{2},j}^{n+1}}{\Delta x} + \frac{W_{i,j+\frac{1}{2}}^{n+1} - W_{i,j-\frac{1}{2}}^{n+1}}{\Delta y}. \tag{2}$$

Depending on the application as well as the calibration data and texture parameters upon model (1), we will have linear or nonlinear diffusion models for image processing (see, e.g., [6,23,30]). As a result of this process the discrete problem (2) would be linear-like $\mathbf{A}^n \mathbf{U}^{n+1} = \mathbf{b}^n$ or nonlinear-like $\mathbf{F}(\mathbf{U}^{n+1}) = 0$ and several interesting methods can be used (see, e.g., [2,6,23,28–30]). We would like to point out at this moment that our contribution relies on the PDE modeling of (1) as well as on the calibration data and texture parameters associated to the pseudo-parabolic modeling in conjunction with a fine tunning of the local descriptors for texture image recognition for the pertinent application under consideration. In summary, we have a family of parameter choice strategies that combines pseudo-parabolic modeling with texture image recognition.

Here, we consider the diffusive flux as $g(x, y, t) \equiv 1$, which results (1) to be a linear pseudo-parabolic model. For a texture image classification based on a pseudo-parabolic diffusion model to be processed, we just consider that each subdomain $\Omega_{i,j}$ corresponds to a pixel with $\Delta x = \Delta y = 1$. As we perform an implicit robust discretization in time (backward Euler), we simply choose the time step $\Delta t = \Delta x$ and the damping coefficient $\tau = 5$. More details can be found in [29]; see also [1, 28].

Therefore, this description summarizes the basic key ideas of our computational PDE modeling approach for texture image classification based on a pseudo-parabolic diffusion model (1).

3   Proposed Methodology

Inspired by ideas presented in [29] and significantly extending comprehension on that study, here we propose the development of a family of images $\{u_k\}_{k=1}^{K}$. These images are obtained by introducing the original image u_0 as initial condition for the 2D pseudo-parabolic numerical scheme presented in Sect. 2. u_k is the numerical solution at each time $t = t_k$. Here, $K = 50$ showed to be a reasonable balance between computational performance and description quality.

Following that, we collected two types of local binary descriptors [13] from each u_k. More exactly, we used sign $LBP\_S_{P,R}^{riu2}$ and magnitude $LBP\_M_{P,R}^{riu2}$ descriptors. In short, the local binary sign pattern $LBP\_S_{P,R}^{riu2}$ for each image pixel with gray level g_c and whose neighbor pixels at distance R have intensities g_p $(p = 1, \cdots, P)$ is given by

$$LBP\_S_{P,R}^{riu2} = \begin{cases} \sum_{p=0}^{P-1} H(g_p - g_c)2^p & \text{if } \mathcal{U}(LBP_{P,R}) \geq 2 \\ P+1 & \text{otherwise,} \end{cases} \quad (3)$$

where H corresponds to the Heaviside step function ($H(x) = 1$ if $x \geq 0$ and $H(x) = 0$ if $x < 0$) and $\mathcal{U}$ is the uniformity function, defined by

$$\mathcal{U}(LBP_{P,R}) = |H(g_{P-1}-g_c)-H(g_0-g_c)|+\sum_{p=1}^{P-1}|H(g_p-g_c)-H(g_{p-1}-g_c)|. \quad (4)$$

Similarly, the magnitude local descriptor is defined by

$$LBP\_M_{P,R}^{riu2} = \begin{cases} \sum_{p=0}^{P-1} t(|g_p - g_c|, C)2^p & \text{if } \mathcal{U}(LBP_{P,R}) \geq 2 \\ P+1 & \text{otherwise,} \end{cases} \quad (5)$$

where C is the mean value of $|g_p - g_c|$ over the whole image and t is a threshold function, such that $t(x, c) = 1$ if $x \geq c$ and $t(x, c) = 0$, otherwise.

Finally, we compute the histogram h of $LBP\_S_{P,R}^{riu2}(u_k)$ and $LBP\_M_{P,R}^{riu2}(u_k)$ for the following pairs of (P, R) values: $\{(1, 8), (2, 16), (3, 24), (4, 24)\}$. The proposed descriptors can be summarized by

$$\mathfrak{D}(u_0) = \bigcup_{\substack{\text{type}=\{S,M\}}} \bigcup_{\substack{(P,R)=\{(8,1), \\ (16,2),(24,3), \\ (24,4)\}}} \bigcup_{k=0}^{K} h(LBP\_type_{P,R}^{riu2}(u_k)). \quad (6)$$

Final descriptors correspond to the concatenated histograms in (6). Figure 1 shows the concatenated histograms for one time ($t = 1$) of the PDE operator. To reduce the dimensionality of the final descriptors, we also apply Karhunen-Loève transform [22] before their use as input to the classifier algorithm. The diagram depicted in Fig. 2 illustrates the main steps involved in the proposed algorithm.

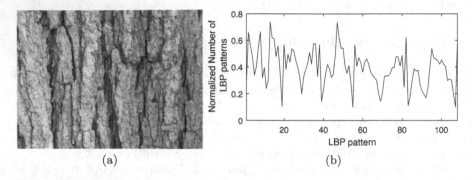

(a)   (b)

Fig. 1. Histograms used to compose the final descriptors obtained from the PDE operator at $t = 1$. (a) Original texture. (b) Normalized histograms.

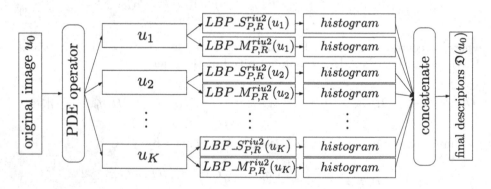

Fig. 2. Main steps of the proposed method.

4   Experiments

The performance of the proposed descriptors is assessed on the classification of two well-established benchmark datasets of texture images, namely, KTH-TIPS-2b [14] and UIUC [18].

KTH-TIPS-2b is a challenging database focused on the real material depicted in each image rather than on the texture instance as most classical databases. In this way, images collected under different configurations (illumination, scale and pose) should be part of the same class. The database comprises a total of 4752 color textures with resolution 200×200 (here they are converted to gray scales), equally divided into 11 classes. Each class is further divided into 4 samples (each sample corresponds to a specific configuration). We adopt the most usual (and most challenging) protocol of using one sample for training and the remaining three samples for testing.

UIUC is a gray-scale texture dataset composed by 1000 images with resolution 256×256 evenly divided into 25 classes. The images are photographed under uncontrolled natural conditions and contain variation in illumination, scale, perspective and albedo. For the training/testing split we also follow the most usual protocol, which consists in half of the images (20 per class) randomly selected for training and the remaining half for testing. This procedure is repeated 10 times to allow the computation of an average accuracy.

For the final step of the process, which is the machine learning classifier, we use Linear Discriminant Analysis [12], given its easy interpretation, absence of hyper-parameters to be tuned and known success in this type of application [29].

5   Results and Discussion

Figures 3 and 4 show the average confusion matrices and accuracies (percentage of images correctly classified) for the proposed descriptors in the classification of KTH-TIPS-2b and UIUC, respectively. The average is computed over all training/testing rounds, corresponding, respectively, to 4 rounds in KTH-TIPS-2b and 10 rounds in UIUC. This is an interesting and intuitive graphical representation of the most complicated classes and the most confusable pairs of classes. While in UIUC there is no pair of classes deserving particular attention (the maximum confusion is of one image), KTH-TIPS-2b exhibits a much more challenging scenario. The confusion among classes 3, 5, 8, and 11 is the most critical scenario for the proposed classification framework. It turns out that these classes correspond, respectively, to the materials "corduroy", "cotton", "linen", and "wool". Despite being different materials, they inevitably share similarities as at the end they are all types of fabrics. Furthermore, looking at the sample from these classes, we can also observe that the attribute "color", that is not considered here, would be a useful class discriminant in that case. In general, the performance of our proposal in this dataset is quite promising and the confusion matrix and raw accuracy confirm our theoretical expectations.

Table 1 presents the accuracy compared with other methods in the literature, including several approaches that can be considered as part of the state-of-the-art in texture recognition. First of all, the advantage over the original Completed Local Binary Patterns (CLBP), whose part of the descriptors are used here as local encoder, is remarkable, being more than 10% in KTH-TIPS-2b. Other advanced encoders based on Scale Invariant Feature Transform (SIFT)

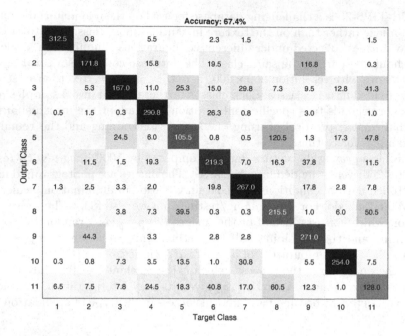

Fig. 3. Average confusion matrix and accuracy for KTH-TIPS-2b.

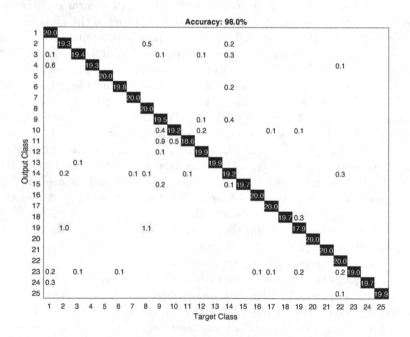

Fig. 4. Average confusion matrix and accuracy for UIUC.

Table 1. Accuracy of the proposed descriptors compared with other texture descriptors in the literature. A superscript [1] in KTHTIPS-2b means training on three samples and testing on the remainder (no published results for the setup used here).

KTH-TIPS-2b		UIUC	
Method	Acc. (%)	Method	Acc. (%)
VZ-MR8 [25]	46.3	RandNet (NNC) [7]	56.6
LBP [21]	50.5	PCANet (NNC) [7]	57.7
VZ-Joint [26]	53.3	BSIF [16]	73.4
BSIF [16]	54.3	VZ-Joint [26]	78.4
LBP-FH [3]	54.6	LBP_{riu2}/VAR [21]	84.4
CLBP [13]	57.3	LBP [21]	88.4
SIFT+LLC [9]	57.6	ScatNet (NNC) [5]	88.6
ELBP [20]	58.1	MRS4 [26]	90.3
SIFT + KCB [8]	58.3	SIFT + KCB [8]	91.4
SIFT + BoVW [8]	58.4	MFS [32]	92.7
LBP_{riu2}/VAR [21]	58.5[1]	VZ-MR8 [25]	92.8
PCANet (NNC) [7]	59.4[1]	DeCAF [8]	94.2
RandNet (NNC) [7]	60.7[1]	FC-CNN AlexNet [9]	91.1
SIFT + VLAD [8]	63.1	CLBP [13]	95.7
ScatNet (NNC) [5]	63.7[1]	SIFT+BoVW [8]	96.1
FC-CNN AlexNet [9]	71.5	SIFT+LLC [9]	96.3
Proposed	**67.4**	**Proposed**	**98.0**

are also outperformed in both datasets (by a large margin in the most challenging textures of KTH-TIPS-2b). SIFT descriptors are complex object descriptors and were considered the state-of-the-art in image recognition for several years. Compared with the most recent CNN-based approaches presented in [9], the results are also competitive. In UIUC, the proposed approach outperforms CNN methods like Deep Convolutional Activation Features (DeCAF) and FC-CNN AlexNet. Here it is worth to observe that FC-CNN AlexNet is not the classical "off-the-shelf" architecture, but an improved version of that algorithm especially adapted for texture images in [9]. These correspond to complex architectures with a high number of layers and large requirements of computational resources and whose results are pretty hard to be interpreted.

Table 2 lists the average computational time of our algorithm compared with some of the methods in Table 1 whose source codes were made available by the authors. This test was carried out over UIUC images (size 640 × 480 pixels). The computational setup corresponds to a potable computer with a CPU Intel Core i7-7700HQ (2.8 GHz), 16 GB of RAM, GPU NVIDIA Geforce GTX 1050Ti, Matlab R2017a 64 bits, Windows 10 Education 64 bits. In general, our method is among the fastest ones in terms of raw computational processing.

Table 2. Average computational time for an UIUC texture image (640 × 480).

Method	Time (s)
CLBP	1.269
LBP	0.713
LBP$_{riu2}$/VAR	1.216
SIFT + BoVW	1.974
SIFT + VLAD	1.406
SIFT+LLC	2.939
SIFT + KCB	5.384
FC-CNN AlexNet	0.579
Proposed	0.406

Generally speaking, the proposed method provided results in texture classification that confirm its potential as a texture image model. Indeed that was theoretically expected from its ability of smoothing spurious noise at the same time that preserves relevant discontinuities on the original image. The combination with a powerful yet simple local encoder like CLBP yielded interesting and promising performance neither requiring large amount of data for training nor advanced computational resources. Another remarkable point was its ability to be competitive with learning-based approaches that rely on transfer learning over huge datasets. Even though this is an interesting approach when we have limited data for training, it may become an infeasible strategy, for example, when the problem involves significant domain shift with respect to general purpose databases like ImageNet. Here our method achieved good generalization on heterogeneous textures without any external pre-training. In general, such great performance combined with the straightforwardness of the model, that allows some interpretation of the texture representation based on local homo/heterogeneous patterns, make the proposed descriptors a candidate for practical applications in texture analysis, especially when we have small to medium datasets and excessively complicated algorithms should be avoided.

6   Conclusions

In this study, we investigated the performance of a nonlinear PDE model (pseudo-parabolic) as an operator for the description of texture images. The operator was applied for a number of iterations (time evolution) and a local encoder was collected from each transformed image. The use of a basic histogram to pooling the local encoders was sufficient to provide competitive results.

The proposed descriptors were evaluated over a practical task of texture classification on benchmark datasets and the accuracy was compared with other approaches from the state-of-the-art. Our method outperformed several other local descriptors that follow similar paradigm and even some learning-based algorithms employing complex versions of convolutional neural networks.

The obtained results confirmed our expectations of a robust texture descriptor, explained by its ability of nonlinearly smoothing out spurious noise and unnecessary details, but preserving relevant information, especially those conveyed by sharp discontinuities. In general, the results and the confirmation of the theoretical formulation suggest the suitability of applying such model in practice, in tasks of texture recognition that require simple models, easy to be interpreted and that do not require much data for training. This is a common situation in areas like medicine and several others.

Finally, such promising results also open space for future works. Particularly, the PDE operator described here can be interpreted as a preconditioner for image analysis algorithms. In this context, it can also be explored, for instance, in combination with the modern deep learning approaches, especially in the typical scenario where deep learning excels, i.e., when there is sufficient amount of data for training and computational power and when model interpretation is not priority. Application to other types of images beyond textures would also be possible in this scenario.

Acknowledgements. J. B. Florindo gratefully acknowledges the financial support of São Paulo Research Foundation (FAPESP) (Grant #2016/16060-0) and from National Council for Scientific and Technological Development, Brazil (CNPq) (Grants #301480/2016-8 and #423292/2018-8). E. Abreu gratefully acknowledges the financial support of São Paulo Research Foundation (FAPESP) (Grant #2019/20991-8), from National Council for Scientific and Technological Development - Brazil (CNPq) (Grant #2 306385/2019-8) and PETROBRAS - Brazil (Grant #2015/00398-0). E. Abreu and J. B. Florindo also gratefully acknowledge the financial support of Red Iberoamericana de Investigadores en Matemáticas Aplicadas a Datos (MathData).

References

1. Abreu, E., Vieira, J.: Computing numerical solutions of the pseudo-parabolic Buckley-Leverett equation with dynamic capillary pressure. Math. Comput. Simul. **137**, 29–48 (2017)
2. Abreu, E., Ferraz, P., Vieira, J.: Numerical resolution of a pseudo-parabolic Buckley-Leverett model with gravity and dynamic capillary pressure in heterogeneous porous media. J. Comput. Phys. **411** (2020). https://doi.org/10.1016/j.jcp.2020.109395. http://www.sciencedirect.com/science/article/pii/S0021999120301698
3. Ahonen, T., Matas, J., He, C., Pietikäinen, M.: Rotation invariant image description with local binary pattern histogram fourier features. In: Salberg, A.-B., Hardeberg, J.Y., Jenssen, R. (eds.) SCIA 2009. LNCS, vol. 5575, pp. 61–70. Springer, Heidelberg (2009). https://doi.org/10.1007/978-3-642-02230-2_7
4. Barros Neiva, M., Guidotti, P., Bruno, O.M.: Enhancing LBP by preprocessing via anisotropic diffusion. Int. J. Mod. Phys. C **29**(08), 1850071 (2018)
5. Bruna, J., Mallat, S.: Invariant scattering convolution networks. IEEE Trans. Pattern Anal. Mach. Intell. **35**(8), 1872–1886 (2013)

6. Catté, F., Lions, P.L., Morel, J.M., Coll, T.: Image selective smoothing and edge detection by nonlinear diffusion. SIAM J. Numer. Anal. **29**(1), 182–193 (1992)
7. Chan, T., Jia, K., Gao, S., Lu, J., Zeng, Z., Ma, Y.: PCANet: a simple deep learning baseline for image classification? IEEE Trans. Image Process. **24**(12), 5017–5032 (2015)
8. Cimpoi, M., Maji, S., Kokkinos, I., Mohamed, S., Vedaldi, A.: Describing textures in the wild. In: Proceedings of the 2014 IEEE Conference on Computer Vision and Pattern Recognition, CVPR 2014, pp. 3606–3613. IEEE Computer Society, Washington, DC (2014)
9. Cimpoi, M., Maji, S., Kokkinos, I., Vedaldi, A.: Deep filter banks for texture recognition, description, and segmentation. Int. J. Comput. Vision **118**(1), 65–94 (2016)
10. Cuesta, C., Pop, I.: Numerical schemes for a pseudo-parabolic burgers equation: discontinuous data and long-time behaviour. J. Comput. Appl. Math. **224**, 269–283 (2009)
11. Dhivyaa, C.R., Sangeetha, K., Balamurugan, M., Amaran, S., Vetriselvi, T., Johnpaul, P.: Skin lesion classification using decision trees and random forest algorithms. J. Ambient Intell. Humaniz. Comput. **1**, 1–13 (2020). https://link.springer.com/article/10.1007/s12652-020-02675-8
12. Fisher, R.A.: The use of multiple measurements in taxonomic problems. Ann. Eugen. **7**(2), 179–188 (1936)
13. Guo, Z., Zhang, L., Zhang, D.: A completed modeling of local binary pattern operator for texture classification. Trans. Image Process. **19**(6), 1657–1663 (2010)
14. Hayman, E., Caputo, B., Fritz, M., Eklundh, J.-O.: On the significance of real-world conditions for material classification. In: Pajdla, T., Matas, J. (eds.) ECCV 2004. LNCS, vol. 3024, pp. 253–266. Springer, Heidelberg (2004). https://doi.org/10.1007/978-3-540-24673-2_21
15. Jain, D.K., Zhang, Z., Huang, K.: Multi angle optimal pattern-based deep learning for automatic facial expression recognition. Pattern Recogn. Lett. **139**, 157–165 (2020)
16. Kannala, J., Rahtu, E.: BSIF: binarized statistical image features. In: ICPR, pp. 1363–1366. IEEE Computer Society (2012)
17. Koenderink, J.J.: The structure of images. Biol. Cybern. **50**(5), 363–370 (1984)
18. Lazebnik, S., Schmid, C., Ponce, J.: A sparse texture representation using local affine regions. IEEE Trans. Pattern Anal. Mach. Intell. **27**(8), 1265–1278 (2005)
19. Lin, J., Pappas, T.N.: Structural texture similarity for material recognition. In: 2019 IEEE International Conference on Image Processing (ICIP), pp. 4424–4428. IEEE International Conference on Image Processing ICIP, Inst Elect & Elect Engineers; Inst Elect & Elect Engineers Signal Proc Soc (2019), 26th IEEE International Conference on Image Processing (ICIP), Taipei, TAIWAN, 22–25 September 2019
20. Liu, L., Zhao, L., Long, Y., Kuang, G., Fieguth, P.: Extended local binary patterns for texture classification. Image Vision Comput. **30**(2), 86–99 (2012)
21. Ojala, T., Pietikäinen, M., Mäenpää, T.: Multiresolution gray-scale and rotation invariant texture classification with local binary patterns. IEEE Trans. Pattern Anal. Mach. Intell. **24**(7), 971–987 (2002)
22. Pearson, F.K.: LIII. On lines and planes of closest fit to systems of points in space. London Edinburgh Dublin Philos. Mag. J. Sci. **2**(11), 559–572 (1901)
23. Perona, P., Malik, J.: Scale-space and edge detection using anisotropic diffusion. IEEE Trans. Pattern Anal. Mach. Intell. **12**(7), 629–639 (1990)
24. Robert Singh, K., Chaudhury, S.: Comparative analysis of texture feature extraction techniques for rice grain classification. IET Image Process. **14**(11), 2532–2540 (2020)

25. Varma, M., Zisserman, A.: A statistical approach to texture classification from single images. Int. J. Comput. Vision **62**(1), 61–81 (2005)
26. Varma, M., Zisserman, A.: A statistical approach to material classification using image patch exemplars. IEEE Trans. Pattern Anal. Mach. Intell. **31**(11), 2032–2047 (2009)
27. Verma, V., Muttoo, S.K., Singh, V.B.: Multiclass malware classification via first- and second-order texture statistics. Comput. Secur. **97**, 101895 (2020)
28. Vieira, J., Abreu, E.: Numerical modeling of the two-phase flow in porous media with dynamic capillary pressure. Ph.D. thesis, University of Campinas, Campinas, SP, Brazil (2018)
29. Vieira, J., Abreu, E., Florindo, J.B.: Texture image classification based on a pseudo-parabolic diffusion model (2020). https://arxiv.org/abs/2011.07173
30. Weickert, J.: A review of nonlinear diffusion filtering. In: ter Haar Romeny, B., Florack, L., Koenderink, J., Viergever, M. (eds.) Scale-Space 1997. LNCS, vol. 1252, pp. 1–28. Springer, Heidelberg (1997). https://doi.org/10.1007/3-540-63167-4_37
31. Witkin, A.P.: Scale-space filtering. In: Proceedings of the Eighth International Joint Conference on Artificial Intelligence, IJCAI 1983, vol. 2, pp. 1019–1022. Morgan Kaufmann Publishers Inc., San Francisco (1983)
32. Xu, Y., Ji, H., Fermüller, C.: Viewpoint invariant texture description using fractal analysis. Int. J. Comput. Vision **83**(1), 85–100 (2009)
33. Zhao, G., Wang, X., Cheng, Y.: Hyperspectral image classification based on local binary pattern and broad learning system. Int. J. Remote Sens. **41**(24), 9393–9417 (2020)

A Study on a Feedforward Neural Network to Solve Partial Differential Equations in Hyperbolic-Transport Problems

Eduardo Abreu[ID] and Joao B. Florindo[✉][ID]

Institute of Mathematics, Statistics and Scientific Computing,
University of Campinas, Rua Sérgio Buarque de Holanda, 651, Cidade Universitária
"Zeferino Vaz" - Distr. Barão Geraldo, Campinas, SP 13083-859, Brazil
{eabreu,florindo}@unicamp.br
http://www.ime.unicamp.br

Abstract. In this work we present an application of modern deep learning methodologies to the numerical solution of partial differential equations in transport models. More specifically, we employ a supervised deep neural network that takes into account the equation and initial conditions of the model. We apply it to the Riemann problems over the inviscid nonlinear Burger's equation, whose solutions might develop discontinuity (shock wave) and rarefaction, as well as to the classical one-dimensional Buckley-Leverett two-phase problem. The Buckley-Leverett case is slightly more complex and interesting because it has a non-convex flux function with one inflection point. Our results suggest that a relatively simple deep learning model was capable of achieving promising results in such challenging tasks, providing numerical approximation of entropy solutions with very good precision and consistent to classical as well as to recently novel numerical methods in these particular scenarios.

Keywords: Neural networks · Partial differential equation · Transport models · Numerical approximation methods for pdes · Approximation of entropy solutions

1 Introduction

In this work, we are interested in the study of a unified approach which combines both data-driven models (regression method by machine learning) and physics-based models (PDE modeling).

Deep learning techniques have been applied to a variety of problems in science during the last years, with numerous examples in image recognition [11], natural language processing [23], self driving cars [7], virtual assistants [13], healthcare

Supported by São Paulo Research Foundation (FAPESP), National Council for Scientific and Technological Development, Brazil (CNPq), and PETROBRAS - Brazil.

M. Paszynski et al. (Eds.): ICCS 2021, LNCS 12743, pp. 398–411, 2021.
https://doi.org/10.1007/978-3-030-77964-1_31

[15], and many others. More recently, we have seen a growing interest on applying those techniques to the most challenging problems in mathematics and the solution of differential equations, especially partial differential equations (PDE), is a canonical example of such task [18].

Despite the success of recent learning-based approaches to solve PDEs in relatively "well-behaved" configurations, we still have points in these methodologies and applications that deserve more profound discussion, both in theoretical and practical terms. One of such points is that many of these models are based on complex structures of neural networks, sometimes comprising a large number of layers, recurrences, and other "ad-hoc" mechanisms that make them difficult to be trained and interpreted. Independently of the approach chosen, the literature of approximation methods for hyperbolic problems primarily concern in the fundamental issues of conservation and the ability of the scheme to compute the correct entropy solution to the underlying conservation laws, when computing shock fronts, in transporting discontinuities at the correct speed, and in giving the correct shape of continuous waves. This is of utmost importance among computational practitioners and theoretical mathematicians and numerical analysts.

Furthermore, with respect to learning-based schemes to solve PDEs in physical models, we have seen little discussion about such procedures on more challenging problems, for instance, such as fractional conservation laws [3], compressible turbulence and Navier-Stokes equations [8], stochastic conservation laws [12] and simulation for darcy flow with hyperbolic-transport in complex flows with discontinuous coefficients [2,10]. Burgers equation has been extensively studied in the literature (see, e.g., [9]). Burgers' equations have been introduced to study different models of fluids. Thus even in the case of classical scalar one-dimensional Burgers equation, where the classical entropy condition (e.g., [16]) singles out a unique weak solution, which coincides with the one obtained by the vanishing viscosity method, there is no rigorous convergence proof for learning-based schemes. See [21] for a recent study of multi-dimensional Burgers equation with unbounded initial data concerning well-posedness and dispersive estimates. Roughly speaking, in solving the Riemann problem for systems of hyperbolic nonlinear equations, we might have nonlinear waves of several types, say, shock fronts, rarefactions, and contact discontinuities [9].

Related to the transport models treated in this work, the purely hyperbolic equation $u_t + H_x(u) = 0$ and the corresponding augmented hyperbolic-parabolic equation $u_t^\epsilon + H_x(u^\epsilon) = \epsilon u_{xx}^\epsilon$, we mention the very recent review paper [6], which discusses machine learning for fluid mechanics, but highlighting that such approach could augment existing efforts for the study, modeling and control of fluid mechanics, keeping in mind the importance of honoring engineering principles and the governing equations [2,10], mathematical [3,21] and physical foundations [8,9] driven by unprecedented volumes of data from experiments and advanced simulations at multiple spatiotemporal scales. We also mention the work [19], where the issue of *domain knowledge* is addressed as a prerequisite essential to gain explainability to enhance scientific consistency from machine learning and foundations of physics-based given in terms of mathematical equations and physical laws. However, we have seen much less discussion on more challenging

PDE modeling problems, like those involving discontinuities and shock' solutions numerical approximation of entropy solutions in hyperbolic-transport problems, in which the issue of conservative numerical approximation of entropy solutions is crucial and mandatory [1,12].

This is the motivation for the study accomplished in this work, where we investigate a simple feed-forward architecture, based on the physics-informed model proposed in [18], applied to complex problems involving PDEs in transport models. More specifically, we analyze the numerical solutions of four initial-value problems: three problems on the inviscid nonlinear Burgers PDE (involving shock wave and smooth/rarefaction fan for distinct initial conditions) and on the one-dimensional Buckley-Leverett equation for two-phase configurations, which is a rather more complex and interesting because it has a non-convex flux function with one inflection point. The neural network consists of 9 stacked layers with tanh activation and geared towards minimizing the approximation error both for the initial values and for values of the PDE functional calculated by automatic differentiation. The achieved results are promising.

Based upon a feedforward neural network approach and a simple algorithm construction, we managed to obtain a significant reduction of the error by simply controlling the input parameters of the simulations for the two fundamental models under consideration, namely, to the Burgers equation (cases rarefaction, shock and smooth) as well as to the Buckley-Leverett problem, respectively. Such results are pretty interesting if we consider the low complexity of the neural model and the challenge involved in these discontinuous cases. It also strongly suggests more in-depth studies on deep learning models that account for the underlying equation. They seem to be a quite promising line to be explored for challenging problems arising in physics, engineering, and many other areas.

What remains of this paper is organized as follows. In Sect. 2, we introduce the key aspects of hyperbolic problems in transport models we are considering in this work, along with a benchmark numerical scheme for comparison purposes with the traditional approach found in the specialized literature. We also offer a brief overview of the relevant approximation results for data-driven models and physics-based models in the context of PDE modeling linked to the feedforward neural network approach. The proposed methodology considered in this work is presented in Sect. 3, considering stable computations and conservation properties of the feedforward neural network approximations. In Sect. 4, we present some numerical experiments to show the efficiency and accuracy of the computed solutions verifying the available theory. Finally, in the last Sect. 5, we present our concluding remarks.

2   Hyperbolic Problems in Transport Models

Hyperbolic partial differential equations in transport models describe a wide range of wave-propagation and transport phenomena arising from scientific and indus-trial engineering area. This is a fundamental research that is in active progress since it involves complex multiphysics and advanced simulations due to a lack of general mathematical theory for closed-analytical solutions. For instance, see the noteworthy book by C. M. Dafermos [9] devoted to the mathematical theory of hyperbolic conservation and balance laws and their generic relations to continuum physics with a large bibliography list as well as some very recent work references cited therein related to recent advances covering distinct aspects, theoretical [3], numerical [5] and applications [8]. In addition, just to name some very recent works, see some interesting results covering distinct aspects, such as, theoretical [3] (uniqueness for scalar conservation laws with nonlocal and nonlinear diffusion terms) and [14] (non-uniqueness of dissipative Euler flows), and well-posedness [21] for multi-dimensional Burgers equation with unbounded initial data, numerical analysis [5] (*a posteriori* error estimates) and numerical computing for stochastic conservation laws [12] and applications [8] (Euler equations for barotropic fluids) and the references cited therein for complimentary discussion to highlight difficulties on the unavailability of universal results for finding the global explicit solution for Cauchy problems to the relevant class of hyperbolic-transport problems involving scalar and systems of conservation laws.

A basic fact of nonlinear hyperbolic transport problems is the possible loss of regularity in their solutions, namely, even solutions which are initially smooth (i.e., initial datum) may become discontinuous within finite time (blow up in finite time) and after this singular time, nonlinear interaction of shocks and rarefaction waves will play an important role in the dynamics. For the sake of simplicity, we consider the scalar 1D Cauchy problem

$$\frac{\partial u}{\partial t} + \frac{\partial H(u)}{\partial x} = 0, \quad x \in \mathbb{R}, \quad t > 0, \qquad u(x,0) = u_0(x), \qquad (1)$$

where $H \in C^2(\Omega)$, $H : \Omega \to \mathbb{R}$, $u_0(x) \in L^\infty(\mathbb{R})$ and $u = u(x,t) : \mathbb{R} \times \mathbb{R}^+ \longrightarrow \Omega \subset \mathbb{R}$. Many problems in engineering and physics are modeled by hyperbolic systems and scalar nonlinear equations [9]. As examples for these equations, just to name a few of relevant situations, we can mention the Euler equations of compressible gas dynamics, the Shallow water equations of hydrology, the Magnetohydrodynamics equations of plasma physics and the Buckley-Leverett scalar equation in petroleum engineering [10] as considered in this work. For this latter model, the flux function is smooth, namely, $H(u) = \frac{u^2}{u^2 + a(1-u)^2}$ in Eq. (1), $0 < a < 1$ ($H(u)$ is not convex with one inflection point, then an associated Riemann problem may be more complicated and the solution can involve both shock and rarefaction waves). Another interesting model is the inviscid Burgers' scalar equation used in many problems in fluid mechanics, where the flux function is $H(u) = u^2/2$ in Eq. (1). A nonlinear phenomenon that arises with the Burgers equation, even for smooth initial data, is the formation of shock, which is a discontinuity that may appear after the finite time. Together these two models [2], the Buckley-Leverett equation and Burgers' equation, are suitable and effective fundamental problems for testing new approximation algorithms to the above mentioned properties as is presented and discussed in the present work.

By using an argument in terms of traveling waves to capture the viscous profile at shocks, one can conclude that solutions of (1) satisfy Oleinik's entropy condition [16], which are limits of solutions $u^\epsilon(x,t) \to u(x,t)$, where $u(x,t)$ is given by (1) and $u^\epsilon(x,t)$ is given by the augmented parabolic equation [17]

$$\frac{\partial u^\epsilon}{\partial t} + \frac{\partial H(u^\epsilon)}{\partial x} = \epsilon \frac{\partial^2 u^\epsilon}{\partial x^2}, \quad x \in \mathbb{R}, \quad t > 0, \qquad u^\epsilon(x,0) = u_0^\epsilon(x), \qquad (2)$$

with $\epsilon > 0$ and the same initial data as in (1).

Thus, in many situations it is of importance to consider and study both hyperbolic-transport problems (1) and (2) and related conservation laws as treated in this work, and some others, of which are described in [3,12]. In this regard, a typical flux function $H(u)$ associated to fundamental prototype models (1) and (2) depends on the application under consideration, for instance, such as modeling flow in porous media [10] and problems in fluid mechanics [2]. Moreover, it is noteworthy that in practice the calibration of function $H(u)$ can be difficult to achieve due to unknown parameters and, thus, for instance, data assimilation can be an efficient method of calibrating reliable subsurface flow forecasts for the effective management of oil, gas and groundwater resources in geological models [22] and PDE models [4,20]. We intend to design a unified approach which combines both PDE modeling and fine tuning machine learning techniques aiming as a fisrt step to an effective tool for advanced simulations related to hyperbolic problems in transport models such as in (1) and (2).

2.1   A Benchmark Numerical Scheme for Solving Model (1)

First, we define a fixed x-uniform grid with a non-constant time step (x_j, t^n), where $(j,n) \in \mathbb{Z} \times \mathbb{N}^0$. In the space coordinates, the cells' middle points are $x_j = jh$, and the cells' endpoints are $x_{j\pm\frac{1}{2}} = jh \pm \frac{h}{2}$. The cells have a constant width $h = x_{j+\frac{1}{2}} - x_{j-\frac{1}{2}} = x_{j+1} - x_j = \Delta x$. Time step $\Delta t^n = t^{n+1} - t^n$ is non-constant subject to some Courant–Friedrichs–Lewy (CFL) stability condition. For simplicity of notation we simply use $k = \Delta t^n$. In order to numerically solve Eq. (1), instead of functions $u(\cdot, t) \in L^p(\mathbb{R})$ for $t \geq 0$, we will consider, for each time level t^n, the sequence $(U_j^n)_{j\in\mathbb{Z}}$ of the average values of $u(\cdot, t^n)$ over the x-uniform grid as follows

$$U_j^n = \frac{1}{h} \int_{x_{j-\frac{1}{2}}}^{x_{j+\frac{1}{2}}} u(x, t^n)\, dx, \qquad (3)$$

for all time steps t^n, $n = 0, 1, 2, \cdots$ and in the cells $[x_{j-\frac{1}{2}}, x_{j+\frac{1}{2}}]$, $j \in \mathbb{Z}$, where for t^0 we have the sequence $(U_j^0)_{j\in\mathbb{Z}}$ as an approximation of the pertinent Riemann data under study. Note that, in Equation (3), the quantity $u(x,t)$ is a solution of (1). The discrete counterpart of the space $L^p(\mathbb{R})$ is l_h^p, the space of sequences $U = (U_j)$, with $j \in \mathbb{Z}$, such that $\|U\|_{l_h^p} = \left(h \sum_{j\in\mathbb{Z}} |U_j|^p \right)^{\frac{1}{p}}$, $1 \leq p < \infty$ (for each time step t^n, as above). Following [1,2], now suppose that the approximate solution U^h has been defined in some strip $\mathbb{R} \times [0, t_n)$, $n \geq 1$. Then we define U^h in $\mathbb{R} \times [t_n, t_{n+1})$ as setting $U^h(x,t)$ constant and equal to U_j^n, by using (3), in the rectangle $(x_{j-1/2}, x_{j+1/2}] \times [t_n, t_{n+1})$ where we see that the Lagrangian-Eulerian numerical scheme applied to (1) reads the conservative method

$$U_j^{n+1} = U_j^n - \frac{k}{h} \left[F(U_j^n, U_{j+1}^n) - F(U_{j-1}^n, U_j^n) \right], \tag{4}$$

with the associated Lagrangian-Eulerian numerical flux [1,2],

$$F(U_j^n, U_{j+1}^n) = \frac{1}{4} \left[\frac{h}{k} (U_j^n - U_{j+1}^n) + 2 \left(H(U_{j+1}^n) + H(U_j^n) \right) \right]. \tag{5}$$

The classical Lax-Friedrichs numerical flux for (4), found elsewhere, is given by:

$$F(U_j^n, U_{j+1}^n) = \frac{1}{2} \left[\frac{h}{k} (U_j^n - U_{j+1}^n) + \left(H(U_{j+1}^n) + H(U_j^n) \right) \right]. \tag{6}$$

Here, both schemes (6) and (5) should follow the stability CFL condition

$$\max_j \left\{ |H'(U_j^n)|, \left| \frac{H(U_j^n)}{U_j^n} \right| \right\} \frac{k}{h} < \frac{1}{2}, \tag{7}$$

for all time steps n, where $k = \Delta t^n$ and $h = \Delta x$, $H'(U_j^n)$ is the partial derivative of H, namely $\frac{\partial H(u)}{\partial u}$ for all U_j^n in the mesh grid.

3   Proposed Methodology

The neural network employed here is based on that described in [18]. There a "physics-informed" neural network is defined to solve nonlinear partial differential equations. That network takes into account the original equation by explicitly including the PDE functional and initial and/or boundary conditions in the objective function and taking advantage of automatic differentiation widely used in the optimization of classical neural networks. It follows a classical feed-forward architecture, with 9 hidden layers, each one with a hyperbolic tangent used as activation function. More details are provided in the following.

The general problem solved here has the form

$$u_t + \mathcal{N}(u) = 0, \qquad x \in \Omega, t \in [0, T], \tag{8}$$

where $\mathcal{N}(\cdot)$ is a non-linear operator and $u(x, t)$ is the desired solution. Unlike the methodology described in [18], here we do not have an explicit boundary condition and the neural network is optimized only over the initial conditions of each problem.

We focus on four problems: the inviscid nonlinear Burgers equation

$$u_t + \left(\frac{u^2}{2} \right)_x = 0, \qquad x \in [-10, 10], \qquad t \in [0, 8], \tag{9}$$

with shock initial condition

$$u(x,0) = 1, x < 0 \text{ and } u(x,0) = 0, x > 0, \tag{10}$$

discontinuous initial data (hereafter *rarefaction fan initial condition*)

$$u(x,0) = -1, x < 0 \text{ and } u(x,0) = 1, x > 0, \tag{11}$$

smooth initial condition

$$u(x,0) = 0.5 + \sin(x), \tag{12}$$

and the two-phase Buckley-Leverett

$$u_t + \left(\frac{u^2}{u^2 + a(1-u)^2}\right)_x = 0, \qquad x \in [-8,8], \qquad t \in [0,8],$$
$$u(x,0) = 1, x < 0 \text{ and } u(x,0) = 0, x > 0. \tag{13}$$

In this problem we take $a = 1$.

For the optimization of the neural network we should define f as the left hand side of each PDE, i.e.,

$$f := u_t + \mathcal{N}(u), \tag{14}$$

such that

$$\mathcal{N}(u) = \left(\frac{u^2}{2}\right)_x \tag{15}$$

in the inviscid Burgers and

$$\mathcal{N}(u) = \left(\frac{u^2}{u^2 + a(1-u)^2}\right)_x \tag{16}$$

in the Buckley-Leverett. Here we also have an important novelty which is the introduction of a derivative (w.r.t. x) in $\mathcal{N}(u)$, which was not present in [18].

The function f is responsible for capturing the physical structure (i.e., selecting the qualitatively correct entropy solution) of the problem and inputting that structure as a primary element of the machine learning problem. Nevertheless, here to ensure the correct entropy solution, we add a small diffusion term to f ($0.01u_{xx}$) for better stabilization, but in view on the modeling problems (1) and (2). It is crucial to mention at this point that numerical approximation of entropy solutions (with respect to the neural network) to hyperbolic-transport problems also require the notion of entropy-satisfying weak solution. The neural network computes the expected solution $u(x,t)$ and its output and the derivatives present in the evaluation of f are obtained by automatic differentiation.

Two quadratic loss functions are defined over f, u and the initial condition:

$$\mathcal{L}_f(u) = \frac{1}{N_f} \sum_{i=1}^{N_f} |f(x_f^i, t_f^i,)|^2,$$
$$\mathcal{L}_u(u) = \frac{1}{N_u} \sum_{i=1}^{N_u} |u(x_u^i, t_u^i) - u^i|^2, \tag{17}$$

where $\{x_f^i, t_f^i\}_{i=1}^{N_f}$ correspond to collocation points over f, whereas $\{x_u^i, t_u^i, u^i\}_{i=1}^{N_u}$ correspond to the initial values at pre-defined points.

Finally, the solution $u(x, t)$ is approximated by minimizing the sum of both objective functions at the same time, i.e.,

$$u(x, t) \approx \arg \min_u [\mathcal{L}_f(u) + \mathcal{L}_u(u)]. \tag{18}$$

Inspired by the great results in [18], here the minimization is performed by the L-BFGS-B optimizer and the algorithm stops when a loss of 10^{-6} is reached. Figure 1 illustrates the evolution of the total loss function (in $\log_{10}$ scale to facilitate visualization) for the inviscid Burgers equation with shock initial condition.

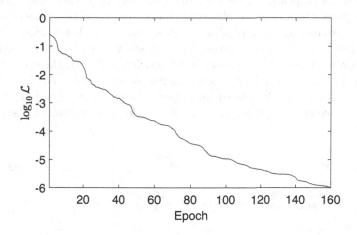

Fig. 1. Loss function evolution (in $\log_{10}$ scale) for the inviscid Burgers equation with shock initial condition.

4   Results and Discussion

In the following we present results for the solutions of the investigated problems obtained by the neural network model. We compare these solutions with two numerical schemes: Lagrangian-Eulerian and Lax-Friedrichs. These are very robust numerical methods with a solid mathematical basis. Here we use one scheme to validate the other. In fact, the solutions obtained by each scheme are very similar. For that reason, we opted for graphically showing curves only for the Lagrangian-Eulerian solution. However, we exhibit the errors of the proposed methodology both in comparison with Lagrangian-Eulerian (EEL) and Lax-Friedrichs (ELF). Here such error corresponds to the average quadratic error, i.e.,

$$ELF(t) = \frac{\sum_{i=1}^{N_u}(u_{NN}(x^i, t) - u_{LF}(x^i, t))^2}{N_u},$$

$$EEL(t) = \frac{\sum_{i=1}^{N_u}(u_{NN}(x^i, t) - u_{LE}(x^i, t))^2}{N_u},$$

(19)

where u_{NN}, u_{LF}, and u_{LE} correspond to the neural network, Lax-Friedrichs, and Lagrangian-Eulerian solutions, respectively. In our tests, we used $N_f = 10^4$ unless otherwise stated, and $N_u = 100$. These parameters of the neural network are empirically determined and are not explicitly related to the collocation points used in the numerical scheme. In fact, theoretical studies on optimal values for these parameters is an open research problem that we also intend to investigate in future works. For the numerical reference schemes we adopted CFL condition 0.4 for Lax-Friedrichs and 0.2 for Lagrangian-Eulerian. We also used $\Delta x = 0.01$.

For the rarefaction case, we observed that using $N_f = 10^4$ collocation points was sufficient to provide good results. In this scenario, we also verified the number of neurons, testing 40 and 60 neurons. Figure 2 shows the obtained solution compared with reference and the respective errors. Interestingly, the error decreases when time increases, which is a consequence of the solution behavior, which becomes smoother (smaller slope) for larger times, showing good accuracy and evidence that we are computing the correct solution in our numerical simulation.

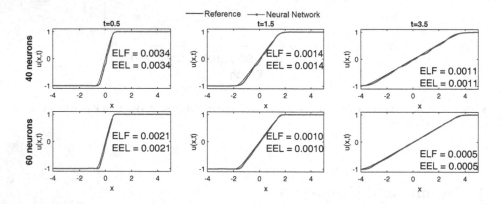

Fig. 2. Burgers: Rarefaction.

Figure 3 illustrates the performance of the neural network model for the inviscid Burgers equation with shock initial condition. Here we had to add a small viscous term $(0.01u_{xx})$ to obtain the entropy solution. Such underlying viscous mechanism did not bring significant reduction in error, but the general structure of the obtained solution is better, attenuating spurious fluctuations around the discontinuities. It was also interesting to see that the addition of more neurons did not reduce the error for this initial condition. This is a typical example of overfitting caused by over-parameterization. An explanation for that is the relative simplicity of the initial condition, assuming only two possible values.

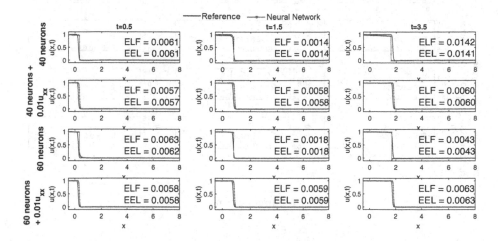

Fig. 3. Burgers: Shock.

Figure 4 depicts the solutions for the smooth initial condition in the inviscid Burgers equation. Here, unlike the previous case, increasing the number of neurons actually reduced the error. Indeed, it turned out better than expected considering that now both initial condition and solution are more complex. Nevertheless, we identified that tuning only the number of neurons was not enough to achieve satisfactory solutions in this situation. Therefore we also tuned the parameter N_f. In particular, we discovered that combining the same small viscous term used for the shock case with $N_f = 10^6$ provided excellent results, with quite promising precision in comparison with our reference solutions.

Another case characterized by solutions with more complex behavior is Buckley-Leverett with shock initial condition (Fig. 5). In this example, similarly to what happened in the smooth case, again the combination of $N_f = 10^6$ with the small viscous term was more effective than any increase in the number of neurons. While the introduction of the small viscous term attenuated fluctuations in the solution when using 40 neurons, at the same time when using $N_f = 10^4$, we observe that increasing the number of neurons causes an increase in the delay between the solution provided by the network and the reference.

Generally speaking, the neural networks studied here were capable of achieving promising results in challenging situations involving different types of discontinuities and nonlinearities. Moreover, our numerical findings might also suggest some good evidence on the robustness of a feedforward neural network as numerical approximation procedure for solving nonlinear hyperbolic-transport problems. Mathematical theoretical foundation of this model is to be pursued in a further work. In particular, the neural networks obtained results pretty close

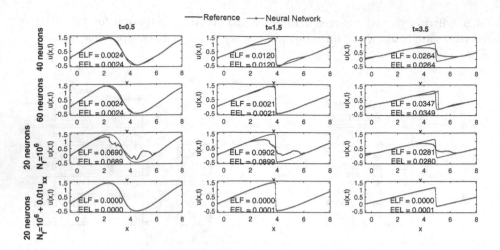

Fig. 4. Burgers: Smooth.

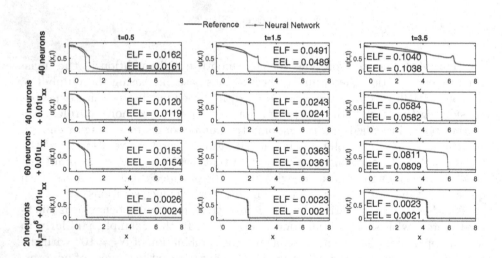

Fig. 5. Buckley-Leverett: Rarefaction + Shock.

to those provided by entropic numerical schemes like Lagrangian-Eulerian and Lax-Friedrichs. Going beyond the analysis in terms of raw precision, these results give us evidences that our neural network model possess some type of entropic property, which from the viewpoint of a numerical method is a fundamental and desirable characteristic.

5   Conclusions

This work presented an application of a feed-forward neural network to solve challenging hyperbolic problems in transport models. More specifically, we solve the inviscid Burgers equation with shock, smooth and rarefaction initial conditions, as well as the Buckley-Leverett equation with classical Riemann datum, which lead to the well-known solution that comprises a rarefaction and a shock wave. Our network was tuned according to each problem and interesting findings were observed. At first, our neural network model was capable of providing solutions pretty similar to those obtained by two numerical schemes used as references: Lagrangian-Eulerian and Lax-Friedrichs. Besides, the general structure of the obtained solutions also behaved as expected, which is a remarkable achievement considering the intrinsic challenge of these problems. In fact, the investigated neural networks showed evidences of an entropic property to the scalar hyperbolic-transport model studies, which is an important attribute of any numerical scheme when dealing with weak solution of scalar conservation laws.

Our approach is substantially distinct from the current trend of merely data-driven discovery type methods for recovery governing equations by using machine learning and artificial intelligence algorithms in a straightforward manner. We glimpse the use of novel methods, fine tuning machine learning algorithms and very fine mathematical and numerical analysis to improve comprehension of regression methods aiming to identify the potential and reliable prediction for advanced simulation for hyperbolic problems in transport models as well as the estimation of financial returns and economic benefits.

In summary, the obtained results share both practical and theoretical implications. In practical terms, the results confirm the potential of a relatively simple deep learning model in the solution of an intricate numerical problem. In theoretical terms, this also opens an avenue for formal as well as rigorous studies on these networks as mathematically valid and effective numerical methods.

Acknowledgements. J. B. Florindo gratefully acknowledges the financial support of São Paulo Research Foundation (FAPESP) (Grant #2016/16060-0) and from National Council for Scientific and Technological Development, Brazil (CNPq) (Grants #301480/2016-8 and #423292/2018-8). E. Abreu gratefully acknowledges the financial support of São Paulo Research Foundation (FAPESP) (Grant #2019/20991-8), from National Council for Scientific and Technological Development - Brazil (CNPq) (Grant #2 306385/2019-8) and PETROBRAS - Brazil (Grant #2015/00398-0). E. Abreu and J. B. Florindo also gratefully acknowledge the financial support of Red Iberoamericana de Investigadores en Matemáticas Aplicadas a Datos (MathData).

References

1. Abreu, E., Matos, V., Pérez, J., Rodríguez-Bermúdez, P.: A class of Lagrangian–Eulerian shock-capturing schemes for first-order hyperbolic problems with forcing terms. J. Sci. Comput. **86**(1), 1–47 (2021). https://doi.org/10.1007/s10915-020-01392-w

2. Abreu, E., Pérez, J.: A fast, robust, and simple Lagrangian-Eulerian solver for balance laws and applications. Comput. Math. Appl. **77**(9), 2310–2336 (2019)
3. Alibaud, N., Andreianov, B., Ouédraogo, A.: Nonlocal dissipation measure and $\mathcal{L}^1$ kinetic theory for fractional conservation laws. Commun. Partial Differ. Equ. **45**(9), 1213–1251 (2020)
4. Berg, J., Nyström, K.: Data-driven discovery of PDEs in complex datasets. J. Comput. Phys. **384**, 239–252 (2019)
5. Bressan, A., Chiri, M.T., Shen, W.: A posteriori error estimates for numerical solutions to hyperbolic conservation laws (2020). https://arxiv.org/abs/2010.00428
6. Brunton, S.L., Noack, B.R., Koumoutsakos, P.: Machine learning for fluid mechanics. Annu. Rev. Fluid Mech. **52**(1), 477–508 (2020)
7. Chen, C., Seff, A., Kornhauser, A., Xiao, J.: DeepDriving: learning affordance for direct perception in autonomous driving. In: 2015 IEEE International Conference on Computer Vision (ICCV), pp. 2722–2730. IEEE (2015)
8. Chen, G.Q.G., Glimm, J.: Kolmogorov-type theory of compressible turbulence and inviscid limit of the Navier-Stokes equations in $\mathcal{R}^3$. Phys. D **400**, 132138 (2019)
9. Dafermos, C.M.: Hyperbolic Conservation Laws in Continuous Physics. Springer, Heidelberg (2016)
10. Galvis, J., Abreu, E., Díaz, C., Pérez, J.: On the conservation properties in multiple scale coupling and simulation for Darcy flow with hyperbolic-transport in complex flows. Multiscale Model. Simul. **18**(4), 1375–1408 (2020)
11. He, K., Zhang, X., Ren, S., Sun, J.: Deep residual learning for image recognition. In: 2016 IEEE Conference on Computer Vision and Pattern Recognition (CVPR), pp. 770–778. IEEE (2016)
12. Hoel, H., Karlsen, K.H., Risebro, N.H., Storrøsten, E.B.: Numerical methods for conservation laws with rough flux. Stoch. Partial Differ. Equ. Anal. Comput. **8**(1), 186–261 (2019). https://doi.org/10.1007/s40072-019-00145-7
13. Kepuska, V., Bohouta, G.: Next-generation of virtual personal assistants (Microsoft Cortana, Apple Siri, Amazon Alexa and Google Home). In: 2018 IEEE 8th Annual Computing and Communication Workshop and Conference (CCWC), pp. 99–103 (2018)
14. Lellis, C.D., Kwon, H.: On non-uniqueness of Hölder continuous globally dissipative Euler flows (2020). https://arxiv.org/abs/2006.06482
15. Litjens, G., et al.: A survey on deep learning in medical image analysis. Med. Image Anal. **42**, 60–88 (2017)
16. Oleinik, O.A.: Discontinuous solutions of nonlinear differential equations. Uspekhi Matematicheskikh Nauk **12**, 3–73 (1957). (Engl. Transl. Trans. Am. Math. Soc. **26**(2), 95–172 (1963)
17. Quinn, B.K.: Solutions with shocks: an example of an L_1-contraction semigroup. Commun. Pure Appl. Math. **24**(2), 125–132 (1971)
18. Raissi, M., Perdikaris, P., Karniadakis, G.: Physics-informed neural networks: a deep learning framework for solving forward and inverse problems involving nonlinear partial differential equations. J. Comput. Phys. **378**, 686–707 (2019)
19. Roscher, R., Bohn, B., Duarte, M.F., Garcke, J.: Explainable machine learning for scientific insights and discoveries. IEEE Access **8**, 42200–42216 (2020). https://doi.org/10.1109/ACCESS.2020.2976199
20. Rudy, S.H., Brunton, S.L., Proctor, J.L., Kutz, J.N.: Data-driven discovery of partial differential equations. Sci. Adv. **3**(4), e1602614 (2017)
21. Serre, D., Silvestre, L.: Multi-dimensional Burgers equation with unbounded initial data: well-posedness and dispersive estimates. Arch. Rational Mech. Anal. **234**, 1391–1411 (2019). https://doi.org/10.1007/s00205-019-01414-4

22. Tang, M., Liu, Y., Durlofsky, L.J.: A deep-learning-based surrogate model for data assimilation in dynamic subsurface flow problems. J. Comput. Phys. **413**(109456), 1–28 (2020)
23. Young, T., Hazarika, D., Poria, S., Cambria, E.: Recent Trends in Deep Learning Based Natural Language Processing. IEEE Comput. Intell. Mag. **13**(3), 55–75 (2018)

Agent-Based Modeling of Social Phenomena for High Performance Distributed Simulations

Mateusz Paciorek$^{(\boxtimes)}$ and Wojciech Turek

AGH University of Science and Technology, Krakow, Poland
{mpaciorek,wojciech.turek}@agh.edu.pl

Abstract. Detailed models of numerous groups of social beings, which find applications in broad range of applications, require efficient methods of parallel simulation. Detailed features of particular models strongly influence the complexity of the parallelization problem. In this paper we identify and analyze existing classes of models and possible approaches to their simulation parallelization. We propose a new method for efficient scalability of the most challenging class of models: stochastic, with beings mobility and mutual exclusion of actions. The method is based on a concept of two-stage application of plans, which ensures equivalence of parallel and sequential execution. The method is analyzed in terms of distribution transparency and scalability at HPC-grade hardware. Both weak and strong scalability tests show speedup close to linear with more than 3000 parallel workers.

Keywords: Agent-based modeling · Social models · Scalability · HPC

1 Introduction

Computer simulation of groups of autonomous social beings, often referred to as Agent-based Modeling and Simulation (ABMS, ABM), is a fast developing domain with impressively broad range of applications. Experts in urban design, architecture, fire safety, traffic management, biology and many other areas use its achievements on a daily basis. Recently, its importance became clearly visible in the face of Covid-19 pandemic, which caused urgent demand for accurate simulations of large and varied populations. Researches continuously work on new modeling methods, models of new problems, general-purpose tools and standards [9], trying to provide useful results. However, reliable simulations of detailed and complex phenomena, observed in the reality by domain experts, require advanced simulation models, large numbers of agents and numerous repetitions. The process of creating, verifying and validating such models is time consuming and requires significant computational effort. The most natural approach is to parallelize the computation and use modern hardware, like GPUs or HPC.

© Springer Nature Switzerland AG 2021
M. Paszynski et al. (Eds.): ICCS 2021, LNCS 12743, pp. 412–425, 2021.
https://doi.org/10.1007/978-3-030-77964-1_32

The considered computational problem typically does not belong to "embarrassingly parallel" class. The model update algorithm must repeatedly modify a single large data structure. If the modifications are performed by parallel processes, the access to the data structure must be managed. Importantly, detailed features of the simulation model strongly influence the complexity of the required synchronization. Efficient and transparent distribution of complex ABMs simulation remains an open problem, especially when HPC-grade systems are considered.

In this paper we discuss the existing approaches to the problem of parallel execution of ABMs. We identify several classes of such models in the context of simulation parallelization and synchronization. The conclusions allowed us to propose a scalable method for modeling and simulation of the most challenging of the identified classes, while removing hidden biases in model update order present in existing approaches. The method presented in Sect. 4 is analyzed in terms of distribution transparency (Sect. 5) and HPC-grade scalability (Sect. 6).

2   Distributed Simulations of Agent-Based Models

Probably the most significant and successful application of software agents paradigm is related to computer simulation. The methodology of Agent-based Modeling (ABM) [9], is a widely used method for simulating groups of autonomous entities in micro-scale. It generalizes many specific areas, like pedestrian dynamics, traffic simulation, building evacuations and others. The simulation based on ABM has been proven superior to approaches based on differential equation models in many areas. The autonomy and differentiation of particular agents can more accurately reflect phenomena observed in the modeled systems.

Simulation of ABMs is computationally expensive. There are several factors which collectively increase the amount of required computations and memory, including agent and environment model complexity, size of the environment and the number of agents. The desire to simulate larger and more accurate models encourage researchers to investigate the methods for parallelization of ABM simulations, which is the main focus of the presented work. The aim is to improve performance by computing model updates in parallel and collecting the same results as in case of sequential execution. The problem of "transparent distribution" is non-trivial due to the need of updating common model state by parallel processes.

The typical approach to the problem is based on division of the environment model [3]. Parallel process compute updates of agents in the assigned fragments of the model and then perform synchronization of results with the processes responsible for adjacent fragments. This synchronization is the key issue – its complexity strongly depends on model characteristics. In the further analysis we will discuss different types of models and existing synchronization strategies.

The majority of ABM simulations uses discrete representation of the environment as a grid of cells, with well defined neighborhood, distance and possible states of cells. Such approach, which is often imprecisely referred to as cellular automaton (CA), significantly simplifies model data structures and update algorithms, while preserving sufficient expressiveness.

The problem of parallel update of CA-based simulations, which is discussed in details in [4], is typically solved by defining overlapping margins along the border of the environment fragments, which are available to two processes responsible for updating the fragments. These margins, often referred to as *buffer zones* or *ghost cells*, are updated by one process and sent to its neighbor for reading the state of "remote cells".

Considering parallel state update we should differentiate two classes of update algorithms, which:

1. always update one cell at a time (classic-CA),
2. update two or more cells at a time (CA-inspired).

If a model belongs to the first class, its simulation is an "embarrassingly parallel" problem – mutual exchange of ghost cells solves the model synchronization problem. However, the second class is required by vast majority of ABM simulations. When an agent "moves" from one cell to another, the state of both cells must change atomically and often the result limits possible moves of other agents in the same simulation step. Migration of an agent to a cell shared with a different process might result in a conflict. The conflict can be avoided if the model is deterministic and each process has sufficient information to ensure move safety [10].

Stochastic models prevent predicting updates performed by a remote process, making the problem of model synchronization really hard. Interestingly, classic-CA models do not expose this issue, which has been shown in [11], where a simulation of city development has been successfully parallelized. Also, models which can accept and handle collisions, can be parallelized relatively easy. For example, the simulation of stochastic model of foraminifera (small, yet mobile sea creatures) provides correct results in distributed configuration [3].

The presented analysis shows that many models can be efficiently parallelized with relatively simple synchronization mechanisms. However, the remaining class of models, which are CA-inspired stochastic models with mutual exclusion of actions, is very important in many applications. The problem has been identified in the domains different from ABM, like simulation of chemical reactions [1].

Synchronization mechanisms for distributed update of the considered class of models have been considered in [2]. The authors propose three different strategies to ensure lack of collisions in decisions of particular agents. Two of them (location-ordered and direction-ordered) divide each simulation step into constant number of sub-steps. In the location-ordered strategy only agents distanced by 2 cells are allowed to move simultaneously. In the direction-ordered strategy agents moving in the same direction are allowed to move at the same time. The third strategy, called trial-and-error, requires collecting plans from all agents and solving conflicts using priorities. Agents which were not able to move can try another destination. The procedure is repeated until all agents move or decide to remain in their original cells. All three synchronization mechanisms provide transparent distribution, all three require repetitive communication between processes during a single simulation step. The authors conclude that the location-ordered and direction-ordered methods are more predictable in terms of computational cost, therefore only these methods have been implemented and tested. The location-ordered strategy has been also described in [5] and used for implementing pedestrian dynamics simulation for multiple GPUs.

The features of the described synchronization strategies will be analyzed and discussed in the next section. The conclusions will formulate basis for a new method proposed in this paper.

3   Discussion on Model Update Distribution

The two methods analyzed and implemented in [2], i.e. location-ordered and direction-ordered, ensure constant synchronization costs and do allow to avoid collisions in agent-based simulations. Their usage is however not without consequences to the behaviors that can be observed in the simulation. A mechanism similar to the location-based method has been implemented in the simulation of emergency evacuation in [7] and proved to introduce patterns in the behavior of evacuated people. Depending on the orientation of the identically shaped and sized junction between a staircase and a building floor, a clear prioritization could be observed. In one case the people inside the staircase were consistently blocked by the inflow of the people evacuating the intersecting floor, while in the other case the flow of the people in the staircase could not be interrupted and the floor occupants were forced to wait for the higher floors to fully evacuate.

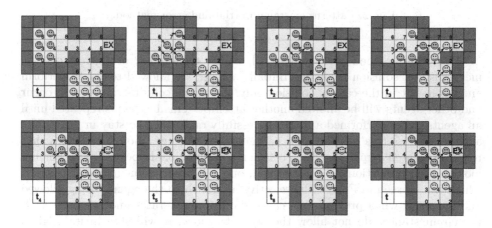

Fig. 1. Patterns emerging in location-ordered method. (Color figure online)

The synthetic example presented in Fig. 1 visualizes the mentioned problem more clearly. In this example the agents try to reach the exit cell marked as *EX*, at which point they are removed from simulation. The arrows show the direction of movements performed between the previous state and the current one. The first scenario, shown in Fig. 1, leads to the situation where the time steps t_6 and t_7 repeat until all agents present in the top left area of the grid (colored blue) are evacuated. No more agents from the bottom area (colored yellow) will be allowed to reach the exit until then. The same is the case for the blue agent at the junction between the two parts of the grid.

This phenomenon will be present in all methods that prioritize some actions above the others and a similar example could be derived for the direction-ordered method. It is important to note that the measurement of the rate of agents reaching the exit cell will not hint at the existence of the problem, as the agents are steadily leaving the grid. Only the observation of more intricate details of the simulation allows to note this behavior, e.g. tracking the distribution of time spent without moving among all the agents. This should not be the case, as the designer of the simulation model will be either unaware of the bias introduced by the method or forced to work around this limitation.

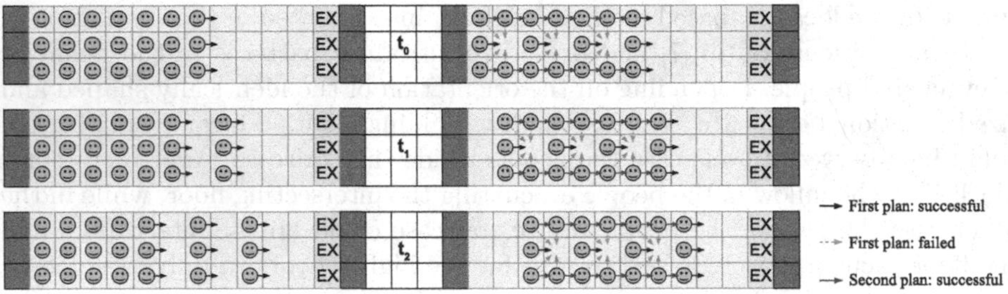

Fig. 2. Patterns emerging in trial-and-error method.

The other method presented in [2], called trial-and-error, introduces the mechanism of collision resolving. Initially, agents are allowed to try to perform an action, but in the case of collision only one of the colliding moves will occur and other agents will be allowed another attempt. This process is repeated until all agents either performed a move successfully or decided to stay in place. The authors dismiss this method as it can generate varying amount of retries that require repeated communications between computational nodes. Taking further look at the implications of this method, one can observe unexpected behavior, which can be seen in Fig. 2. The selective approach to the agents being allowed another move when previous one caused collision generates another pattern. If the circumstances do not allow the agent to move, it will stay in its original cell (Fig. 2, left). However, if the agent is allowed to attempt to move, even if it collides, it can then exploit the new movement possibility generated by the previous batch of moves. As shown in the example (Fig. 2, right), majority of agents are allowed to take action that was unavailable to them in the initial state of the grid. As a result, the example presents two configurations of agents that will result almost 1 : 2 difference in the throughput of the corridor.

This analysis of the existing approaches to the synchronization and collision-handling methods yields several requirements for the approach that could be deemed transparent for the model, while retaining good scalability:

- The amount of the communication related to the decision-making and collision handling should be predictable—as suggested in [2].
- No prioritization of moves or agents should be present unless explicitly implemented in the model.
- All agents should have equal capabilities of observing environment, i.e. either all decisions are based only on the initial state or all agents can track intermediate changes in environment and decide accordingly.

These conclusions led us towards a new method of model update distribution. This method, which is the main contribution of this paper, will be described in the next section.

4   Distributed Model Update Algorithm

To satisfy requirements outlined in previous section, a new method has been designed. As only the method of decision-making and conflict-resolving is the main focus of this work, several important assumptions are made to facilitate the understanding the method without unnecessary insights into the other parts of the simulation system:

- Prior to the main body of the simulation the simulation state is divided into parts that are then assigned to their respective "owner" computational nodes.
- The mapping from the identifier of the cell to the owner node is known to each node.
- Although the examples shown in the following sections use the Moore neighborhood, this method can be easily adapted to allow agents to interact with more distant cells.

The crucial concept introduced in the method is a *plan*. One or more *plan* is produced by each agent in each simulations step. Each *plan* consists of the following information:

- *action*—the change the algorithm intends to apply to the state of the handled cell or its neighbor.
- *consequence*—the optional change to the state of handled cell that is required to accompany the application of the action.
- *alternative*—the optional change to the state of handled cell that is required to accompany the rejection of the action.

None of these is guaranteed to be applied to the simulation state. Instead, the proposed distributed simulation update algorithm, after collecting *plans* from all agents, decides which *actions*, *consequences* and *alternatives* are executed.

A good example of this mechanism might be a model in which agents track their age (measured in iterations) and traverse the cells of the grid. The plan of an agent would be following: the action is inserting a copy of the agent with increased age into neighboring cell, the consequence is removing the agent from the original cell, and the alternative is replacing the old agent with its copy with increased age. If a plan is accepted, the result will be the application of action and consequence (add aged agent in neighboring cell, remove from the original cell). If a plan is rejected, the result will be application of alternative (keep aged agent in the original cell).

The possibility of multiple plans might be further explained by extending the example model with reproduction: if agent fulfills some criteria, it is allowed to create a new agent of the same type in adjacent cell. In such case, another planof the same agent could comprise of action of placing a new agent in the target cell, with empty consequence and alternative.

The important assumptions, emerging from this concept, are following:

- No two plans created by the same agent can contradict or depend on each other, as their acceptation or rejection is resolved independently—e.g. the agent cannot try to move into two different cells.
- Consequences and alternatives are not subject to rejection and must always be applicable.

The core of the method is the course of each simulation iteration, which has been described in Algorithm 1. The *exchange* executed in lines 6, 16 and 24 is realized by grouping the exchanged elements by their cell identifiers and sent to the owners of these identifiers.

There are a few model-specific components, marked **bold** in the Algorithm 1:

- A component responsible for the creation of the initial grid state.
- *createPlans* - creation of the plans basing on the cell and its neighbors.
- *isApplicable* - accepting or rejecting the plan basing on the target cell.
- *apply* - application of the given plan to the given cell.

The actual model part of the implementation is unable to determine whether any given neighbor cell is owned by the current node or is the "ghost cell". The same lack of knowledge pertains to the origin of any plan that is being resolved. Therefore there can be no influence of the distribution of the grid on the decision making algorithm itself.

The only part of the algorithm affected by the distribution and communication between computational nodes is the necessity of the exchange of the plans (Algorithm 1, line 6) if the plan would affect the cell owned by another computational node. The plans are not resolved immediately, which results in all of them being handled in the same point in the iteration. Therefore the only possible effect of the distribution would be the change in the order of their resolving (e.g. locally created plans first, then plans from other nodes in order dictated by the order of communication events). However, this possibility is eliminated by the introduction of the obligatory randomization of the order of plans handling (Algorithm 1, line 7).

As a side note, if the model explicitly requires ordering of the plans application, the randomization step can be easily replaced by sorting (preceded by shuffling, if the order is not total, to ensure no bias).

Algorithm 1: Steps of the single iteration of simulation.

```
   // plans creation step
 1 plans ← emptyList;
 2 foreach cell in localCells do
 3     cellNeighbors ← getCellNeighbors(cell);
 4     plansForCell ← createPlans(cell, cellNeighbors);
 5     plans.append(plansForCell);

 6 localPlans ← exchange(plans);
 7 shuffledLocalPlans ← shuffle(localPlans);

   // actions application step
 8 reactions ← emptyList;
 9 foreach plan in shuffledLocalPlans do
10     targetCell ← localCells.getCell(plan.getAction().getTargetId());
11     if isApplicable(targetCell, plan) then
12         apply(targetCell, plan.getAction());
13         reactions.append(plan.getConsequence());
14     else
15         reactions.append(plan.getAlternative());

16 localReactions ← exchange(reactions);

   // reactions application step
17 foreach reaction in localReactions do
18     targetCell ← localCells.getCell(reaction.getTargetId());
19     apply(targetCell, reaction);

   // "ghost cells" update step
20 edgeCells ← emptyList;
21 foreach cell in localCells do
22     if isEdgeCell(cell) then
23         edgeCells.append(cell);

24 ghostCells ← exchange(edgeCells);
```

---

5   Transparency of Simulation Distribution

The method described in previous section has been added to the framework discussed in detail in [3]. To ensure the method does fulfill the transparency of distribution requirements presented in Sect. 3, the experimental verification was designed and executed.

The exemplary model implemented using the new method was a simple predator-prey scenario involving rabbits and lettuce, similar to the one described in [3]. Each cell is either occupied by a lettuce agent, a rabbit agent, or is an empty cell. In each iteration each agent can perform an action. Lettuce is allowed to grow by creating new lettuce agent in a random neighboring cell, limited by the growing interval. Rabbit is allowed to take one of the three actions, depending on the energy parameter of the agent: if below zero, the rabbit dies (is removed from the grid); if above reproduction threshold, the rabbit reproduces by creating a new rabbit agent in a random neighboring cell; otherwise, the rabbit moves towards the lettuce, expending some of its energy.

In addition to the energy of the rabbit agent, both agent types track their "lifespan", i.e. the number of iterations since their creation. No two agents of the same type can occupy the same cell. If any action leads to the situation where agent of one type would be in the same cell as the agent of the other type, the lettuce is consumed (removed from the grid) and the rabbit increases its energy level. The reproduction threshold, energy gained from consumption, energy cost of movement, initial energy of rabbits and lettuce growing interval are the main parameters of the simulation and can be adjusted.

The implemented system collects metrics during the simulation execution. In the case of the mentioned simulation, the metrics collected for each iteration are following (naming uses camel case notation mirroring the one used in the implementation):

- rabbitCount - the number of the rabbits present on the grid.
- rabbitReproductionsCount - the number of rabbit reproductions.
- rabbitTotalEnergy - the total energy of all rabbits present on the grid.
- rabbitDeaths - the number of rabbit deaths.
- rabbitTotalLifespan - the cumulative lifespan of rabbits that died.
- lettuceCount - the number of the lettuces present on the grid.
- consumedLettuceCount - the number of the lettuces consumed.
- lettuceTotalLifespan - the cumulative lifespan of consumed lettuces.

The HPC system used to run the experiments was a Prometheus supercomputer located in the AGH Cyfronet computing center in Krakow, Poland. According to the TOP500 list, as of November 2020 it is 324th fastest supercomputer. Prometheus is a peta-scale (2.4 PFlops) cluster utilizing HP Apollo 8000 nodes with Xeon E5-2680v3 CPUs working at 2.5 GHz. Each of the nodes connected via InfiniBand FDR network has 24 physical cores (53,604 cores total).

5.1   Stochastic Experiments

The first set of experiments was performed within 1000 iterations each and using the constant size of the grid: 480×500 cells. To observe the influence of the changes in parameters on the simulation course and results, four batches of experiments were conducted, each using different set of parameters, later referred to as "variants" numbered 0–3. The detailed description of the parameters and the variants can be found in [8] (where the "variant 0" is referred to as "default"). Each variant was then executed in different degrees of distribution - using 1, 2, 4, 6, 8, 10, 20, 40, 60, 80 and 100 nodes, 24 cores each. Each unique combination of the variant and the distribution degree was executed 10 times to obtain sample of size necessary for the variance analysis used in later steps.

It is clearly visible that the simulation model displays high variance in the results, even within the identical configurations. As is visible in Fig. 3, which

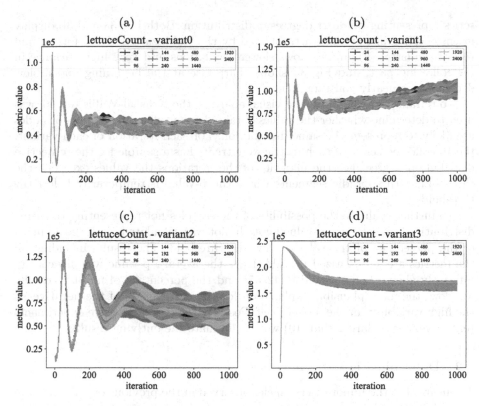

Fig. 3. Comparison of lettuceCount metric over time in stochastic execution. Each plot represents different model parameters configuration. Series correspond to means with standard deviations collected for different distribution degree.

Table 1. Summary of average p-values and percentage of p-values below threshold

	Variant 0		Variant 1		Variant 2		Variant 3	
	avg. p	< 0.05	avg. p	< 0.05	avg. p	< 0.05	avg. p	< 0.05
rabbitCount	0.47	4.2%	0.42	3.9%	0.48	1.8%	0.44	1.4%
rabbitReproductionsCount	0.47	2.9%	0.44	4.2%	0.48	2.7%	0.45	1.1%
rabbitTotalEnergy	0.47	3.9%	0.42	4.1%	0.48	2.5%	0.44	1.4%
rabbitDeaths	0.48	3.0%	0.44	4.7%	0.48	2.7%	0.45	2.7%
rabbitTotalLifespan	0.48	3.3%	0.44	3.9%	0.48	3.0%	0.46	3.2%
lettuceCount	0.55	2.4%	0.44	5.7%	0.52	0.1%	0.47	1.8%
consumedLettuceCount	0.47	2.8%	0.42	4.5%	0.48	3.0%	0.44	1.6%
lettuceTotalLifespan	0.50	3.2%	0.45	7.3%	0.47	5.2%	0.53	1.5%

represents the change of one of the metrics (*lettuceCount* - chosen randomly) in the time, the bands created by the means are noticeably wide and do not overlap perfectly. However, each of the variants presents some tendency obeyed by all series representing different degrees of distribution. Both Fig. 3a and 3b display intense oscillations at the start, followed by the stabilization of the former and rise of the latter. Figure 3c shows slower oscillations and very high variance in the following part, and Fig. 3d shows sharp rise at the beginning and a slow decrease to a nearly constant value.

To refrain from the imprecise visual analysis, the Kruskal-Wallis test [6] was used to determine whether the values obtained from different numbers of nodes are likely to represent the same distribution. For each iteration of each variant, the 10 collected series of each metric were treated as a sample for the respective distribution degree. For the majority of the iterations the values exceeded the threshold, with sporadic segments where the p-value is temporarily below the threshold.

To further evaluate the possibility of the samples not representing the same distributions, the average p-value for each plot was calculated, with the addition of the percentage of the values below the threshold. The results including all the metrics are summarized in Table 1. As the average p-value was always significantly above the threshold (0.42–55) and the percentage of negative results was low, the most plausible explanation for the occurrences of the negatives is the high variability of the model itself. Possibly the more extensive experiments (e.g. sample sizes larger than 10) would eliminate the outlying results.

5.2 Deterministic Experiments

To ensure that the minor discrepancies observed in the previous experiments are a result of the high sensibility of the model to the random decisions, another set of experiments was conducted. The general course and parameters of the experiment remained identical to the previous one, but the random factors were completely eliminated from the model and the method. Randomness in the moves of agents was replaced by the direction prioritization, while the randomized traversal of the plans was substituted with sorting. Instead of 10 times, each instance was executed only once, as there is no explicit randomness present in the system.

In each instance all collected data differed only at the least significant decimal digits due to differences in rounding, as the metrics are collected and saved separately by each core. This behavior is a result of the aggregation and formatting of the metrics, and has no effect on the course of simulation. No other differences were present. The comparison footage[1] of sample 100×100 cells experiment executed on single core and on 100 cores (10×10 cores, 10×10 cells per core) is available for additional visual verification of distribution transparency.

6   Scalability Experiments

As the crucial role of the implemented method is related to the handling of the distribution and the coordination of the communication between computational

---

[1] https://youtu.be/9W-zmyQo-K8.

nodes, the scalability tests were necessary. Tests were conducted using the same implementation and model as described in previous section.

6.1 Weak Scaling

The first scalability experiments were the weak scaling tests. The size of the problem was variable to achieve constant size per core—1e4 cells per core. The experiment was executed 5 times on each of the following numbers of nodes: 1, 2, 4, 6, 8, 10, 20, 40, 60, 80, 100 and 120, which translates to the number of cores ranging from 24 to 3600. Therefore, the final sizes of the grids varied from 2.4e5 cells to 3.6e7 cells. Each run consisted of 1000 iterations, the first 100 of which were omitted in final calculations to exclude any outlying results caused by the initialization of the simulation system. The execution times measured on each core were averaged to achieve a single time value for each run.

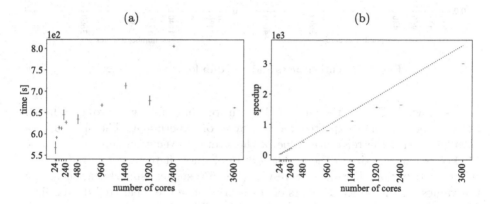

Fig. 4. Execution times and speedup for weak scaling. (Color figure online)

The results were grouped by the number of cores and aggregated into means and standard deviations, which are presented in Fig. 4. The first plot (Fig. 4a) presents the execution times of the iterations 100–1000 for each number of cores. The vertical lines show the standard deviation of the measurements. Additionally, the speedup adjusted to weak scaling experiments (i.e. with corrections for the changing problem size) was calculated. The average execution time on 1 node was used as the reference time. This metric is shown in Fig. 4b, with the ideal linear speedup marked with red dotted line. After the initial nearly-perfect scalability, the values became lower than the ideal. However, the deviation from the ideal was expected due to the non-negligible amounts of communication.

6.2 Strong Scaling

The second scalability experiments were the strong scaling tests. The problem size was kept constant—2.4e7 cells. Due to the memory limitations, the lower numbers of nodes were not included, and the experiment was executed on 10, 20, 40, 60, 80, 100, 150 and 200 nodes, that is 240–4800 cores. The effective problem size for each core ranged from 1e5 cells to 5e3 cells. All other aspects—number of iterations, data collection and aggregation—were identical to the weak scaling experiments.

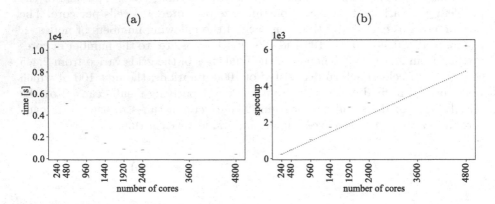

Fig. 5. Execution times and speedup for strong scaling.

As shown in Fig. 5a, the measured execution times follow the roughly hyperbolic shape, which is expected for this type of experiment. The speedup was calculated with the reference value of the average execution time on 10 nodes (240 cores), as the size of the problem did not allow it to be computed in more coarse distribution. The speedup results (Fig. 5b) suggest superlinear scaling, as the values for the larger numbers of cores are above the ideal (red dotted line). The explanation for such results can be the alleviation of the pressure on the memory, which were the reason the experiments with less than 10 nodes were not feasible. As the system is running on JVM, the memory management can introduce additional overhead when the problem parts are large in comparison to the available memory.

7   Conclusions and Further Research

In this work we analyzed the issues present in popular approaches to the problem of ABM simulation distribution. The most important ones concerned the introduction of undesired—and, in some cases, difficult to detect—changes to the simulation model. We proposed the new approach to this problem that ensures no changes to the model, unless the simulation designer intends to introduce them. This solution proved to retain the high scalability expected from the efficient distribution method, while displaying no influence on the simulation outcomes, regardless of the degree of distribution. The method has an additional advantage of flexibility of the neighborhood definition - the approaches described in Sect. 3 would require more synchronization cycles to achieve the same.

To further explore the new solution, we intend to focus on adapting models based on real life scenarios. Our concurrent work suggests that it is possible to use this method in the microscopic pedestrian dynamics simulation of epidemic spread in urban environment. Another direction of study possible due to the qualities of the proposed solution is the adaptation of models that require interaction with non-adjacent cells (outside of Moore neighborhood).

Acknowledgements. The research presented in this paper was partially supported by the funds of Polish Ministry of Science and Higher Education assigned to AGH University of Science and Technology. This research was supported in part by PLGrid Infrastructure.

References

1. Bezbradica, M., Crane, M., Ruskin, H.J.: Parallelisation strategies for large scale cellular automata frameworks in pharmaceutical modelling. In: 2012 International Conference on High Performance Computing & Simulation (HPCS), pp. 223–230. IEEE (2012)
2. Bowzer, C., Phan, B., Cohen, K., Fukuda, M.: Collision-free agent migration in spatial simulation. In: Proceedings of 11th Joint Agent-Oriented Workshops in Synergy (JAWS 2017), Prague, Czech (2017)
3. Bujas, J., Dworak, D., Turek, W., Byrski, A.: High-performance computing framework with desynchronized information propagation for large-scale simulations. J. Comput. Sci. **32**, 70–86 (2019)
4. Giordano, A., De Rango, A., D'Ambrosio, D., Rongo, R., Spataro, W.: Strategies for parallel execution of cellular automata in distributed memory architectures. In: 2019 27th Euromicro International Conference on Parallel, Distributed and Network-Based Processing (PDP), pp. 406–413 (2019)
5. Kłusek, A., Topa, P., Wąs, J., Lubaś, R.: An implementation of the social distances model using multi-GPU systems. Int. J. High Perform. Comput. Appl. **32**(4), 482–495 (2018)
6. Kruskal, W.H., Wallis, W.A.: Use of ranks in one-criterion variance analysis. J. Am. Stat. Assoc. **47**(260), 583–621 (1952)
7. Paciorek, M., Bogacz, A., Turek, W.: Scalable signal-based simulation of autonomous beings in complex environments. In: Krzhizhanovskaya, V.V., et al. (eds.) ICCS 2020. LNCS, vol. 12139, pp. 144–157. Springer, Cham (2020). https://doi.org/10.1007/978-3-030-50420-5_11
8. Paciorek, M., Bujas, J., Dworak, D., Turek, W., Byrski, A.: Validation of signal propagation modeling for highly scalable simulations. Concurr. Comput. Pract. Exp,. e5718 (2020). https://doi.org/10.1002/cpe.5718
9. Railsback, S.F., Grimm, V.: Agent-Based and Individual-based Modeling: A Practical Introduction. Princeton University Press, Princeton (2019)
10. Turek, W.: Erlang-based desynchronized urban traffic simulation for high-performance computing systems. Futur. Gener. Comput. Syst. **79**, 645–652 (2018)
11. Xia, C., Wang, H., Zhang, A., Zhang, W.: A high-performance cellular automata model for urban simulation based on vectorization and parallel computing technology. Int. J. Geogr. Inf. Sci. **32**(2), 399–424 (2018)

Automated Method for Evaluating Neural Network's Attention Focus

Tomasz Szandała[✉] ⓘ and Henryk Maciejewski ⓘ

Wroclaw University of Science and Technology, Wroclaw, Poland
{Tomasz.Szandala,Henryk.Maciejewski}@pwr.edu.pl

Abstract. Rapid progress in machine learning and artificial intelligence (AI) has brought increased attention to the potential security and reliability of AI technologies. This paper identifies the threat of network incorrectly relying on counterfactual features that can stay undetectable during validation but cause serious issues in life application. Furthermore, we propose a method to counter this hazard. It combines well-known techniques: object detection tool and saliency map obtaining formula to compute metric indicating potentially faulty learning. We prove the effectiveness of the method, as well as discuss its shortcomings.

Keywords: Deep learning · Xai · Convolutional neural networks · Region proposal · Object detection

1 Introduction

Object recognition networks are trained to minimize the loss on a given training dataset and subsequently evaluated with regard to the testing dataset. With this approach, we expect the model will learn its own correlation between images and their labels. However, the actual model reliability and robustness highly depend on the choice of the unskewed and heterogenous training dataset. Otherwise, we may encounter a latent correlation that is present in our training set but is not a reliable indicator in the real world.

Such unsought correlation can be color yellow for a cab or biased background representing a typical environment for classified objects like a desert for the camel or sea for the ship. Multiple works [10, 11] prove that background bias can have a serious impact on classifier reliability and even state-of-the-art models are not free of relying on latent correlations [16]. Yet these defects are recognizable merely by manual examination. While for several classes it is achievable, for a wider range of types it can be challenging.

In this paper we aim to provide a solution to detect and avert learning misguided correlations. For this purpose, we choose a pre-trained tool, Detectron2 [2, 6], to indicate potential Regions of Interest (ROI) on a given image. Secondly, we process images through our to-inspect network and thus obtain saliency maps with the GradCAM technique [3] for the most accurate class. Since the GradCAM result is a matrix that assigns potential importance for a given classification to areas of the image, we can calculate the importance value inside the detected ROI and divide it by the entire image's importance.

© Springer Nature Switzerland AG 2021
M. Paszynski et al. (Eds.): ICCS 2021, LNCS 12743, pp. 426–436, 2021.
https://doi.org/10.1007/978-3-030-77964-1_33

The obtained ratio is considered a fitting metric for a given classification. The higher it is, the more certain we can be that the network has learned the actual object, not the background.

During our research, we have performed 4 experiments. In the first one, we have provided the evidence for the problem. Secondly, we have formulated the method details and tested it on an ImageNet's subset of 9 classes. This proved the effectiveness of the proposed method and, due to the small set, we could easily verify its effectiveness. In the third attempt, we moved to the entire ImageNet set and pretrained VGG-11 from Model Zoo [4]. In this phase, we used the method to determine several classes with potential faulty correlations. We have further investigated them and in the final experiment we used the possessed knowledge to create a successful adversarial attack on a presumably robust network [5].

2 State of the Art

Image backgrounds are a natural source of correlation between images and their labels in object recognition. Indeed, prior work has shown that models may use backgrounds in classification [14, 16–19], and suggests that even human vision makes use of image context for scene and object recognition [20].

Moreover, Xiao et al. [16] have demonstrated that standard models not only use but require backgrounds for correctly classifying large portions of test sets. They have also studied the impact of backgrounds on classification for a variety of classifiers, and found that more accurate models tend to simultaneously exploit background correlations more and have greater robustness to changes in the image background.

The researchers from Zhang's team [7] created a CNN that classifies an X-ray image of the chest as "high disease probability" based not on the actual appearance of the disease but rather on an "L" metal token placed on the patient's left shoulder. The key is that this "L" token is placed directly on the patient's body only if he is lying down, and the patient will only lie down on the X-rays if they are too weak to stand. Therefore, CNN has learned the correlation between "metal L on the shoulder" and "patient is sick" - but we expect CNN to be looking for real visual signs of the disease, not metal tokens.

The final example: CNN learns [13] to classify an image as a "horse" based on the presence of a source tag in the lower-left corner of one-fifth of the horse's images in the data set. If this "horse source tag" is placed on a car image, the network classifies the image as "horse".

3 Method

The proposed method is based on three simple concepts. The first is to determine what network should consider as a discriminative feature. Secondly, we have to go through and discover what the network is actually learning. And finally, we need to evaluate how much of the learned correlations are tied to the expected area of the image.

3.1 ROI Generation

The first phase is the generation of the Regions-of-Interest (ROI). For this purpose, we can employ an expert that will manually indicate ROI, but in this paper we utilized the Detectron2 framework [6]. It is a Facebook AIResearch's next-generation software system that implements state-of-the-art object detection algorithms. It is also a common practice to use a base model pre-trained on a large image set (such as ImageNet [9]) as the feature extractor part of the network. Detectron2 framework allows us to obtain coordinates of two points that mark the top left and bottom right of the ROI. For the purpose of this research, we did not put attention to which class object has been recognized.

Fig. 1. Detectron2 processed image with highlighted 3 regions of interests

Detectron2 produces images with highlighted ROIs (see Fig. 1) and returns a tuple for each image that contains 4 coordinates: x and y for the top left corner and x and y for the bottom right corner of the ROI.

3.2 Saliency Map Generation

Saliency map generation is the subsequent stage, although it can be performed in parallel to ROIs generation if we have sufficient computing power. For this purpose we have used the popular GradCAM method [3]. To obtain the class-discriminative localization map, Grad-CAM computes the gradient of y^c (score for class c) with respect to feature maps A of a convolutional layer. These gradients flowing back are global-average-pooled to obtain the importance weights for pixel k with respect to class c: α^c_k (Eq. 1).

$$\alpha_k^c = \overbrace{\frac{1}{Z}\sum_i\sum_j}^{\text{global average pooling}} \underbrace{\frac{\partial y^c}{\partial A_{ij}^k}}_{\text{gradients via backprop}} \tag{1}$$

The importance weights create a saliency map as a matrix of values between 0.0 to 1.0 which corresponds to the importance of a particular pixel. Graphical representation of this map is a picture with colors from blue (lowest importance) to red (highest importance).

3.3 Fit Metric Calculation

The final step is to compute how many classification-important pixels are inside ROI in relation to all important pixels in the image. In the proposed method we sum values of saliency map inside ROI and divide them by the sum of values over the entire map. If there are more ROIs detected we calculate the ratio for each separately and in the end choose the highest value.

$$m_{fit} = \frac{\sum_{x,y}^{roi} \alpha_k^c}{\sum_{x,y}^{image} \alpha_k^c} \tag{2}$$

This proportion gives us fit measurement. The higher value we obtain for a given class, the higher confidence that the network has learned the object, not the background.

4 Experiments

Our research has been divided into 4 linked experiments.

4.1 Huskies and Wolves Discrimination Using Resnet Architecture

In the first study, we are trying to reproduce the Singh's et al. [13] experiment about discrimination of wolves and huskies. They have noticed that their network has learned snow as a feature strictly correlated with wolves' class.

In our attempt, we have collected 40 images of each specimen and taught the network as a binary classifier. This time attention has been paid to ensure that wolves are not only displayed in the winter background. The precision achieved was satisfactory. However, this time the visualization methods showed that most of the wolf images were taken in their natural habitat, which usually contained bushes, trees and branches, and these were the criteria for distinguishing the wolf from the husky.

The features that determine wolves according to the network are leafless trees and bushes (Fig. 2.). Indeed, most of our wolves are pictured in their natural habitat and often with some trees in the background. On the other hand, the huskies are all beasts without any branches.

This research supports the thesis that networks may learn incorrect correlations even when the training set appears to be hardened against this setback. Nonetheless, when there are only 2 classes the flaw can be easily exposed using known visualization techniques.

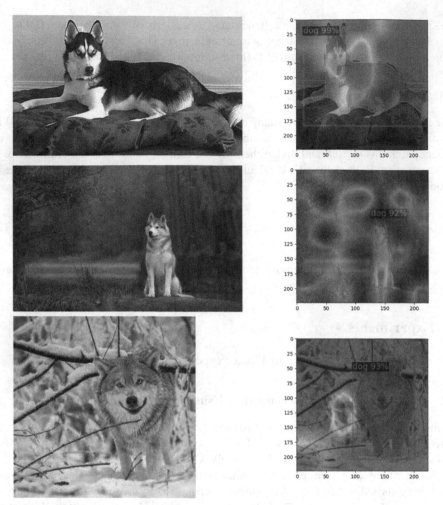

Fig. 2. Top-down: husky classified as husky and its saliency map, husky classified as wolf due to trees and branches, wolf classified as wolf due to branches

4.2 Re-trained Resnet50 on Imagenet's Subset with 9 Classes

For this research, we have utilized transfer learning [12] of ResNet50 and expect it to recognize 9 subjectively chosen classes. The subset ImageNet-9 consists of classes: bird, camel, dog, fish, insect, musical_instrument, ship, snowboard, wheeled_vehicle. Each class is often associated with certain environments like camel-desert, insect-plants, ship-water, snowboard-mountains, etc. Each class has been represented by approximately 500 images. It appears that transfer learning is quite an efficient method since 8 object types were recognized correctly as the actual object. Only one class, snowboard, took our attention. It appears that the presence of the sky implicated this recognition.

Using CAM-in-ROI method we have noticed that only 28% of significant pixels were inside ROI's rectangle, while for e.g. camels this value oscillated around 74%.

Table 1. ROI-fitting value and percentage of correctly classified images over ImageNet-9. Note that images where ROI has been not found were excluded

Class	Average fitting	Model accuracy
Bird	77.75%	88.25%
Camel	74.20%	22.70%
Dog	85.28%	96.18%
Fish	67.87%	62.82%
Insect	66.31%	77.10%
Musical_instrument	59.26%	81.80%
Ship	68.19%	88.08%
Snowboard	28.01%	82.28%
Wheeled_vehicle	83.77%	94.26%

Table 1 shows average importance, according to GradCAM, found inside ROI. The most outstanding is the snowboard class where less than one-third of saliency fits inside indicated ROI. Human conducted scrutiny revealed that for this certain class a top area associated with the sky is signal classification as the snowboard (Fig. 3).

Fig. 3. Images classified as snowboard and their respective saliency maps with ROIs

It is worth noticing that despite a low fitting ratio for the snowboard, the model's classification accuracy does not differ significantly from other classes. This leads to a conclusion that incorrect correlation may be undetectable by standard means.

This proof of concept demonstrated the practicality of the proposed method on a small subset, where a human validation could be done. The method has correctly identified the snowboard classification as defective and other correlation as satisfactory.

4.3 Pretrained VGG-11 for Object Classification

Xiao et al. stated that state-of-the-art pretrained networks appear to be immune to background skewed learning [16]. We have decided to scrutinize this statement by applying CAM-in-ROI measurement to the samples of all classes found in the ImageNet dataset and their classification using VGG architecture. We have prepared a batch of approximately 50 images of each class listed in the ImageNet set. Chosen results are displayed in Table 2.

Table 2. ROI-fitting value over full ImageNet. Only selected classes were displayed

Class	Average fitting
Japanese_spaniel	96.73%
Persian_cat	93.04%
Black_and_gold_garden_spider	11.08%
Valley	7.75%
Lakeside	7.50%
Seashore	2.46%
Bell_cote	1.53%
Alp	1.12%
Breakwater	0.68%
Mosque	0.25%
Rapeseed	0.00%

Some of the classes, like seashore, alp or lakeside can be considered object-less, since Detectron2 hardly ever recognized any object on the picture. Or there could have been highlighted an object that has only subsidiary importance. As we may see on the Fig. 4: ROI targets a person, while the network focuses, correctly, on the surroundings and classifies the image as a valley. Furthermore, some objects like bell cote are not known to the Detectron2. These issues caused many false positives in our results, although these may be overcome by applying different, more compelling ROI detecting tools.

Further we must acknowledge one more setback: our technique will turn out ineffective for detecting incorrect perception of the object. Namely like the aforementioned association of the cab object with yellow [21].

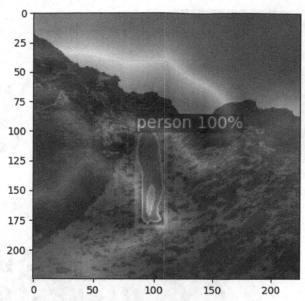

Fig. 4. An image classified correctly as a valley, despite the person detected in the foreground. Note that a tiny saliency region is included inside ROI thus giving as non-zero fitting measure

Nonetheless, our method has greatly limited the number of classes that require a closer look. Therefore we had to manually inspect only the lowest fitting classes. By reviewing images from the lowest fits we can highlight one distinctive class: black_and_gold_garden_spider. The black-and-gold-garden-spider often appears on the spider web background. While ROI indicates the insect quite well, the saliency map marks the spider web as the most significant area on the image, therefore resulting in only 11% average fitting. This leads us to the conclusion that the network assigns the label black_and_gold_garden_spider to an image that displays an object on a spider's web background.

4.4 Mocking Spider Image

Enhanced with knowledge about background influence on a certain class we can forge an image that has a spider's web in the background and something else in the foreground. We have downloaded a web image as well as a deer's head and merged them into one picture. Before the fusion, each picture was separately correctly classified: web as web and deer as a hartebeest (Fig. 5). The blended image, as expected, has been classified as black-and-gold-garden-spider.

By knowing the flaws in the network's reasoning, we were able to conduct an adversarial attack on it. Its success ultimately proves the potency of the proposed method.

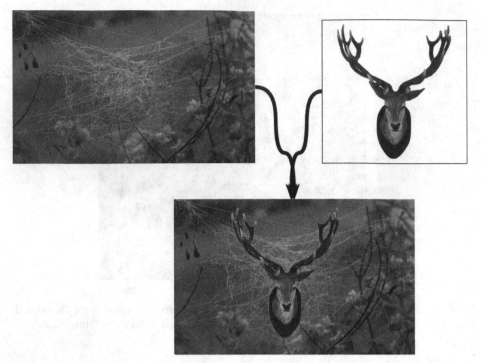

Fig. 5. Source images of spider's web and a deer's head that, according to results from 3rd experiment, should result in classification as black_and_gold_garden_spider

5   Conclusion

Presented evidence proves the existence of the serious problem in deep learning when a network can accidentally learn an incorrect correlation for a given class. Pretrained VGG-11 network from PyTorch's Model Zoo is flawed for at least one class: the black-and-gold-garden-spider.

The proposed method allows neural network practitioners to recognize incorrect correlations picked by the model and thus counter them by e.g. enhancing our training set.

Despite the named drawbacks, the proposed method might be a milestone in improving convolutional neural networks. It allows user to indicate classes that may lack legitimate feature selection and therefore cause complication in real-life applications.

6   Future Works

The proposed method is currently a proof of concept. It relies on the GradCAM saliency map and detectron2 framework trained on the COCO dataset. Possible improvements consist of choosing a better ROI detection method that knows more types of objects. Moreover, a more complex ROI area could be used instead of a simple rectangular space.

Consequently, the saliency maps generating method can be changed. Possible alternatives are GradCAM++ [3], occlusion maps [8] or excitation backpropagation [7].

Last but not least a refining the formulation of fitting metric can be changed. Currently, it is a naive ratio of in-ROI to entire pixels saliency values. A more accurate measure might rely on integrals computation.

References

1. Krizhevsky, A., Sutskever, I., Hinton, G.E.: ImageNet classification with deep convolutional neural networks. Commun. ACM **60**(6), 84–90 (2017)
2. Vung, P., Pham, C., Dang, T.: Road damage detection and classification with detectron2 and faster R-CNN. arXiv preprint arXiv:2010.15021 (2020)
3. Chattopadhay, A., et al.: Grad-cam++: generalized gradient-based visual explanations for deep convolutional networks. In: 2018 IEEE Winter Conference on Applications of Computer Vision (WACV). IEEE (2018)
4. Simonyan, K., Zisserman, A.: Very deep convolutional networks for large-scale image recognition. arXiv preprint arXiv:1409.1556 (2014)
5. Dinh, V., Ho, L.S.T.: Consistent feature selection for analytic deep neural networks. arXiv preprint arXiv:2010.08097 (2020)
6. Detectron2,   Detectron2   modelzoo.   https://github.com/facebookresearch/detectron2/ Accessed 16 Oct 2020
7. Zhang, J., Bargal, S.A., Lin, Z., Brandt, J., Shen, X., Sclaroff, S.: Top-down neural attention by excitation backprop. Int. J. Comput. Vis. **126**(10), 1084–1102 (2017). https://doi.org/10. 1007/s11263-017-1059-x
8. Shokoufandeh, A., Marsic, I., Dickinson, S.J.: View-based object recognition using saliency maps. Image Vis. Comput. **17**(5–6), 445–460 (1999)
9. Deng, Dong, W. Socher, R., Li, L.-J., Li, K., Li, F.-F.: ImageNet: a large-scale hierarchical image database. In: 2009 IEEE Conference on Computer Vision and Pattern Recognition, pp. 248–255. IEEE (2009)
10. Nguyen, A., Yosinski, J., Clune, J.: Deep neural networks are easily fooled: high confidence predictions for unrecognizable images. In: Proceedings of the IEEE Conference on Computer Vision and Pattern Recognition (2015)
11. Engstrom, L., et al.: Adversarial robustness as a prior for learned representations. arXiv preprint arXiv:1906.00945 (2019)
12. Schott, L., et al.: Towards the first adversarially robust neural network model on MNIST. arXiv preprint arXiv:1805.09190 (2018)
13. Ribeiro, M.T., Singh, S., Guestrin, C.: Why should I trust you?: explaining the predictions of any classifier. In: Proceedings of the 22nd ACM SIGKDD International Conference on Knowledge Discovery and Data Mining. ACM (2016)
14. Lapuschkin, S., et al.: Unmasking Clever Hans predictors and assessing what machines really learn. Nat. Commun. **10**(1), 1096 (2019)
15. Zech, J.R., et al.: Confounding variables can degrade generalization performance of radiological deep learning models. arXiv preprint arXiv:1807.00431 (2018)
16. Xiao, K., et al.: Noise or signal: the role of image backgrounds in object recognition. arXiv preprint arXiv:2006.09994 (2020)
17. Zhang, J., et al.: Local features and kernels for classification of texture and object categories: a comprehensive study. Int. J. Comput. Vis. **73**(2), 213–238 (2007)

18. Zhu, Z., Xie, L., Yuille, A.L.: Object recognition with and without objects. arXiv preprint arXiv:1611.06596 (2016)
19. Sagawa, S., et al.: Distributionally robust neural networks for group shifts: on the importance of regularization for worst-case generalization. arXiv preprint arXiv:1911.08731 (2019)
20. Torralba, A.: Contextual priming for object detection. Int. J. Comput. Vis. **53**(2), 169–191 (2003)
21. Geirhos, R., et al.: ImageNet-trained CNNs are biased towards texture; increasing shape bias improves accuracy and robustness. arXiv preprint arXiv:1811.12231 (2018)

Machine Learning Control Design for Elastic Composite Materials

Sebastián Ossandón[1]([✉]), Mauricio Barrientos[1], and Camilo Reyes[2]

[1] Instituto de Matemáticas, Pontificia Universidad Católica de Valparaíso,
Blanco Viel 596, CerroBarón, Valparaíso, Chile
sebastian.ossandon@pucv.cl
[2] Facultad de Ingeniería, Ciencia y Tecnología, Universidad Bernardo O'Higgins,
Avenida Viel 1497, Santiago, Chile

Abstract. A novel numerical method, based on a machine learning approach, is used to solve an inverse problem involving the Dirichlet eigenfrequencies for the elasticity operator in a bounded domain filled with a composite material. The inhomogeneity of the material under study is characterized by a vector which is designed to control the constituent mixture of homogeneous elastic materials that compose it. Using the finite element method, we create a training set for a forward artificial neural network, solving the forward problem. A forward nonlinear map of the Dirichlet eigenfrequencies as a function of the vector design parameter is then obtained. This forward relationship is inverted and used to obtain a training set for an inverse radial basis neural network, solving the aforementioned inverse problem. A numerical example showing the applicability of this methodology is presented.

Keywords: Machine learning · Composite materials · Inverse problems · Eigenfrequencies of the elasticity operator · Finite element method

1 Introduction

Inverse problems applied to elasticity are usually promoted by the need to capture relevant information concerning the features and design parameters of the elastic materials under study. There has been a great body of research related to this problem, which the following ones can be mentioned:

- Reconstruction of inclusions (see [2] and [4])
- Non-destructive evaluation for mechanical structures (see [21])
- Parametric identification for elastic models: Lamé coefficients, elastic moduli, Poisson's ratio, mass density or wave velocity (see [3])
- Reconstruction of residual stresses (see [6])
- Model updating: dynamic control for mechanical structures ([9])

European Union's Horizon 2020 research and innovation programme under the Marie Sklodowska-Curie grant agreement No. 777778: MATHROCKS PROJECT.

© Springer Nature Switzerland AG 2021
M. Paszynski et al. (Eds.): ICCS 2021, LNCS 12743, pp. 437–451, 2021.
https://doi.org/10.1007/978-3-030-77964-1_34

On the other hand, the use of spectral analysis to address elasticity problems is not new, for example, Babuška [5], Zienkiewicz [37,38], Oden [24] and Boffi [10,12], have analyzed the calculation of eigenfrequencies and eigenfunctions using this numerical technique. Also, see Sun and Zhou [31] and references therein. Forward numerical simulations of composites materials with local and well-defined inhomogeneities have been widely applied through the Finite Element Method (FEM), see for example Hassell and Sayas [20], Xu et al. [32] and Rodrigues et al. [29]. However applications of this method for composite (or inhomogeneous) materials is more complex. One of the reasons is that in anisotropic and composite (or inhomogeneous) materials, the models and the result interpretation are not easy to obtain using FEM, see Yan [33], Eigel and Peterseim [16], Choi [14] and Zhou et al. [36]. Consequently, FEM is not able to give an easy way to control the constituent mixture of composite materials.

Nowadays, machine learning algorithms (based on Artificial Neural Network (ANN)) are widely used in many problems. In Griffiths et al.[19] the solution of cornea curvature using a meshless method is discussed. In Liu et al. [22] a deep material network, is developed based on mechanistic homogenization theory of representative volume element and advanced machine learning techniques. The use of neural networks to solve the Stokes and Navier Stokes forward problems is studied in Baymani et al. [7,8]. The results show that the neural network has higher accuracy than classical methods. In Ossandón and Reyes [25], Ossandón et al. [26,28] the researchers solve the inverse eigenfrequency problems for the linear elasticity operator, the anisotropic Laplace operator, and the Stokes operator, respectively. Moreover, Ossandón et al. [27] solve an inverse problem to calculate the potential coefficients associated with the Hamiltonian operator in quantum mechanics.

In this article we are interested, in applying a machine learning approach, to obtain a vector of design parameters $\alpha = (\alpha_1, \alpha_2, \cdots, \alpha_m)^{\mathbf{T}} \in \mathbb{R}^m$, which characterizes the thicknesses between the interfaces (we assume m interfaces) of each elastic homogeneous material that make up the composite material under study, as a function of eigenfrequencies of the elasticity operator. We note that α is a vector of design parameters which control the constituent mixture of the composite material. The methodology proposed is based on the design of two ANNs (forward and inverse ANNs). The proposed ANNs are multilayered Radial-Basis Function (RBF) networks, and they are chosen due to the nature of the problem that is analyzed and the features exhibited by the neural network (see Schilling et al. [30] and Al-Ajlouni et al. [1]). Our main goal pursued with this article is to prove the effectiveness of the machine learning approach, evaluating its speed and accuracy in comparison with an approach based only on FEM. Therefore given a desired spectral behavior of the analyzed material, a specific composition for the mixture can be determined calculating α. The successful application of this methodology depends on the ability to face real inverse problem, specifically acquiring the eigenfrequencies associated with the elastic composite materials under study. In practice, using devices (or mechanisms) that use piezoelectric transducers, we can measure both the eigenfrequencies and the design parameters. Usually, the functioning of these mechanisms is based on resonance techniques such as resonant ultrasound spectroscopy (see [34,35]).

The article is organized as follows: first we start with a machine learning analysis to solve general inverse problems which is described in Sect. 2. The constitutive relations for elastic composite materials are described in Sect. 3, on the other hand in Sect. 4 we introduce the corresponding forward and inverse problems, and its respective solutions are displayed in Sects. 5 and 6. In Sect. 7, a numerical example is exposed, and finally the conclusions follow in Sect. 8. Figure 1 shows an example of the type of composite material this article is analyzing, which is composed by homogenous domains.

Fig. 1. Elastic composite material.

2   Solving Inverse Problems Using Machine Learning Algorithms

Mathematically, solving an inverse problem consists of estimating a signal P from data observations M where:

$$\mathcal{F}_T(P) = M, \qquad P \in \mathbb{X} \quad \text{and} \quad M \in \mathbb{Y}. \tag{1}$$

In Eq. (1), $\mathbb{X}$ and $\mathbb{Y}$ are vector spaces with norm, and $\mathcal{F}_T : \mathbb{X} \to \mathbb{Y}$ is called the forward operator. Let us observe that the forward operator has been parametrized by T. For example in some non-destructive evaluation problems, T denotes a well-known trajectory composed of several transmitter and receiver locations (sensor locations).

Machine learning methodology can be used in order to obtain (approximately) a non-linear mapping $\mathcal{F}_{T,\theta}^{\dagger} : \mathbb{Y} \to \mathbb{X}$ satisfying:

$$\mathcal{F}_{T,\theta}^{\dagger}(M) \approx P_{\text{seek}}, \qquad M \in \mathbb{Y} \quad \text{and} \quad P_{\text{seek}} \in \mathbb{X}. \tag{2}$$

Let us observe that $\mathcal{F}_{T,\theta}^{\dagger}$ (which is an ANN) has been parametrized also by T and by $\theta \in \Theta$. The main idea when we use machine learning algorithms is to make a choice of an optimal parameter $\theta^* \in \Theta$ given some training data. This optimal parameter can be obtained by minimizing a functional to control the quality of a learned $\mathcal{F}_{T,\theta}^{\dagger}$. In other words, we propose solving the inverse problem using machine learning algorithms amounts to learn $\mathcal{F}_{T,\theta}^{\dagger}$ from data such that it approximates an inverse of $\mathcal{F}_T$.

We assume that the metric spaces $\mathbb{X}$ and $\mathbb{Y}$ are endowed with the respective norms: $\| \cdot \|_{\mathbb{X}} : \mathbb{X} \to \mathbb{R}$ and $\| \cdot \|_{\mathbb{Y}} : \mathbb{Y} \to \mathbb{R}$.

2.1 Supervised Learning

In supervised learning the input/output data set are known. The optimization (minimization) problem that we propose to solve is:

$$\min_{\theta \in \Theta} \left\{ \frac{1}{2} \| \mathcal{F}_T \circ \mathcal{F}^{\dagger}_{T,\theta} \circ \mathcal{F}_T(\mathrm{P}) - \mathrm{M} \|_{\mathbb{Y}}^2 + \beta \mathcal{R}_T(\mathrm{P}) \right\}. \tag{3}$$

where the functional $\mathcal{R}_T : \mathbb{X} \to \mathbb{R}$ is known as a regularization functional that penalizes unfeasible solutions. The parameter β measures this penalization.

2.2 Unsupervised Learning

In the case of unsupervised learning the output data set is unknown. For his problem we propose to minimize:

$$\min_{\theta \in \Theta} \left\{ \frac{1}{2} \| \mathcal{F}_T \circ \mathcal{F}^{\dagger}_{T,\theta} \circ \mathcal{F}_T(\mathrm{P}) - \mathcal{F}_T(\mathrm{P}) \|_{\mathbb{Y}}^2 + \beta \mathcal{R}_T(\mathrm{P}) \right\}, \tag{4}$$

Let us notice that in many cases it can be quite complicated to use directly $\mathcal{F}_T$ in the optimization algorithm, which can be a minimizing numerical method. In these cases, it is convenient to also use an approximation (another ANN) $\mathcal{F}_{T,\theta_1}$ for $\mathcal{F}_T$. In addition, if this new network (called forward ANN) is well trained, we can greatly reduce the calculation time involved in solving many times many forward problems.

Finally, we must indicate that since in general the forward problems are well-posed, the use of the operator $\mathcal{F}_T \circ \mathcal{F}^{\dagger}_{T,\theta} \circ \mathcal{F}_T$ in the optimization problem limits the existence of unfeasible solutions, which together with the use of an additional regularization functional $\mathcal{R}_T$ could be suitable for choosing a "good physical" solution for the inverse problem.

3 Constitutive Relations in Elasticity

Let $\Omega \subset \mathbb{R}^l$ ($l = 2$ or 3) be a nonempty, open, convex and bounded domain, with its regular boundary Γ, filled with an elastic composite material characterized by its Lamé functions $\gamma_\alpha(\boldsymbol{x})$, $\mu_\alpha(\boldsymbol{x})$ ($\boldsymbol{x} \in \Omega$). In this work, $\boldsymbol{\alpha}$ is a designed vector of parameters used to modify the spectral response (eigenfrequencies and eigenfunctions) associated to an elastic material through the control of its constituent mixture.

We denote the elastic composite tensor, for an isotropic medium, by $C^{\alpha}_{ijrq}(\boldsymbol{x}) = \gamma_\alpha(\boldsymbol{x})\delta_{ij}\delta_{rq} + \mu_\alpha(\boldsymbol{x})(\delta_{jr}\delta_{iq} + \delta_{iq}\delta_{ir})$ ($1 \leqslant i,j,r,q \leqslant l$), where $\delta_{.,.}$ is the Kronecker delta function. We remark that this tensor is positive definite.

The generalized Hooke's law, relating the mechanical displacements u $(u(x))$ with the associated stress tensor σ^α $(\sigma^\alpha(x, u) = \gamma_\alpha(x)(\nabla \cdot u)\mathcal{I} + 2\mu_\alpha(x)\mathcal{E}(u))$ is given by:

$$\sigma_{ij}^\alpha(x, u) = \sigma_{ji}^\alpha(x, u) = \sum_{r,q} C_{ijrq}^\alpha(x)\mathcal{E}_{rq}(u), \tag{5}$$

where:

$$\mathcal{E}_{ij}(u) = \frac{1}{2}(\partial_j u_i + \partial_i u_j) \tag{6}$$

is the strain tensor.

4   The Forward and Inverse Problems

4.1   The Forward Problem

As a forward problem, we consider the following eigenfrequency problem: given $\alpha \in \mathbb{R}_+^m$, find $\lambda \in \mathbb{R}$ and the non-null valued functions u which are the solutions of:

$$\begin{cases} -\text{div}(\gamma_\alpha(x)(\nabla \cdot u)\mathcal{I} + 2\mu_\alpha(x)\mathcal{E}(u)) = \lambda u & \text{for} \quad x \in \Omega, \\ \qquad\qquad\qquad\qquad\qquad\qquad\quad u = 0 & \text{for} \quad x \in \Gamma. \end{cases} \tag{7}$$

where $\mathcal{I}$ is the identity matrix of size $l \times l$.

We note (see [5]) that the only non-null solutions of (7) are a countable pair sequence $\{(\lambda_n, u_n)\}_{n \geqslant 1}$ of eigenfrequencies and eigenfunctions.

We define the function $\mathcal{F}_N$ associated with (7):

$$\mathcal{F}_N : \mathbb{R}_+^m \to \mathbb{R}^N, \quad \overrightarrow{\lambda}^N := (\lambda_1, \lambda_2, \cdots, \lambda_N)^\mathbf{T} = \mathcal{F}_N(\alpha). \tag{8}$$

We remark that the function $\mathcal{F}_N$ $(N \in \mathbb{N})$ solves the forward problem associated to (7).

4.2   The Inverse Problem

We consider the following inverse problem associated to (7):

Find $\alpha \in \mathbb{R}_+^m$ such that

$$\begin{cases} -\text{div}(\gamma_\alpha(x)(\nabla \cdot u_n^d)\mathcal{I} + 2\mu_\alpha(x)\mathcal{E}(u_n^d)) = \lambda_n^d u_n^d & \text{in} \quad \Omega, \\ \qquad\qquad\qquad\qquad\qquad\qquad\qquad\qquad u_n^d = 0 & \text{on} \quad \Gamma, \end{cases} \tag{9}$$

where the given pair $\left\{\lambda_n^d, u_n^d\right\}_n$, with $n \in \mathbb{N}$ and $n \leq N < +\infty$, characterizing the desired spectral behavior.

We define the function $\mathcal{F}_N^{-1}$, which is the inverse function of $\mathcal{F}_N$, associated to (9):

$$\mathcal{F}_N^{-1} : \mathbb{R}^N \to \mathbb{R}_+^m, \quad \alpha = \mathcal{F}_N^{-1}(\vec{\boldsymbol{\lambda}}^N). \tag{10}$$

We remark that the function $\mathcal{F}_N^{-1}$ ($N \in \mathbb{N}$) solves the inverse problem associated to (9).

5   Solution of the Forward Problem

5.1   Variational Formulation

We define the functional space

$$\mathbf{V} = \mathbf{U} = \left\{ \boldsymbol{v} = (v_1, v_2, \cdots, v_l) \in [H^1(\Omega)]^l; \quad v_i = 0 \quad \text{on} \quad \Gamma, \quad 1 \leqslant i \leqslant l \right\}, \tag{11}$$

associated with the norm $\|\boldsymbol{v}\|_{1,\Omega}^2 = (\sum_{i=1}^{l} \|v_i\|_{1,\Omega}^2)^{1/2}$.

The corresponding variational form of Eq. (7) is given by:

$$a_\alpha(\boldsymbol{u}, \boldsymbol{v}) := \int_\Omega (\gamma_\alpha(\boldsymbol{x})(\nabla \cdot \boldsymbol{u})(\nabla \cdot \boldsymbol{v}) + 2\mu_\alpha(\boldsymbol{x})\mathcal{E}(\boldsymbol{u}) : \mathcal{E}(\boldsymbol{v}))dx = \lambda \int_\Omega \boldsymbol{u} \cdot \boldsymbol{v} dx. \tag{12}$$

Thus the weak formulation for the eigenfrequency problem in elasticity, considering homogeneous boundary conditions, is given by: *Find* $(\lambda, \boldsymbol{u}) \in (\mathbb{R}, \mathbf{U})$ *such that*

$$a_\alpha(\boldsymbol{u}, \boldsymbol{v}) = \lambda(\boldsymbol{u}, \boldsymbol{v})_{0,\Omega} \quad \forall \boldsymbol{v} \in \mathbf{V}, \tag{13}$$

where $(\cdot, \cdot)_{0,\Omega}$ is the inner product of $[L^2(\Omega)]^l$.

5.2   Discretization

To obtain the discrete form of the variational formulation (13), the approach is based on $\mathbb{P}_k$-Lagrange Finite Element ($k \geqslant 1$) in Ω is used.

Let $\{\mathcal{T}_h\}_{h>0}$ be a regular mesh discretizing Ω (see Ciarlet [15]), composed by triangles T_i ($i = 1, ..., M_h$) of diameter h_{T_i}, such that $h := \sup_{T_i \in \mathcal{T}_h} h_{T_i}$ measures the size of the mesh $\mathcal{T}_h$. Furthermore, we consider the finite element space $\mathbf{V}_h \subset \mathbf{V}$ of piecewise polynomials $\mathbb{P}_k$ ($k \geqslant 1$).

Let $(\lambda_h, \boldsymbol{u}_h) \in (\mathbb{R}, \mathbf{V}_h)$ be the eigenpair solution to the discrete weak form of (13). The Rayleigh quotient for each discrete eigenfrequency λ_h is given by:

$$\lambda_h = \frac{a_\alpha(\boldsymbol{u}_h, \boldsymbol{u}_h)}{(\boldsymbol{u}_h, \boldsymbol{u}_h)_{0,\Omega}} = \mathcal{F}_N(\boldsymbol{\alpha}). \tag{14}$$

Finally, let us mention that the Babuska-Brezzi condition (see [5, 10, 11, 13, 15, 23]) satisfied by the approximation spaces, ensures the wellposedness of the discrete weak form of (13).

6 Solution of the Inverse Problem Using Machine Learning

A the machine learning approach can be used to solve the inverse problem previously discussed. Afterwards a the machine learning algorithm can reconstruct the nonlinear inverse mapping $\mathcal{F}_{\theta,N}^{\dagger} : \mathbb{R}^N \to \mathbb{R}_+^m$ that approximate $\mathcal{F}_N^{-1}$ above defined. Let $\mathcal{F}_{\theta,N}^{\dagger}$ be an inverse Artificial Neural Network (ANN) with activation functions defined by a Radial Based Function (RBF). The structure will consist in one input layer containing n neurons, one hidden layer containing s neurons and an output layer containing m neurons. Let us notice that θ is a vector parameter associated with the specific network topology. The use of this type of ANN is treated by Girosi et al. [17] and Girosi and Poggio [18]. In these articles, the authors analyze, in detail, the regularization features of an RBF ANN.

To perform the construction of the above inverse network (which solves the inverse problem), we define a forward RBF ANN $\mathcal{F}_{\theta_1,N} : \mathbb{R}_+^m \to \mathbb{R}^N$, used as an approximation of the direct operator $\mathcal{F}_N$, with one hidden layer containing s_1 neurons and one output layer containing N neurons. Let us remark that, we use the forward network $\mathcal{F}_{\theta_1,N}$ instead of $\mathcal{F}_N$ to accelerate the calculation of eigenfrequencies. In this case θ_1 is a vector parameter associated to the topology of $\mathcal{F}_{\theta_1,N}$. An optimal estimation for θ_1 is given by:

$$\theta_1^* = \min_{\theta_1} \left\{ \frac{1}{2} \| \mathcal{F}_N(\alpha) - \mathcal{F}_{\theta_1,N}(\alpha) \|_{\mathbb{R}^N}^2 \right\}. \tag{15}$$

Once calculated θ_1^*, we can obtain an optimal estimation for θ using:

$$\theta^* = \min_{\theta} J(\theta) = \min_{\theta} \left\{ \frac{1}{2} \| \mathcal{F}_{\theta_1^*,N}(\alpha) - \mathcal{F}_{\theta_1^*,N} \circ \mathcal{F}_{\theta,N}^{\dagger} \circ \mathcal{F}_{\theta_1^*,N}(\alpha) \|_{\mathbb{R}^N}^2 \right\}. \tag{16}$$

Obtaining θ_1^* and θ^* is usually known as the training process associated with the forward and inverse networks respectively. Let us notice that the training process can be performed, for example, using a backpropagation algorithm (see [17,18]): starting with an initial parameter vector θ^0, the training algorithm iteratively decreases the mean square error updating θ, where each iteration is given as follows:

$$\theta^{i+1} = \theta^i - \epsilon \mathbf{L} \cdot \frac{\partial J(\theta^i)}{\partial \theta^i}, \tag{17}$$

where ϵ controls the length of the update increment and $\mathbf{L}$ is a matrix that defines the backpropagation algorithm to be used.

Finally, we remark that the regularization functional $\mathcal{R}_T$ for our specific problem is not required because $\mathcal{F}_{\theta_1^*,N} \circ \mathcal{F}_{\theta^*,N}^\dagger \circ \mathcal{F}_{\theta_1^*,N}$ (from its definition) only allows the "good physical" solution.

7   Numerical Result

Let us consider the problem (7), where $\Omega =]1,3[\times]1,2[\subset \mathbb{R}^2$ ($\boldsymbol{x} = (x,y)^\top$) and $\Gamma = \overline{\Omega} - \Omega$. A representative diagram showing the domain of our example is the Fig. 2. Figure 3 shows the function that models the interfaces between each of the homogenous domains. In this example $\alpha = (\alpha_1, \alpha_2)^\mathbf{T} = (\alpha, \alpha)^\mathbf{T}$.

Let us assume that $\gamma_j = \frac{\nu_j E_j}{(1+\nu_j)(1-2\nu_j)}, \mu_j = \frac{E_j}{2(1+\nu_j)}$ ($1 \leqslant j \leqslant 3$), where E_j is the Young's modulus and ν_j is the Poisson's ratio for the elastic homogeneous material of the domain Ω_j (see Fig. 2). Table 1 shows the values for E_j and ν_j ($1 \leqslant j \leqslant 3$) used in our example.

Table 1. Coefficients used in this numerical example.

Material	E (Young's modulus) GPa	ν (Poisson's ratio)
Cooper ($j = 1$)	124	0.34
Stainless steel ($j = 2$)	200	0.30
Aluminum ($j = 3$)	79	0.35

Let us define:

$$
\tilde{\Omega}_{\alpha,1}(\boldsymbol{x}) = \begin{cases} 1 & \text{if} \quad 1 < x \leqslant 1.5 - \dfrac{\alpha}{2}, 1 < y < 2.0, \\[2mm] \dfrac{1.5 + \frac{\alpha}{2} - x}{\alpha} & \text{if} \quad 1.5 - \dfrac{\alpha}{2} < x \leqslant 1.5 + \dfrac{\alpha}{2}, 1 < y < 2.0, \quad (18) \\[2mm] 0 & \text{if} \quad 1.5 + \dfrac{\alpha}{2} < x < 3.0, 1 < y < 2.0, \end{cases}
$$

$$
\tilde{\Omega}_{\alpha,2}(\boldsymbol{x}) = \begin{cases} 0 & \text{if} \quad 1 < x \leqslant 1.5 - \dfrac{\alpha}{2}, 1 < y < 2.0, \\[2mm] \dfrac{x - 1.5 + \frac{\alpha}{2}}{\alpha} & \text{if} \quad 1.5 - \dfrac{\alpha}{2} < x \leqslant 1.5 + \dfrac{\alpha}{2}, 1 < y < 2.0, \\[2mm] 1 & \text{if} \quad 1.5 + \dfrac{\alpha}{2} < x \leqslant 2.5 - \dfrac{\alpha}{2}, 1 < y < 2.0, \quad (19) \\[2mm] \dfrac{2.5 + \frac{\alpha}{2} - x}{\alpha} & \text{if} \quad 2.5 - \dfrac{\alpha}{2} < x \leqslant 2.5 + \dfrac{\alpha}{2}, 1 < y < 2.0, \\[2mm] 0 & \text{if} \quad 2.5 + \dfrac{\alpha}{2} < x < 3.0, 1 < y < 2.0, \end{cases}
$$

and

$$
\tilde{\Omega}_{\alpha,3}(x) = \begin{cases} 0 & \text{if} & 1 < x \leqslant 2.5 - \dfrac{\alpha}{2}, 1 < y < 2.0, \\[2mm] \dfrac{x - 2.5 + \frac{\alpha}{2}}{\alpha} & \text{if} & 2.5 - \dfrac{\alpha}{2} < x \leqslant 2.5 + \dfrac{\alpha}{2}, 1 < y < 2.0, \\[2mm] 1 & \text{if} & 2.5 + \dfrac{\alpha}{2} < x < 3.0, 1 < y < 2.0. \end{cases} \quad (20)
$$

Note that $\sum_{j=1}^{3} \tilde{\Omega}_{\alpha,j}(x) = 1$.

The relationships $\tilde{\Omega}_{\alpha,j}(x), 1 \leqslant j \leqslant 3$ are functions related to sets $\Omega_j, 1 \leqslant j \leqslant 3$ (see Fig. 3) where a1 $= 1.5 - \frac{\alpha}{2}$, a2 $= 1.5 + \frac{\alpha}{2}$, a3 $= 2.5 - \frac{\alpha}{2}$ and a4 $= 2.5 + \frac{\alpha}{2}$ respectively. These functions (which are a partition of the unity) are employed with the purpose to model the constituent mixture of homogeneous materials in order to obtain composite materials with new physical properties. It is important to remark that inhomogeneities considered in this example only depend only on the x variable and do not depend on y variable (see Fig. 2).

We also define the variable Lamé coefficients characterizing the inhomogeneities of elastic materials understudy

$$
\gamma_\alpha(x) = \frac{\sum_{j=1}^{3} \gamma_j \tilde{\Omega}_{\alpha,j}(x)}{\sum_{j=1}^{3} \tilde{\Omega}_{\alpha,j}(x)} \quad \text{and} \quad \mu_\alpha(x) = \frac{\sum_{j=1}^{3} \mu_j \tilde{\Omega}_{\alpha,j}(x)}{\sum_{j=1}^{3} \tilde{\Omega}_{\alpha,j}(x)}, \quad (21)
$$

In this example, the thickness $\alpha =$ a2$-$a1 $=$ a4$-$a3 is the design parameter used to control the constituent mixture of the composite material. Let us observe that:

$$
\lim_{\alpha \to 0} \gamma_\alpha(x) = \sum_{j=1}^{3} \gamma_j 1_{\Omega_j}(x) \quad \text{and} \quad \lim_{\alpha \to 0} \mu_\alpha(x) = \sum_{j=1}^{3} \mu_j 1_{\Omega_j}(x) \quad (22)
$$

where

$$
1_{\Omega_1}(x) = \begin{cases} 1 & \text{if} & 1 < x < 1.5, y \in]1,2[, \\ 0.5 & \text{if} & x = 1.5, y \in]1,2[, \\ 0 & \text{elsewhere}, \end{cases} \quad (23)
$$

$$
1_{\Omega_2}(x) = \begin{cases} 1 & \text{if} & 1.5 < x < 2.5, y \in]1,2[, \\ 0.5 & \text{if} & x = 1.5, y \in]1,2[, \\ 0.5 & \text{if} & x = 2.5, y \in]1,2[, \\ 0 & \text{elsewhere}, \end{cases} \quad (24)
$$

and

$$
1_{\Omega_3}(x) = \begin{cases} 1 & \text{if} & 2.5 < x < 3.0, y \in]1,2[, \\ 0.5 & \text{if} & x = 2.5, y \in]1,2[, \\ 0 & \text{elsewhere}. \end{cases} \quad (25)
$$

The purpose of this example is to obtain numerically a specific composition for the mixture (calculating α) with a desired spectral behavior of the composite material (solve the inverse problem).

The forward RBF ANN $\mathcal{F}_{\theta_1, N}$ is trained with data provided by simulations obtained using FEM ($\mathbb{P}_2$ finite elements): $\alpha^{(i)} = 0.4 - 0.02(i - 1)$, $(\overrightarrow{\lambda}^N)^{(i)} = \mathcal{F}_N(\alpha^{(i)})$, $1 \leqslant i \leqslant N_1$, being in this case $N_1 = 20$. We use Eq. (15), applying the backpropagation algorithm for the training data, in order to obtain the optimal vector parameter θ_1^* (associated with the forward RBF ANN topology) and performing the approximation for $\mathcal{F}_N$. After training the forward network, we use it to simulate a larger amount of data $N_2 \gg N_1$, with $\alpha^{(i)} = 0.4 - 0.002(i - 1)$ ($1 \leqslant i \leqslant N_2 = 200$), to obtain a set of training data for the construction of the inverse RBF ANN $\mathcal{F}_{\theta, N}^{\dagger}$. In this case we also use the backpropagation algorithm in the training process, using Eq. (16) to obtain the optimal vector parameter θ^* (associated with the inverse RBF ANN topology). Finally, we have the inverse network trained and we can use it to solve numerically the inverse problem. Figure 4 shows a comparison of the evolution, when the value of the simulated data is $N_s = 2000$ and $\alpha^{(i)} = 0.4 - 0.0002(i - 1)$, $1 \leqslant i \leqslant N_s$, of the first 4 Dirichlet eigenfrequencies: 1) directly calculated using the FEM and 2) calculated from the forward RBF ANN with the data input obtained from the ANN: $\mathcal{F}_{\theta_1^*, N} \circ \mathcal{F}_{\theta^*, N}^{\dagger} \circ \mathcal{F}_{\theta_1^*, N}(\alpha^{(i)})$. As seen in this figure, the Dirichlet eigenfrequencies calculated using ANNs approximate quite well the calculated Dirichlet eigenfrequencies using FEM. However when $\alpha \to 0$ the accuracy of the approximation decreases. This problem can be solved increasing the size of the data set as α converges to zero.

The computational performance is summarized in Table 2. The merit figure used are the mean squared error (MSE) and the computational time (using ANN: CT ANN) vs FEM: CT FEM), required for the simulations of our example. Let us notice that CT ANN is calculated taking into account the computational time required to obtain the training data set, using the FEM results, which are needed by the forward RBF ANN. Let us observe the CT ANN compared with the CT FEM (CT ANN $\ll$ CT FEM), remarking the good computational performance attained using the MSE. It is important to remark that the difference between the data set prepared using FEM ($N_1 = 20$) and the data set used for the simulation ($N_s = 2000$) in this example, implies the difference between the computational times CT ANN and CT FEM, and also the MSE. It is possible to improve the MSE by increasing N_1, implying a longer training time for the forward RBF ANN, and thus increasing the CT ANN.

The computer, used to obtain the results shown in this section, has a 2.4 GHz Intel Core Duo processors with 3 GB of RAM.

Table 2. Computational performance summary.

N_s	MSE: $\dfrac{1}{N_s}\displaystyle\sum_{i=1}^{N_s}(\alpha^{(i)} - \widehat{\alpha}^{(i)})^2$	CT ANN (s)	CT FEM (s)
2000	$1.8440e - 06$	65.9468	5330.69

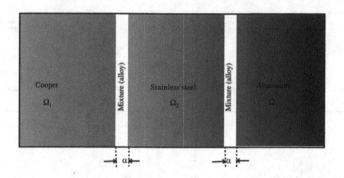

Fig. 2. Composite material used in the numerical example.

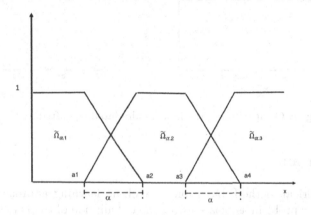

Fig. 3. Functions used in the numerical example.

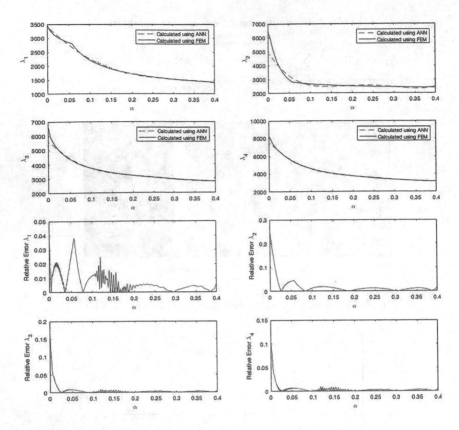

Fig. 4. Comparison of the first 4 calculated eigenfrequencies.

8 Conclusion

A novel numerical method, based on a machine learning approach, is used to solve an inverse problem associated with the calculation of the Dirichlet eigenfrequencies for the elasticity operator in a bounded domain filled with a composite material. The numerical results shows that the calculation, using RBF ANNs, of the vector of design parameters $\boldsymbol{\alpha} = (\alpha_1, \alpha_2)^{\mathbf{T}} = (\alpha, \alpha)^{\mathbf{T}}$ (in this case α is the thickness between the interfaces of each homogeneous material that compose the material used in the numerical example) from the eigenfrequencies of the elastic composite material, has a relatively negligible error and clearly the time consumption performance shows very important improvements compared to a more traditional approach based only on FEM (see Table 2). In summary, we have proved the effectiveness of a method that can be used as a control design tool: given a desired spectral response, we can control (motivated for the design) the constituent mixture of an elastic composite material. The method improves time performance without compromising the accuracy of the numerical results.

Finally, as a consequence of the notable improvements in the time calculation of our methodology, it can be used, in future works, to design real-time controllers for the mixture of composite materials.

Acknowledgements. S. Ossandón acknowledges support from the European Union's Horizon 2020 research and innovation programme under the Marie Sklodowska-Curie grant agreement No. 777778: MATHROCKS PROJECT.

References

1. Al-Ajlouni, A.-F., Schilling, R.-J., Harris, S.-L.: Identification of nonlinear discrete-time systems using raised-cosine radial basis function networks. Int. J. Syst. Sci. **35**(4), 211–221 (2004)
2. Alves, C.-J.S., Antunes, P.-R.S.: The method of fundamental solutions applied to the calculation of eigenfrequencies and eigenmodes of $2D$ simply connected shapes. CMC-Comput. Mater. Con. **2**(4), 251–265 (2005)
3. Ammari, H., Kang, H., Nakamura, G., Tanuma, K.: Complete asymptotic expansions of solutions of the system of elastostatics in the presence of an inclusion of small diameter and detection of an inclusion. J. Elast. **67**, 97–129 (2002)
4. Andrieux, S., Ben Abda, A., Bui, H.-D.: Sur l'identification de fissures planes via le concept d'écart à la réciprocité en élasticité. C.R. Acad. Sci. Paris, Série II **324**, 1431–1438 (1997)
5. Babuška, I., Osborn, J.-E.: Eigenvalue Problems, Handbook of Numerical Analysis: Finite Element Methods (Part 1), vol. 2, Ciarlet, P.G., Lions, J.L. (eds.). North-Holland, Amsterdam (2000)
6. Ballard, P., Constantinescu, A.: On the inversion of subsurface residual stresses from surface stress measurements. J. Mech. Phys. Solids **42**, 1767–1788 (1994)
7. Baymani, M., Effati, S., Kerayechian, A.: A feed-forward neural network for solving Stokes problem. Acta Appl. Math. **116**(1), 55–64 (2011)
8. Baymani, M., Effati, S., Niazmand, H., Kerayechian, A.: Artificial neural network method for solving the Navier-Stokes equations. Neural Comput. Appl. **26**, 765–773 (2015)
9. Ben Abdallah, J.: Inversion gaussienne appliquée à la correction paramétrique de modèles structuraux. Ph.D. thesis, Ecole Polytechnique, Paris, France (1995)
10. Boffi, D.: Finite element approximation of eigenvalue problems. Acta Numerica **19**, 1–120 (2010)
11. Boffi, D., Brezzi, F., Fortin, M.: Mixed finite element methods and applications. Springer Series in Computational Mathematics, vol. 44. Springer, Heidelberg (2013). https://doi.org/10.1007/978-3-642-36519-5
12. Boffi, D., Gastaldi, L.: Some remarks on finite element approximation of multiple eigenvalues. Appl. Numer. Math. **79**, 18–28 (2014)
13. Brezzi, F., Fortin, M.: Mixed and Hybrid Finite Elements Methods. Springer, Berlin (1991). https://doi.org/10.1007/978-1-4612-3172-1
14. Choi, H.-J.: A numerical solution for the inhomogeneous Dirichlet boundary value problem on a nonconvex polygon. Appl. Math. Comput. **341**, 31–45 (2019)
15. Ciarlet, P.-G.: The Finite Element Method for Elliptic Problems. North-Holland, Amsterdam (1978)
16. Eigel, M., Peterseim, D.: Simulation of composite materials by a nertwork FEM with error control. Comput. Methods Appli. Math. **15**(1), 21–37 (2014)

17. Girosi, F., Jones, M., Poggio, T.: Regularization theory and neural networks architectures. J. Neural Comput. **7**, 219–269 (1995)
18. Girosi, F., Poggio, T.: A theory of networks for approximation and learning, MIT Artificial Intelligence Laboratory, A.I. Memo No. 1140, C.B.I.P Paper No. 31 (1989)
19. Griffiths, G.-W., Plociniczak, L., Schiesser, W.-E.: Analysis of cornea curvature using radial basis functions - part I: methodology. Comput. Biol. Med. **77**, 274–284 (2016)
20. Hassell, M.-E., Sayas, F.-J.: A fully discrete BEM-FEM scheme for transient acoustic waves. Comput. Methods Appl. Mech. Eng. **309**, 106–130 (2016)
21. Leonard, K.-R., Malyarenko, E.-V., Hinders, M.-K.: Ultrasonic Lamb wave tomography. Inverse Prob. **18**, 1795–1808 (2002)
22. Liu, Z., Wu, C.-T., Koishi, M.: A deep material network for multiscale topology learning and accelerated nonlinear modeling of heterogeneous materials. Comput. Methods Appl. Mech. Eng. **345**, 1138–1168 (2019). https://doi.org/10.1016/j.cma.2018.09.020
23. Mercier, B., Osborn, J., Rappaz, J., Raviart, P.-A.: Eigenvalue approximation by mixed and hybrid methods. Math. Comp. **36**, 427–453 (1981)
24. Oden, J.-T., Reddy, J.-N.: An Introduction to the Mathematical Theory of Finite Elements. Wiley, New York (1976)
25. Ossandón, S., Reyes, C.: On the neural network calculation of the Lamé coefficients through eigenvalues of the elasticity operator. C. R. Mecanique **344**, 113–118 (2016)
26. Ossandón, S., Reyes, C., Reyes, C.-M.: Neural network solution for an inverse problem associated with the Dirichlet eigenvalues of the anisotropic Laplace operator. Comput. Math. Appl. **72**, 1153–1163 (2016)
27. Ossandón, S., Reyes, C., Cumsille, P., Reyes, C.-M.: Neural network approach for the calculation of potential coefficients in quantum mechanics. Comput. Phys. Commun. **214**, 31–38 (2017)
28. Ossandón, S., Barrientos, M., Reyes, C.: Neural network solution to an inverse problem associated with the eigenvalues of the Stokes operator. C. R. Mecanique **346**, 39–47 (2018)
29. Rodrigues, D.-E.S., Belinha, J., Pires, F.-M.A., Dinis, L.-M.J.S., Natal Jorge, R.-M.: Homogenization technique for heterogeneous composite materials using meshless methods. Eng. Anal. Boundary Elem. **92**, 73–89 (2018)
30. Schilling, R.-J., Carroll Jr., J.-J., Al-Ajlouni, A.-F.: Approximation of nonlinear systems with radial basis function neural networks. IEEE Trans. Neural Netw. **12**(1), 1–15 (2001)
31. Sun, J., Zhou, A.: Finite Element Methods for Eigenvalue Problems. CRC Press Taylor & Francis Group, Boca Raton (2016)
32. Xu, W., Xu, B., Guo, F.: Elastic properties of particle-reinforced composites containing nonspherical particles of high packing density and interphase: DEM-FEM simulation and micromechanical theory. Comput. Methods Appl. Mech. Eng. **326**, 122–143 (2017)
33. Yan, X.: Finite element modeling of consolidation of composite laminates. Acta Mechanica Sinica **22**(1), 62–67 (2006)
34. Zadler, B.-J.: Properties of elastic materials using contacting and non-contacting acoustic spectroscopy. Ph.D. thesis, Colorado School of Mines, Golden, Colorado, USA (2005)
35. Zadler, B.-J., Scales, J.-A.: Monitoring crack-induced changes in elasticity with resonant spectroscopy. J. Appl. Phys. **104**(2), 023536 (2008)

36. Zhou, L., Ren, S., Liu, C., Ma, Z.: A valid inhomogeneous cell-based smoothed finite element model for the transient characteristics of functionally graded magneto-electro-elastic structures. Compos. Struct. **208**, 298–313 (2019)

37. Zienkiewicz, O.-C.: Origins, milestones and directions of the finite element method a personal view. In: Ciarlet, P.G., Lions, J.L. (eds.) Handbook of Numerical Analysis: Techniques of Scientific Computing (Part 2), vol. 5. North-Holland, Amsterdam (1997)

38. Zienkiewicz, O.-C.: The Finite Element Method, 5th edn. McGraw-Hill, New York (2000)

Optimize Memory Usage in Vector Particle-In-Cell (VPIC) to Break the 10 Trillion Particle Barrier in Plasma Simulations

Nigel Tan[1]($\boxtimes$), Robert Bird[2], Guangye Chen[2], and Michela Taufer[1]

[1] University of Tennessee, Knoxville, TN 37919, USA
[2] Los Alamos National Laboratory, Los Alamos, NM 87545, USA

Abstract. Vector Particle-In-Cell (VPIC) is one of the fastest plasma simulation codes in the world, with particle numbers ranging from one trillion on the first petascale system, Roadrunner, to ten trillion particles on the more recent Blue Waters supercomputer. As supercomputers continue to grow rapidly in size, so too does the gap between compute capability and memory capability. Current memory systems limit VPIC simulations greatly as the maximum number of particles that can be simulated directly depends on the available memory. In this study, we present a suite of VPIC memory optimizations (i.e., particle weight, half-precision, and fixed-point optimizations) that enable a significant increase in the number of particles in VPIC simulations. We assess the optimizations' impact on a GPU-accelerated Power9 system. Our optimizations enable a 31.25% reduction in memory usage and up to 40% increase in the number of particles.

Keywords: Particle-In-Cell method · Mixed-precision · Fixed-point · Plasma physics

1 Introduction

Vector Particle-In-Cell (VPIC) is a high-performance particle-in-cell code that models plasma phenomena such as magnetic reconnection, fusion, solar weather, and laser-plasma interaction [2]. VPIC is one of the fastest PIC codes in the world and has performed some of the largest plasma simulations in history, ranging from one trillion particles on the first petascale system, Roadrunner [3], to ten trillion particle simulations on the more recent Blue Waters supercomputer [5]. VPIC simulations use large numbers of particles (i.e., on the order of trillions of particles [3]) to model real world phenomena. As we move the VPIC code from CPUs to accelerators (i.e., GPUs), the number of particles in VPIC simulations become limited by memory rather than compute capabilities; modern CPUs can access up to 4 TB of memory while GPUs (such as Nvidia A100) are limited to 80 GB, a factor of 50 difference. Moreover, data movement between CPUs and GPUs is costly, with PCIe 4.0 limited to 32 GB/s in one direction. Specialized protocols and hardware have

© Springer Nature Switzerland AG 2021
M. Paszynski et al. (Eds.): ICCS 2021, LNCS 12743, pp. 452–465, 2021.
https://doi.org/10.1007/978-3-030-77964-1_35

been developed to help address the issue; NVLink 3.0 achieves up to 300 GB/s in one direction [10]. Hardware and software techniques for maximizing communication efficiency are a major ongoing field of research [18]. At the code level, running VPIC on modern supercomputers with accelerators, while scaling up the number of particles, requires rethinking how the code uses memory.

In this paper, we introduce a new particle representation and develop a suite of optimizations for VPIC's particle storage format (i.e., particle weight, half-precision position, and fixed-point position optimizations) to tackle the memory usage problem. Our new particle representation reduces particle memory usage by up to 31.25%. We demonstrate that our optimizations enable significantly larger simulations, and that the optimized simulations produce accurate scientific results. Section 2 introduces VPIC's particle representation and workflow; Sect. 3 describes our method to increase particle count by reducing memory usage; Sect. 4 presents our performance and accuracy tests; Sect. 5 provides an overview of existing plasma simulation codes; and Sect. 6 summarizes the key results and introduces future work.

2 VPIC Particles and Workflow

VPIC is a high-performance implementation of the particle-in-cell method used for plasma simulations [11]. VPIC operates by defining a simulation space divided into a grid of cells and modeling particle movement across the cells. In other words, particles are distributed across an n-dimensional (n-D) space that is decomposed into an n-D grid. The resolution of the grid determines cell size and the maximum time step length. Figure 1a shows an example of a 2-D grid. Each simulated particle is a macroparticle (Fig. 1b) with a defined weight (i.e., the number of real particles modeled by each macroparticle).

The grid resolution and the number of particles both affect simulation accuracy and computational costs. Fine resolution grids better approximate a continuous n-D space. However, such fine resolution grids increase computational costs due to field operations, and the time step size must shrink accordingly to ensure numerical stability. Thus the number of time steps increases for the same period of simulated time, which further increases computational costs. Increasing the number of particles can also improve accuracy by more closely modeling real world plasma phenomena. High particle count primarily affects the particle advance stage with computational costs scaling linearly with the number of particles. In standard PIC simulations, the cell size is close to the Debye length (which is the smallest physical length scale in a plasma), and the time step is set as large as possible. Between the number of time steps and particle count, increasing particle count is generally preferred, as increasing the number of time steps incurs higher computational and communication costs.

A VPIC simulation is an iterative process across a defined number of time steps: each iteration has four key stages (as shown in Fig. 1c). First, the electric and magnetic fields are gathered from the grid points to each particle's location (interpolate fields). Second, particles move around based on the forces calculated from the electric and magnetic fields (advance particles). Third, the current generated by the particles' movements is scattered for each cell (accumulate currents). Last, electric and magnetic fields are advanced based on the accumulated current (advance EM fields), and the next iteration starts. For large, particle-heavy simulations, the first three stages take most of the simulation time, as each of these stages must operate on all the particles. In the last stage, VPIC advances the electric and magnetic fields at each grid point. The field advance stage is cheaper compared to prior stages since the number of particles tends to greatly outnumber the number of grid points. Furthermore, past I/O studies using VPIC noted that particles are responsible for the vast majority of memory usage. The trillion particle run in [6] required only 80 GB of storage for the electric and magnetic fields compared to the approximately 30 TB necessary for the particles. The 375-fold difference in memory requirement demonstrates that particle storage is a bottleneck for large scale VPIC simulations.

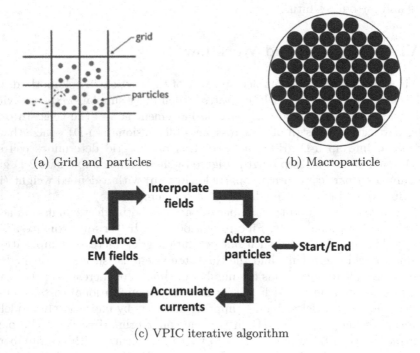

(a) Grid and particles   (b) Macroparticle

(c) VPIC iterative algorithm

Fig. 1. VPIC features: grid and particles (a); concept of macroparticle (b); and VPIC stages (c).

The particle position can be stored in two different ways based on the coordinate system: globally and locally. Global coordinates of particles are calculated and stored based on their absolute position in the n-D space. In Fig. 2a the global positions of the particles are depicted in a 2-D space as (dx, dy). Alternatively, particle positions can be stored in local coordinates with each particle storing its cell index and position within the cell. In Fig. 2b), each particle position in the 2-D space is defined as $(i * H_x + dx, j * H_y + dy)$, where (i, j) is the cell index, (H_x, H_y) is the cell size, and (dx, dy) is the local position of the particle. This representation requires extra space for the cell index but allows VPIC to represent each particle position more accurately. This is because the distribution of floating-point values follows an almost logarithmic distribution [16]. Values further away from zero are less precise, and approximately half of all floating-point values exist in $[-1, 1]$. Thus, local coordinates within the smaller cell have a more even and dense distribution of values to represent a particle position. The original VPIC code uses local coordinates and single-precision for storing particle positions.

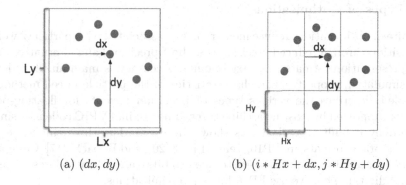

(a) (dx, dy) (b) $(i * Hx + dx, j * Hy + dy)$

Fig. 2. Global and local particle coordinates.

3   Increase Particle Count by Reducing Memory Usage

In VPIC simulations, the larger the number of particles, the more physically accurate the simulations and the greater the memory usage. Each particle is a macroparticle of 32 bytes (as shown in Fig. 3) that comprises 3 floats for position, 3 floats for momentum, 1 integer for the cell index, and 1 float for weight. There is a major disparity in memory usage between particle data and all the other data used in the code. Figure 4 shows an example of disparity for a VPIC simulation of 512 particles per cell; the particles are responsible for over 90% of the total memory usage. Larger VPIC simulations have thousands (or more) of particles per grid cell [3], making particles the primary focus when it is time to optimize memory usage. For instance, the simulations described in [1], which study laser-plasma modeling, have up to 65,536 particles per grid cell.

```
struct particle {
  float w;              // Weight
  float dx, dy, dz;     // Position
  int i;                // Cell index
  float ux, uy, uz;     // Momentum
};
```

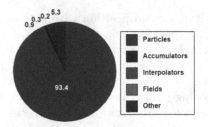

Fig. 3. Original VPIC particle data structure.

Fig. 4. Breakdown of memory usage for a simulation using 512 particles per cell.

3.1 Types of Optimizations

Our suite of optimizations reduce memory usage associated with particles' weight and position through reduced precision (i.e., half-precision) and alternative number representation formats (i.e., fixed-point). Accuracy is maintained by leveraging simulation properties and characteristics of the particle-in-cell method.

Table 1 describes the various types of precision formats for floating-point numbers supported by existing architectures. The original VPIC code uses single-precision by default (i.e., float), as shown in Fig. 3. The three candidates for reduced storage formats are FP16, TensorFloat [20], and Bfloat16 [17]. Compared to FP16, TensorFloat requires more storage (19 bits), and Bfloat16 loses precision (decimal digits). Thus we use FP16 for our optimizations.

Table 1. Floating-point formats with their numerical representations.

Precision	Sign	Exponent	Fraction	Decimal digits
Double (FP64)	1	11	52	$\approx$15.9
Single (FP32)	1	8	23	$\approx$7.2
Half (FP16)	1	5	10	$\approx$3.3
TensorFloat	1	8	10	$\approx$3.3
Bfloat16	1	8	7	$\approx$2.4

3.2 Particle Weight

In VPIC, the macroparticle's weight represents the number of real particles modeled by the simulated particle. The particle weight either changes within a limited range or does not change at all during a simulation. We use this property to optimize memory usage by adjusting how weight is stored.

When the particle weight varies but has a limited range of values, we can replace the single-precision weight with a 16-bit integer denoting a multiple of a known base weight. Alternatively, the 16-bit integer can be an index for a lookup table of particle weights. Both methods allow 65,536 different weights while reducing particle memory by 6.25%. We call this optimization short weight (SW).

When the particle weight remains constant for all particles of the same species throughout a simulation, particle weight can be removed entirely and replaced with a per species constant weight. This solution reduces memory usage by 12.5% over original VPIC. We call this optimization constant weight (CW). Figure 5 shows the reduction in memory usage for both optimizations (i.e., SW and CW).

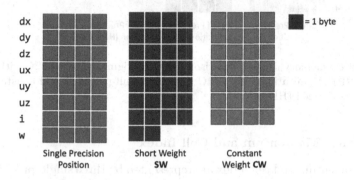

Fig. 5. Particle memory usage for default single-precision VPIC and VPIC with our short weight (SW) and constant weight (CW) optimizations.

3.3   Particle Position

The position of a particle in the original VPIC code is represented by three float values in single-precision (32 bits). We optimize the code by switching the representation to half-precision (16 bits). Figure 6 shows that by deploying half-precision we can reduce memory usage by 18.75% compared to the original VPIC and 31.25% when combined with the constant weight optimization (or CW). For a fine resolution grid, half-precision can produce simulations of sufficient accuracy while enabling larger simulations, as we will show in Sect. 4.

3.4   From Half-Precision to Fixed-Point

Half-precision particle position may incur an accuracy penalty in comparison to the default single-precision. Particle positions in each cell in VPIC using local particle coordinates are normalized to $[-1, 1]$, and thus we can use 16-bit fixed-point numbers instead of half-precision to minimize the loss in accuracy [7]. Fixed-point numbers allow us to maximize the number of bits used for precision. The fixed-point $Qm.n$ format specifies m bits for the integer and n bits for the fractional portion. We use the $Q1.14$ fixed-point format for storing position. One bit is for the sign, one bit for the integer portion, and the remaining 14 bits for the fractional portion. The $Q1.14$ format uses the same amount of memory as half-precision but has an additional four bits (approximately one decimal digit) for improving accuracy.

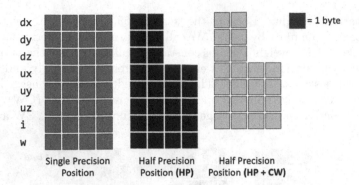

Fig. 6. Particle memory usage comparison between original VPIC, VPIC with our half-precision (HP) optimization, and VPIC with both half-precision and constant weight optimizations applied (HP+CW).

3.5 Particle Momentum and Cell Index

Particle momentum and cell index are represented by three single-precision floats and a 32-bit integer respectively. Momentum is left unchanged in this work. Momentum values are normalized to c (speed of light). Initial tests indicate that switching the momentum values from single-precision to half-precision would result in insufficient accuracy. Particle cell index determines which cell the particle resides in and is kept at the default 32-bit integer. A short 16-bit integer has insufficient range for large-scale simulations.

4 Performance and Accuracy

We present four test scenarios. The first test models laser-plasma interaction and is used for measuring runtime performance and memory usage of the original VPIC and the optimized version. The remaining three tests constitute a set of simple benchmark problems for analyzing the impact of our optimizations on the VPIC's numerical accuracy. All the tests were conducted on a four-node, GPU-accelerated IBM Power9 system. Each node has 155 GB of memory and 32 cores with 128 threads; two nodes host two Nvidia Tesla V100 GPUs each for a total of 4 GPUs. A single GPU is used, which is sufficient for testing and simplifies both data collection and analysis.

4.1 Performance

For testing runtime and memory performance, we have four problem sizes designed to use the GPU's full 16 GB memory capacity. Problem size is determined by the total number of particles in the system. The base case requires 16 GB to run using the original single-precision VPIC. The remaining cases increase in size until only our optimized versions can successfully run. Runtime tests are repeated 10 times and the average runtime is used to compare configurations. Memory usage is measured using the Space Time Stack tool provided by the Kokkos Tools profiling utilities [12]. The tool tracks Kokkos allocations and the maximum memory usage on the GPU.

We test VPIC scalability in terms of the number of particles using simulations modeling laser-plasma interactions. Figure 7 shows the scalability for the original VPIC code and when using our four optimizations (i.e., with short particle weight (SW), with constant particle weight (CW), using half-precision particle position (HP), and using fixed-point particle position (FP)). Missing columns indicate failed simulations that crash due to insufficient memory. We control problem size by adjusting the number of particles per cell (Nppc). The grid resolution and number of time steps are kept constant.

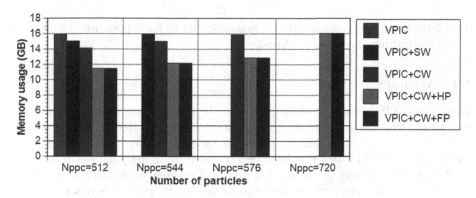

Fig. 7. Memory usage in GB of original VPIC and VPIC with our optimizations. Number of particles is proportional to the number of particles per cell (Nppc).

Our optimizations demonstrate a significant reduction in memory usage, as shown in Fig. 7. The optimizations to particle position (HP and FP) provide the greatest reduction in memory usage; when combined with constant weight (CW), they can increase the total number of particles by a factor of up to 40%. Figure 8 demonstrates that our optimizations also have minimal effect on runtime performance. Specifically, no negative effect is observed on runtime performance; in fact runtime can improve thanks to the hardware needing to move less data through its memory system and between processors. In other worlds, the optimized VPIC minimizes the amount of data movement between CPU and GPU. This results in a relatively small improvement in performance that largely cancels out the additional overhead from converting data between different formats when entering and exiting the advance particle stage (i.e., from single-precision

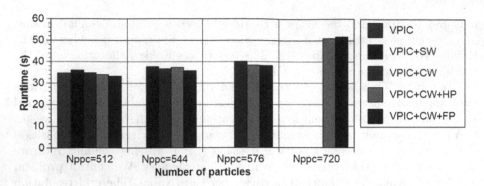

Fig. 8. Runtime in seconds of original VPIC and VPIC with our optimizations.

to half-precision and vice versa, from floating-point to fixed-point and vice versa) where the optimizations are deployed.

4.2   Accuracy

Measuring VPIC accuracy is difficult due to the lack of rigorous error quantification for the PIC method. Numerical accuracy testing is conducted with three different benchmark problems. In each test, we compare the original single-precision VPIC against VPIC with our half-precision and 16-bit fixed-point optimizations.

The first benchmark problem is a simple 1-D problem with periodic boundary conditions and an arbitrary number of particles N_p. Each particle has a constant weight $\frac{1}{N_p}$ and an initial momentum of $V_0 = 0$. The initial potential and electric field are given by

$$\phi_i(x) = \begin{cases} x(1 - x_i),\, 0 < x < x_i \\ x_i(1 - x),\, x_i < x < 1 \end{cases}$$

$$\phi_b(x) = \frac{x^2 - x}{2}, \quad \phi(x) = \sum_{i=1}^{N_p} \phi_i(x)w_i + \phi_b(x), \quad E(x) = -\frac{\partial \phi}{\partial x}$$

This test has an analytical solution for particle position described by

$$x_i = \frac{1}{N_p}\left[\left(\sum_k x_{k0}\right) - \frac{N_p + 1}{2} + i\right] - \frac{1}{N_p}\left[\sum_k (x_{k0} - x_{i0}) - \frac{N_p + 1}{2} + i\right]\cos(t)$$

Using the analytical solution we can track errors in an individual particle's position over time. For simplicity, we set the number of particles to two.

Table 2 describes the relative memory usage, and accuracy for each particle position representation. The 16-bit fixed-point setting guarantees an additional decimal digit of precision compared to 16-bit floating point.

Table 2. Relative memory usage, run-times, and accuracy for each potential particle position representation.

Format	Space reduction (# times)	Accuracy (Digits)
32-bit floating-point	1.00×	7.225
16-bit floating-point	0.81×	3.311
16-bit fixed-point	0.81×	4.515

This two-particle test problem can demonstrate the effects of our optimizations combined with grid resolution on particle position accuracy. Relative position error is tracked across 1,000,000 time steps with two different grid resolutions (i.e., 10,000 and 6,000 cells). The accuracy results for the two resolutions are shown in Fig. 9a and Fig. 9b respectively. Both the 16-bit floating-point and the 16-bit fixed-point formats trade storage space for precision as shown theoretically in Table 2: 16-bit floating-point maintains 3.311 decimal digits of precision while 16-bit fixed-point maintains 4.515 digits, a significant drop from the 7.225 digits for 32-bit floating-point. Figure 9a demonstrates that a sufficiently high resolution grid can make up for the drop in accuracy when using 16-bit floating-point or fixed-point for particle position.

It is important to note that grid resolution, time step length, and simulation length are directly related. High resolution grids cause the time step length to shorten, which in turn causes the number of time steps and the overall runtime to increase for a fixed simulation length. As a result, most simulations keep grid resolution as coarse as possible to minimize the number of simulated time steps. Striking a balance between grid resolution and accuracy is important. Figure 9b shows the same test with a lower resolution grid (i.e., 6,000). The effect of grid resolution on accuracy is clearly shown by the 16-bit floating-point format failing to maintain sub 10% relative error in this test. The fixed-point format has a notable increase in accuracy over half-precision and is comparable to the 32-bit single-precision floating-point; both 16-bit fixed-point and 32-bit floating-point achieve sub 10% relative error throughout the test Our results indicate that 16-bit fixed-point not only enables larger simulations than 32-bit single-precision but also improves accuracy over half-precision.

The second benchmark problem models the two-stream instability, which is an electrostatic instability commonly seen in plasma physics. Two counter streaming electron beams move in a stationary ion background [22]. The streams are vulnerable to electrostatic perturbations that result in charge bunching behavior [21]. The simulation is based on the two-stream instability deck described in [9]. The two-stream test focuses on verifying the conservation of momentum. In Fig. 10a, we observe how the conservation of momentum is maintained extremely well by all three variations of VPIC for the first 600 time steps. After 600 time steps, errors for fixed-point start growing while the errors for 32-bit and 16-bit floating-point remain low for a few dozen time steps before

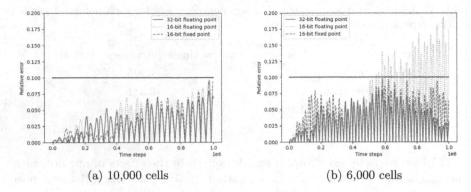

(a) 10,000 cells   (b) 6,000 cells

Fig. 9. Accuracy comparison VPIC simulations with different grid resolution. Particle weight is kept constant and particle position is shown using 32-bit floating-point, 16-bit floating-point, and 16-bit fixed-point.

growing and eventually plateauing. Half-precision and 16-bit fixed-point notably plateau with lower relative error than single-precision.

The third benchmark problem simulates the Weibel instability [15]. Similar to the two-stream instability, the Weibel instability can arise from counter streaming beams. Unlike the two-stream instability, the Weibel instability has electromagnetic perturbations that result in elongated structures in the plasma (filamentation). The process converts the kinetic energy of the particle beams into magnetic field energy [19]. The purpose of this test is to verify that energy is conserved with the particle format optimizations applied. Figure 10b clearly shows that total energy is conserved very well with less than 0.1% relative error across 10 million time steps for the Weibel instability. The curves for single-precision, half-precision, and fixed-point are tightly grouped together. Thus with our optimizations applied, conservation of energy is upheld with minimal variation compared to original VPIC.

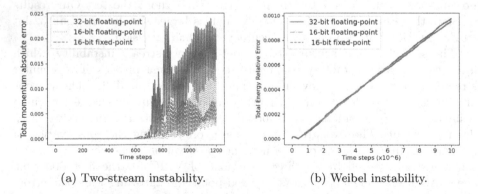

(a) Two-stream instability.   (b) Weibel instability.

Fig. 10. Conservation of momentum relative error for the two-stream instability (a) and conservation of energy relative error for the Weibel instability.

5   Related Work

Current state-of-the-art plasma simulation codes include VPIC, OSIRIS [13], WarpX [23], and PIConGPU [4]. These codes have all been used to run plasma simulations with approximately 10 trillion particles at the largest scale [5,8,14]. Both WarpX and PIConGPU use a similar particle representation format as VPIC. All three codes use single-precision for performance reasons. To the best of our knowledge, no other particle-in-cell code has investigated the use of half-precision or fixed-point representation for optimizing particle format.

6   Conclusions and Future Work

This paper demonstrates how the combination of constant weight and lower precision position in the VPIC code reduces memory usage by up to 31.25% and enables up to 40% increase in the number of particles using the same amount of memory for particle-heavy simulations. The optimizations improve performance by reducing the amount of data movement, which compensates for any additional operations introduced. Unlike other work in mixed-precision algorithms, our optimizations use reduced precision for storage rather than accelerating computation. We also show that our optimizations not only greatly increase simulation scale on memory constrained hardware, but can also achieve similar accuracy as the original single-precision VPIC. The fixed-point optimizations, in particular, show great promise thanks to the higher precision compared to half-precision. The higher precision allows for lower grid resolutions which ultimately decreases the number of time steps in the simulation. It is important to note that our optimizations may require changes to simulation parameters, namely the grid resolution. A grid resolution that produces accurate results for single-precision VPIC may not be accurate enough with half-precision or fixed-point. Such simulations may need to be adjusted to fully benefit from our optimizations.

Future work includes studying the algorithmic changes necessary to enable lower precision storage for particle momentum and the impact of these changes on both scalability and accuracy. We also plan to develop heuristics and methodologies to help physicists use our suite of optimizations. Adjustments to simulation parameters (e.g., grid resolution, step size and number of steps) are necessary to maintain accuracy. Different classes of plasma simulations also need to be studied to better understand which classes benefit from our optimizations the most and what changes to simulation settings are required to use our optimizations.

Acknowledgments. Work performed under the auspices of the U.S. DOE by Triad National Security, LLC, and Los Alamos National Laboratory (LANL). This work was supported the LANL ASC and Experimental Sciences programs. The UTK authors acknowledge the support of LANL under contract #578735 and IBM through a Shared University Research Award. LA-UR-21-21297.

References

1. Arber, T., et al.: Contemporary particle-in-cell approach to laser-plasma modelling. Plasma Phys. Control. Fus. **57**(11), 113001 (2015)
2. Bowers, K.J., Albright, B., Yin, L., Bergen, B., Kwan, T.: Ultrahigh performance three-dimensional electromagnetic relativistic kinetic plasma simulation. Phys. Plasmas **15**(5), 055703 (2008)
3. Bowers, K.J., Albright, B.J., Bergen, B., Yin, L., Barker, K.J., Kerbyson, D.J.: 0.374 pflop/s trillion-particle kinetic modeling of laser plasma interaction on road-runner. In: SC 2008: Proceedings of the 2008 ACM/IEEE Conference on Super-computing, pp. 1–11. IEEE (2008)
4. Burau, H., et al.: PIConGPU: a fully relativistic particle-in-cell code for a GPU cluster. IEEE Trans. Plasma Sci. **38**(10), 2831–2839 (2010)
5. Byna, S., Sisneros, R., Chadalavada, K., Koziol, Q.: Tuning parallel I/O on blue waters for writing 10 trillion particles. Cray User Group (CUG) (2015)
6. Byna, S., et al.: Parallel I/O, analysis, and visualization of a trillion particle simula-tion. In: SC 2012: Proceedings of the International Conference on High Performance Computing, Networking, Storage and Analysis, pp. 1–12. IEEE (2012)
7. Catrina, O., Saxena, A.: Secure computation with fixed-point numbers. In: Sion, R. (ed.) FC 2010. LNCS, vol. 6052, pp. 35–50. Springer, Heidelberg (2010). https://doi.org/10.1007/978-3-642-14577-3_6
8. Chandrasekaran, S., et al.: Running PIConGPU on summit: CAAR: preparing PIConGPU for frontier at ORNL. In: 4th OpenPOWER Academia Discussion Group Workshop (2019)
9. Chen, G., Chacón, L., Yin, L., Albright, B.J., Stark, D.J., Bird, R.F.: A semi-implicit, energy-and charge-conserving particle-in-cell algorithm for the relativistic Vlasov-Maxwell equations. J. Comput. Phys. **407**, 109228 (2020)
10. Choquette, J., Gandhi, W.: Nvidia A100 GPU: Performance & innovation for GPU computing. In: 2020 IEEE Hot Chips 32 Symposium (HCS), pp. 1–43. IEEE Com-puter Society (2020)
11. Dawson, J.M.: Particle simulation of plasmas. Rev. Modern Phys. **55**(2), 403 (1983)
12. Edwards, H.C., Trott, C.R., Sunderland, D.: Kokkos: enabling manycore per-formance portability through polymorphic memory access patterns. J. Parallel Distrib. Comput. **74**(12), 3202–3216 (2014). https://doi.org/10.1016/j.jpdc.2014.07.003. http://www.sciencedirect.com/science/article/pii/S0743731514001257. Domain-Specific Languages and High-Level Frameworks for High-Performance Computing
13. Fonseca, R.A., et al.: OSIRIS: a three-dimensional, fully relativistic particle in cell code for modeling plasma based accelerators. In: Sloot, P.M.A., Hoekstra, A.G., Tan, C.J.K., Dongarra, J.J. (eds.) ICCS 2002. LNCS, vol. 2331, pp. 342–351. Springer, Heidelberg (2002). https://doi.org/10.1007/3-540-47789-6_36
14. Fonseca, R.A., et al.: Exploiting multi-scale parallelism for large scale numeri-cal modelling of laser wakefield accelerators. Plasma Phys. Control. Fus. **55**(12), 124011 (2013)
15. Fried, B.D.: Mechanism for instability of transverse plasma waves. Phys. Fluids **2**(3), 337–337 (1959)
16. Goldberg, D.: What every computer scientist should know about floating-point arithmetic. ACM Comput. Surv. (CSUR) **23**(1), 5–48 (1991)
17. Kalamkar, D., et al.: A study of bfloat16 for deep learning training. arXiv preprint arXiv:1905.12322 (2019)

18. Li, A., Song, S.L., Chen, J., Li, J., Liu, X., Tallent, N.R., Barker, K.J.: Evaluating modern GPU interconnect: PCIe, NVLink, NV-SLI, NVSwitch and GPUDirect. IEEE Trans. Parallel Distrib. Syst. **31**(1), 94–110 (2019)
19. Morse, R., Nielson, C.: Numerical simulation of the Weibel instability in one and two dimensions. Phys. Fluids **14**(4), 830–840 (1971)
20. NVIDIA Corporation: Nvidia A100 tensor core GPU architecture. Technical report (2020). https://www.nvidia.com/content/dam/en-zz/Solutions/Data-Center/nvidia-ampere-architecture-whitepaper.pdf
21. Stix, T.H.: Waves in plasmas. Springer (1992)
22. Thode, L., Sudan, R.: Two-stream instability heating of plasmas by relativistic electron beams. Phys. Rev. Lett. **30**(16), 732 (1973)
23. Vay, J.L., et al.: Warp-X: a new exascale computing platform for beam-plasma simulations. Nucl. Instrum. Methods Phys. Res. Sect. A Acceler. Spectr. Detect. Assoc. Equip. **909**, 476–479 (2018)

Deep Learning for Prediction of Complex Geology Ahead of Drilling

Kristian Fossum[1]([✉]), Sergey Alyaev[1], Jan Tveranger[1], and Ahmed Elsheikh[2]

[1] NORCE Norwegian Research Centre, Bergen, Norway
krfo@norceresearch.no

[2] School of Energy, Geoscience, Infrastructure and Society, Heriot-Watt University, Edinburgh, UK

Abstract. During a geosteering operation the well path is intentionally adjusted in response to the new data acquired while drilling. To achieve consistent high-quality decisions, especially when drilling in complex environments, decision support systems can help cope with high volumes of data and interpretation complexities. They can assimilate the real-time measurements into a probabilistic earth model and use the updated model for decision recommendations.

Recently, machine learning (ML) techniques have enabled a wide range of methods that redistribute computational cost from on-line to off-line calculations. In this paper, we introduce two ML techniques into the geosteering decision support framework. Firstly, a complex earth model representation is generated using a Generative Adversarial Network (GAN). Secondly, a commercial extra-deep electromagnetic simulator is represented using a Forward Deep Neural Network (FDNN).

The numerical experiments demonstrate that the combination of the GAN and the FDNN in an ensemble randomized maximum likelihood data assimilation scheme provides real-time estimates of complex geological uncertainty. This yields reduction of geological uncertainty ahead of the drill-bit from the measurements gathered behind and around the well bore.

Keywords: Geosteering · Machine learning · Deep neural network · Generative Adversarial Network · Ensemble randomized maximum likelihood

1 Introduction

The process of drilling wells for hydrocarbon production represents a major cost in petroleum reservoir development. However, drilling of new wells is necessary to increase the total oil recovery. To maximize the value for each drilled well it is necessary to optimize the placement of the well within the reservoir structure. An optimally placed well will mobilize more of the petroleum resources, and reduce the need for injected water – reducing the environmental impact of oil production.

© Springer Nature Switzerland AG 2021
M. Paszynski et al. (Eds.): ICCS 2021, LNCS 12743, pp. 466–479, 2021.
https://doi.org/10.1007/978-3-030-77964-1_36

To place a well in its optimal position, operators apply geosteering. Here, the well trajectory is adjusted while drilling in response to real-time measurement of the geology surrounding the drill bit. The value of geosteering has been well documented in the literature [1,17,19].

The main objective with geosteering is to utilize the information in the measurements to make optimal decisions. Hence, geosteering can be seen as a sequential decision process under uncertainty and should be treated in a probabilistic framework [20]. Recently, a workflow based on the Ensemble Kalman Filter (EnKF) [12] has been employed to condition the geological model on measurements acquired while drilling [9,22]. In the EnKF the uncertainty is represented by an ensemble of equiprobable realizations. This workflow has then been combined with a global optimization method and applied as a Decision Support System (DSS) [3].

The DSS framework provides high quality decisions on synthetic cases, but practical challenges should be addressed for it to be applicable to real operations [3]. Firstly, to our knowledge, there is no studies which combine the ensemble update with a commercial tool for simulation of measurements. Secondly, the earth model utilized in the published studies does not represent a realistic geological complexity. Conceptually, it is easy to insert any numerical model for simulating the measurements into the DSS workflow. Similarly, there is nothing that prohibits the use of complex earth models. However, as the complexity increase, the numerical run-time also increases hindering real-time performance. Moreover, complex earth models can not typically be represented using a Gaussian distribution, and consequently EnKF updates will not retain the geological complexity [28].

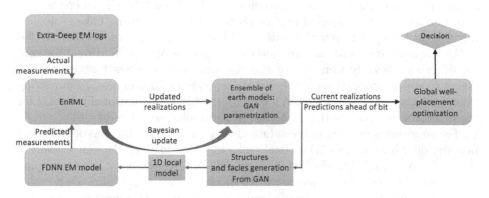

Fig. 1. The proposed DSS workflow. Green boxes highlight the new elements introduced in this paper. (Color figure online)

In this paper, we introduce important elements to make the DSS better suited to real operations, see Fig. 1. The main novelty in our approach is to introduce a machine-learning method to represent both the earth model and the forward model of extra-deep borehole electromagnetic (EM) measurements. Within this setting we demonstrate that a real-time ensemble-based inversion can indeed predict the distribution of non-trivial geology ahead of bit.

To construct a reference earth model we generate realizations of a fluvial geological environment using a commercial software. These realizations are then sub-sampled to form a training dataset for the offline training of a Generative Adversarial Network (GAN). The GAN is then used, online, to generate plausible geological realizations from a low-dimensional Gaussian input vector. The earth modeling is described in Sect. 2. For modeling the extra-deep EM measurements we use a forward deep neural network (FDNN) trained on a dataset generated using a commercial simulator (Sect. 3). In Sect. 4 we discuss the exact and the fast data assimilation (DA) methods. The numerical results, showing the applicability of our proposed method are given in Sect. 5. Finally, we summarize and conclude the paper in Sect. 6.

2   Earth Modeling Using GAN

GANs are a class of unsupervised machine learning methods which can learn to generate new formatted data with the same statistics as the training set. Motivated by successful applications of GANs for modeling channelized structures for reservoir simulation [7,8], we use a GAN for efficient earth modeling.

The GAN consists of two deep neural networks (DNNs): a generator and a discriminator. The generator takes a random Gaussian low dimensional vector as input and generates a realization of formatted data: geological realization. The discriminator takes the formatted data and gives a probability of it being 'real', i.e., belonging to the training set. During training the DNNs contest each other in a min-max game. They are trained simultaneously. On each training step the generator creates (fake) geological realization from random vectors. Fake geological realizations are combined with random samples of real earth model and are fed to the discriminator. The loss function for the generator is proportional to number of 'fakes' correctly identified by the discriminator. The loss function for the discriminator is proportional to the total misjudged data samples. In our study we use an adapted Wasserstein GAN with hierarchical convolutional networks for the generator and the discriminator, see [4] for implementation details.

For geosteering we want to reproduce likely geological realizations of facies and porosity distributions on a 2D vertical geological section along the well to identify the oil-bearing sands ahead of bit. For training of GAN we use a large (compare to the area of prediction) reference earth model, which should provide a realistic test case for the present study in terms of scale and actual geological features and properties. The reference earth model is constructed using a commercial software that models a synthetic structural framework, a facies model set-up derived from outcrop analogue data, and synthetic petrophysical properties of individual facies derived from published literature. The resulting model measures 4000 m × 1000 m × 200 m (xyz) with cell dimensions set to $10 \times 10 \times 0.5$ m, yielding a regular corner-point grid of size $400 \times 100 \times 400$, see Fig. 2.

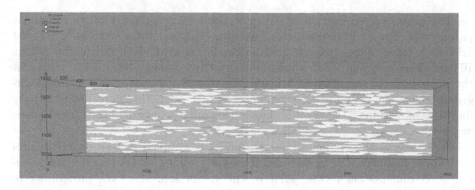

Fig. 2. The original earth model generated by the commercial tool.

The constructed facied model represents a low net/gross fluvial depositional system. It was chosen since it provides complex 3D architectures comprising a limited number of facies, which form contrasting geometries, see Fig. 2. Input numbers for statistical generation of facies and geometries are derived from a well-documented outcrop of the Cretaceous lower Williams Fork Formation (Mesa Verde Group) at Coal Canyon, Colorado, USA [26, 27, 30].

Key parameters of the facies model set-up are listed in Table 1. The model is not intended as a rendering of the outcrop itself and is consequently simplified compared to descriptions of the original outcrop [11, 24, 25, 27]. The model contains three facies: Background/shale, Channels and Crevasse splays. The probability distribution of channel width in the model is adapted to include "narrow channel bodies", and stacking of channels accounts for multi-story channels which comprise more than 80% of the observed channel bodies. The flow direction of the channel system is set towards $45 \pm 10°$. No trends were used to condition the spatial distribution of channels.

Table 1. Parameter settings for facies models.

Volumetric fraction	Value	Tolerance	Comments
Channel system volume fraction	0.3	0.05	
Channel positioning	1		No trends
Crevasse volume fraction	0.1	0.03	Of channel system vol. frac.

Channel geometry	Value	SD	Min.	Max.
Thickness	4.2	1.5		
Width	155	50	20	500
Correlation W/T	36			
Amplitude	400	50		
Sinuosity	1.3			
Azimuth	45	10		

Form/repulsion	Setting
Cross-section geometry	Parabolic, basic variability
Channel form	Rigid
Repulsion	None

The geological realization is parameterized by a vector of 60 independent parameters. For each 60-dimensional vector, the generator outputs a 64×64 grid with three values in each grid block. For a grid block (with dimensions 10.0 m along-well and 0.5 m thickness) the three values, 'channels', represent the probability of the grid-block belonging to the respective facies class: Background/Channel/Crevasse. Our generator is also predicting porosity/resistivity distribution within the geo-bodies, but in this initial study only the facies classes are used.

For training, the original 3D earth model is sampled as 64×64 2D images with three channels. The facies index from the training set is converted into one-hot three-dimensional vector. That is, the vector represents the probability of facies: the value of the true index is set to one and other channels to zero. During evaluation, the resistivity of the facies with highest probability is applied.

3   LWD Neural Network

To maintain real-time performance of a data assimilation workflow the forward model should be fast and support batch, preferably parallel execution. Proprietary forward models provided by measurement instrument vendors provide the most accurate results, but they are often not sufficiently fast, and not always optimized for batch execution. In [2], the authors developed a DNN approximation of such a forward model [29], which we abbreviate FDNN.

The model approximates the output of the ultra-deep electromagnetic wellbore logging instrument. The instrument is configured to transmit four shallow and nine pairs of deep directional measurements, and has sensitivity to boundaries up to 30 m to the side from the well bore. We emphasize that the tool provides information around, but not ahead of the drilling position. An illustration of the deep measurements is provided in Fig. 3.

The input to the FDNN model is a layered geological media with up to three boundaries above and below the measurement instrument as well as the resistivity values of all seven layers. In this study we assume that the layer resistivity is isotropic and that the well is aligned with the horizontal axis.

We produce one synthetic set of measurements for every horizontal position of the gridded model which we 'drill' through. First, we choose most probable facies for each 'pixel' and substitute it with the corresponding resistivity value:

1. Background, $R = 220.0 \, \Omega\text{m}$;
2. Channel, $R = 3.6 \, \Omega\text{m}$;
3. Crevasse, $R = 4.1 \, \Omega\text{m}$.

Second, we find boundaries between layers composed of pixels with equal resistivities and use them as the input to the forward model.

4   Data Assimilation

In the DSS for geosteering [3], one uses data assimilation to condition the earth model to measurements made while drilling. The fundamental idea is that if a

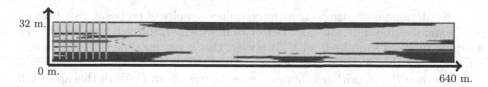

Fig. 3. Resistivity of earth model plotted in 1:2 aspect ratio. The green lines shows the measurements and their extent illustrate the maximum sensitivity depth. The full red line is the drilled well, and the dashed red lines indicate the potential for geosteering. (Color figure online)

poorly known earth model can be made consistent with measurements it will provide more accurate forecasts, and, hence, provide a better basis for decisions.

In this paper, the emphasis is placed on the data assimilation part of the DSS. Especially data assimilation utilizing an efficient neural network model for the synthetic logs, and an efficient GAN-generator for representing complex earth models. To check whether this setup provides useful conditioned models we consider two data assimilation algorithms. Firstly, the Markov Chain Monte Carlo (MCMC), which is considered as a gold standard method for sampling. Secondly, the ensemble randomized maximum likelihood (EnRML), an approximate method suitable for DSS.

4.1 MCMC

A reliable method for sampling from a complex posterior distribution is the MCMC technique. MCMC relates to the general framework of methods introduced in [23] and [18] for Monte Carlo (MC) integration. One designs a Markov chain that produce samples from the desired posterior distribution, and subsequently utilize these samples for MC estimation. In this section, the adaptive Metropolis-Hastings method, utilized in the numerical study, is introduced. For more information on MCMC we refer the reader to [6], and references therein.

Suppose we want samples from the un-normalized posterior distribution F, which is the general case with the Bayesian method where the normalizing factor often is very difficult to calculate. Assume that the current element is m, and that the chain proposes a move to m^*, with conditional probability density $q(m^*|m)$. The move is performed with probability

$$b(m, m^*) = min(1, r(m, m^*)) \tag{1}$$

where the Hastings ratio is defined as

$$r(m, m^*) = \frac{F(m^*) q(m|m^*)}{F(m) q(m^*|m)}. \tag{2}$$

If the move is not made $m^* = m$. This is the basis for the Metropolis-Hastings method, and it can be shown that the method will generate samples from the posterior distribution F.

The Metropolis-Hastings algorithm requires a choice of proposal distribution, and some distributions will work better than others. Intuitively, one would like to draw proposal samples from F. However, this is not possible since we cannot sample from this distribution. However, one idea is to consider the previous samples from the algorithm as approximate samples from F. With this approach proposal samples are drawn from

$$m^* \sim (1 - \beta) \mathcal{N} \left(m, \left(\frac{2.38^2}{N_m} \right) \tilde{C}_m \right) + \beta \mathcal{N} (m, Q_m), \qquad (3)$$

where $\tilde{C}_m$ is the empirical covariance matrix calculated utilizing all the preceding iterations of the Markov Chain, Q_m is some fixed non-singular matrix and $0 < \beta < 1$. Note that $\beta = 1$ until C_m is well defined. This sampling method was applied in [13,14]. It is well known that the MCMC requires a certain burn-in period, since the initial samples are not from the posterior distribution. Hence, it is necessary to monitor the convergence of the method. In this work, convergence is monitored by assessing the maximum root statistic of the multivariate potential scale reduction factor [5].

4.2   EnRML

The EnRML [16] has recently become one of the most successful methods for automatic history matching of petroleum reservoirs. The EnRML is based on minimization of an objective function using the ensemble approximation of the sensitivity matrix. Hence, the EnRML can be formulated in many different ways. In this study we utilize the approximate form of the Levenberg-Marquardt method, introduced in [10].

Iteration number i of the Levenberg-Marquardt method is given as

$$\delta m_i = - \left[(1 + \lambda_i) C_m^{-1} + G_i^T C_d^{-1} G_i \right]^{-1} \qquad (4)$$

$$\times \left[C_m^{-1} (m_i - m_{prior}) + G_i^T C_d^{-1} (g(m) - (d_{obs} + \epsilon)) \right] \qquad (5)$$

where λ_i is the Levenberg-Marquardt multiplier, G is the sensitivity of data to the parameters, and $\epsilon \sim \mathcal{N}(0, C_d)$ is a realization of the measurement observation noise.

In the ensemble framework, we approximate C_m and G using the ensemble. To this end we define

$$\tilde{G} = C_{sc}^{1/2} \Delta d (\Delta m)^{-1} \qquad (6)$$

$$\tilde{C}_m = \Delta m \Delta m^T \qquad (7)$$

where

$$\Delta m = [m_1, \ldots, m_j, \ldots, m_N] \left(I_N - \frac{1}{N} 11^T \right) / \sqrt{N - 1}, \qquad (8)$$

$$\Delta d = C_{sc}^{-1/2} [g(m_1), \ldots, g(m_j), \ldots, g(m_N)] \left(I_N - \frac{1}{N} 11^T \right) / \sqrt{N - 1}, \qquad (9)$$

N denotes the ensemble size, and C_{sc} is a diagonal matrix for scaling the data, typically containing the measurement variance on the diagonal. We get the approximate version of the Levenberg-Marquardt update equation by inserting ensemble approximations of G and C_m, neglecting the updates from the model mismatch term, substituting the prior precision matrix C_m^{-1} with $\tilde{C}_{m_i}^{-1}$, and rewriting the equation using the Sherman-Woodbury-Morrison matrix inversion formula [15]

$$\delta m_i = -\tilde{C}_{m_i} \tilde{G}_i^T \left[(1 + \lambda_i) C_d + \tilde{G}_i \tilde{C}_{m_i} \tilde{G}_i^T \right]^{-1} (g(m) - (d_{obs} + \epsilon)). \tag{10}$$

The update equation is simplified by calculating the truncated singular value decomposition of Δd

$$\Delta d = U_p S_p V_p^T, \tag{11}$$

where the subscript p indicates the number of singular values that are kept. In this work, we define p such that the cumulative sum of the p singular values equals 99% of the cumulative sum of all the singular values. Further, to allow for correlated measurement errors, we substitute C_D with the ensemble approximation $\tilde{C}_D$

$$\tilde{C}_d = \Delta \epsilon \Delta \epsilon^T, \tag{12}$$

where

$$\Delta \epsilon = [\epsilon_1, \ldots, \epsilon_j, \ldots, \epsilon_N] \left(I_N - \frac{1}{N} 11^T \right) / \sqrt{N-1}. \tag{13}$$

Inserted into (10) gives

$$\delta m_i = -\Delta m_i V_p \left[(1 + \lambda_i) S_p^{-1} U_p^T C_{scl}^{-1/2} \Delta \epsilon \Delta \epsilon^T C_{scl}^{-1/2} U_p S_P^{-T} + I \right]$$
$$(U_p S_p^{-1})^T C_{sc}^{-1/2} (g(m) - (d_{obs} + \epsilon)) \tag{14}$$
$$= -\Delta m_i V_p Z \left[(1 + \lambda_i) \zeta + I \right]^{-1} (U_p S_p^{-1} Z)^T C_{sc}^{-1/2} (g(m) - (d_{obs} + \epsilon)),$$

where Z and ζ are the eigenvectors and eigenvalues of $S_p^{-1} U_p^T C_{scl}^{-1/2} \Delta \epsilon \Delta \epsilon^T C_{scl}^{-1/2} U_p S_P^{-T}$.

The iterative scheme is run until it is converged. Here, we consider the method to be converged when the relative difference in the data misfit is below a given threshold.

5   Numerical Results

We, throughout, utilize the generative neural network, introduced in Sect. 2, to represent the poorly known earth model. Hence, realizations of the earth model is generated by applying the generative network to parameters sampled from the multivariate standard distribution, $m \sim \mathcal{N}(0, C_m)$. The numerical investigation considers a well drilled horizontally, approximately, in the center of the model and through the 9 first grid-cells. For each drilled grid-cell we simulate measurements using the model introduced in Sect. 3.

Throughout the investigation, we consider a diagonal C_m with elements equal to 1×10^{-6}. Figure 4 shows two random earth model realizations from the prior model. From the figure, we observe that this setup provides significant variation in the earth model.

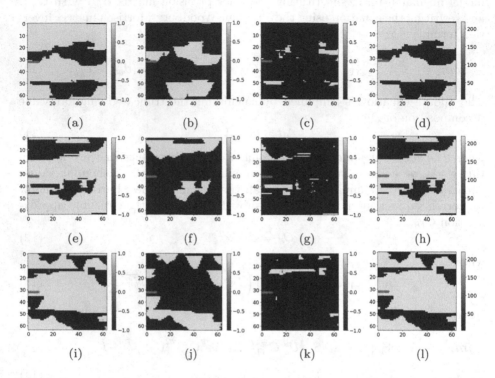

(a)   (b)   (c)   (d)

(e)   (f)   (g)   (h)

(i)   (j)   (k)   (l)

Fig. 4. Rows 1–3 show three realizations from the prior ensemble. Each row shows the three output channels (e.g. fig a–c) of GAN corresponding to probability of facies (Background/Channel/Crevasse), and the derived resistivity image (e.g. fig. d). Red stars indicate measurement position. (Color figure online)

The synthetic true earth model is also drawn from the prior model. All 3 channels and the derived resistivity of the synthetic truth is illustrated in Fig. 5. The true observations are simulated using the true earth model.

For each of the 9 measurements positions the measurement standard deviation is given as 5% of the measurement value. In addition, we let the measurements at each position be correlated. Assuming equidistance between measurements, the correlation length equals 10 times the inter-measurement distance. We further assume that the measurements at two different well positions are uncorrelated.

We conduct two numerical experiments. Firstly, we conduct a MCMC run to properly characterize the posterior distribution. Here, 8 chains are run in parallel for 10^6 iteration. At that point the chains were converged, and we extract samples

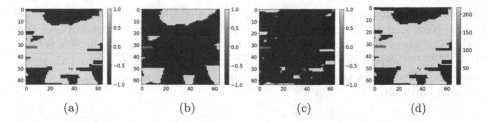

(a)   (b)   (c)   (d)

Fig. 5. (a)–(c): Facies probabilities from GAN for the synthetic truth. (d): Derived resistivity for the synthetic truth. Red stars indicate measurement positions.

from the posterior by, for each of the 8 chains, removing the first half of the chain and retaining every 100 iteration from the second half of the chain. Hence, leaving 4×10^4 samples from the posterior distribution. Secondly, we estimate the posterior distribution using the EnRML method introduced in Sect. 4.2. Due to the fast simulation time we utilized an ensemble size of $N = 500$, and in addition we applied the correlation based localization technique introduced in [21]. The method is allowed to iterate until the relative improvement of the updates is less than 1%. When showing the numerical results we will only plot the values of the derived resistivity.

5.1   MCMC

Based on the posterior realizations obtained by the MCMC, we calculate the posterior mean and posterior standard deviation, shown in Fig. 6 (a) and (b). In Fig. 6 (c) and (d) we plot two random realizations from the posterior.

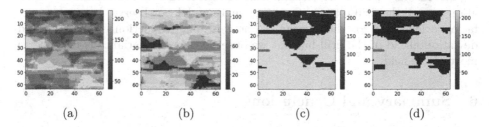

(a)   (b)   (c)   (d)

Fig. 6. (a): MCMC mean of the resistivity. (b): MCMC standard deviation of the resistivity. (c): Resistivity for MCMC posterior realization 1. (d): Resistivity for MCMC posterior realization 2. Red stars indicate measurement positions. (Color figure online)

The MCMC result illustrates that the generative neural network can be utilized for data assimilation, and be successfully conditioned to measurements utilizing a neural network proxy model. The model standard deviation is significantly reduced close to the drill bit, and also ahead of the drill bit position.

Moreover, the mean value shows that the correct resistivity is identified in these areas. Note that the posterior still has significant variance in most parts of the field. The areas directly around the drill bit have not obtained sufficient reduction of the standard deviation. We do not properly understand why this is so, however, it indicates that there measurements are less sensitive to the region near the drill bit.

5.2   EnRML

After the EnRML has converged we calculate the mean and standard deviation from the ensemble. These are approximations to the true posterior mean and standard deviations, and are shown in Fig. 7 (a) and (b). In addition, the results from two random realizations are illustrated in Fig. 7 (c) and (d).

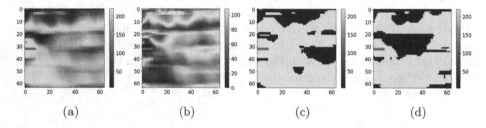

(a)   (b)   (c)   (d)

Fig. 7. (a): Mean resistivity from the EnRML. (b): Standard deviation of the resistivity from the EnRML. (c): Resistivity for EnRML realization 1. (d): Resistivity for EnRML realization 2. Red stars indicate measurement positions. (Color figure online)

The results for the EnRML show a significant reduction in the standard deviation around the drill bit, and from the mean model we observe that the correct facies is identified for these regions. There is some reduction of the standard deviation ahead of the drill bit, but we observe that the variance is retained in most of the field.

6   Summary and Conclusions

In this paper, we have demonstrated that two essential parts, the earth model and the simulated log, of an ensemble based DSS system can be substituted with neural networks. For the earth model we utilize the GAN trained with images from a realistic geological setting, for the simulated log we use a deep neural network trained using a large set of simulations from a commercial tool. The setup redistributes the computational cost from on-line to off-line calculations, enabling complex earth models utilizing simulated logs with high accuracy to be used in the real time DSS. The numerical results illustrate that DSS, equipped with the GAN, provides good predictions ahead of drilling when conditioning to only measurements with sideways sensitivity.

The proposed approach has many beneficial factors. Firstly, a GAN provides large flexibility for defining the geological setting. Here, we consider three different facies, but one can easily imagine the inclusion of features like faults and pinch-outs as well as smoothly-varying properties. Secondly, we only need to condition a few parameters with Gaussian distribution to the measurements, which is very beneficial for the ensemble based DA approach. Thirdly, since we are utilizing a neural network model to generate the simulated log the computational cost of simulating a single ensemble member is very low. Hence, the proposed approach can utilize a large ensemble for the DA part.

The numerical experiments illustrated that the setup provides a reasonable posterior distribution, and we can estimate this using the EnRML approximation. In this study we have only considered a single part of the DSS, namely the conditioning of the earth model to measured data. However, due to the promising results, the developments shown in this paper will integrate with the framework developed in [3]. Hence, allowing DSS under much more complex geological setting. Demonstrating the complete DSS with the proposed setup is left for future work.

Acknowledgments. The authors are supported by the research project 'Geosteering for IOR' (NFR-Petromaks2 project no. 268122) which is funded by the Research Council of Norway, Aker BP, Equinor, Vår Energi and Baker Hughes Norway.

References

1. Al-Fawwaz, A., et al.: Increased net to gross ratio as the result of an advanced well placement process utilizing real-time density images. In: IADC/SPE Asia Pacific Drilling Technology Conference and Exhibition, pp. 151–160. Society of Petroleum Engineers (2004). https://doi.org/10.2118/87979-MS. http://www.onepetro.org/doi/10.2118/87979-MS
2. Alyaev, S., et al.: Modeling extra-deep EM logs using a deep neural network. arXiv preprint arXiv:2005.08919 (2020). Accepted in SEG Geophysics
3. Alyaev, S., Suter, E., Bratvold, R.B., Hong, A., Luo, X., Fossum, K.: A decision support system for multi-target geosteering. J. Petrol. Sci. Eng. **183**, 106381 (2019). https://doi.org/10.1016/j.petrol.2019.106381. https://doi.org/10.1016/j.petrol.2019.106381linkinghub.elsevier.com/retrieve/pii/S0920410519308022
4. Arjovsky, M., Chintala, S., Bottou, L.: Wasserstein GAN. arXiv (2017). http://arxiv.org/abs/1701.07875
5. Brooks, S.P., Gelman, A.: General methods for monitoring convergence of iterative simulations. J. Comput. Graph. Stat. **7**(4), 434 (1998). https://doi.org/10.2307/1390675. http://www.tandfonline.com/doi/abs/10.1080/10618600.1998.10474787www.jstor.org/stable/1390675?origin=crossref
6. Brooks, S.P., Gelman, A., Jones, G.L., Meng, X.L. (eds.): Handbook of Markov Chain Monte Carlo. Chapman and Hall/CRC, Boca Raton (2011)
7. Chan, S., Elsheikh, A.H.: Parametric generation of conditional geological realizations using generative neural networks. Comput. Geosci. **23**(5), 925–952 (2019). https://doi.org/10.1007/s10596-019-09850-7. http://link.springer.com/10.1007/s10596-019-09850-7

8. Chan, S., Elsheikh, A.H.: Parametrization of stochastic inputs using generative adversarial networks with application in geology. Front. Water **2**, 1–21 (2020). https://doi.org/10.3389/frwa.2020.00005. https://www.frontiersin.org/article/10.3389/frwa.2020.00005/full

9. Chen, Y., Lorentzen, R.J., Vefring, E.H.: Optimization of well trajectory under uncertainty for proactive geosteering. SPE J. **20**(02), 368–383 (2015). https://doi.org/10.2118/172497-PA. https://onepetro.org/SJ/article/20/02/368/206467/Optimization-of-Well-Trajectory-Under-Uncertainty

10. Chen, Y., Oliver, D.S.: Levenberg-Marquardt forms of the iterative ensemble smoother for efficient history matching and uncertainty quantification. Comput. Geosci. (2013). https://doi.org/10.1007/s10596-013-9351-5. http://link.springer.com/10.1007/s10596-013-9351-5

11. Cole, R.D., Cumella, S.: Sand-body architecture in the lower Williams Fork Formation (Upper Cretaceous), Coal Canyon, Colorado, with comparison to the Piceance Basin subsurface. Mt. Geol. **42**, 85–107 (2005)

12. Evensen, G.: Sequential data assimilation with a nonlinear quasi-geostrophic model using Monte Carlo methods to forecast error statistics. J. Geophys. Res. **99**(C5), 10143 (1994). https://doi.org/10.1029/94JC00572. http://doi.wiley.com/10.1029/94JC00572

13. Fossum, K., Mannseth, T.: Parameter sampling capabilities of sequential and simultaneous data assimilation: II. Statistical analysis of numerical results. Inverse Prob. **30**(11), 114003 (2014). https://doi.org/10.1088/0266-5611/30/11/114003. http://stacks.iop.org/0266-5611/30/i=11/a=114003?key=crossref.2e2654e480ea9cb9eab6d8c245d1dc46

14. Fossum, K., Mannseth, T.: Assessment of ordered sequential data assimilation. Comput. Geosci. **19**(4) (2015). https://doi.org/10.1007/s10596-015-9492-9. http://link.springer.com/10.1007/s10596-015-9492-9

15. Golub, G.H., Van Loan, C.F.: Matrix Computations. Johns Hopkins series in the mathematical sciences, The Johns Hopkins University Press, Baltimore (1983)

16. Gu, Y., Oliver, D.S.: An iterative ensemble Kalman filter for multiphase fluid flow data assimilation. SPE J. **12**(4), 438–46 (2007). https://doi.org/10.2118/108438-PA. http://www.spe.org/ejournals/jsp/journalapp.jsp?pageType=Preview&jid=ESJ&mid=SPE-108438-PA&pdfChronicleId=090147628014cce3

17. Guevara, A.I., Sandoval, J., Guerrero, M., Manrique, C.A.: Milestone in production using proactive azimuthal deep-resistivity sensor combined with advanced geosteering techniques: Tarapoa Block, Ecuador. In: SPE Latin America and Caribbean Petroleum Engineering Conference, vol. 2, pp. 1508–1520. Society of Petroleum Engineers (2012). https://doi.org/10.2118/153580-MS. http://www.onepetro.org/doi/10.2118/153580-MS

18. Hastings, W.K.: Monte Carlo sampling methods using Markov chains and their applications. Biometrika **57**(1), 97 (1970). https://doi.org/10.2307/2334940. http://www.jstor.org/stable/2334940?origin=crossref

19. Janwadkar, S., et al.: Reservoir-navigation system and drilling technology maximize productivity and drilling performance in the granite wash, US midcontinent. SPE Drill. Compl. **27**(01), 22–31 (2012). https://doi.org/10.2118/140073-PA. https://onepetro.org/DC/article/27/01/22/198159/Reservoir-Navigation-System-and-Drilling

20. Kullawan, K., Bratvold, R., Bickel, J.E.: A decision analytic approach to geosteering operations. SPE Drill. Compl. **29**(01), 36–46 (2014)

21. Luo, X., Bhakta, T., Nævdal, G.: Correlation-based adaptive localization with applications to ensemble-based 4D-seismic history matching. SPE J. **23**(2), 396–427 (2018). https://doi.org/10.2118/185936-PA

22. Luo, X., et al.: An ensemble-based framework for proactive geosteering. In: SPWLA 56th Annual Logging Symposium 2015 (2015)

23. Metropolis, N., Rosenbluth, A.W., Rosenbluth, M.N., Teller, A.H., Teller, E.: Equation of state calculations by fast computing machines. J. Chem. Phys. **21**(6), 1087 (1953). https://doi.org/10.1063/1.1699114. http://link.aip.org/link/JCPSA6/v21/i6/p1087/s1&Agg=doi

24. Panjaitan, H.: Sand-body dimensions in outcrop and subsurface, Lower Williams Fork Formation, Piceance Basin, Colorado. Master's thesis, Colorado School of Mines (2006)

25. Pranter, M.J., Cole, R.D., Panjaitan, H., Sommer, N.K.: Sandstone-body dimensions in a lower coastal-plain depositional setting: Lower Williams Fork formation, Coal Canyon, Piceance Basin, Colorado. Am. Assoc. Petrol. Geol. Bull. **93**(10), 1379–1401 (2009). https://doi.org/10.1306/06240908173

26. Pranter, M.J., Hewlett, A.C., Cole, R.D., Wang, H., Gilman, J.: Fluvial architecture and connectivity of the Williams Fork Formation: use of outcrop analogues for stratigraphic characterization and reservoir modelling. Geol. Soc. Lond. Spec. Publ. **387**(1), 57–83 (2014). https://doi.org/10.1144/sp387.1

27. Pranter, M.J., Sommer, N.K.: Static connectivity of fluvial sandstones in a lower coastal-plain setting: an example from the Upper Cretaceous lower Williams Fork Formation, Piceance Basin. Colorado. AAPG Bull. **95**(6), 899–923 (2011). https://doi.org/10.1306/12091010008

28. Sebacher, B., Stordal, A.S., Hanea, R.: Bridging multipoint statistics and truncated Gaussian fields for improved estimation of channelized reservoirs with ensemble methods. Comput. Geosci. 341–369 (2015). https://doi.org/10.1007/s10596-014-9466-3. http://link.springer.com/10.1007/s10596-014-9466-3

29. Sviridov, M., et al.: New software for processing of LWD extradeep resistivity and azimuthal resistivity data. SPE Reserv. Eval. Eng. **17** (2014). https://doi.org/10.2118/160257-PA

30. Trampush, S.M., Hajek, E.A., Straub, K.M., Chamberlin, E.P.: Identifying autogenic sedimentation in fluvial-deltaic stratigraphy: evaluating the effect of outcrop-quality data on the compensation statistic. J. Geophys. Res. Earth Surface **122**(1), 91–113 (2017). https://doi.org/10.1002/2016JF004067

Biomedical and Bioinformatics Challenges for Computer Science

Controlling Costs in Feature Selection: Information Theoretic Approach

Paweł Teisseyre[1,2](✉) [ID] and Tomasz Klonecki[1] [ID]

[1] Institute of Computer Science, Polish Academy of Sciences, Warsaw, Poland
{Pawel.Teisseyre,Tomasz.Klonecki}@ipipan.waw.pl
[2] Faculty of Mathematics and Information Science, Warsaw University
of Technology, Warsaw, Poland

Abstract. Feature selection in supervised classification is a crucial task in many biomedical applications. Most of the existing approaches assume that all features have the same cost. However, in many medical applications, this assumption may be inappropriate, as the acquisition of the value of some features can be costly. For example, in a medical diagnosis, each diagnostic value extracted by a clinical test is associated with its own cost. Costs can also refer to non-financial aspects, for example, the decision between an invasive exploratory surgery and a simple blood test. In such cases, the goal is to select a subset of features associated with the class variable (e.g., the occurrence of disease) within the assumed user-specified budget. We consider a general information theoretic framework that allows controlling the costs of features. The proposed criterion consists of two components: the first one describes the feature relevance and the second one is a penalty for its cost. We introduce a cost factor that controls the trade-off between these two components. We propose a procedure in which the optimal value of the cost factor is chosen in a data-driven way. The experiments on artificial and real medical datasets indicate that, when the budget is limited, the proposed approach is superior to existing traditional feature selection methods. The proposed framework has been implemented in an open source library (Python package: https://github.com/kaketo/bcselector).

Keywords: Cost sensitive feature selection · Information theory · Mutual information

1 Introduction

Feature selection in supervised classification is a crucial task in many biomedical applications. Feature selection improves the comprehensibility of the considered model and allows to discover the relationship between features and the target variable. Most importantly, it helps to build models with better generalization and larger predictive power [7]. Last years have witnessed a rapid and substantial advancement of feature selection methods coping with the high dimensionality of data. However, most existing methods usually assume that all features have

© Springer Nature Switzerland AG 2021
M. Paszynski et al. (Eds.): ICCS 2021, LNCS 12743, pp. 483–496, 2021.
https://doi.org/10.1007/978-3-030-77964-1_37

the same cost, which may be inappropriate as in some situations the acquisition of feature values is costly. For example, in a medical diagnosis, obtaining some information is inexpensive (e.g., gender or age of the patient), but each diagnostic value extracted by a clinical test is associated with its own cost. In general, feature costs may also correspond to non-financial factors, for example time or difficulty in obtaining administrative data (e.g., due to privacy reasons) [18]. Other examples of feature costs include risk associated with certain diagnostic examinations (such as general anesthesia [9], diagnostic X-rays [6]). Finally, the costs may correspond to a choice of diagnostic procedure, e.g., the decision between invasive exploratory surgery and a simple blood test. Ignoring the costs may lead to choosing features that yield a powerful model, but the model cannot be used in practice as high cost is incurred in the prediction [21]. In such cases, it may be better to have a model with an acceptable classification performance, but a much lower cost.

In this work, we focus on a model-free feature selection approach based on the information theory, which has several important advantages. First, it avoids reliance on a particular classification model which allows to find all features associated with the class variable, not only those which are indicated by the employed model. Information theoretic methods, unlike some classical approaches (e.g., logistic regression with lasso regularization), are able to detect both linear and non-linear dependencies between features and class variables. Moreover, some advanced criteria are able to discover interactions between features as well as to take the redundancy of features into account. The information theoretic approach is versatile as it can be used for both classification and regression tasks, i.e., nominal and quantitative class variables, as well as for any type of features. Finally, information theoretic filter methods are usually computationally much faster than their model-based counterparts (such as lasso or random forest variable importance measures). Methods from the latter group require fitting complex classification models, which may be challenging for datasets having a large number of features.

We propose a novel greedy feature selection method that takes into account information on feature costs. In each step of the proposed procedure, we select a feature that maximizes the proposed criterion. Our criterion consists of two components describing the feature relevance and its cost, respectively. The first term is an approximation of the conditional mutual information (CMI) between a candidate feature and a target variable under the condition of already selected features. The approximation of CMI is divided by a second term which is proportional to the cost of the candidate feature. Moreover, we introduce a cost factor that controls the trade-off between feature relevance (measured by CMI) and its cost. We argue that the cost factor plays an important role in cost sensitive feature selection, although it is neglected in most related methods. In particular, its choice should depend on the assumed budget. What distinguishes our idea from previous related methods is a data-driven method of choosing the optimal value of the cost factor.

The paper is organized as follows. We discuss related work in Sect. 2 and introduce the basic concepts of the information theory in Sect. 3. In Sect. 4, we introduce the proposed method and discuss the results of the experiments in Sect. 5. Section 6 concludes the paper.

2   Related Work

Feature selection methods based on the information theory have attracted significant attention in recent years. Various criteria have been proposed ranging from a simple MIM filter [11] (involving the computation of the mutual information between a class variable and a candidate feature) to more powerful methods like CIFE [12], JMI [22] or IIFS [15] that take into account high-order interactions between features as well as possible redundancies between features. We refer to review articles [19] and [3]. In the latter one, the authors analysed dozens of feature selection methods both theoretically and experimentally. Most of the information theoretic methods only produce a ranking of the features and do not select a subset of relevant ones (see however [13]).

In the machine learning literature, there are some attempts to include cost information in the feature selection. The task is challenging as it is necessary to find a trade-off between the feature relevance and its cost. The method most related to our approach was proposed by [2] in which the popular information theoretic filter mRMR was modified by adding a penalty for the feature cost to the term describing the feature relevance. In our contribution, we propose a more general framework in which feature relevance can be measured by any approximation of conditional mutual information. Moreover, the method proposed in [2] lacks the choice of cost factor parameter; our method aims to fill this gap. Another related method has been described in the recent paper [8]. In this approach, the feature relevance is measured using an increase of the Akaike Information Criterion (AIC). The feature relevance term is simply divided by the cost. Unlike in our method, there is no cost factor. Importantly, the method is based on a parametric model whose quality is measured using the AIC, whereas in our approach we consider a more flexible, model-free criterion which is able to detect non-linear dependencies among variables. There are also some attempts to modify existing classification methods in which the feature selection is embedded in the base learner. For example, [23] proposed a random forest-based feature selection algorithm that incorporates the feature cost into the base decision tree construction process. In particular, when constructing a base tree, a feature is randomly selected with a probability inversely proportional to its associated cost. Although the method is appealing, it is not clear how to control the trade-off between the feature relevance and its cost and how to optimize the prediction performance within the assumed budget. Davis et al. [5] present a cost sensitive modification of the ID3 decision tree algorithm. They propose a new cost sensitive feature selection criterion that maximizes the information gain while minimizing the cost. The modification of the lasso method for logistic regression was considered in [17]. The authors introduced a regularization term which depends on the feature costs.

3   Background

In this section, we review the basic concepts used in the information theory: the mutual information and the conditional mutual information [4], which are necessary to introduce a general framework of feature selection. First, we discuss some notations. We consider a target class variable Y and a vector of features $\mathbf{X} = (X_1, \ldots, X_p)$, where p is the number of all considered features. In addition, we denote by $\mathbf{X}_S$ a subvector of $\mathbf{X}$ corresponding to a subset of some features $S \subseteq \{1, \ldots, p\}$.

3.1   Mutual Information

The mutual information (MI) is the basic measure of dependence between two variables. MI between the class variable Y and a candidate feature X_k is defined as

$$I(Y, X_k) = H(Y) - H(Y|X_k),$$

where $H(Y)$ is the entropy of the class variable and $H(Y|X_k)$ is the conditional entropy. MI is a popular non-negative measure of association and equals 0 only if Y and X_k are independent. The MI can be also interpreted as the amount of uncertainty in the class variable which is removed by knowing the other variable X_k. In this context, it is often called information gain. In the context of feature selection, the MI is used to assess the individual relevance of the feature X_k, i.e., it measures marginal dependence between Y and X_k. Estimation of the MI is a challenging problem, especially in the case of continuous features [14]. In our experiments, we discretize all continuous features and use a plug-in estimator of the entropy in which the probabilities are estimated by fractions.

3.2   Conditional Mutual Information and Its Approximations

The conditional mutual information (CMI) is a crucial concept in the feature selection [3]. Most feature selection methods based on the information theory are forward sequential procedures that start from an empty set of features and, in each step, add a new feature from a set of candidate features. The CMI is used to measure how the candidate feature is associated with the class variable conditioned on the already selected features. The CMI is defined as

$$I(Y, X_k|\mathbf{X}_S) = H(Y|\mathbf{X}_S) - H(Y|X_k, \mathbf{X}_S),$$

where Y is a class variable, X_k is a candidate feature and $\mathbf{X}_S$ is a vector of features corresponding to already selected features. Importantly, it may happen that the candidate feature X_k is associated with the class variable, i.e., $I(Y, X_k) > 0$ but it is redundant when considering together with features S. The simplest example is the situation when S contains a copy of X_k. Another interesting situation is the case when $I(Y, X_k) = 0$ and $I(Y, X_k|\mathbf{X}_S) > 0$, i.e., there is no marginal effect of X_k, but the interaction between X_k and features from the set S exists. Estimation of the CMI is a very challenging problem, even for a moderate size of conditioning the set S and it becomes practically infeasible for the larger S. To overcome this problem, various approximations of CMI have been proposed, resulting in different feature selection criteria. We refer to [3] and [10] which clarify when various feature selection criteria can be indeed seen as approximations of the CMI. In the following, we briefly review the most popular ones. The simplest approximation is known as MIM (mutual information maximization) criterion defined simply as $I_{mim}(Y, X_k|\mathbf{X}_S) = I(Y, X_k)$, which totally ignores the conditioning set. The other popular method is MIFS (Mutual Information Feature Selection) proposed in [1]

$$I_{mifs}(Y, X_k|\mathbf{X}_S) = I(X_k, Y) - \sum_{j \in S} I(X_j, X_k), \tag{1}$$

in which the first term $I(X_k, Y)$ describes the feature relevance and the second is penalty enforcing low correlations with features already selected in S. Brown et al. [3] have shown that if the selected features from S are independent and class-conditionally independent given any unselected feature X_k then CMI reduces to so-called CIFE criterion [12]

$$I_{cife}(Y, X_k|\mathbf{X}_S) = I(X_k, Y) + \sum_{j \in S} [I(X_j, X_k|Y) - I(X_j, X_k)]. \tag{2}$$

Note that CIFE criterion is much more powerful than MIM and MIFS as it takes into account possible interactions of order 2 between candidate feature X_k and features selected in the previous steps. There are also criteria that take into account higher-order interactions, see e.g., [15] and [20].

4 Controlling Costs in Feature Selection

4.1 Problem Statement

In this section, we describe an information theoretic framework for feature selection. It has to be recalled that Y is a class variable which is predicted using features $X_1, \ldots, X_p$. We assume that there are costs $c_1, \ldots, c_p \in (0, 1]$ associated with features $X_1, \ldots, X_p$. It can be denoted that by $C(S) = \sum_{j \in S} c_j$ a cost associated with a feature subset $S \subseteq \{1, \ldots, p\}$. The total cost is $TC = \sum_{j=1}^{p} c_j$. In addition, the assumption that we have a total admissible budget $B \leq TC$, can be made. The goal is to find a subset of features that allows to predict the class variable accurately within the assumed total budget B. The budget is a user-based parameter that can be manipulated according to current needs. Within an information theoretic framework, the problem can be stated as

$$S_{\text{opt}} = \arg \max_{S:C(S) \leq B} I(Y, \mathbf{X}_S), \tag{3}$$

i.e., we aim to select a feature subset S_{opt} that maximizes joint mutual information between the class variable Y and the vector $\mathbf{X}_S$ within the assumed budget.

4.2   Greedy Forward Selection

Note that the number of possible subsets in the problem (3) may grow exponentially, which means that it is possible to solve it only for a small or moderate number of features. Moreover, the estimation of $I(Y, \mathbf{X}_S)$ is a challenging problem when S is large. In this work, we consider a sequential forward search which starts from an empty set of features and in each step adds a feature from a set of candidate features. We first describe the algorithm which finds the optimal feature subset within the assumed budget B, for the fixed value of a cost factor r (see Algorithm 1 for a detailed description). The core element of our algorithm is a cost sensitive criterion of adding a candidate feature. In the i-th step, from a set of candidate features $F_i(r)$ (see Algorithm 1), we select the feature with index $k_i(r)$ such that

$$k_i(r) = \arg\max_{k \in F_i(r)} \frac{I_{\text{approx}}(Y, X_k | \mathbf{X}_{S_{i-1}(r)})}{c_k^r}, \tag{4}$$

where I_{approx} is one of the approximations of CMI (see Sect. 3.2 for examples), $S_{i-1}(r)$ is a set of features selected in the previous step. Note that criterion (4) can be written in the alternative form $\arg\max_{k \in F_i(r)} [\log I_{\text{approx}}(Y, X_k | \mathbf{X}_{S_{i-1}(r)}) - r \log c_k]$. The first term corresponds to the relevancy of the candidate feature, whereas the second term is a penalty for its cost. We aim to select a candidate feature that maximizes the conditional mutual information with the class variable given already selected features, but at the same time we try to minimize the cost.

Algorithm 1: Finding the optimal subset for the fixed cost factor r

> **Input**   : $Y, \mathbf{X} = (X_1, \ldots, X_p), r, B$
> $S_0(r) = \emptyset$, $F_1(r) = \{1, \ldots, p\}$
> $I_{\text{approx-cum}}(r) = 0$
> **for** $i = 1, \ldots, p$ **do**
> > $k_i(r) = \arg\max_{k \in F_i(r)} \frac{I_{\text{approx}}(Y, X_k | \mathbf{X}_{S_{i-1}(r)})}{c_k^r}$
> > **if** $C(S_{i-1}(r) \cup k_i(r)) \leq B$ **then**
> > > $S_i(r) := S_{i-1}(r) \cup k_i(r)$
> > > $F_{i+1}(r) := F_i(r) \setminus k_i(r)$
> > > $I_{\text{approx-cum}}(r) = I_{\text{approx-cum}}(r) + I_{\text{approx}}(Y, X_{k_i(r)} | \mathbf{X}_{S_{i-1}(r)})$
> > > $S(r) = S_i(r)$
> > **else**
> > > $S(r) = S_{i-1}(r)$
> > > break for loop
> > **end**
> **end**
> **Output**   : $S(r)$, $I_{\text{approx-cum}}(r)$

The cost factor r controls the trade-off between the relevancy of the candidate feature and its cost. Indeed, for $r = 0$, the cost c_k is ignored, whereas for larger r, the cost term plays a more important role. An interesting question arises: how to choose the optimal value of the parameter? In related papers, it is often stated that cost factors should be specified by the user according to his needs, see e.g., [2]. However, in practice, it is not clear how to select the optimal r. We argue that the choice of r should depend on the assumed budget B. Indeed, when B is large, say it is close to a total cost TC, then there is no need to take costs into account, so r should be close to zero. On the other hand, if B is small, then we need to take more into account the costs in order to fit into assumed budget, so r should be large. In order to find the feature subset corresponding to the optimal r, we propose the following procedure, described by the Algorithm 2. We run Algorithm 1 for different values of r, ranging between 0 and certain value $r_{\max}$. For each r we calculate the cumulative increments of the CMI related to the added candidate features. Finally, we choose r_{opt} corresponding to the largest cumulative increment and the feature subset corresponding to r_{opt}. The value $r_{\max}$ is chosen in the following way. Let $I_{\max} := \max_k I(X_k, Y)$ and $I_{\min} := \min_k I(X_k, Y)$ be the maximal and the minimal MIs, respectively. Next, let $c_{(1)} \leq c_{(2)} \leq \ldots \leq c_{(p)}$ be the feature costs sorted in ascending order. For $r = r_{max}$, we should select the cheapest feature regardless of its relevance. In particular, we could potentially have $I_{\max}/(c_{(2)}^{r_{\max}}) \leq I_{\min}/(c_{(1)}^{r_{\max}})$, as the cheapest feature with cost $c_{(1)}$ should be selected regardless of the value of mutual information. Using the above equation, we define $r_{\max} := \log(I_{\max}/I_{\min})/\log(c_{(2)}/c_{(1)})$. The number of values in the grid $0, \ldots, r_{\max}$ depends on the user preferences. For a denser grid, the optimal value of r can be chosen more precisely, but at the same time the computational cost of the procedure increases.

Algorithm 2: Finding the optimal feature subset with cost factor optimization

---

Input   : $Y, \mathbf{X}, B$
for $r = 0, \ldots, r_{\max}$ **do**
  | Run Algorithm 1 to obtain $S(r)$ and $I_{\mathrm{approx\text{-}cum}}(r)$
end
$r_{\mathrm{opt}} := \arg\max_{r=0,\ldots,r_{\max}} I_{\mathrm{approx\text{-}cum}}(r)$
Output   : $S(r_{\mathrm{opt}})$

---

5 Experiments

The main goal of the experiments was to compare the proposed cost sensitive feature selection procedure with traditional feature selection that ignores information about feature costs (we used a standard sequential forward search with CIFE criterion as a representative traditional method). Regarding the proposed cost sensitive approach, we used the greedy procedure described in Algorithm 1 in which the conditional mutual information was approximated with CIFE criterion (2). We used the logistic regression model to calculate the ROC AUC score for the selected set of features. Moreover, the cost factor r was selected using the Algorithm 2. We performed experiments on both artificial and real medical datasets. The proposed framework has been implemented in a publicly available Python package https://github.com/kaketo/bcselector.

5.1 Artificial Dataset

The advantage of using an artificially generated dataset is that we can easily control the relationship between the feature relevancy and its cost. Below we present a method of generating the artificial data. We consider p original features with a cost equal to 1. The additional features are obtained from the original features by adding noise. The cost of additional features is inversely proportional to the variance of the noise. The above framework mimics a real scenario. For example, in a medical diagnosis we can perform the expensive diagnostic test which yields the accurate value of the feature or alternatively we can choose the cheaper diagnostic test which gives an approximate value of the feature. As an example, one may consider the medical ultrasonography (USG): the 3D scans are more effective and precise than traditional 2D scans, but they are also more expensive; the 2D scan can be regarded as an approximation of the 3D scan.

Generation of Artificial Data

1. Generate p independent random variables $X_1, \cdots, X_p \sim N(0, 1)$ of size n. Let $x_i^{(j)}$ be the i-th value of j-th feature. We set $c_1 = c_2 = \cdots = c_p = 1$.
2. For each observation $i = 1, \ldots, n$, calculate the following term:

$$\sigma_i = \frac{e^{\sum_{j=1}^p x_i^{(j)}}}{1 + e^{\sum_{j=1}^p x_i^{(j)}}}.$$

3. Generate target variable $Y = \{y_1, \cdots, y_n\}$, where y_i is drawn from the Bernoulli distribution with the success probability σ_i.
4. Generate p noisy random features $e_1, \cdots, e_p$, where $e_j \sim N(0, \sigma)$.
5. Create additional p noisy features, defined as: $X_j' := X_j + e_j$. For each noisy feature we assign cost $c_j' = \frac{1}{\sigma+1}$.
6. Steps 4 - 5 are repeated for different values of σ and finally we obtain $(k+1) \times p$ features, where k is a number of repetitions of steps 4 - 5.

We present the illustrative example for $n = 10000$, $p = 4$ and $k = 4$. This setting yields 20 features in total (4 original and 16 noisy features). Noisy features were generated for four values of σ, randomly selected from $[1, 10]$. Figure 1 (top-left panel) shows the mutual information between considered features and the target variable. It is important to note that the mutual information for noisy features is always lower than for the original features. The left-bottom panel presents the costs of the considered features; note that noisy features have much lower costs than the original features. In the right panel we present the averaged results of 10 trials of feature selection performed for various fractions of the

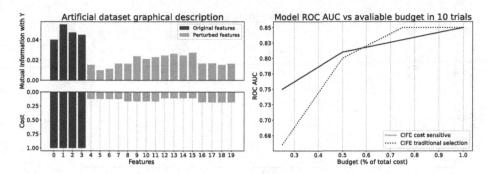

Fig. 1. Feature selection for artificial dataset.

total cost. On OX axis, the budgets are described as a percent of the total cost. On OY axis, we can see the ROC AUC score of the logistic regression model built on the selected features within the assumed budget. We can see that until 60% of the total cost, cost sensitive method performs better. This is due to the fact that, in this case, traditional methods can only use a fraction of all original features (say 1 or 2 out of 4 original features) within the assumed budget, which deteriorates the predictive performance of the corresponding classification model. On the other hand, the cost sensitive method aims to replace the original features by their cheaper counterparts, which allows to achieve higher accuracy of the corresponding model. When the budget exceeds 60% of the total cost, the traditional feature selection method tends to perform better than the proposed method, which is associated with the fact that, in this case, traditional methods include all original features (i.e., those which constitute the minimal set allowing for accurate prediction of the target variable) which results in a large predictive power of the corresponding model. For a larger budget, cost sensitive methods include both noisy features as well as the original ones. The noisy features become redundant when considering together with the original ones. This results in slightly lower prediction accuracy of the corresponding model. As expected, the cost sensitive methods are worth considering when the assumed budget is limited.

Figure 2 visualizes the selection of the cost factor r described by the Algorithm 2, for one trail. Vertical dashed lines correspond to the optimal parameter values for different values of the budget.

5.2   MIMIC-II Dataset

We performed an experiment on the publicly available medical database MIMIC-II [16] which provides various medical data about patients from the intensive care unit and their diseases. We randomly selected 6500 patients and chose hypertension disease as the target variable. We used 33 variables which refer to basic medical interviews and results of various medical tests. The costs of the features are provided by the experts and they are based on the prices of

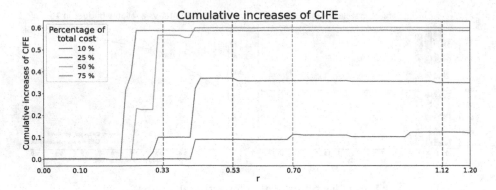

Fig. 2. Artificial dataset. Cumulative increases of CIFE for different values of r and different budgets. Vertical lines correspond to maximum of the curves.

diagnostics tests in laboratories. We used the cost data described in [17]. Before running the algorithm, the original costs are normalized in such a way that $c_j \in (0, 1]$. It should be noted here that in most countries the relations between the prices of different diagnostic tests are similar.

Figure 3 depicts the values of mutual information between considered features and the target variable as well as the costs of the features. Features are sorted according to the increasing cost. Values of the first four features (Marital status, Admit weight, Gender and Age), which are based on basic interviews with patients, are really cheap to collect. Note that the variable *Age* is highly correlated with the class variable although it has low cost. Therefore, we can expect that this feature will be selected as relevant by both traditional and cost sensitive methods. Values of the remaining features are possible to obtain using various medical tests. We can distinguish three groups of features: results of blood tests, blood pressure measurements and urine analysis. There are two conspicuous features: *NBP systolic* (number 14) and *urea nitrogen in serum or plasma* (number 23), both of them are moderately correlated with the target variable, but their cost is rather high.

Figure 4 visualizes the results of feature selection for various budgets for traditional and cost sensitive methods. The figure shows how the ROC AUC depends on the number of features used to train the model. The parameter r is calculated for each budget, therefore the sets of selected features may be different for different values of budget B. Observe that the variable *age* is selected as the most relevant feature in all cases. This can be easily explained as *age* has small cost and high mutual information with the target variable. The first discrepancy between the methods can be seen in the second step, where the traditional method selects expensive *urea nitrogen in serum* and the cost sensitive method selects *weight* which is really cheap and has a positive value of the MI. In the next steps, the cost sensitive algorithm favors cheap features with moderate value of the MI, which explains why *mean heart rate* or *mean platelet volume in blood* are selected. The most important observation is that the cost sensitive feature

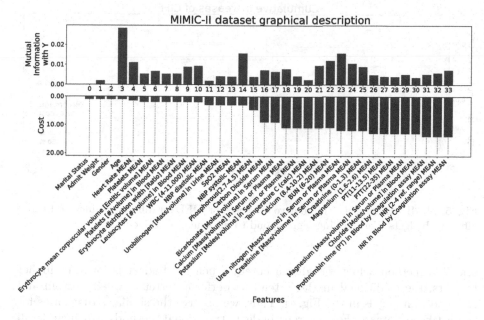

Fig. 3. MIMIC-II dataset. Basic characteristics of features.

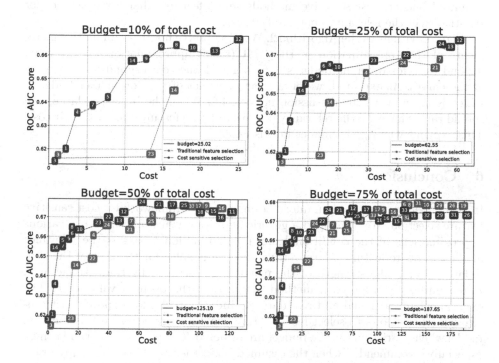

Fig. 4. Feature selection for MIMIC-II dataset.

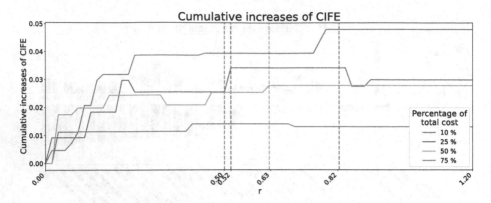

Fig. 5. MIMIC-II dataset. Cumulative increases of CIFE for different values and r and different budgets. Vertical lines correspond to maximum of the curves.

selection method achieves higher accuracy when the budget is low. For higher budgets, the traditional methods tend to perform better (see left-bottom and right-bottom panels in the Fig. 4). Thus, we observe the similar situation as for the artificial dataset. For a larger budget, traditional methods can include all relevant features, which results in a large predictive power of the model. For a limited budget, cost sensitive methods select features that serve as cheaper substitutes of the relevant expensive features.

Figure 5 visualizes the Algorithm 2. We can observe how the cumulative increments of the approximation of the CMI (CIFE approximation in this case) depend on r for different budgets. The vertical lines correspond to maximum of the curves. As expected, we obtain larger values of r_{opt} for smaller budgets, which is in line with the discussion in Sect. 4.2. When the budget is large, one should rather focus on the relevancy of the candidate features and not their cost. This explains why r_{opt} is smaller in this case.

6   Conclusions

In this paper, we proposed an information theoretic framework for cost sensitive feature selection. We developed a generic algorithm which allows to use various approximations of the conditional mutual information to assess the relevance of the candidate feature. Moreover, we use the penalty for the cost of the candidate feature. The strength of the penalty is controlled by the cost factor r. Importantly, we proposed a method of choosing the optimal value of r. The experiments on artificial and real datasets indicate that the proposed cost sensitive method allows to select features that yield a more accurate classification model when restrictions on the budget are imposed. The proposed method can be especially recommended when the assumed budget is low. There are many interesting issues left for future research. In this work, we assumed that each feature has equal extraction cost. However, in many medical applications, features are extracted in groups rather than individually, that is, the feature extraction cost is common for a whole group of features and one pays to simultaneously select all features belonging to such group instead of a single feature at a time. It would be interesting to adapt our method to such a case. Another interesting problem is to consider many target variables (e.g., many diseases) simultaneously, which in the machine learning community is known as a multilabel classification problem. In such cases, it is challenging to approximate the conditional mutual information as instead of a single variable Y we consider a multivariate variable $\mathbf{Y} = (Y_1, \ldots, Y_K)$.

References

1. Battiti, R.: Using mutual information for selecting features in supervised neural-net learning. IEEE Trans. Neural Netw. **5**(4), 537–550 (1994)
2. Bolón-Canedo, V., Porto-Díaz, I., Sánchez-Maroño, N., Alonso-Betanzos, A.: A framework for cost-based feature selection. Pattern Recogn. **47**(7), 2481–2489 (2014)
3. Brown, G., Pocock, A., Zhao, M.J., Luján, M.: Conditional likelihood maximisation: a unifying framework for information theoretic feature selection. J. Mach. Learn. Res. **13**(1), 27–66 (2012)
4. Cover, T.M., Thomas, J.A.: Elements of Information Theory (Wiley Series in Telecommunications and Signal Processing). Wiley-Interscience, Hoboken (2006)
5. Davis, J.V., Ha, J., Rossbach, C.J., Ramadan, H.E., Witchel, E.: Cost-sensitive decision tree learning for forensic classification. In: Fürnkranz, J., Scheffer, T., Spiliopoulou, M. (eds.) ECML 2006. LNCS (LNAI), vol. 4212, pp. 622–629. Springer, Heidelberg (2006). https://doi.org/10.1007/11871842_60
6. Hall, E.J., Brenner, D.J.: Cancer risks from diagnostic radiology. Br. J. Radiol. **81**(965), 362–378 (2008)
7. Hastie, T., Tibshirani, R., Friedman, J.: The Elements of Statistical Learning: Data Mining, Inference and Prediction. Springer Series in Statistics, Springer, New York (2009). https://doi.org/10.1007/978-0-387-84858-7
8. Jagdhuber, R., Lang, M., Stenzl, A., Neuhaus, J., Rahnenfuhrer, J.: Cost-constrained feature selection in binary classification: adaptations for greedy forward selection and genetic algorithms. BMC Bioinform. **21**(2), 307–333 (2020)
9. Lagasse, R.S.: Anesthesia safety: model or myth?: A review of the published literature and analysis of current original data. Anesthesiol. J. Am. Soc. Anesthesiol. **97**(6), 1609–1617 (2002)
10. Lazecka, M., Mielniczuk, J.: Analysis of information-based nonparametric variable selection criteria. Entropy **22**(9), 974 (2020)
11. Lewis, D.D.: Feature selection and feature extraction for text categorization. In: Proceedings of the Workshop on Speech and Natural Language, HLT 1991, pp. 212–217. Association for Computational Linguistics (1992)
12. Lin, D., Tang, X.: Conditional infomax learning: an integrated framework for feature extraction and fusion. In: Leonardis, A., Bischof, H., Pinz, A. (eds.) ECCV 2006. LNCS, vol. 3951, pp. 68–82. Springer, Heidelberg (2006). https://doi.org/10.1007/11744023_6
13. Mielniczuk, J., Teisseyre, P.: Stopping rules for mutual information-based feature selection. Neurocomputing **358**, 255–271 (2019)

14. Paninski, L.: Estimation of entropy and mutual information. Neural Comput. **15**(6), 1191–1253 (2003)
15. Pawluk, M., Teisseyre, P., Mielniczuk, J.: Information-theoretic feature selection using high-order interactions. In: Nicosia, G., Pardalos, P., Giuffrida, G., Umeton, R., Sciacca, V. (eds.) LOD 2018. LNCS, vol. 11331, pp. 51–63. Springer, Cham (2019). https://doi.org/10.1007/978-3-030-13709-0_5
16. Saeed, M., et al.: Multiparameter intelligent monitoring in intensive care II: a public-access intensive care unit database. Crit. Care Med. **39**(5), 952–960 (2011)
17. Teisseyre, P., Zufferey, D., Słomka, M.: Cost-sensitive classifier chains: selecting low-cost features in multi-label classification. Pattern Recogn. **86**, 290–319 (2019)
18. Turney, P.D.: Types of cost in inductive concept learning. In: Proceedings of the 17th International Conference on Machine Learning, ICML 2002, pp. 1–7 (2002)
19. Vergara, J.R., Estévez, P.A.: A review of feature selection methods based on mutual information. Neural Comput. Appl. **24**(1), 175–186 (2014)
20. Vinh, N., Zhou, S., Chan, J., Bailey, J.: Can high-order dependencies improve mutual information based feature selection? Pattern Recogn. **53**, 46–58 (2016)
21. Xu, Z.E., Kusner, M.J., Weinberger, K.Q., Chen, M., Chapelle, O.: Classifier cascades and trees for minimizing feature evaluation cost. J. Mach. Learn. Res. **15**(1), 2113–2144 (2014)
22. Yang, H.H., Moody, J.: Data visualization and feature selection: new algorithms for non Gaussian data. Adv. Neural. Inf. Process. Syst. **12**, 687–693 (1999)
23. Zhou, Q., Zhou, H., Li, T.: Cost-sensitive feature selection using random forest: selecting low-cost subsets of informative features. Knowl. Based Syst. **95**, 1–11 (2016)

How Fast Vaccination Can Control the COVID-19 Pandemic in Brazil?

Rafael Sachetto Oliveira[1]([✉]) [iD], Carolina Ribeiro Xavier[1],
Vinícius da Fonseca Vieira[1], Bernardo Martins Rocha[2][iD], Ruy Freitas Reis[2],
Bárbara de Melo Quintela[2], Marcelo Lobosco[2],
and Rodrigo Weber dos Santos[2][iD]

[1] Universidade Federal de São João del-Rei, São João del-Rei, MG, Brazil
{sachetto,carolinaxavier,vinicius}@ufsj.edu.br
[2] Universidade Federal de Juiz de Fora, Juiz de Fora, MG, Brazil
{bernardomartinsrocha,ruyfreitas,barbara,marcelo.lobosco,
rodrigo.weber}@ufjf.edu.br

Abstract. The first case of Corona Virus Disease (COVID-19) was registered in Wuhan, China, in November 2019. In March, the World Health Organization (WHO) declared COVID-19 as a global pandemic. The effects of this pandemic have been devastating worldwide, especially in Brazil, which occupies the third position in the absolute number of cases of COVID-19 and the second position in the absolute number of deaths by the virus. A big question that the population yearns to be answered is: When can life return to normal? To address this question, this work proposes an extension of a SIRD-based mathematical model that includes vaccination effects. The model takes into account different rates of daily vaccination and different values of vaccine effectiveness. The results show that although the discussion is very much around the effectiveness of the vaccine, the daily vaccination rate is the most important variable for mitigating the pandemic. Vaccination rates of 1M per day can potentially stop the progression of COVID-19 epidemics in Brazil in less than one year.

Keywords: COVID-19 · Vaccination · SIRD

1 Introduction

The first case of Corona Virus Disease (COVID-19) was registered in Wuhan, China, in November 2019. Quickly, the fast spread of the virus in the Chinese city was characterized as an epidemic, and in February 2020, 8 countries already had cases of the disease. Deeply concerned both by the alarming levels of spread and severity of the disease and the alarming levels of inaction, the World Health Organization (WHO) declared COVID-19 as a global pandemic in March 2020 [18].

Until February 3rd 2021, COVID-19 had already infected more than 104.221M people globally, 9.283M only in Brazil. The global number of deaths

Supported by UFSJ, UFJF, Capes, CNPq and Fapemig.

has reached more than 2.262M, 226,309 only in Brazil[1]. In this context, the world population's best hope is mass vaccination, which is a big challenge for the pharmaceutical industry and national health systems.

According to Li *et al.* [10], vaccines, in a general context and not only for COVID-19, will have prevented 69 million (95% confidence interval) deaths between 2000 and 2030, showing that vaccination is fundamental to mitigate the effect of infectious diseases. The conclusion of the tests for the first vaccines for COVID-19 took place in December 2020. As soon as national regulatory agencies approved them, the vaccines began to be applied to the population according to each country's national immunization plan. By February 3rd 2021, 83.83M vaccines were administrated globally, 1.09% of its population. In Brazil, the number of vaccine shots administrated is 2.52M, 1.19% of its population. Israel is already experiencing a reduction in hospitalization and transmission rates after the vaccination of 3.3M people (38.11% of its population)[2]. It is known that a rapid vaccination is essential to mitigate the spread of the disease, but the limitations imposed by the productive capacity and a clear definition of a logistics plan for the distribution of the vaccines reduces the potential of application of vaccines in the population, especially in lower-income countries. Particularly in Brazil, by keeping the current rate of vaccinated people per day, it would take more than two years to get 70% of the population vaccinated.

Many researchers have been carrying out studies to model the behavior of viruses computationally. In [12], the authors present a study showing that real-time vaccination following an outbreak can effectively mitigate the damage caused by an infectious disease using a stochastic SIR model. They analyzed the trade-offs involving vaccination and time delays to determine the optimal resource allocation strategies. The earlier a population undergoes mass vaccination, the more effective the intervention, and fewer vaccines will be required to eradicate the epidemic. In [1], the authors estimated the number of clinical cases, hospitalizations, and deaths prevented in the United States that were directly attributable to the 2009–2010 A(H1N1)pdm09 virus vaccination program. They show that early vaccination improves the efficacy of immunization significantly.

There is also a study [11] where the authors try to determine an optimal control strategy for vaccine administration for COVID-19 considering real data from China. This problem was solved using multi-objective and mono-objective Differential Evolution. The mono-objective optimal control problem considers minimizing the number of infected individuals during the treatment. On the other hand, the multi-objective optimal control problem considers minimizing the number of infected individuals and the prescribed vaccine concentration during the treatment.

Many works have dedicated efforts to adjust models for short-term prediction of virus behavior, such as [4], which calibrates a model to estimate the pandemic's behavior in the State of Mato Grosso and Brazil as a whole. To accomplish this objective, the authors proposed an exponential model and time series forecasting

---

[1] Available at https://data.humdata.org/dataset/novel-coronavirus-2019-ncov-cases.
[2] Available at https://ourworldindata.org/covid-vaccinations.

techniques. In [14], the use of a simple mathematical model is proposed, based on the classical SIRD model to adjust and predict the pandemics behavior in the three countries: Brazil, Italy, and Korea, which are examples of very different scenarios and stages of the COVID-19 pandemic [13].

In this work, we extend the model proposed in [14] by adding a new time-dependent variable called immunization rate. This variable is calculated as a linear function that grows as time passes by and depends on the vaccination rate (people/day) and the vaccine effectiveness for simulating different vaccination scenarios. In this context, the main goal of this work is to show that by considering the immunization rate, we can better simulate the evolution of the number of infections and deaths for COVID-19 and forecast when we can expect to overcome the pandemic, considering different scenarios that take into account different vaccine efficiency and vaccination rates. Moreover, this study can raise awareness that even with the vaccination, it will take some time to alleviate the non-pharmacological measures (social distancing, use of masks, among others) to prevent the spread of the disease. Our results show that although the discussion is very much around the effectiveness of the vaccines, the daily vaccination rate is the most important variable for mitigating the pandemic. In addition, our results suggest that Brazil's vaccination should be done more quickly to make it possible to overcome the pandemic.

2 Materials and Methods

2.1 Mathematical Model

The model used here is an extension of the model presented in [14] which is based on the classic compartmental SIRD model [2,5–8] and was kept as simple as possible to reduce the number of unknown parameters to be estimated. The modified model to include the vaccination scheme is described by the following set of equations:

$$
\begin{cases}
\frac{dS}{dt} = -\frac{\alpha(t)}{N}(S - v(t)S)I, \\
\frac{dI}{dt} = \frac{\alpha(t)}{N}(S - v(t)S)I - \beta I - \gamma I, \\
\frac{dR}{dt} = \gamma I, \\
\frac{dD}{dt} = \beta I, \\
I_r = \theta I, \\
R_r = \theta R, \\
C = I_r + R_r + D,
\end{cases}
\tag{1}
$$

where S, I, R, D, I_r, R_r, and C are the variables that represent the number of individuals within a population of size N that are susceptible, infected, recovered, dead, reported as infected, reported as recovered, and total confirmed cases, respectively. The term $\alpha(t) = a(t)b$ denotes the rate at which a susceptible individual becomes infected, where $a(t)$ denotes the probability of contact and b the rate of infection. See [14] for further details on the model.

The new term $v(t)$ denotes the rate of immunization and is described by the following equation:

$$v(t) = \begin{cases} 0, & \text{if } t < (t_{vs} + t_{im}), \\ 1 - v_e v_r t, & \text{otherwise.} \end{cases} \qquad (2)$$

where t_{vs} is the day when the vaccination program starts; t_{im} is the immunization delay, i.e., the time between the vaccination and the acquired immunization; v_e is the vaccine effectiveness (in %), and v_r is the vaccination rate (the percent of the population vaccinated per day). To simplify the model and due to the lack of data for the Covid-19 vaccines, the following hypotheses are considered: 1) In the lack of numbers for vaccine effectiveness, we use the disclosed range of reported vaccine efficacy (between 50% and 90%); 2) We consider the optimistic scenario, reported in some recent studies [9,17], where once immunized an individual when in contact with the virus has a significant reduction in the virus load, i.e., once immunized an individual is no longer infectious; 3) we do not directly consider the differences of vaccine effectiveness after a booster dose, i.e., the need of a second dose is implicitly captured by the parameter t_{im}, the immunization delay.

2.2   Numerical Simulations

We started our investigation by adjusting the model variables with available public data for Brazil, using the 78 days and the original model, described in Sect. 2.1. The variable adjustments were made using Differential Evolution (DE), using the same approach presented in [14]. Then we applied the modified model (including the vaccination scheme), simulating one year (365 days) using the adjusted parameters and the configurations described in the scenarios below.

- **Scenario 1**: Fixing vaccination rate at 100k p/d and varying vaccine efficacy from 0.5 to 0.9, with an increment of 0.1.
- **Scenario 2**: Fixing vaccination rate at 2M p/d and varying vaccine efficacy from 0.5 to 0.9, with an increment of 0.1.
- **Scenario 3**: Fixing vaccine efficacy at 0.5 and using vaccination rates of 100k, 1M and 2M p/d.
- **Scenario 4**: Fixing vaccine efficacy at 0.9 and using vaccination rates of 100k, 1M and 2M p/d.

To execute the simulations, we consider that the vaccine takes 28 days to trigger the immune response. We also limit the maximum population to be vaccinated to 109.5 Million, which corresponds to the number of people for vaccination, according to the Brazilian Immunization Program [15].

A global sensitivity analysis of the modified mathematical model for vaccination was carried out using Sobol indices [16]. In particular, in this study, only the vaccination parameters such as immunization time, the vaccine efficacy, and vaccination rate were analyzed, since the remaining parameters were investigated previously [14].

3   Results

After applying the methodology described in Sect. 2, we were able to simulate the proposed scenarios considering different vaccination and efficacy rates. The results of the numerical simulations are depicted in Figs. 1, 2, 3 and 4, that illustrate the evolution of active cases and deaths in Brazil. First, it is important to notice some important information that is common to all simulation scenarios. If no vaccination scheme is adopted, we would have approximately 276.614 active cases and 540.674 total deaths after the 365-time span of the simulations. The particular results for each of the proposed scenarios are presented in Sects. 3.1 to 3.4. A deeper discussion on the results is presented in Sect. 4.

3.1   Scenario 1

Figure 1 shows the evolution of active cases and deaths in Brazil, considering the simulation of the numerical method presented in Sect. 2 using 365 days, with a fixed vaccination rate of 100 thousand people per day and varying the vaccine efficacy rate. From Scenario 1, we observe that the number of active cases after the 365 days would vary from 148.000 with a vaccine efficacy of 50% to 85.135 for a 90% effective vaccine. The total number of deaths would stand between 496.988 and 469.493 considering vaccine efficacy between 50% and 90%, respectively, indicating a reduction in the total number of deaths between 8% and 13%.

3.2   Scenario 2

The results of the numerical simulation of Scenario 2, which considers a fixed vaccination rate of 2 million people per day, Brazilian theoretical vaccine production capacity, and a varied vaccine efficacy rate, are depicted in Fig. 2, which shows the variation of active cases and total deaths in 365 days. The number of active cases would vary from 8.387 with a vaccine efficacy of 50% to 201 for a 90% effective vaccine, and the total deaths would stay between 349.564 and 314.201 considering a vaccine efficacy between 50% and 90% respectively. Compared to the number of deaths observed in the simulation without vaccination, the number of deaths considering Scenario 2 represents a decrease of approximately 35% for a 50% effective vaccine and almost 42% for a vaccine with a 90% efficacy rate after 365 days of simulation. It is also important to notice a drastic drop in the number of active cases after day 110, considering Scenario 2 for all the simulated vaccination and vaccine efficacy rates, which reflects a substantial reduction in the expected number of deaths on the following days of the simulation.

3.3   Scenario 3

Differently from the previous scenarios, where the vaccination rate is fixed, in Scenario 3, we investigate the vaccination effects with three different rates (100k,

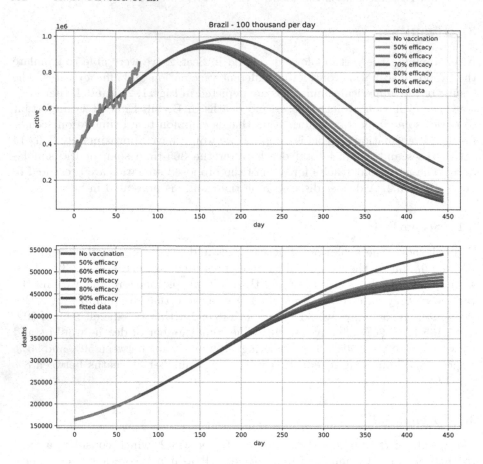

Fig. 1. Fixing vaccination rate at 100k p/d and varying vaccine efficacy from 0.5 to 0.9

1M, and 2M people per day), but fixing the efficacy rate as 50%. Figure 3 presents the number of active cases and the total of deaths for the numerical simulation considering Scenario 3 in the 365 days. The resulting curves for the 1M and 2M people per day simulations show a drastic reduction in the number of active cases around day 110, and in Fig. 2 which represents the results for Scenario 2. A less drastic reduction is observed in the number of active cases for the parameterization with a vaccination rate of 100k people per day, which causes the number of deaths to be increasingly high, even after the 365 days.

3.4   Scenario 4

In Scenario 4, the vaccine efficacy rate is fixed, and the vaccination rate is varied in 100k, 1M, and 2M people per day, as in Scenario 3. However, the numerical simulations consider a vaccine efficacy of 90%. Figure 4 shows the number of

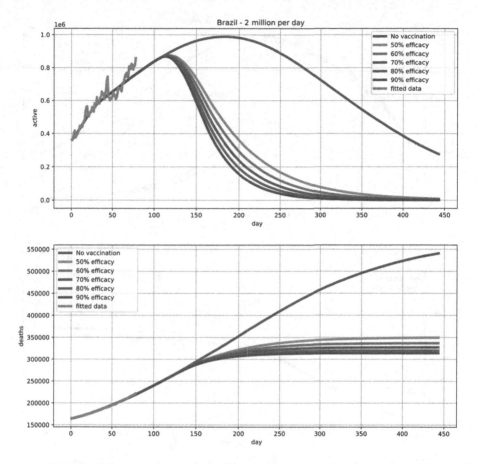

Fig. 2. Fixing vaccination rate at 2M p/d and varying vaccine efficacy from 0.5 to 0.9

active cases and the number of deaths observed for the simulation of the proposed model in 365 days. The peaks of the curves of active cases considering the vaccination of 1M and 2M people per day occur around day 120, while the peak of the curve for the vaccination of 100k people per day occurs after day 150. Moreover, for the curves that represent the total number of deaths considering the vaccination of 1M and 2M people per day, we can observe a steady state after the 365 days time span, while for the vaccination of 100k people per day, the number of deaths still increases even after the 365 days of simulation.

3.5   Sensitivity Analysis

To better understand how the different parameters considered impact the number of active cases and the number of deaths for the model presented in Sect. 2, this section presents a Sensitivity Analysis (SA) carried out for immunization delay, vaccine efficacy and, vaccination rate. The interval for immunization delay

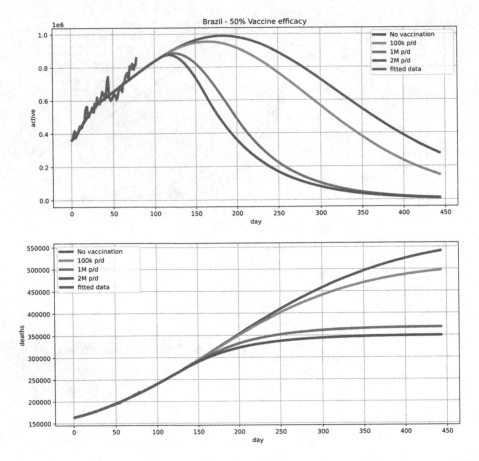

Fig. 3. Fixing vaccine efficacy at 0.5 and using vaccination rates of 100k, 1M and 2M p/d

considers the bounds $[14, 42]$, representing a standard 28 ± 14 days immunization time. The bounds for the vaccine efficacy, $[0.5, 0.9]$, are the same values used for the experiments, as reported in Sect. 2.2. The bounds for vaccination rate, $[0.000545107, 0.009371513]$, consider the proportion of the Brazilian population regarding the vaccination rates of 100k and 2M people per day, also reported in Sect. 2.2.

First, it is essential to mention that the results observed for the curves representing the Sensitivity Analysis regarding the number of active cases and the number of deaths are very similar. Thus, to avoid redundancy, we chose to omit the results of the Sensitivity Analysis for the number of deaths. The Sensitivity Analysis results based on main Sobol indices regarding the number of active cases are presented in Fig. 5, which reports the sensitivities in the period after the beginning of vaccination. One can observe that the immunization delay plays an essential role during the vaccination program, whereas the vaccination rate

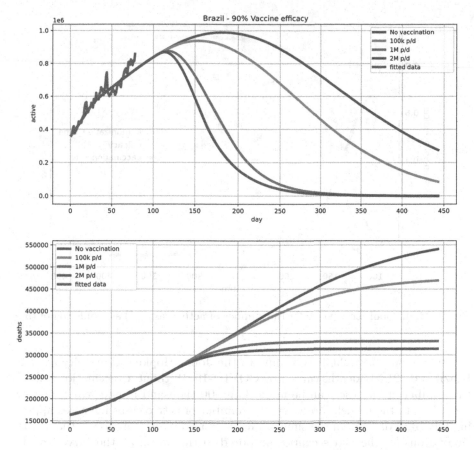

Fig. 4. Fixing vaccine efficacy at 0.9 and using vaccination rates of 100k, 1M and 2M p/d

becomes the more relevant parameter to control the number of active cases in the long term. We can also note that the efficacy of the vaccine appears as the least important parameter for controlling the number of active cases.

4 Discussion

By analyzing the results presented in Sects. 3.1 to 3.4, it can be seen that, according to the simulations performed in this work, with a vaccination rate of 100 thousand people per day, the disease would not be fully controlled after the 365 time span, regardless of the vaccine efficacy as the number of active cases would still be high (85.135 active cases), even with a 90% effective vaccine. By vaccinating the population at a rate of 2M p/d, the active cases would vary from 8.387, for a 50% effective vaccine to 201, for a vaccine with 90% of efficacy. These results indicate that, at this rate, the pandemic would be controlled, after one year of vaccination, regardless of the vaccine efficacy.

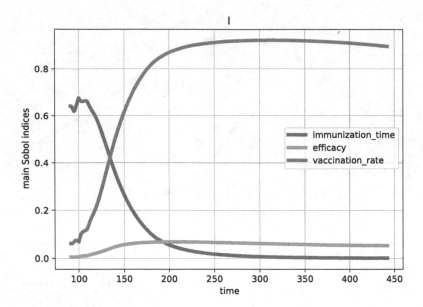

Fig. 5. Main Sobol sensitivities for the number of active cases (I) as a function of time.

By January 28th, 2021, according to the Johns Hopkins University data [3], Brazil had 221.547 deaths caused by COVID-19. Without a vaccination scheme, the number of fatalities can be more than 500 thousand in one year from now, according to the model. Moreover, the number of active cases increases approximately until day 180 of the simulations. It is also interesting to see that, in the simulations for the four scenarios proposed in this work, all the curves overlap until at least day 100, indicating that, in the short-term, the choice of a vaccination scheme, or the lack of it, is irrelevant. However, it is worth observing that, in the long-term, the adoption of immunization policies could substantially reduce the number of deaths due to COVID-19. Even considering a more conservative vaccination scheme with the vaccination of 100k people per day and a 50% effective vaccine, a similar rate to that currently adopted in Brazil, represented in Scenario 1 (Sect. 3.1), the model predicts 43.686 fewer deaths. The data provided by our simulations also suggests that a solid vaccination scheme could potentially control the pandemics in Brazil in 356 days, and thousands of lives could be saved. If a more assertive vaccination scheme is adopted, considering the vaccination of 2M people per day, taking advantage of the immunization potential of the Brazilian health system, we would have 191.110 less deaths with a 50% effective vaccine, and 226.473 lives could be saved if a 90% effective vaccine was adopted, according to the numerical simulations in the 365 days.

The experiments conducted in this work allow us to better understand an essential issue in the definition of a vaccination scheme for COVID-19: the impact of the vaccination rate and vaccine efficacy in the mitigation of the pandemic. By analyzing the simulation Scenarios 1 (Sect. 3.1) and 2 (Sect. 3.2), where the vaccination rate is fixed, it can be observed that the curves that represent the results of the numerical simulation behave very similarly, although the number of deaths and active cases decreases as the vaccine efficacy increases. This suggests that the significance of the vaccine efficacy is limited to define the success of a global immunization effort. When we consider the scenarios where the vaccination rate varies (Scenarios 3 and 4, presented in Sects. 3.3 and 3.4 respectively), we can observe a clear distinction between the curve that represent the results of the simulation with a vaccination rate of 100k people per day and the curves for 1M and 2M people per day. According to the simulations performed in this work, regardless of the vaccine efficacy considered (50% for Scenario 3 and 90% for Scenario 4), the vaccination of 1M and 2M can mitigate the spreading of the virus after 365 days, unlike Scenario 1, with the vaccination of 100k people per day, where the curves that represent the number of deaths still increases after 365 days. The observation of the high relevance of the vaccination rate in the success of the immunization efforts for COVID-19 is corroborated by the Sensitivity Analysis (SA), presented in Sect. 3.5. The Sobol indices observed for the immunization delay, vaccination rate, and vaccine efficacy as a function of time show that, in the long term, the vaccination rate is the most relevant parameter to control the number of active cases and, accordingly, the number of deaths, considering the numerical simulations conducted in this work.

5   Limitations and Future Works

The model considered in this work, as well as the model in which it is based [14], is able to adjust well to real data, as presented in Sect. 3. However, it shows a series of limitations, which we intend to discuss in this section.

As the first case of COVID-19 occurred not much more than one year ago, some limitations of our model relate to uncertainties regarding the disease's characteristics, which only recently has been investigated. Currently, there is no consensus on how often reinfections of COVID-19 can occur and how does a recurrent occurrence of the disease could affect the dynamics of its transmission. We still do not understand how different Corona Virus variations can impact the transmission rate in a population and the severity of COVID-19.

Other limitations are imposed by the complex behavior of a population, especially facing a pandemic, due to explicit public policies and guidelines or the self-organization of people. As the pandemic progresses, people harden or soften the social distancing and the adoption of measures like the use of masks and hands sanitation; restaurants, schools, and business places for other economic activities close and open as the occupancy of hospitals change. The variations of the social dynamics are not considered by our model and represent a limitation of our work. Our model also considers the population to be homogeneously spread in the space and a homogeneous contact between people and, these aspects should be considered in future works if we want our model to be more realistic.

The definition of public policies regarding the immunization strategy also involves the stratification of the population in different levels of priority to be vaccinated, according to the risk of severity of COVID-19, in case of infection. For instance, in Brazil, the first individuals to be vaccinated are the most vulnerable to COVID-19 infection, which means that if this kind of information is incorporated into our model in future works, the number of deaths could be reduced in our simulations. To keep it simple, we are also disregarding that a vaccine's efficacy to avoid infections is frequently lower than its efficacy in preventing more severe occurrences of the disease, sometimes almost avoiding the evolution of patients' clinical conditions to death or even the need to use respirators.

Moreover, as more vaccines are developed and approved by regulatory agencies, public and private organizations negotiate their acquisition and improve the strategies for the logistics and the administration of vaccine shots, changing the vaccination rate, which, in this work, remains fixed as a function of time.

As reported in Sect. 2, the model proposed in this work is intended to be kept as simple as possible. The limitations here described do not prevent us to better understand the effect of the adoption of different vaccination schemes over time, and we intend to consider and solve them in future works.

6   Conclusions

This work presents an extension of the model proposed in Reis *et al.* [14], which simulates the propagation of COVID-19 in a homogeneous population by incorporating the immunization of a portion of the population as a function of time. The model parameters were adjusted to the real data of 78 days before the start of vaccination in Brazil via differential evolution regarding the number of active cases and the number of deaths. Different values for the daily vaccination rate and vaccine efficacy were tested to simulate Brazil's pandemic effects in 365 days. From the numerical simulation results, we investigated the parameters set that would better control the pandemic and how long it would take for a significant reduction in the number of deaths by varying these values. In this context, the observation of the resulting curves from the simulations combined with a Sensitivity Analysis of the parameters allows us to verify that the vaccination rate is much more important than the vaccine's efficacy to mitigate the pandemic.

One of the main issues in the discussion regarding vaccines' acquisition for implementing public policies for immunization is its reported efficacy. However, according to the results of the numerical simulations and the Sensitivity Analysis conducted in this work, considering the parameters adopted for the vaccination and vaccine efficacy rates, it is clear that this should be a secondary topic to be discussed, as the vaccination rate is much more relevant for mitigating COVID-19 than the vaccine efficacy. Thus, considering that the primary purpose of an immunization policy is rapid mitigation of the disease and a substantial reduction in the number of deaths, more attention should be given to adopting vaccines that allow a logistic that enables universal and very fast vaccination of the population.

References

1. Borse, R.H., et al.: Effects of vaccine program against pandemic influenza A (H1N1) virus, United States, 2009–2010. Emerg. Infect. Diseas. **19**(3), 439 (2013)
2. Diekmann, O., Heesterbeek, J.: Mathematical Epidemiology of Infectious Diseases: Model Building, Analysis and Interpretation. Wiley Series in Mathematical & Computational Biology, Wiley, Hoboken (2000). https://books.google.com.br/books?id=5VjSaAf35pMC
3. Dong, E., Du, H., Gardner, L.: An interactive web-based dashboard to track COVID-19 in real time. Lancet Infect. Diseas. (2020). https://doi.org/10.1016/S1473-3099(20)30120-1
4. Espinosa, M.M., de Oliveira, E.C., Melo, J.S., Damaceno, R.D., Terças-Trettel, A.C.P.: Prediction of COVID -19 cases and deaths in Mato Grosso state and Brazil. J. Health Biol. Sci. 1–7 (2020). https://doi.org/10.12662/2317-3076jhbs.v8i1.3224.p1-7.2020
5. Hethcote, H.W.: The mathematics of infectious diseases. SIAM Rev. **42**(4), 599–653 (2000). https://doi.org/10.1137/S0036144500371907
6. Keeling, M.J., Rohani, P.: Modeling Infectious Diseases in Humans and Animals. Princeton University Press, Princeton (2011). https://doi.org/10.1111/j.1541-0420.2008.01082_7.x
7. Kermack, W.O., McKendrick, A.G.: Contributions to the mathematical theory of epidemics–I. Bull. Math. Biol. **53**(1), 33–55 (1991). https://doi.org/10.1007/BF02464423
8. Kermack, W.O., McKendrick, A.G., Walker, G.T.: A contribution to the mathematical theory of epidemics. Proc. R. Soc. Lond. Ser. A, Containing Papers of a Mathematical and Physical Character **115**(772), 700–721 (1927). https://doi.org/10.1098/rspa.1927.0118. Publisher: Royal Society
9. Levine-Tiefenbrun, M., et al.: Decreased SARS-CoV-2 viral load following vaccination. medRxiv (2021)
10. Li, X., et al.: Estimating the health impact of vaccination against ten pathogens in 98 low-income and middle-income countries from 2000 to 2030: a modelling study. Lancet **397**(10272), 398–408 (2021)
11. Libotte, G.B., Lobato, F.S., Platt, G.M., Silva Neto, A.J.: Determination of an optimal control strategy for vaccine administration in Covid-19 pandemic treatment. Comput. Methods Programs Biomed. **196** (2020). https://doi.org/10.1016/j.cmpb.2020.105664
12. Nguyen, C., Carlson, J.M.: Optimizing real-time vaccine allocation in a stochastic sir model. PloS one **11**(4) (2016)
13. Reis, R.F., et al.: The quixotic task of forecasting peaks of Covid-19: rather focus on forward and backward projections. Front. Public Health (2021). https://doi.org/10.3389/fpubh.2021.623521
14. Reis, R.F., et al.: Characterization of the COVID-19 pandemic and the impact of uncertainties, mitigation strategies, and underreporting of cases in South Korea, Italy, and Brazil. Chaos, Solitons Fractals **136** (2020). https://doi.org/10.1016/j.chaos.2020.109888
15. da Saúde, M.: Preliminary vaccination plan against Covid-19 foresees four phases (2021). https://www.gov.br/saude/pt-br/assuntos/noticias/vacinacao-contra-a-covid-19-sera-feita-em-quatro-fases. Accessed 3 Feb 2021. (in portuguese)
16. Sobol, I.M.: Global sensitivity indices for nonlinear mathematical models and their Monte Carlo estimates. Math. Comput. Simul. **55**(1–3), 271–280 (2001). https://doi.org/10.1016/S0378-4754(00)00270-6

17. Voysey, M., et al.: Single dose administration, and the influence of the timing of the booster dose on immunogenicity and efficacy of ChAdOx1 nCoV-19 (AZD1222) vaccine. Preprints with The Lancet (2021)
18. World Health Organization: WHO timeline - Covid-19 - 27 April 2020 (2020). https://www.who.int/news/item/27-04-2020-who-timeline--covid-19. Accessed 03 Feb 2021

Uncertainty Quantification of Tissue Damage Due to Blood Velocity in Hyperthermia Cancer Treatments

Bruno Rocha Guedes, Marcelo Lobosco⑩, Rodrigo Weber dos Santos⑩, and Ruy Freitas Reis$^{(\boxtimes)}$⑩

Departamento de Ciência da Computação, Universidade Federal de Juiz de Fora, Juiz de Fora, MG, Brazil
`ruy.reis@ufjf.br`

Abstract. In 2020, cancer was responsible for almost 10 million deaths around the world. There are many treatments to fight against it, such as chemotherapy, radiation therapy, immunotherapy, and stem cell transplant. Hyperthermia is a new treatment that is under study in clinical trials. The idea is to raise the tumour temperature to reach its necroses. Since this is a new treatment, some questions are open, which can be addressed via computational models. Bioheat porous media models have been used to simulate this treatment, but each work adopts distinct values for some parameters, such as the blood velocity, which may impact the results obtained. In this paper, we carefully perform an uncertainty quantification analysis due to the uncertainties associated with estimating blood velocity parameters. The results of the *in silico* experiments have shown that considering the uncertainties presented in blood velocity, it is possible to plan the hyperthermia treatment to ensure that the entire tumour site reaches the target temperature that kills it.

Keywords: Hyperthermia · Uncertainty quantification · Porous media · Mathematical modeling · Computational physiology

1 Introduction

According to the National Cancer Institute (NCI), in 2018, there were 18,1 million new cases of cancer registered and 9,5 million deaths worldwide caused by cancer [3]. As projected in early 2020 by a Brazilian non-governmental organisation, Oncoguia Institute, between 2020 and 2022 Brazil will register a total of 625,000 new cases of cancer per year. Considering all types of cancers, the most frequents in Brazil are non-melanoma skin cancer (177 thousand cases), breast and prostate cancer (66 thousand cases each), colon and rectum (41 thousand cases), lung (30 thousand cases), and stomach (21 thousand cases) [1]. According to the NCI, numerous treatments were created to treat cancer, for example, chemotherapy, radiation therapy, bone marrow transplant, and hyperthermia treatment [2].

© Springer Nature Switzerland AG 2021
M. Paszynski et al. (Eds.): ICCS 2021, LNCS 12743, pp. 511–524, 2021.
https://doi.org/10.1007/978-3-030-77964-1_39

The hyperthermia treatment is considered a non-invasive method [16]. The basic idea of this technique is to increase the temperature of the target above a threshold aiming to destroy tumour cells. Besides, it is desired to minimise the damage to healthy tissue. Therefore, a consolidated technique to heat living tissues is based on the properties of magnetic nanoparticles. These particles produce heat when submitted to an alternating magnetic field through Néel relaxation and/or Brownian motion [17]. Due to biocompatibility, iron oxides Fe_3O_4 (magnetite) and $\gamma - Fe_2O_3$ (maghemite) are normally employed to hyperthermia treatment [18]. Furthermore, the nanoparticles can reach the target by the bloodstream or directly in the tumour site [16]. This study focuses on direct injection due to the facility to handle different tumour sizes and shapes.

In the context of this study, a mathematical model is employed to describe the behaviour of the tumour temperature over time due to hyperthermia treatment. Several models were developed and applied to describe the dynamics of heat in living tissues [13,14,21,25,26]. Among them, this work employed a porous media approach described in [14]. This model describes bioheat according to the properties of the living tissue such as density, specific heat, thermal conductivity, metabolism, blood temperature and velocity. Concerning blood velocity, the scientific literature describes that cancer growth can stimulate the angiogenesis of capillaries in the tumour site [9,33]. Furthermore, it depends on the cancer type, stage, location and other specificities [12,33]. So, this study chooses to describe the blood velocity as a density probability function based on the velocities found in the scientific literature intending to quantifying their influence on the bioheat dynamics.

To approach the employed mathematical model, this study uses an explicit strategy of the finite volume method (FVM) to obtain the numerical solution of the time-dependent partial differential equation [6,30]. Moreover, the uncertainty quantification (UQ) technique was employed to quantify the blood velocity influence in the numerical output. UQ is a technique used to quantify uncertainties in real and computational systems. To incorporate the real variability and stochastic behaviour in systems, it is used statistical distributions as model inputs, with the goal validate and reduce the uncertainties of those parameters. Besisdes, UQ aims to quantify the influence of the variability in the parameter to the model output [28,29,32]

Several studies adopt the uncertainty quantification technique to propagate uncertainties of model parameters on both ordinary [4,15,22–24] and partial differential equations [5,10,19,20,31]. Concerning bioheat models, Fahrenholtz [7] employed generalised polynomial chaos to quantify the influence of model parameters of Pennes Bioheat model for planning MR-guided laser-induced thermal therapies (MRgLITT). Iero [11] also analyse the influence of uncertainties in hyperthermia treatment in patient-specific breast cancer using a modified version of Pennes model. In our study, we consider the propagation of the uncertainties related to the blood velocity in porous media and their impacts on a cancer hyperthermia treatment via magnetic nanoparticles.

We organise this paper as follows. Section 2 describes the mathematical model and numerical techniques employed in this study. The results are presented in Sect. 3 and discussed in Sect. 4. Finally, Sect. 5 presents the conclusions and plans for future work.

2 Materials and Methods

In this study, the porous media equation is used in the representation of bioheat transference over the tumour cells [14,25]. Uncertainties in the blood velocity in the tumour/healthy tissues were considered due to the study presented in [12] which calculates the velocity of the blood on tumour and healthy tissues. The values of this parameter are based on a uniform distribution, which randomly chooses a value in a given interval. The interval used for the blood velocity in tumour and healthy tissues was defined by the velocity presented in [12].

2.1 Mathematical Model

Let's consider that the modelled tissue containing the tumour is represented by an open bounded domain $\Omega \subset \mathbb{R}^2$ and $I \in (0, t_f] \subset \mathbb{R}^+$ is the simulated time. So, the bioheat model considering a medium porous approach can be defined as:

$$\begin{cases} \sigma \frac{\partial T}{\partial t} = \nabla \cdot \kappa \nabla T + \beta \frac{\rho_b u_b \varepsilon c_{pb}}{\delta} (T_b - T) + (1 - \varepsilon)(Q_m + Q_r), & \text{in } \Omega \times I, \\ \kappa \nabla T \cdot n = 0, & \text{on } \partial\Omega \times I, \\ T(\cdot, 0) = 37, & \text{in } \Omega, \end{cases}$$

(1)

where $\sigma = [\rho c_p (1 - \varepsilon) + \rho_b c_{pb} \varepsilon]$, $\kappa = (1 - \epsilon)k_t + \epsilon k_b$. ρ, c_p and k_t are density, specific heat and thermal conductivity of the tissue, respectively. ρ_h, c_{ph}, k_b and u_b are density, specific heat, thermal conductivity and average velocity of the blood, respectively. ϵ, δ, β are porosity of the medium, average distance between the transverse blood vessels and a correction factor, respectively. T_b is the blood temperature. Q_m is the metabolic heat source, Q_r heat generated from the hyperthermia treatment, T_0 is the initial temperature, and the heat generated from hyperthermia treatment is given by:

$$Q_r(\boldsymbol{x}) = \sum_{i=1}^{n} A_i e^{-r(\boldsymbol{x}_i^2)/r_{0,i}^2},$$

(2)

where n is the number of injections, A is maximum heat generation rate, $r(\boldsymbol{x}_i^2)$ is the distance from the injection point and r_0 is the radius of coverage of hyperthermia. The mathematical model of bioheat transference used was based on a previous study [25].

2.2   Numerical Method

The finite volume method was employed to solve Eq. (1) [30]. The domain $\Omega \cup \partial\Omega$ is discretized into a set of nodal points defined by $S = \{(x_i, y_j); i = 0, .., N_x; j = 0, ..., N_y\}$ with N_x and N_y being the total number of points spaced with length Δx and Δy, respectively. Moreover, the time domain I is partitioned into N equal time intervals of length Δt, i.e. $t_n = n\Delta t$ for $n \in [0, N]$. Finally, the fluxes at the volumes surfaces for each nodal point were approximated by a central difference, and, to obtain an explicit method, the forward difference was employed for the time stepping. These numerical approaches ensure the first-order accuracy for time and a second-order for space. Thus, Eq. (1) is numerically computed by:

$$
\begin{aligned}
T_{i,j}^{n+1} = {} & \frac{\Delta t}{\rho_c c_b \epsilon + \rho c(1-\epsilon)} \Big\{ \frac{[\kappa_{i+\frac{1}{2},j}(T_{i+1,j}^n - T_{i,j}^n) - \kappa_{i-\frac{1}{2},j}(T_{i,j}^n - T_{i-1,j}^n)]}{\Delta x^2} \\
& + \frac{[\kappa_{i,j+\frac{1}{2}}(T_{i,j+1}^n - T_{i,j}^n) - k_{i,j-\frac{1}{2}}(T_{i,j}^n - T_{i,j-1}^n)]}{\Delta y^2} \\
& + \beta_{i,j} \frac{p_b u_b \epsilon c_b}{\delta_{i,j}} (T_b - T) + (1-\epsilon)(Q m_{i,j} + Q r) \Big\} + T_{i,j}^n
\end{aligned}
\tag{3}
$$

where the thermal conductivity function is discontinuous, therefor the harmonic mean $\kappa_{i+\frac{1}{2},j} = \frac{2\kappa_{i,j}\kappa_{i+1,j}}{\kappa_{i,j}+\kappa_{i+1,j}}$ is adopted to ensure the flux continuity.

2.3   Uncertainty Quantification

In real-world, uncountable uncertainties are intrinsic to nature. In view of this, the uncertainty quantification (UQ) is employed for quantitative characterisation and uncertainties reduction on computational and real world applications, with the goal to determine how likely outcomes of models which uses stochastic behaviour. With the objective of including stochastic behaviour we employ UQ via Monte Carlo (MC) simulation [27]. The deterministic model described in Eq. (1) is solved several times considering the u_b parameter as a probability density function (PDF) to stochastic responses. MC draws samples of the u_b and evaluates the model for each of them. The results are used to provide statistical properties for the quantities of interest (QoI).

Consider that Y is the model solution for all space and time points. Furthermore, consider one or more uncertain input parameter Q and the known parameters K:

$$
Y = f(x, y, t, Q, K).
\tag{4}
$$

Both Q and K are treated as sets of model parameters, and Y is computed using FVM described in the prior section. The uncertainty in this problem comes from the parameters in Q, which are assumed to have a known probability density function ϕ_Q. So, several samples are evaluated considering different values of Q according to ϕ_Q. Finally, the set of model results Y are used to evaluate the statistical properties of each QoI.

In this study, we perform the MC simulation aided by Chaospy library [8]. Chaospy is an open-source library designed to be used in the Python programming language, which provides functions for performing uncertainty quantification. Through this library, it is possible to create probabilistic distribution and samples variables to approach numerically the uncertainty quantification [8].

3 Results

This section presents the UQ analysis of temperature dynamics and tissue damage considering u_b as a PDF. To perform the UQ analysis we consider $10,000$ Monte Carlo iteration employing u_b as a uniform distribution. So, it was evaluated the mean and confidence interval (CI) in two different points on the domain, (one inside the tumour and another outside) over all time steps. Moreover, we present the mean of the final temperature distribution in the entire domain. Finally, it was estimated the amount of necrosis.

The values of parameters used in Eq. (1) to perform the Monte Carlo simulation are given by Table 1, where $X \sim F^{-1}(a, b)$ is a uniform distribution that generate a random value between a and b. The values of other parameters were based on a previous work [25].

Table 1. Model parameters values employed in Eq. (1).

Parameter	Description	Unit	Muscle	Tumour
c_p	Specific heat	(J/Kg^oC)	4200.0	4200.0
c_{pb}	Specific heat of blood	(J/Kg^oC)	4200.0	4200.0
k_t	Thermal conductivity of tissue	(W/m^oC)	0.51	0.56
k_b	Thermal conductivity of blood	(W/m^oC)	0.64	0.64
p	Density	(Kg/m)	1000.0	1000.0
p_b	Density of blood	(Kg/m)	1000.0	1000.0
u_b	Blood velocity	(mm/w)	$X \sim F^{-1}(1.5, 2.5)$	$X \sim F^{-1}(1.5, 4)$
δ	Average distance between blood vessels	(mm)	6.0	4.0
ϵ	Porosity of the medium	–	0.02	0.02
β	Correction factor	–	0.1	0.1
Q_m	Metabolic heat source	W/m	420.0	4200.0

Two distinct scenarios were considered. The first one (Scenario 1) considers that magnetic nanoparticles are applied in a single point inside the tumour. The second scenario (Scenario 2) considers the application of magnetic nanoparticles in four distinct points. The domain is $x \in [0.0, 0.1]$ m and $y \in [0.0, 0.1]$ m, with the tumour located in $(x, y) \in [0.04, 0.06] \times [0.04, 0.06]$.

3.1   Scenario 1

In the first scenario, we consider only one injection site in the middle of the tumour, using the injection parameters given by Table 2.

Table 2. Parameters values for hyperthermia treatment with one injection site.

Position (m)	A	r_0
$(0.05, 0.05)$	$1.9 \times 10^6 \, \text{W/m}^3$	3.1×10^{-3}

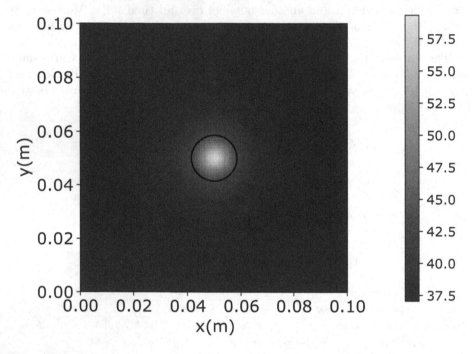

Fig. 1. Temperature distribution at $t = 5,000$ s employing Eq. (3) with a tumour located in $(x, y) \in [0.04, 0.06] \times [0.04, 0.06]$ and one injection of nanoparticles as specified in Table 2. The solid black contour represents the temperature of $43 \, ^\circ\text{C}$, highlighting the necrosis area.

The solution of Eq. (1) at $t = 5,000 \, \text{s}$, using the method described by Eq. (3), is shown in Fig. 2. This result shows the steady state temperature distribution. It is possible to observe an exponential decay in the temperature as the distance to the injection site increases.

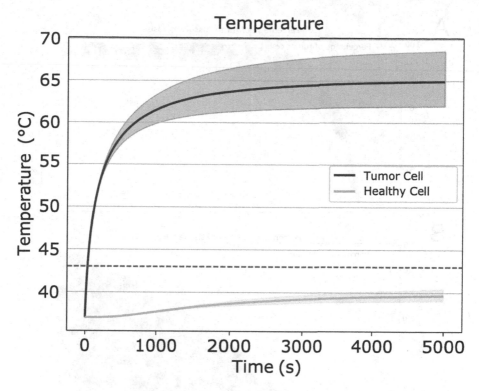

Fig. 2. Evolution of temperature determined by the Eq. (3) with one injection of nanoparticles located in the position presented in Table 2. The solid lines represents the mean temperature, shaded regions represents the 95% confidence interval and the red dashed line represents the temperature where the necroses occurs. The blue line and blue shaded represent the temperature of the tumour tissue located on the point $x = 0.05$ m and $y = 0.05$ m. The orange line and orange shaded represent the temperature of a healthy tissue located on the point $x = 0.065$ m and $y = 0.065$ m.

Figure 2 presents the temperature evolution in both health and tumour tissues. For health tissue we consider the point seated in $(0.065$ m; 0.065 m$)$ and for tumour tissue a point located at $(0.05$ m; 0.05 m$)$. It can be observed that the temperature inside the tumour is much higher than $43\,°C$, which causes its necrosis, while the temperature on the health point does not reaches $43\,°C$.

Figure 3A represents the evolution of tumour tissue necrosis over time. The percentage of tumour tissue killed due to the hyperthermia was computed considering all tumour points whose temperature was above $43\,°C$. This figure shows that up to 100% of the tumour is destroyed by the end of the simulation.

Figure 3B represents the evolution of healthy tissue necrosis over time. This value was computed counting the number of healthy cell destroyed by a elevation of temperature above $43\,°C$. This figure shows that a small number of health tissue, 3.5%, were damaged due to the hyperthermia treatment.

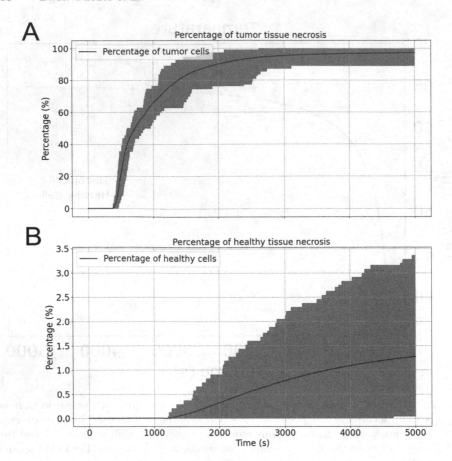

Fig. 3. (A) Percentage of tumour tissue killed by hyperthermia, and (B) Percentage of healthy tissue killed by hyperthermia. In both cases, one injection of nanoparticles was administered in the position given by Table 2. The percentage of tumour tissue killed by hyperthermia was computed at each time step considering all tumour cells whose temperature was above 43 °C. The solid line represents the mean and shaded regions represents the 95% confidence interval. The same method was applied to the healthy tissue.

3.2   Scenario 2

For the second scenario, we assume four injection sites in distinct positions. Table 3 presents the parameters values for this scenario.

The temperature distribution presented on Fig. 4 shows the solution of Eq. (3) on the final time step, *i.e.* 5,000 s.

Figure 5 shows the evolution of temperature on a point inside the tumour (0.05 m, 0.05 m) and one point outside it (0.065 m, 0.065 m) in a simulation using four injection sites as described in Table 3. This figure shows that the temperature inside the tumour rises above 43 °C, causing the tumour tissue to

Table 3. Parameters values for hyperthermia treatment with four injection sites.

Position (m)	A	r_0
$(0.045, 0.045)$	$0.425 \times 10^6 \text{ W/m}^3$	3.1×10^{-3}
$(0.045, 0.055)$	$0.425 \times 10^6 \text{ W/m}^3$	3.1×10^{-3}
$(0.055, 0.045)$	$0.425 \times 10^6 \text{ W/m}^3$	3.1×10^{-3}
$(0.055, 0.055)$	$0.425 \times 10^6 \text{ W/m}^3$	3.1×10^{-3}

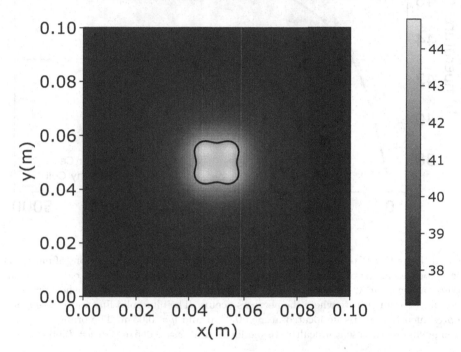

Fig. 4. Temperature distribution at $t = 5,000$ s employing Eq. (3) with a tumour located in $(x, y) \in [0.04, 0.06] \times [0.04, 0.06]$ and four injections of nanoparticles as specified in Table 3. The solid black contour represents the temperature of $43\,^\circ$C, highlighting the necrosis area.

necrosis, while the temperature outside the tumour does not. Therefore, the healthy tissue located on the point outside the tumour does not reach $43\,^\circ$C.

The percentage of tumour tissue killed due to hyperthermia with four injection site locations is shown in Fig. 6A. This result presents the percentage of tumour tissue killed over time, showing that at the end of the simulation, the expected value of tumour necrosis reaches 100%.

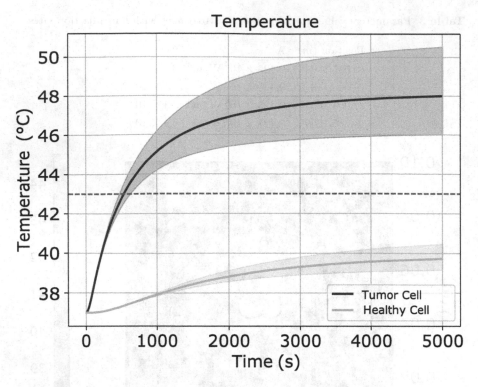

Fig. 5. Evolution of temperature determined by Eq. (3) with four injections of nanoparticles located in the positions presented in Table 3. The solid lines represents the mean temperature, shaded regions the 95% confidence interval and the red dashed line represents the temperature where the necroses occurs. The blue solid line and blue shaded represent a tumour tissue located at $x = 0.05$ m and $y = 0.05$ m. The orange line and orange shaded represent a healthy tissue located at $x = 0.065$ m and $y = 0.065$ m.

The consequences of hyperthermia treatment on healthy tissue are shown in Fig. 6B, which presents the percentage of healthy tissue killed due to the elevation of temperature above 43 °C. Only 3% of healthy tissue suffered necrosis at the end of the simulation of the hyperthermia treatment.

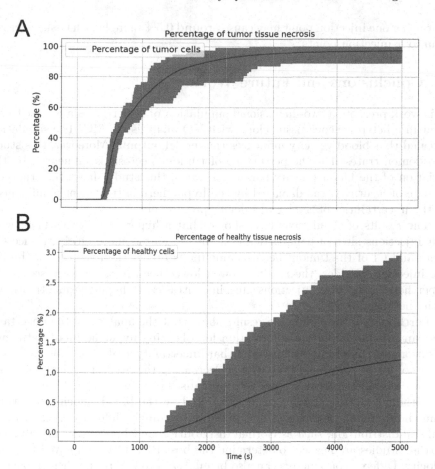

Fig. 6. (A) Percentage of tumour tissue killed by hyperthermia, and (B) Percentage of healthy tissue killed by hyperthermia. In both cases, four injection of nanoparticles was administered in the position given by Table 3. The percentage of tumour tissue killed by hyperthermia was computed at each time step considering all tumour cells whose temperature was above 43 °C. The solid line represents the mean and shaded regions represents the 95% confidence interval. The same method was applied to the healthy tissue.

4   Discussion

From the numerical results, it is possible to conclude that in both cases, with one and four injections sites on the tumour, the temperature rises above 43 °C at the tumour site which leads to its necrosis. Particularly Figs. 3A and 6A shows that the tumour is destroyed by the treatment. Comparing the different strategies for application points, one can observe that using one injection points leads to higher temperatures than four injection points. Besides, Figs. 3B and 6B show that using one injection point may cause around 0.5% more health tissue necrosis than four injection points.

5   Conclusions and Future Works

This work presents a two-dimensional simulation of hyperthermia cancer treatment in a heterogeneous tissue along with UQ analysis via MC. The simulation revealed that blood velocity influences the model outputs. Moreover, this study also demonstrates that the percentage of tumour necrosis is influenced by the variation of the blood velocity parameter. Also, this study shows that the percentage of healthy tissue damaged by the hyperthermia treatment is influenced by the uncertainties of the blood velocity parameter.

The results of simulations have shown that a hyperthermia treatment with four injections sites of nanoparticles on the tumour is more efficient, once the mean and CI of the tumour necrosis amount is higher than the case with only one injection. Besides, the results showed lower healthy tissue necrosis on the hyperthermia with four injections sites, in contrast with hyperthermia treatment with one injection site.

Furthermore, the simulation results show that the analysis of the uncertainties can provide important information for planning hyperthermia treatments. Additionally, this study demonstrates that uncertainty analysis can be an important tool for *in silico* medicine, mainly decreasing the need of clinical trial with animals as well as cohort studies with humans.

Although the goal of this paper is to analyse the blood velocity parameter from the mathematical model, it is possible to employ different types of probabilistic distribution, such as Normal distribution and LogNormal distribution. Further studies are needed to determine the best PDF to be employed to blood velocity. Different parameters can also be analyzed to determine their impact on the results of the mathematical model simulation. Furthermore, we intend to use real data from hyperthermia treatment and compare it with the *in silico* trials along with parameters of real patients to perform patient-specific experiments.

Acknowledgments. The authors would like to express their thanks to CAPES, CNPq, FAPEMIG and UFJF for funding this work.

References

1. Cancer estimates in brazil. http://www.oncoguia.org.br/conteudo/estimativas-no-brasil/1705/1/. Accessed 12 Feb 2020
2. Comprehensive cancer information about treatments. https://www.cancer.gov/about-cancer/treatment/types. Accessed 12 Feb 2020
3. Comprehensive cancer information and statistics. https://www.cancer.gov/about-cancer/understanding/statistics. Accessed 12 Feb 2020
4. Bhaumik, P., Ghosal, S.: Efficient Bayesian estimation and uncertainty quantification in ordinary differential equation models. Bernoulli **23**(4B), 3537–3570 (2017). https://doi.org/10.3150/16-BEJ856

5. Campos, J.O., Sundnes, J., Dos Santos, R.W., Rocha, B.M.: Effects of left ventricle wall thickness uncertainties on cardiac mechanics. Biomech. Model. Mechanobiol. **18**(5), 1415–1427 (2019)
6. Eymard, R., Gallouët, T., Herbin, R.: Finite volume methods. Handbook Numer. Analy. **7**, 713–1018 (2000)
7. Fahrenholtz, S.J., Stafford, R.J., Maier, F., Hazle, J.D., Fuentes, D.: Generalised polynomial chaos-based uncertainty quantification for planning MRgLITT procedures. Int. J. Hyperthermia **29**(4), 324–335 (2013)
8. Feinberg, J., Langtangen, H.P.: Chaospy: an open source tool for designing methods of uncertainty quantification. J. Comput. Sci. **11**, 46–57 (2015)
9. Hicklin, D.J., Ellis, L.M.: Role of the vascular endothelial growth factor pathway in tumor growth and angiogenesis. J. Clin. Oncol. **23**(5), 1011–1027 (2005)
10. Hurtado, D.E., Castro, S., Madrid, P.: Uncertainty quantification of 2 models of cardiac electromechanics. Int. J. Numer. Meth. Biomed. Eng. **33**(12) (2017)
11. Iero, D.A.M., Crocco, L., Isernia, T.: Thermal and microwave constrained focusing for patient-specific breast cancer hyperthermia: a robustness assessment. IEEE Trans. Antennas Propag. **62**(2), 814–821 (2014). https://doi.org/10.1109/TAP.2013.2293336
12. Ishida, H., Hachiga, T., Andoh, T., Akiguchi, S.: In-vivo visualization of melanoma tumor microvessels and blood flow velocity changes accompanying tumor growth. J. Appl. Phys. **112**(10), 104703 (2012)
13. Jiji, L.M.: Heat transfer in living tissue. In: Heat Conduction, pp. 302–346. Springer, Heidelberg (2009)
14. Khaled, A.R., Vafai, K.: The role of porous media in modeling flow and heat transfer in biological tissues. Int. J. Heat Mass Trans. **46**(26), 4989–5003 (2003)
15. d.L.e Silva, L., Xavier, M.P., dos Santos, R.W., Lobosco, M., Reis, R.F.: Uncertain quantification of immunological memory to yellow fever virus. In: 2020 IEEE International Conference on Bioinformatics and Biomedicine (BIBM), pp. 1281–1288 (2020). https://doi.org/10.1109/BIBM49941.2020.9313282
16. Minkowycz, W., Sparrow, E.M., Abraham, J.P.: Nanoparticle Heat Transfer and Fluid Flow, vol. 4. CRC Press, Boca Raton (2012)
17. Moros, E.: Physics of Thermal Therapy: Fundamentals and Clinical Applications. CRC Press, Boca Raton (2012)
18. Moroz, P., Jones, S., Gray, B.: Magnetically mediated hyperthermia: current status and future directions. Int. J. Hyperthermia **18**(4), 267–284 (2002)
19. Nardini, J.T., Bortz, D.: The influence of numerical error on parameter estimation and uncertainty quantification for advective PDE models. Inverse Prob. **35**(6), 065003 (2019)
20. Osnes, H., Sundnes, J.: Uncertainty analysis of ventricular mechanics using the probabilistic collocation method. IEEE Trans. Biomed. Eng. **59**(8), 2171–2179 (2012)
21. Pennes, H.H.: Analysis of tissue and arterial blood temperatures in the resting human forearm. J. Appl. Physiol. **1**(2), 93–122 (1948)
22. Reagan, M.T., 4, H.N.N., Debusschere, B.J., Maître, O.P.L., Knio, O.M., Ghanem, R.G.: Spectral stochastic uncertainty quantification in chemical systems. Combust. Theory Model. **8**(3), 607–632 (2004). https://doi.org/10.1088/1364-7830/8/3/010
23. Reis, R.F., et al.: Characterization of the covid-19 pandemic and the impact of uncertainties, mitigation strategies, and underreporting of cases in South Korea, Italy, and Brazil. Chaos, Solitons Fractals **136**, 109888 (2020). https://doi.org/10.1016/j.chaos.2020.109888, https://www.sciencedirect.com/science/article/pii/S0960077920302885

24. Reis, R.F., et al.: The quixotic task of forecasting peaks of covid-19: Rather focus on forward and backward projections. Front. Public Health **9**, 168 (2021). https://doi.org/10.3389/fpubh.2021.623521
25. Reis, R.F., dos Santos Loureiro, F., Lobosco, M.: Parameters analysis of a porous medium model for treatment with hyperthermia using OpenMP. J. Phys. Conf. Ser. **633**, 012087 (2015)
26. Reis, R.F., dos Santos Loureiro, F., Lobosco, M.: 3D numerical simulations on GPUs of hyperthermia with nanoparticles by a nonlinear bioheat model. J. Comput. Appl. Math. **295**, 35–47 (2016)
27. Rubinstein, R.Y., Kroese, D.P.: Simulation and the Monte Carlo method, vol. 10. John Wiley & Sons, Hoboken (2016)
28. Saltelli, A., et al.: Global Sensitivity Analysis: the Primer. John Wiley & Sons, Hoboken (2008)
29. Sullivan, T.J.: Introduction to Uncertainty Quantification, vol. 63. Springer, Heidelberg (2015)
30. Versteeg, H.K., Malalasekera, W.: An Introduction to Computational Fluid Dynamics: the Finite Volume Method. Pearson Education, London (2007)
31. Wu, J.L., Michelén-Ströfer, C., Xiao, H.: Physics-informed covariance kernel for model-form uncertainty quantification with application to turbulent flows. Comput. Fluids **193**, 104292 (2019)
32. Xiu, D.: Numerical Methods for Stochastic Computations: a Spectral Method Approach. Princeton University Press, New Jersey (2010)
33. Zhou, W., Chen, Z., Zhou, Q., Xing, D.: Optical biopsy of melanoma and basal cell carcinoma progression by noncontact photoacoustic and optical coherence tomography: in vivo multi-parametric characterizing tumor microenvironment. IEEE Trans. Med. Imaging **39**(6), 1967–1974 (2019)

EEG-Based Emotion Recognition – Evaluation Methodology Revisited

Sławomir Opałka⦿, Bartłomiej Stasiak(✉)⦿, Agnieszka Wosiak⦿,
Aleksandra Dura⦿, and Adam Wojciechowski⦿

Institute of Information Technology, Łódź University of Technology,
ul. Wólczańska 215, 93-005 Łódź, Poland
bartlomiej.stasiak@p.lodz.pl

Abstract. The challenge of EEG-based emotion recognition had inspired researchers for years. However, lack of efficient technologies and methods of EEG signal analysis hindered the development of successful solutions in this domain. Recent advancements in deep convolutional neural networks (CNN), facilitating automatic signal feature extraction and classification, brought a hope for more efficient problem solving. Unfortunately, vague and subjective interpretation of emotional states limits effective training of deep models, especially when binary classification is performed basing on datasets with non-bimodal distribution of emotional state ratings. In this work we revisited the methodology of emotion recognition, proposing to use regression instead of classification, along with appropriate result evaluation measures based on mean absolute error (MAE) and mean squared error (MSE). The advantages of the proposed approach are clearly demonstrated on the example of the well-established and explored DEAP dataset.

Keywords: EEG · Emotion recognition · Regression · Classification · CNN

1 Introduction

In recent years, electroencephalography (EEG) became an emerging source of bioelectrical signals providing invaluable information concerning various aspects of human brain activity. Among them, emotional states and their manifestations in the EEG signal still hide many secrets and raise expectations in several research domains: medicine, psychology and human computer interaction.

EEG signal reflects a complex brain activity spectrum, requiring advanced signal processing methods and feature extraction methodologies to be meaningfully interpreted. Although recent advancements in deep learning techniques have revealed a potential for EEG signal classification in several domains [1], emotion analysis still seems to be one of the most challenging ones, especially due to the subjective evaluation process, intrinsic noise and acquisition channels

© Springer Nature Switzerland AG 2021
M. Paszynski et al. (Eds.): ICCS 2021, LNCS 12743, pp. 525–539, 2021.
https://doi.org/10.1007/978-3-030-77964-1_40

crosstalk [2]. Data acquisition for emotion recognition tasks typically involves eliciting specific emotions in subjects, e.g. by watching video clips, appropriately selected by experts. EEG is recorded during these sessions and emotion self-assessment is usually conducted after each video clip.

The current research on deep learning has shown outstanding results in computer vision and image processing [4]. Due to the nature of the EEG signal on one hand and the fundamental principles and characteristics of deep learning tools and methods on the other hand, it can be expected that these methods will become the mainstream research technique for EEG signal processing in the near future [5]. Especially the deep learning approach for the EEG-based emotion recognition problem seems to leave considerable room for improvement.

The main motivation for our contribution regards the problem of ambiguities inherent to the self-assessment (ground-truth ratings) of the subject's emotional state and their impact on the machine learning strategies applied. We propose a novel approach for emotional state recognition with a convolutional neural network (CNN), which is based on regression rather than on binary classification. Our evaluation methodology puts more emphasis on the self-assessment nuances and it also lets us interpret the results in a more suitable and understandable manner.

2   Previous Work

One of the first studies, originating the discussion on emotional state classification and becoming an inspiration for the research community was published by Koelstra et al. [6]. The authors have introduced a Database for Emotion Analysis using Physiological signals (DEAP) which encompasses a set of psycho-physiological parameters acquired from users who were watching specially selected music video clips.

DEAP dataset has originated numerous experiments on quantitative classification of emotional dimensions, namely *valence* (quality of emotion from unpleasant to pleasant) and *arousal* (emotion activation level from inactive to active) in accordance with Russell's valence-arousal scale [3], as well as *dominance* (from a helpless and weak feeling to an empowered feeling) and personal impressions encoded in *liking* parameter.

Extensive review of DEAP-based experiments was published by Roy et al. [8]. The authors concluded that the EEG signal suffers from considerable limitations that hinder its effective processing and analysis. Due to low signal-to-noise ratio (SNR), non-stationary characteristics and high inter-subject variability, signal classification becomes a big challenge for real-life applications.

An important source of inspiration for our research was the paper by Craik et al. [9]. The authors have reported a systematic review on EEG-related tasks classification. One of the reviewed aspects concerned input formulation for a CNN-based deep learning solution to the emotional state classification problem, which became a supportive context for the present study. Despite of the fact that most of the authors addressed the problem of signal artifact removal, inherent signal characteristics are hardly addressed. It is difficult to identify persistent noisy channels or noise that is sparsely presented in multi-channel recordings. Manually processed data was highly subjective and rendering it was difficult for other researches to reproduce the procedures. Surprisingly, one of the main findings reported by Craik et al. [9] was that – according to the authors' best knowledge – there were no studies demonstrating that deep learning can achieve results comparable with classification methods based on handcrafted features. Especially convolutional neural networks, although popular in image processing domain, can hardly be found in the domain of EEG-based emotional state recognition. In fact, just a few authors examined CNN architecture with frequency domain EEG spectrograms prepared as an input [10–13]. They have concentrated on motor impairment [10], mental workload [11,13] and sleep stage scoring [12], rather than on emotional state analysis/classification.

There were also several neural architectures containing convolutional layers, which were employed for examining the DEAP dataset and which might therefore be treated as a meaningful reference. In their both works Li et al. [14,15] proposed hybrid neural architectures interpreting wavelet-derived features, with the CNN output connected to LSTM RNN modules, but they achieved low classification accuracy. On the other hand, noticeable difference in classification accuracy could be attributed to alternative input formulation. Yanagimoto et al. [16] directly used signal values as inputs into a neural network, while Qiao et al. [17] and Salama et al. [18] converted the input data into Fourier maps and 3D grids respectively, considerably improving the classification accuracy. Nevertheless, CNN application for processing spectrogram-based EEG data in the frequency domain seems to be a quite novel strategy of outstanding and unexplored potential.

In the context of valence-arousal dimensions and deep learning-based classification, Lin et al. [19] proposed an interesting multi-modal approach for the DEAP dataset, achieving for bi-partitioned classification of valence and arousal 85.5% and 87.3% respectively. It must be noted, however, that the obtained accuracy is greater than many other DEAP-based experiments, mainly due to considering all the available physiological signals rather than just EEG.

Frequency-domain representation of the EEG signal, collected with numerous electrodes (32 channels in the case of DEAP) and split into frequency subbands, puts high demands on the neural network model, which must deal with this highly-dimensional input data in conditions of the limited number of subjects, stimuli, recording time, etc. Feature extraction is therefore an important processing step applied by most authors to reduce the complexity of the neural network and to let it learn effectively with low generalization error. It should be noted that data dimensionality reduction may also be obtained by appropriate selection of the EEG electrodes, e.g. on the basis of channel cross-correlation.

An extensive and detailed analysis of the most suitable approach for EEG signal feature extraction was provided by Nakisa et al. [20]. The authors considered different time-domain, frequency-domain and time-frequency domain features in the context of the selected datasets [6,7]. Four-class (High/Low

Valence/Arousal) emotion classification process, based on a probability neural network (PNN) and 30 features selected with 5 alternative evolutionary computation algorithms, provided average accuracy reaching 65% within 100 iterations. Supportive conclusion regarded DEAP-oriented EEG channels selection. The authors noticed that FP1, F7, FC5, AF4, CP6, PO4, O2, T7 and T8 were the most relevant electrodes in the context of emotion recognition. The above findings and the current progress in deep learning-based feature extraction and classification [21] led to an observation that automatically selected features from a limited number of channels (9 for DEAP) may result in more spectacular results, particularly given that traditional classifiers like SVM-related ones reach 78% for bi-partitioned valence and arousal classes [22] on 16 frequency and time domain features and respectively 63% and 58% for 3-partitioned valence and arousal classes [23] on 11 frequency and time domain features. High accuracy of non-deep learning approaches was also reported by other authors [24].

Inherent emotional dimensions in the DEAP dataset were subjectively quantified by subjects, who assigned numerical ratings between 1 and 9. However, most of the authors aggregated emotional dimensions into two groups: high (greater or equal 5) and low (lower than 5) within each dimension respectively. In consequence, the mean values of valence and arousal within four quadrants on the Low-High/Arousal-Valence plane were as follows: Low-Arousal-Low-Valence (LALV: 4.3 ± 1.1; 4.2 ± 0.9), High-Arousal-Low-Valence (HALV: 5.7 ± 1.5; 3.7 ± 1.0), Low-Arousal-High-Valence (LAHV: 4.7 ± 1.0; 6.6 ± 0.8), High-Arousal-High-Valence (HAHV: 5.9 ± 0.9; 6.6 ± 0.6). It must be noted that these means were relatively close to each other (although statistically different), as compared to the whole range of the possible values. Additionally, considering the fact that the authors of DEAP selected only 40 video clips – out of initial 120 – namely the most extreme ones in the context of valence/arousal, unambiguous quantitative differentiating of emotional dimensions becomes a real challenge requiring in-depth analysis.

The above findings, addressing the ambiguity in emotional state classification, suggest revisiting deep learning-based approaches. Specifically, the *quantization* of the emotional dimensions and reformulating the problem from binary (low-high) classification into the *regression* task will be covered in the following part of this paper.

3   Materials and Methods

The DEAP database [6], introduced in the previous section, was published by Queen Mary University of London for the purpose of emotion analysis on the basis of physiological signals, including EEG. The data consists of recordings of 32 participants who were watching music videos. Forty 1-min long videos were carefully selected to induce emotions falling into 4 general categories: LAHV, LALV, HAHV and HALV, as described above. The recorded data comprises 32 EEG channels conforming to the international 10–20 system and 8 additional channels representing various physiological signals, including EOG (horizontal

and vertical eye movements), EMG (activity of zygomaticus major and trapez-
ius muscles), GSR, respiration, blood volume pressure and temperature. In our
research only the EEG channels were used.

Every participant rated every video he/she had watched, in terms of four
distinct emotional qualities: valence, arousal, dominance and liking. Each rating
was expressed as a real number from the range $[1, 9]$. The two first qualities
(valence, arousal) defined our research goal: to predict the participant's rating
on the basis of the EEG signal recording.

3.1 EEG Data Preprocessing

The EEG signal recorded from a participant watching one film comprises 8064
time-domain samples per electrode, which corresponds to 63 s at sampling frequency 128 Hz. Power spectral density is computed for each frame of size 128
samples (one second, *von Hann* window applied) with 50% overlap (hop-size: 64
samples). This yields a spectrogram with 125 time points (frames) and frequency
resolution 1 Hz. The frequencies below 4 Hz and above 45 Hz are rejected (in fact,
they are already filtered out in the originally preprocessed DEAP dataset). The
spectrogram values are then scaled logarithmically so that they fit within $[0, 1]$
range. The first 5 frames are rejected, and the spectrogram is rescaled along
the frequency axis (antialiasing filter applied) to the final shape of 20×120
(frequency $\times$ time). An example is presented in Fig. 1.

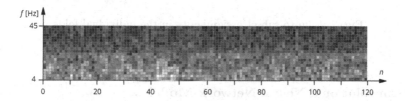

Fig. 1. Spectrogram of the first EEG channel (Fp1 electrode) recorded for the first
participant watching the first video (n denotes the frame index)

Each spectrogram is cut into chunks of 10 frames each (which corresponds to
ca 5 s) with overlap of 50%. This yields 23 chunks for a single spectrogram. The
corresponding chunks coming from all the 32 electrodes (i.e. the chunks representing the same time range within all 32 spectrograms) are grouped together,
forming the single *object* (input tensor) to be recognized by the network. Every
recording of a single film watched by a single user is therefore represented by
separate 23 fragments (input tensors) of size $32 \times 20 \times 10$ (electrodes $\times$ frequency bands $\times$ time frames). All these 23 fragments have the same target value
(defining our training goal) which is simply equal to the participant's rating of
the film under consideration. Each tensor is an individual input object for a
convolutional neural network described in the next section.

We decided to apply a 4-fold crossvalidation scheme with fixed division into the training, testing and validation subsets. Every experiment was based on the EEG data from 40 films of *a single participant only*. It was therefore repeated for all 32 participants individually (and for each of the 4 folds) and averaged results are reported (Sect. 4).

Considering the data from a single participant, 10 films were included in the testing set, another 10 films – in the validation set and the remaining 20 films were used for training. Complete films were always used, i.e. we did not mix fragments (input tensors) from different films. In Fig. 2 the assignments of individual films to particular subsets (in each of the 4 folds) are shown in the bottom 4 rows. Ts denotes the testing set, Vd – the validation set and the blank fields indicate the films used for training. The reason for these particular assignments is explained in the upper part of the table in Fig. 2. Most of the first 20 films have on average higher valence ratings (H) than the last 20 films (L). As for the arousal, the first and the last 10 films tend to be rated higher than the middle 20. Therefore, the chosen assignment yields more balanced testing set (and also the validation and the training ones), containing 5 H's and 5 L's both for valence and arousal, irrespective of the fold number (although the ratings of individual participants may occasionally deviate from this simple H/L distinction).

Film idx	1	2	3	4	5	6	7	8	9	10	11	12	13	14	15	16	17	18	19	20	21	22	23	24	25	26	27	28	29	30	31	32	33	34	35	36	37	38	39	40
Valence	H	H	H	H	H	H	H	H	H	H	H	H	H	H	H	H	H	H	H	H	L	L	L	L	L	L	L	L	L	L	H	H	H	H	H	H	H	H	H	H
Arousal	H	H	H	H	H	H	H	H	H	H	L	L	L	L	L	L	L	L	L	L	L	L	L	L	L	L	L	L	L	L	H	H	H	H	H	H	H	H	H	H
Fold 1	Ts	Ts	Ts	Ts	Ts						Vd	Vd	Vd	Vd	Vd						Ts	Ts	Ts	Ts	Ts						Vd	Vd	Vd	Vd	Vd					
Fold 2						Ts	Ts	Ts	Ts	Ts						Vd	Vd	Vd	Vd	Vd						Ts	Ts	Ts	Ts	Ts						Vd	Vd	Vd	Vd	Vd
Fold 3	Vd	Vd	Vd	Vd	Vd						Ts	Ts	Ts	Ts	Ts						Vd	Vd	Vd	Vd	Vd						Ts	Ts	Ts	Ts	Ts					
Fold 4						Vd	Vd	Vd	Vd	Vd						Ts	Ts	Ts	Ts	Ts						Vd	Vd	Vd	Vd	Vd						Ts	Ts	Ts	Ts	Ts

Fig. 2. Film-to-subset assignments in the individual folds

3.2 Convolutional Neural Network Model

Having analyzed the extensive review presented by Roy et al. [8], we decided to use a simple, yet effective CNN architecture shown in Fig. 3.

Fig. 3. Applied model architecture

The architecture is implemented as Sequential model with *Keras* interface for *TensorFlow* library. Apart from the dropout layers, there are some additional elements (not shown) aimed at generalization properties enhancement: L2 kernel regularizer (regularization coefficient: 0.01) in both convolutional layers and a GaussianNoise layer ($\sigma = 1.5$) applied before the first convolutional layer.

4 Experiment Objectives and Design

Polarity inference is the most fundamental task in many emotion recognition or sentiment analysis problems. *Does she like me or not? Are they interested or bored?* We often tend to ignore the possible shades of gray between the extremes. The DEAP dataset construction principles seem to support this view, provided that the 40 films had been deliberately selected (out of the initial collection of 120 videos) to maximize the strength of the elicited emotions. More precisely, the database contains these films, which happened to lie closest to the 4 extreme corners in the valence-arousal 2D space, as rated by at least 14 volunteers per film in a preliminary step of video material selection [6].

Using the collected videos in the actual experiments, targeted at emotion recognition from the physiological signals, follows naturally the same principle. A typical approach found in most research works, including also the original paper by Koelstra et al. [6], aims at *classification* of the videos in two classes: low and high, with respect to any of the four aforementioned emotional qualities. For example, if the valence rating of a film exceeds 5, the film is automatically included in the "high valence" class ("low valence" in the opposite case). This approach seems natural, straightforward and valid. However, taking into account the actual data collected from the subjects participating in DEAP database construction, it is probably overly simplistic, as we demonstrate below.

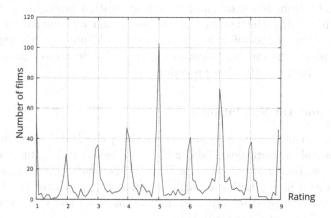

Fig. 4. Histogram of *valence* ratings

The first thing to consider is the histogram of the participants' ratings (Fig. 4). Its unusual shape results from the fact, that the participants usually tried to give integer scores, although the available input method was based on a continuous scale (ranging from 1.0 to 9.0, as mentioned before). Now we can clearly see the problem of "neutral responses": over 100 ratings were very close to 5 or – if we take a slightly broader tolerance – 221 ratings fell within the range [4.5, 5.5]. Considering the total number of ratings (1280 = 40 films evaluated by 32 users), this accounts for over 17% ratings which were probably intended to mean "no opinion". Splitting the data with a hard threshold of 5 into positive/negative valence ratings will inevitably lead to significant confusion, irrespective of the particular machine learning or classification approach.

Moreover, assigning the same class ("low") for films labelled "1" and "4", while separating the "4"s ("low") from the "6"s ("high") seems also arguable. Although the emotional valence rating is highly subjective and it probably does not follow any simple linear scale or distance measure suggested by Fig. 4, predicting the actual *rating* instead of the "low-high" quantization seems much more appropriate.

Finally, we have to take into account that the participant's rating is the result of an intrinsic decision-making process based not only on purely emotional reactions but on many other premises as well. They may include prior knowledge and personal attitude towards the video content, the general worldview, the social, political and cultural background and – last but not least – the comparison with the previously watched videos and the ratings given. These factors may easily change the final rating of the current video within certain limits, independently on the actual emotions deducible from the recording of the physiological signals. This change may be relatively small in terms of sheer numbers, but in some cases it may easily shift the film from the "low" to the "high" class, or *vice versa*.

This, again, supports the claim that *prediction of the participant's rating* (i.e. *regression*) should be the preferred approach to the analysis of the DEAP dataset (and other collections based on similar data acquisition principles). It should be noted that increasing the number of classes, as an alternative to regression, would probably not be as effective, because we could not directly represent the relations between consecutive classes on the ordinal scale in our machine learning approach (and in the evaluation procedures).

4.1   Experimental Validation

In a single experiment (for a single participant), the training set for the convolutional neural network described in Sect. 3.2 included 460 input objects (20 films × 23 input tensors), according to Sect. 3.1. For each dimension of the input tensor, the mean μ_{train} and standard deviation σ_{train} within the training set were computed and used for data normalization of all the three sets: training (Tr), testing (Ts) and validation (Vd):

$$Tr_{norm} = \frac{Tr - \mu_{train}}{\sigma_{train}} \tag{1}$$

$$Ts_{norm} = \frac{Ts - \mu_{train}}{\sigma_{train}} \tag{2}$$

$$Vd_{norm} = \frac{Vd - \mu_{train}}{\sigma_{train}} \tag{3}$$

The goal of the supervised training was to obtain proper regression, i.e. to minimize the mean squared error (MSE) between the output of the last layer (a single neuron with a unipolar sigmoidal activation function, Sect. 3.2) and the target participant's rating value (ground-truth). The target value was divided by

10.0 to make it fit within the range [0.1, 0.9]. The network was trained with Adam optimizer [8]. After numerous preliminary experiments, the maximum number of epochs and the batch size were set to 600 and 100, respectively. The validation dataset (Sect. 3.1) was used to select the best model in terms of validation MSE minimization.

Apart from this, two additional experiments were done. In one of them ("regression_opt"), no validation set was used (or, more precisely, it was merged with the training set for the total number of 30 training films) and the optimal model was selected on the basis of the MSE value for the testing set. In this case, the obtained results may be interpreted as the theoretical maximum that might be reached, provided that the optimal stopping criterion is known beforehand. The other experiment is based on high/low classification instead of regression, so it is basically a reference for comparison of the results. In this case, all the participant's ratings and the network outputs are thresholded (below 0.5 → 0; greater or equal 0.5 → 1). It is worth to note, however, that this binary representation of the targets is used for the training process only. The CNN, once trained, is tested and evaluated in the same way as in the case of the regression-based experiments, as described in the following section.

4.2   Evaluation Metrics

Following the non-binary formulation of the training target, also the evaluation methods, used for the analysis of the testing set results, should be defined in a more "fine-grained" way. Mean square error (MSE) and mean absolute error (MAE) between the CNN outputs and the participant's ratings seem to be reasonable measures, telling us how big the discrepancy is on average. We also used a standard binary classification metric (CLS) based on thresholding both the outputs and targets with the fixed threshold of 0.5 and simply counting the objects for which the match occurred.

These three measures (MSE, MAE and CLS) were computed in two ways: for individual input tensors representing the film fragments (23 independent results for a single film) and for the whole films. In the latter case, the arithmetic mean of the outputs obtained for all the 23 input objects representing the same film was treated as the final prediction and compared with the target.

As an additional form of result presentation, we computed the percentage of films within a given range of MAE values (with respect to the true participant's rating). This computation was done on the whole-film basis, as defined above.

Apart from the objective evaluation measures, individual results were also carefully inspected and verified manually. This allowed us to detect a significant problem, responsible for degradation of the results in many cases. It occurred especially for these users and folds for which most of the ratings in the training set were close to the middle 5. In such cases the network tended to get stuck in the local minimum, producing non-diversified results, also close to 5 (Fig. 5, left). This problem may be viewed as a direct consequence of resigning from the "hard" binary classification approach which had forced the network to decide on either the high or low value at the output.

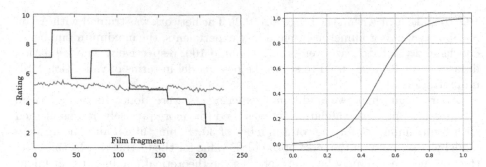

Fig. 5. Left: participant's ratings (black) and network output (grey) for all 230 fragments from the test set (user 10, fold 2; each sequence of 23 consecutive fragments represents one film); **Right**: the sigmoidal function used to transform the target values in order to force the network to produce more diversified output values. (Color figure online)

As a remedy, a special scaling of the targets in the training set, with a sigmoidal function, was applied (Fig. 5, right). This "soft alternative" to the binary classification drew the target values more to the extremes, encouraging the network to leave the "mid-level comfort zone" during training, while preserving the relative order of the rating values. The result for the same user/fold pair is presented in Fig. 6. Naturally, it should be remembered that training the network with target values transformed with a sigmoidal function requires that in the testing phase its outputs are transformed with the inverse function, before any comparison or evaluation is performed.

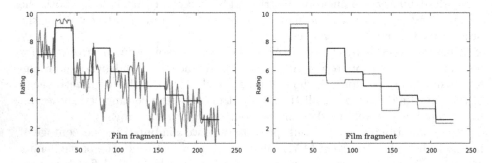

Fig. 6. Left: training result for the same dataset as in Fig. 5, but with the targets transformed with the sigmoidal function; **Right**: the same result but with the mean output ratings per film

4.3   Results

All the training sessions were independently done for the valence and for the arousal ratings. In every experiment, the training was repeated three times and the mean values of the evaluation metrics are reported in Table 1 (valence) and Table 2 (arousal).

Table 1. Results (valence)

	Film fragments			Whole films		
	MSE	MAE	CLS	MSE	MAE	CLS
Classification	0.128	0.298	58.9%	0.096	0.252	60.0%
Regression	0.052	0.181	59.2%	0.046	0.172	60.2%
Regression_opt	0.038	0.153	67.5%	0.032	0.143	71.4%

Table 2. Results (arousal)

	Film fragments			Whole films		
	MSE	MAE	CLS	MSE	MAE	CLS
Classification	0.134	0.307	59.6%	0.106	0.268	60.9%
Regression	0.050	0.173	60.4%	0.046	0.165	62.2%
Regression_opt	0.035	0.145	68.2%	0.031	0.137	70.3%

Figures 7 and 8 reveal how many films were rated sufficiently close to the ground-truth in terms of MAE. For example, the last column in the last group of Fig. 7 tells us that the absolute difference between the CNN output and the participant's rating was less than 4.5 in 98.8% of all the films. It is worth noting that this result (and all other presented results) is averaged over individual results from 384 training sessions (3 repetitions × 32 participants × 4 folds) and that the CNN output is in fact the mean of 23 outputs for 23 different fragments of the same film.

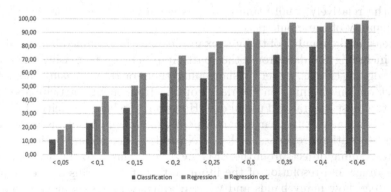

Fig. 7. Percentage of films within a given MAE range (valence)

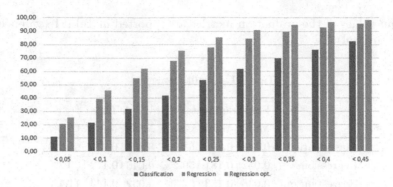

Fig. 8. Percentage of films within a given MAE range (arousal)

4.4 Discussion

The obtained results clearly show the advantages of the proposed approach. Although the binary classification accuracy (CLS) is only slightly better for the regression-based model than for the classification-based one (60.2% vs 60% and 62.2% vs 60.9% for the valence and arousal, respectively), the MSE and MAE values are definitely smaller (roughly two times smaller). In fact, this outcome is understandable when we take into consideration that the thresholding, used for producing the binary class labels, inevitably leads to discarding all the nuances present in the participants' ratings. Nevertheless, the practical usefulness of the obtained small MSE and MAE values seems quite clear, when we note, for example, that over two-thirds (three-fourths) of the films will get the arousal rating prediction within ± 2 (± 2.5) from the ground-truth, if we apply our regression-based CNN training approach.

Considering the binary classification accuracy itself, it has to be agreed upon that the obtained result, slightly exceeding 60% (or 70% for the optimally determined early stopping criterion), is not very impressive. One potential reason for that is the relatively small training set, especially when we consider the huge dimensionality of the input space.

Increasing the number of training examples may be obtained in several ways, e.g. by including the data from other participants or by increasing the number of spectrogram chunks, either by shortening them or by increasing the overlap. The first option (training on the data coming from many participants) would also be the most general and useful one. However, the heterogeneity of the EEG characteristics among the participants, poses significant problems in obtaining good generalization properties of the CNN models. The second option (generating more training objects from the EEG signal) is related to the question of the optimal range and resolution of the input data both in terms of the frequency content (e.g. how many bands and what frequency range should be analyzed) and the temporal characteristics (e.g. duration of the analyzed film fragments).

Instead of increasing the training set size, we may also search for dimensionality reduction. The EEG signals coming from adjacent electrodes are usually quite significantly correlated, and some EEG channels may be more useful in emotion analysis that the others. Similarly, some frequency ranges might probably be excluded from the input data or, at least, represented with decreased resolution. These are just a few examples of the research directions that will be considered in our future work.

5 Conclusion

In the presented work, we compared two Russell's emotional state evaluation methodologies in the task of valence/arousal prediction on the basis of the EEG signal. We confronted state-of-the-art binary classification with our regression-based approach. Our motivation was supported by the detailed analysis of the representative DEAP dataset, highlighting the pitfalls and difficulties resulting from simple high/low label assignment. We also proposed new evaluation metrics (MSE/MAE) conforming to the reformulated emotion recognition task. Subject-oriented experimental evaluation of the proposed methodology, based on a convolutional neural network trained on EEG signal spectrograms, revealed the improvement in the obtained results, both in terms of the new metrics and binary classification accuracy. The CNN trained to perform the regression task yielded much higher target rating prediction rates (with respect to the binary classification), with the difference reaching 26.2 percentage points (arousal) and 19.5 percentage points (valence), for MAE tolerance range of ± 0.2.

Future works will concentrate on analysis of automatic vs handcrafted EEG signal feature selection, including both EEG channels and frequency range selection. We will also investigate potential solutions for improvement of the generalization properties of the proposed CNN model.

As a final conclusion, we encourage the research community to revise the evaluation methodology used in emotional state recognition tasks and to consider regression as more appropriately reflecting the subjective nature of emotional state ratings reported by the users.

References

1. Opałka, S., Stasiak, B., Szajerman, D., Wojciechowski, A.: Multi-channel convolutional neural networks architecture feeding for effective EEG mental tasks classification. Sensors **18**, 3451 (2018)
2. Teplan, M.: Fundamentals of EEG measurement. Meas. Sci. **2**, 1–11 (2002)
3. Russell, J.A.: A circumplex model of affect. J. Pers. Soc. Psychol. **39**(6), 1161–1178 (1980)
4. Lazarek, J., Pryczek, M.: A review of point cloud semantic segmentation methods. J. Appl. Comput. Sci. **26**(2), 99–105 (2018)
5. Li, G., Lee, C.H., Jung, J.J., Youn, Y.C., Camacho, D., Deep learning for EEG data analytics: a survey. Concurrency Comput. **32**(18), e5199 (2020)

6. Koelstra, S., et al.: DEAP: a database for emotion analysis using physiological signals. IEEE Trans. Affect. Comput. **3**, 18–31 (2012). https://doi.org/10.1109/T-AFFC.2011.15

7. Soleymani, M., Lichtenauer, J., Pun, T., Pantic, M.: A multimodal database for affect recognition and implicit tagging. IEEE Trans. Affect. Comput. **3**(1), 42–55 (2012). https://doi.org/10.1109/T-AFFC.2011.25

8. Roy, Y., Banville, H., Albuquerque, I., Gramfort, A., Falk, T.H., Faubert, J.: Deep learning-based electroencephalography analysis: a systematic review. J. Neural Eng. **16**(5), 051001 (2019)

9. Craik, A., He, Y., Contreras-Vidal, J.L., Deep learning for electroencephalogram (EEG) classification tasks: a review. J. Eng. **16**(3), 031001 (2019)

10. Vrbancic, G., Podgorelec, V.: Automatic classification of motor impairment neural disorders from EEG signals sing deep convolutional neural networks. Elektron. Elektrotech. **24**, 3–7 (2018)

11. Kuanar, S., Athitsos, V., Pradhan, N., Mishra, A., Rao, K.R.: Cognitive analysis of working memory load from EEG, by a deep recurrent neural network. In: 2018 IEEE International Conference on Acoustics, Speech and Signal Processing (ICASSP) (2018)

12. Vilamala, A., Madsen, K.H., Hansen, L.K.: Deep convolutional neural networks for interpretable analysis of EEG sleep stage scoring. In: 2017 IEEE 27th International Workshop on Machine Learning for Signal Processing (MLSP) (2017)

13. Jiao, Z., Gao, X., Wang, Y., Li, J., Xu, H.: Deep Convolutional Neural Networks for mental load classification based on EEG data. Pattern Recog. **76**, 582–95 (2018)

14. Li, X., Song, D., Zhang, P., Yu G., Hou, Y., Hu, B.: Emotion recognition from multi-channel EEG data through convolutional recurrent neural network. In: 2016 IEEE International Conference on Bioinformatics and Biomedicine, pp. 352–359 (2016)

15. Li, Y., Huang, J., Zhou, H., Zhong, N.: Human emotion recognition with electroencephalographic multidimensional features by hybrid deep neural networks. Appl. Sci. **7**, 1060 (2017)

16. Yanagimoto, M., Sugimoto, C.: Recognition of persisting emotional valence from EEG using convolutional neural networks. In: 2016 IEEE 9th International Workshop on Computational Intelligence and Applications, pp. 27–32 (2016)

17. Qiao, R., Qing, C., Zhang, T., Xing, X., Xu, X.: A novel deep-learning based framework for multi-subject emotion recognition. In: 2017 4th International Conference on Information, Cybernetics and Computational Social SystemsICCSS, pp. 181–185 (2017)

18. Salama, E.S., El-khoribi, R.A., Shoman, M.E., Shalaby, M.A.: EEG-based emotion recognition using 3D convolutional neural networks. Int. J. Adv. Comput. Sci. Appl. **9**, 329–37 (2018)

19. Lin, W., Li, C., Sun, S.: Deep convolutional neural network for emotion recognition using EEG and peripheral physiological signal. In: Zhao, Y., Kong, X., Taubman, D. (eds.) ICIG 2017. LNCS, vol. 10667, pp. 385–394. Springer, Cham (2017). https://doi.org/10.1007/978-3-319-71589-6_33

20. Nakisa, B., Rastgoo, M.N., Tjondronegoro, D., Chandran, V.: Evolutionary computation algorithms for feature selection of EEG-based emotion recognition using mobile sensors. Expert Syst. Appl. **93**, 143–155 (2018)

21. Asghar, M.A., et al.: EEG-based multi-modal emotion recognition using bag of deep features: an optimal feature selection approach. Sensors **19**(23), 5218 (2019)

22. Yin, Z., Wang, Y., Liu, L., Zhang, W., Zhang, J. Cross-subject EEG feature selection for emotion recognition using transfer recursive feature elimination. Front. Neurorobotics **11**, 19 (2017)
23. Menezes, M.L.R., et al.: Towards emotion recognition for virtual environments: an evaluation of EEG features on benchmark dataset. Pers. Ubiq. Comp. 1–11 (2017)
24. Wang, X.-W., Nie, D., Lu, B.-L.: Emotional state classification from EEG data using machine learning approach. Neurocomputing **129**, 94–106 (2014)

Modeling the Electromechanics
of a Single Cardiac Myocyte

Anna Luisa de Aguiar Bergo Coelho[1], Ricardo Silva Campos[1,4],
João Gabriel Rocha Silva[1,2], Carolina Ribeiro Xavier[3],
and Rodrigo Weber dos Santos[1(✉)]

[1] Federal University of Juiz de Fora, Juiz de Fora, Brazil
rodrigo.weber@ufjf.edu.br
[2] Federal Institute of Education, Science and Technology of Mato Grosso,
Pontes e Lacerda, Brazil
[3] Federal University of São João del Rei, São João del-Rei, Brazil
[4] Faculdade Metodista Granbery, Juiz de Fora, Brazil

Abstract. The synchronous and proper contraction of cardiomyocytes
is essential for the correct function of the whole heart. Computational
models of a cardiac cell may spam multiple cellular sub-components,
scales, and physics. As a result, they are usually computationally expensive. This work proposes a low-cost model to simulate the cardiac
myocyte's electromechanics. The modeling of action potential and active
force is performed via a system of six ordinary differential equations.
Cardiac myocyte's deformation that considers details of its geometry is
captured using a mass-spring system. The mathematical model is integrated in time using Verlet's method to obtain the position, velocity, and
acceleration of each discretized point of the single cardiac myocyte. Our
numerical results show that the obtained action potential, contraction,
and deformation reproduces very well physiological data. Therefore, the
low-cost mathematical model proposed here can be used as an essential
tool for the correct characterization of cardiac electromechanics.

Keywords: Mass-spring systems · Eletromechanical coupling ·
Cardiac myocyte

1 Introduction

Cardiac diseases are still the first cause of death in the world, taking an estimated
17.9 million lives per year, according to World Health Organization. Modeling
this organ is a complex task that begins with the model of a single cardiac
myocyte. Computational models of a cardiac cell may spam multiple cellular
sub-components, scales, and physics. In general, robust models with partial differential equations are used for the electrical action potential propagation and
the coupling to the mechanical deformation, i.e., the myocyte contraction [7].

Supported by UFJF, UFSJ, Capes, CNPq (under grant 153465/2018-2) and Fapemig.

The finite Element Method is widely used in solving these equations, but with high computational costs [3].

For tackling the computational costs, the work of *Silva et al.* [10] proposed simplified models that reproduce the most important cardiac mechanics features. These models capture how action potential influences active force, i.e., the so-called electromechanical coupling. These models are based on few differential equations and have low computational costs without losing quality to reproduce the physiological phenomena.

Besides a simple active force model, it is also necessary to implement the passive mechanical model, responsible for restoring a contracted cell to its initial configuration. In this case, a common choice is a mass-spring system (MSS), which represents elastic materials by a finite set of masses connected by springs. The ability to simulate the elastic behavior of bodies in real-time made MSSs of great interest in computer graphics due to its simple formulation and computational performance [1,6]. They are used in animations and virtual reality applications, especially in simulations of surgeries and biologic tissues [8].

A related work that simulates the heart mechanic with mass-spring systems is presented by Weise *et al.* [12]. It proposes a discrete reaction-diffusion-mechanics model, where Hooke's law describes the elastic properties of the material. The model was used for studying heart phenomena such as the effect of mechano-electrical feedback.

Another works in this field use simplified methods based on cellular automata for simulating the action potential propagation and the active force application. A example of this kind of method is the *Campos et al.* [4], that proposes a meshless simulator, where the 3D geometries are split in a discrete set of masses, connected by springs. Its simple implementation resulted in very fast execution times. Later, *Campos et al.* [5] proposed a more robust cardiac electromechanic simulator, able to handle more complex geometries, with a discretization based on tetrahedrons. It also contains more realistic features, such as volume preservation and anisotropy controlling in a mass spring system. The model reproduced a cycle of contraction and relaxation of a human left ventricle.

In this sense, this work proposes a new tool for cardiac myocyte electromechanics simulation, aiming for low computational costs and correct physiological results. The action potential and active force are modeled by a system of six ordinary differential equations. The active force is responsible for contracting the cell, and then a passive force acts for bringing the cell to its initial configuration. The passive force is modeled by a mass-spring system. The cardiomyocite shape is obtained via a confocal microscopy, which we discretize by a set of point masses connected by springs, in a irregular mesh fashion. Our equations are integrated in time using Verlet's method to obtain the position, velocity, and acceleration of mass point.

2   Mass-Spring Systems

Mass-spring models have a simple formulation and fast execution time, making them a suitable altenative for modeling elastic materials without the need of higher computational resources. In such systems, masses are connected to their neighbors by springs. Forces can be applied to the system deforming its spatial distribution.

The cardiac tissue does not have a linear stress-strain relation but considering small scale deformations, its contraction can be aproximated by a linear model, using springs. The springs of the system will try to bring the system back to its initial configuration after contraction. The proposed model is described below.

2.1   Mathematical Model

Considering one mass unity as a rigid body, the following Ordinary Diferential Equations can be deduced to integrate trajectory and velocity, according to classical mechanics.

$$F = ma \tag{1}$$

$$v = \frac{\partial x}{\partial t} \tag{2}$$

$$a = \frac{\partial v}{\partial t} \tag{3}$$

This formulation can be manipulated in order to isolate derivatives and then obtaining a linear system with two ODE's.

$$\frac{\partial x}{\partial t} = v \tag{4}$$

$$\frac{\partial v}{\partial t} = \frac{F}{m} \tag{5}$$

The force can be categorized in two different types: passive and active forces. Passive forces are made from particles to its neighboors through the springs when its position changes. The active force is applied as a load external to the system, in this case, modeling the active tension generated by the action potential in the cell. In order to avoid the rigid-body displacement of translation, the mass center of the cell is fixed. Considering only one mass linked to a wall by a spring, as depicted in Fig. 1a, the force can be calculated with Hooke's Law. The force exerted will be f_p and the reaction of the wall will be f_r.

$$f_p = -f_r = k(x_L - x_0) \tag{6}$$

We arranged our spring and masses in a regular grid, where masses are connected in a Moore Neighborhood fashion, as depicted in Fig. 1b. In this manner, the total force exerted in a mass will be the sum of contributions of the neighborhood composed by 8 other masses. Therefore, the total force will be the sum of passive and active forces applied externally to the system. Equation 7 was

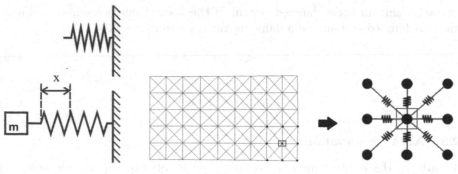

(a) Mass coupled to a spring and its deformation.

(b) Neighborhood of a mass.

Fig. 1. A simple mass-spring system and our mass-spring configuration.

used to calculate passive forces, that are proportional to the deformation of the spring.

$$f_p = -[k_{elas}(|l| - r)]\frac{l}{|l|} \qquad (7)$$

In addition to passive forces and active forces, a viscous damping force is considered. This damping force is calculated as shown in the Eq. 8.

$$f_p = \left[k_{damping}\frac{(\dot{l} \cdot l)}{|l|} \right] \frac{l}{|l|} \qquad (8)$$

After system initialization, a loop starts over a period of time. At each iteration, the sum of passive, active, external and damping forces that a mass receives at a given moment is calculated.

2.2 Critical Damping

The spring mass system without the use of damping results in a state in which the energy never dissipates. A factor called Damping Ratio denoted by ξ can be used to calculate damping in a MSSs:

$$\xi = \frac{k_{damping}}{k_{dcritical}}. \qquad (9)$$

Considering the equation of motion in the formulation described by Eq. 10, the critical damping coefficient will be equal to the relation in Eq. 11. If damping below the critical value is used, the damping ratio will be less than 1, characterizing an under-damped system. If the above value is used, we will have an over-damped system and a damping ratio greater than 1.

$$m\ddot{x} + k_d\dot{x} + k_e x = 0 \tag{10}$$

$$k_{dcritical} = 2\sqrt{km} \tag{11}$$

2.3   Area Preservation

To adjust the cardiac myocyte contraction to obtain a more physiological behavior a force of area preservation is added, i.e., to reproduce the quasi-incompressible feature of the cell.

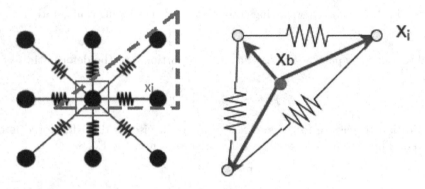

(a) A mass and one adjacent surface area.

(b) Force applied to preserve the area.

Fig. 2. Surface area preserving.

For each mass, two masses in the neighborhood are visited, forming a triangle, as showed in Fig. 2. The baricenter of the formed triangle is calculated using the mean value between the three masses coordinates, displayed in Eq. 12.

$$x_b = \frac{1}{3}\sum_{j=1}^{3} x_j \tag{12}$$

The direction in wich force is applied is calculated through a vector between the baricenter and the node receiving the preservation force. The force applied will be proportional to a preservation area constant and the area of the triangle formed by the nodes:

$$f_{prev} = -k_{prev}area\frac{(x_i - x_b)}{||x_i - x_b||}. \tag{13}$$

2.4   Verlet's Numerical Method

Verlet's numerical method was used to solve the mathematical model in order to obtain its position and velocity through time:

Verlet's method can be deduced using Taylor's Series Expansion for progressive and regressive aproximations:

$$X_{t_{n+1}} = X_{t_n} + V_{t_n}\Delta t + \frac{h^2}{2}\frac{F_n}{m} + \frac{h^3}{6}X_{t_n}^{(3)} + O(h^4), \tag{14}$$

$$X_{t_{n-1}} = X_{t_n} - V_{t_n}\Delta t + \frac{h^2}{2}\frac{F_n}{m} - \frac{h^3}{6}X_{t_n}^{(3)} + O(h^4). \tag{15}$$

Adding the two expansions we obtain:

$$X_{t_{n+1}} = 2X_{t_n} + h^2\frac{F_n}{m} - X_{t_{n-1}} + O(h^4). \tag{16}$$

The velocity of the mass can be obtained by Finite Diferences Method in its centered aproximation:

$$V_{t_{n+1}} = \frac{V_{t_{n+1}} - V_{t_{n-1}}}{2h} + O(h^2). \tag{17}$$

Therefore, this method aproximates the position with an error order of h^4 and velocity of h^2.

3   Coupled Eletromechanical Model

For the mass system to model the cardiac myocyte with its contraction characteristics and properties, the applied force must follow a cell active tension that is associated to a cell action potential. To model the cell action potential, the Minimal Model proposed in [2] was used and adjusted to reproduce the model described in [11]:

$$\frac{du}{dt} = -(J_{fi} + J_{so} + J_{si}) \tag{18}$$

$$\frac{dv}{dt} = (1 - H(u - \theta_v))(v_\infty - v)/\tau_v^- - H(u - \theta_v)v/\tau_v^+ \tag{19}$$

$$\frac{dw}{dt} = (1 - H(u - \theta_w))(w_\infty - w)/\tau_w^- - H(u - \theta_w)w/\tau_w^+ \tag{20}$$

$$\frac{ds}{dt} = ((1 + tanh(k_s(u - u_s)))/2 - s)/\tau_s \tag{21}$$

$$J_{fi} = -vH(u - \theta_v)(u - \theta_v)(u_u - u)/\tau_{fi} \tag{22}$$

$$J_{so} = (u - u_o)(1 - H(u - \theta_w))/\tau_o + H(u - \theta_w)/\tau_{so} \tag{23}$$

$$J_{si} = -H(u - \theta_w)WS/\tau_{si} \tag{24}$$

$$\tau_v^- = (1 - H(u - \theta_v^-))\tau_{v1}^- + H(u - \theta_v^-)\tau_{v2}^- \tag{25}$$

$$\tau_w^- = \tau_{w1}^- + (\tau_{w2}^- - \tau_{w1}^-)(1 + tanh(k_w^-(u - u_w^-)))/2 \tag{26}$$

$$\tau_{so} = \tau_{so1} + (\tau_{so2} - \tau_{so1})(1 + tanh(k_{so}(u - u_{so})))/2 \tag{27}$$

$$\tau_s = (1 - H(u - \theta_w))\tau_{s1} + H(u - \theta_w)\tau_{s2} \tag{28}$$

$$\tau_o = (1 - H(u - \theta_o))\tau_{o1} + H(u - \theta_o)\tau_{o2} \tag{29}$$

$$v_\infty = \begin{cases} 1, u < \theta_v^- \\ 0, u \geq \theta_v^- \end{cases} \tag{30}$$

$$w_\infty = (1 - H(u - \theta_o))(1 - u/\tau_{w\infty}) + H(u - \theta_o)w_\infty^*, \tag{31}$$

where $u_o, u_u, \theta_v, \theta_w, \theta_v^-, \theta_o, \tau_{v1}^-, \tau_v^+, \tau_{w1}^-, \tau_{w2}^-, k_w^-, u_w^-, \tau_w^+, \tau_{fi}, \tau_{o1}, \tau_{o2}$, $\tau_{so1}, \tau_{so2}, k_{so}, u_{so}, \tau_{s1}, \tau_{s2}, k_s, u_s, \tau_{si}, \tau_{w\infty}, w_\infty^*$ are 28 adjustable parameters of the model with values reported in [2].

The model presented in [10] proposes two ODE's to capture the cell's active tension triggered by an action potential:

$$\frac{dTa_i}{dt} = c_0(k(V) - Ta_i) \tag{32}$$

$$\frac{dTa}{dt} = \epsilon_1(V, Ta_i)(Ta_i - Ta) \tag{33}$$

$$k(V) = \frac{1}{\sigma\sqrt{2\pi}} e^{\frac{-1}{2}(\frac{V-1}{\sigma})^2} \tag{34}$$

$$\epsilon_1(V) = \begin{cases} x_1 \text{ para V} > x_2 \text{ e } Ta_i < x_3 \\ c_0 \qquad \text{otherwise.} \end{cases} \tag{35}$$

The parameters used were adjusted using a Genetic Algorithm as described before in [10]. The parameters of the coupled electromechanical model are presented in Table 1.

The active tension obtained in Eq. 33 was multiplied by a factor of 85.

Table 1. Parameters of the coupled model.

u_o	0.529297	τ_{w2}^-	62.9688	k_{so}	0.253711	c_0	0.0166016
θ_v	0.0673828	u_w^-	58.0469	τ_{s1}	2.36621	x_1	0.0001
θ_w	0.00195313	τ_w^+	0.59668	τ_{s2}	11.4453	x_2	0.78
θ_v^-	0.0976563	τ_{fi}	273.633	k_s	2.25586	x_3	0.2925
θ_o	0.932618	τ_{o1}	0.644532	u_s	0.903321		
τ_{v1}^-	57.7148	τ_{o2}	477.344	τ_{si}	1.76816		
τ_v^+	1101.56	τ_{so1}	14.1992	$\tau_{w\infty}$	0.785157		
τ_{w1}^-	1.96973	τ_{so2}	25.4492	w_∞^*	0.500977		

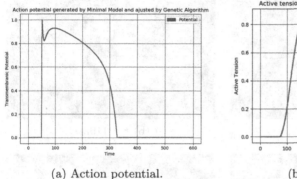

(a) Action potential.

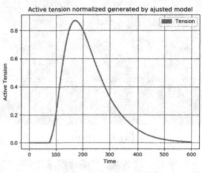

(b) Active tension.

Fig. 3. Action potential and active tension.

4   Results

4.1   Computational Aspects

A C++ code was developed to simulate the coupling of units of masses with springs in horizontal, vertical, and diagonal links, considering the force each mass applies to its neighbors. The tool allows the representation of irregular geometries. This is essential to reproduce the complex geometry of single cardiac myocytes, as presented in Fig. 4b, which was obtained in the laboratory. A similar numerical mesh is also presented in Fig. 4a.

4.2   Numerical Results

We tested our model by simulating a cycle of contraction and relaxation. The results for action potential and active tension are presented in Fig. 3b. The active force reaches its peak around 170 ms, where it applies 100% of stress. After that, it returns to 0% of stress.

The active tension drives the mass-spring system which results in the contraction and relaxation of the single cardiac myocyte, as presented in Fig. 5.

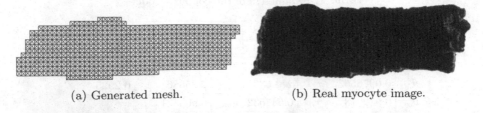

(a) Generated mesh.   (b) Real myocyte image.

Fig. 4. A real myocyte image and its corresponding mesh.

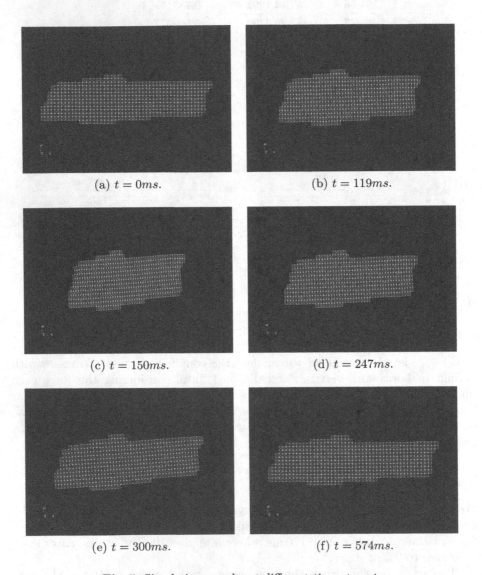

(a) $t = 0ms$.   (b) $t = 119ms$.

(c) $t = 150ms$.   (d) $t = 247ms$.

(e) $t = 300ms$.   (f) $t = 574ms$.

Fig. 5. Simulations results at different time-steps t.

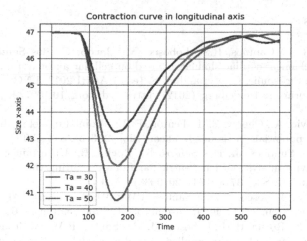

Fig. 6. Contraction curve

Figure 6 shows the shortening of the myocyte in the x-direction. We applied three different values of active stress, $T_a = 30$ kPa, $T_a = 40$ kPa and $T_a = 50$ kPa causing a maximum deformation of 12%. The cell responded to active stress as expected, achieving its maximum contraction when stress is maximum.

5   Conclusion

In this work, we present a low-cost model to simulate the electromechanics of a single cardiac myocyte. The modeling of action potential and active force was performed via a system of six ordinary differential equations. Cardiac myocyte's deformation that considers details of its complex geometry was captured using a mass-spring system with an irregular mesh. The mathematical model was integrated in time using Verlet's method to obtain the position, velocity, and acceleration of each discretized point of the single cardiac myocyte. Our numerical results show that the obtained shortening reproduces very well physiological data. The comparison was made considering measures in [9]. In this work, the measured contraction displayed a shortening of 8% to 10% of the cell volum. Therefore, the low-cost mathematical model proposed here can be used as a tool to help the characterization of cardiac electromechanics.

As future work, we intend to perform a sensitivity analysis in order to evaluate the significance of each parameter in the simulations. We also intend to quantitatively compare our simulations to experimental data.

References

1. Amorim, R.M., Campos, R.S., Lobosco, M., Jacob, C., dos Santos, R.W.: An electro-mechanical cardiac simulator based on cellular automata and mass-spring models. In: Sirakoulis, G.C., Bandini, S. (eds.) ACRI 2012. LNCS, vol. 7495, pp. 434–443. Springer, Heidelberg (2012). https://doi.org/10.1007/978-3-642-33350-7_45
2. Bueno-Orovio, A., Cherry, E.M., Fenton, F.H.: Minimal model for human ventricular action potentials in tissue. J. Theor. Biol. **253**(3), 544–560 (2008)
3. Campos, J., Sundnes, J., Dos Santos, R., Rocha, B.: Uncertainty quantification and sensitivity analysis of left ventricular function during the full cardiac cycle. Philos. Trans. R. Soc. **378**(2173), 20190381 (2020)
4. Campos, R.S., Lobosco, M., dos Santos, R.W.: A GPU-based heart simulator with mass-spring systems and cellular automaton. J. Supercomputing **69**(1), 1–8 (2014)
5. Campos, R.S., Rocha, B.M., Lobosco, M., dos Santos, R.W.: Multilevel parallelism scheme in a genetic algorithm applied to cardiac models with mass-spring systems. J. Supercomputing **73**(2), 609–623 (2017)
6. Kot, M., Nagahashi, H., Szymczak, P.: Elastic moduli of simple mass spring models. Vis. Comput. **31**(10), 1339–1350 (2014). https://doi.org/10.1007/s00371-014-1015-5
7. Oliveira, R.S., et al.: Ectopic beats arise from micro-reentries near infarct regions in simulations of a patient-specific heart model. Sci. Rep. **8**(1), 1–14 (2018)
8. Pappalardo, O., et al.: Mass-spring models for the simulation of mitral valve function: looking for a trade-off between reliability and time-efficiency. Med. Eng. Phys. **47**, 93–104 (2017)
9. Philips, C.M., Duthinh, V., Houser, S.R.: A simple technique to measure the rate and magnitude of shortening of single isolated cardiac myocytes. IEEE Trans. Biomed. Eng. **10**, 929–934 (1986)
10. Silva, J.G.R., Campos, R.S., Xavier, C.R., dos Santos, R.W.: Simplified models for electromechanics of cardiac myocyte. In: Gervasi, O., et al. (eds.) ICCSA 2020. LNCS, vol. 12249, pp. 191–204. Springer, Cham (2020). https://doi.org/10.1007/978-3-030-58799-4_14
11. ten Tusscher, K.H., Noble, D., Noble, P.J., Panfilov, A.V.: A model for human ventricular tissue. Am. J. Physiol.-Heart Circulatory Physiol. **286**(4), H1573–H1589 (2004)
12. Weise, L.D., Nash, M.P., Panfilov, A.V.: A discrete model to study reaction-diffusion-mechanics systems. Plos One **6**(7), e21934 (2011)

Towards Mimetic Membrane Systems in Molecular Dynamics: Characteristics of *E. Coli* Membrane System

Mateusz Rzycki[1,2](✉) ⓘ, Sebastian Kraszewski[2] ⓘ, and Dominik Drabik[2,3] ⓘ

[1] Department of Experimental Physics, Faculty of Fundamental Problems of Technology,
Wroclaw University of Science and Technology, Wroclaw, Poland
`mateusz.rzycki@pwr.edu.pl`
[2] Department of Biomedical Engineering, Faculty of Fundamental Problems of Technology,
Wroclaw University of Science and Technology, Wroclaw, Poland
[3] Laboratory of Cytobiochemistry, Faculty of Biotechnology, University of Wroclaw, Wroclaw,
Poland

Abstract. Plenty of research is focused on the analysis of the interactions between bacteria membrane and antimicrobial compounds or proteins. The hypothesis of the research is formed according to the results from the numerical models such as molecular docking or molecular dynamics. However, simulated membrane models often vary significantly from the real ones. This may lead to inaccurate conclusions. In this paper, we employed molecular dynamic simulations to create a mimetic *Escherichia coli* full membrane model and to evaluate how the membrane complexity may influence the structural, mechanical and dynamical mainstream parameters. The impact of the O-antigen region presence in the outer membrane was also assessed. In the analysis, we calculated membrane thickness, area per lipid, order parameter, lateral diffusion coefficient, interdigitation of acyl chains, mechanical parameters such as bending rigidity and area compressibility, and also lateral pressure profiles. We demonstrated that outer membrane characteristics strongly depend on the structure of lipopolysaccharides, changing their properties dramatically in each of the investigated parameters. Furthermore, we showed that the presence of the inner membrane during simulations, as it exists in a full shell of *E. coli*, significantly changed the measured properties of the outer membrane.

Keywords: Molecular dynamics · Mimetic systems · Lipid membrane model

1 Introduction

Escherichia coli is one of the most frequently investigated bacteria being responsible for common infections among humans and animals [1, 2]. This strain is widely used for antimicrobial studies [3–5]. It belongs to the Gram-negative ones which membranes consist of the outer (OM) and the inner membrane (IM) separated by the periplasm. OM is an asymmetric bilayer primarily composed of lipopolysaccharides (LPS) in the top leaflet and phospholipids (PL) in the bottom one [6]. It serves as a protective shield

© Springer Nature Switzerland AG 2021
M. Paszynski et al. (Eds.): ICCS 2021, LNCS 12743, pp. 551–563, 2021.
https://doi.org/10.1007/978-3-030-77964-1_42

preventing the entry of toxic compounds e.g. antibiotics [7, 8]. The LPS is composed of three segments: lipidA-the hydrophobic fatty surface forming the base of the top OM leaflet, a phosphorylated, highly anionic core and an O-antigen unit composed of sugar chains performing a hydrophilic surface [7, 9]. IM has a dynamic structure mostly formed by phosphatidylethanolamine (PE), phosphatidylglycerol (PG) and cardiolipin (CL) [10, 11].

In computational studies on bacteria membranes, methods such as molecular dynamics are often employed [12]. This allows observation of the behavior of studied molecules up to the atomic level. However, many modelled systems are simplified and limited to one particular membrane even for Gram-negative bacteria [13–16]. Piggot *et al.* performed the analysis on the base model of OM *E. coli* with embedded FecA protein, presenting the LPS structure on the upper leaflet, while the lower one was composed of PE and PG [15]. A similar model was proposed by Wu *et al.* a few years earlier where a couple of LPS structures were studied [13]. One of the most comprehensive approaches was delivered by Hwang *et al.* where the two – inner and outer membranes were separately modelled and analyzed [14]. While those models are quite close to reality, they may significantly differ from the real OM/IM bacterial membrane or even experimental models. This may result in influencing the outcome results.

In this work, we investigate the changes in membrane properties with the increasing complexity of the systems to better reflect bacterial membrane and to draw attention that both OM and IM should not be studied separately for better biological context. For this purpose, we created five bacterial membrane models based on the composition of *E. coli*. We started from simple pure IM and pure OM (with and without O-antigen units used). To better reflect the natural conditions we created whole OM/IM bacteria membrane system (first without O-antigen units and latter with O-antigen units). Each of the systems was analyzed in detail to characterize the topological and mechanical properties of the membranes in investigated systems and to present the influence on how structural complexity can affect membrane behavior.

2   Methods

The all-atom models of the membranes were generated using CHARMM-GUI membrane builder [17]. The IM model consisted of 80% PYPE, 15% PYPG, 5% PVCL2 [10, 11, 18]. The lipid bilayer was solvated with TIP3P water molecules (100 water molecules per lipid) and 240 mM NaCl were added based on literature data [19]. Final IM configuration included: 256 PYPE, 48 PYPG, 16 PVCL2, 276 Na$^+$, 196 Cl$^-$ and 32000 TIP3P molecules.

The OM models were composed of 75% PYPE and 25% PYPG in the upper leaflet and 100% LPS in the lower one [13, 20]. The type1 of lipidA, R1 core and repeating units of O6-antigen were included in the LPS sequence. The length of the O6-antigens was adapted based on results published by Wu *et al.* [13, 21]. The number of LPS molecules and phospholipids was equally adjusted to the total lipid area occupied on each leaflet. The same procedure of solvation and ion addition was employed as before, except for Ca^{2+} ions, which were automatically added based on LPS length to neutralize the system. Afterward, molecular dynamics (MD) simulations of two asymmetric LPS based bilayers were performed. The upper leaflet contained: lipidA, R1 core and 2 repeating units of O6-antigen (OMA) and lipidA, R1 core (OM0). Final OMA and OM configuration included: 52 LPS, 120 PYPE, 40 PYPG, 260 Ca^{2+}, 232 Na^+, 112 Cl^-, 28964 TIP3, and 60 LPS, 144 PYPE, 48 PYPG, 300 Ca^{2+}, 168 Na^+, 120 Cl^- and 27856 TIP3P molecules respectively. Three-dimensional periodic boundary conditions were applied to deal with potential energy disruption due to the origin of cell discontinuity.

MD simulations of pure membranes were performed using the GROMACS (version 2020.4) package with the CHARMM36 force field [22, 23]. Each system was first minimized using the steepest descent algorithm for energy minimization. Calculations were carried out in the NPT ensemble (constant Number of particles, Pressure and Temperature) using a Nose-Hoover thermostat at T = 303.15 K and semi-isotropic coupling with Parrinello-Rahman barostat at p = 1bar. The long-run production was conducted for at least 300 ns using the leap-frog integrator. Chemical bonds between hydrogen and heavy atoms were constrained to their equilibrium values by the LINCS algorithm, while long-range electrostatic forces were evaluated using the particle mesh Ewald (PME) method, which allowed us to employ the integration timestep of 2 fs.

The complete bacterial membrane models: LIPA (IM + OMA) and LIPO (IM + OM0) have been constructed by assembling IM and OM separated by a small water slab (2.4 nm, 4140 water molecules) imitating the periplasm. The minimization procedure and NPT ensemble were carried out according to the same protocol as described above. We analyzed the last 10 ns of all simulations using a combination of GROMACS tools, self-made MATLAB (The MathWorks, Natick, MA) scripts, VMD and VMD's dedicated plugins such as MEMBPLUGIN 1.1 [24] for interdigitation calculation.

The order parameter of the acyl chains was obtained using:

$$S_{CH} = \frac{3}{2}\langle cos^2\,\theta \rangle - \frac{1}{2} \tag{1}$$

where θ for a particular carbon atom is the angle between the bilayer normal and carbon-hydrogen bond.

The diffusion was calculated in the Diffusion Coefficient Tool [25] from the slope of the mean-squared displacement (MSD) curve through Einstein's relation. For the computation accuracy, only phosphorous atoms of all lipids were taken into account.

$$D(\tau) = \frac{M(\tau)}{2E\tau} \tag{2}$$

where $M(\tau)$ – is the MSD at a range of lag time τ and E represents the dimensionality (two in our case - XY).

Lateral Pressure profiles (LPPs) were computed using a custom version of GROMACS-LS [26]. The obtained beforehand trajectories were adapted to comply with the software requirements, thus the calculations of the PME electrostatic forces were settled to cutoff. We also adjusted the cutoff to 2.2nm according to Vanegas *et al.* [26, 27]. The lateral component of pressure tensor $(P_L(z) = 0.5 \times (P_{xx}(z) + P_{yy}(z)))$ and the normal component $(P_N = P_{zz})$ are computed from the GROMACS-LS output. Finally, LPPs $\pi(z)$ was determined from:

$$\pi(z) = P_L(z) - P_N. \tag{3}$$

Bending rigidity was determined using the real space fluctuation method [28]. Briefly, a probability distribution for both tilt and splay is determined for all lipids over the last

10 ns of simulation. Tilt is defined as an angle between the lipid director (vector between lipid head – the midpoint between C2 and P atoms – and lipid tail – the midpoint between last carbon atoms) and bilayer normal. Lipid splay is defined as divergence of an angle formed by the directors of neighboring lipids providing that they are weakly correlated. Area compressibility was determined using a method developed by Doktorova *et al.* [29]. Briefly, a real-space analysis of local thickness fluctuations is sampled from the simulations.

Determined parameters' statistical significance was performed using one-way ANOVA significance test with Tukey post hoc test in Origin 2018 (OriginLabs) software.

3   Results

Each of the investigated systems was characterized thoroughly. In the Fig. 1 and Fig. 2 we present LIPA and LIP0 systems in detail, including their density profiles and graphical representation. Structural, stress and mechanical parameters of lipid membranes were determined. We studied whether simplification of biological membranes, which is common for numerical simulations, is feasible. We assume that significant differences between the systems may result in the different occurrence of biological phenomena.

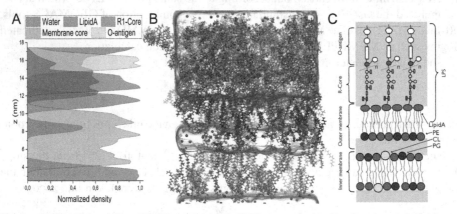

Fig. 1. A) Density profile of LIPA regions. B) LIPA system visualization (several lipids are hidden for clarity). The Inner membrane, outer membrane, lipidA, R-Core, O-antigen region, water, calcium ions, sodium together with chlorine ions have been colored orange, gray, red, magenta, pink, azure, yellow and dark blue, respectively. C) Graphical representation of created LIPA system. (Color figure online)

For the structural aspect of the membrane, we decided to perform a standard analysis with a couple more comprehensive parameters afterward. To understand and characterize the molecular effect of LPS on the membrane and/or additional membrane in complex systems, we determined different bilayer properties such as membrane thickness (MT), area per lipid (APL), order parameter, interdigitation and lateral diffusion. Membrane thickness was determined between phosphorus atoms, while the area per lipid was determined using Voronoi tessellation. The results of structural characteristics are presented in Table 1. All of the parameters were statistically distinct.

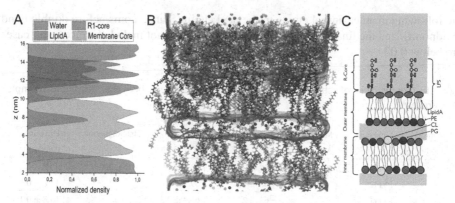

Fig. 2. A) Density profile of LIP0 regions. B) LIP0 system visualization (several lipids are hidden for clarity). The inner membrane, outer membrane, lipidA, R-Core, water, calcium ions, sodium together with chlorine ions have been colored in orange, ice blue, red, magenta, azure, yellow and dark blue, respectively. C) Graphical representation of created LIP0 system. (Color figure online)

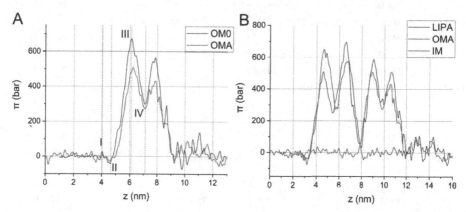

Fig. 3. Lateral pressure profile of A) OM0 and OMA, B) LIPA system with corresponding pure bilayer components OMA and IM.

Further structural characterization was enhanced with a description of lipid behavior in the systems. For this purpose, lateral mobility, which is usually described by the lateral diffusion coefficient, was investigated. The diffusion coefficient from the 2D mean square displacement equation was calculated. Obtained values were statistically significantly different between the investigated systems. Finally, the determination of acyl chain interdigitation to assess interactions between the leaflets itself was performed. The parameter allows estimating whether O-antigens presence may influence the interactions between the leaflets. As before the differences in values of interdigitation were statistically significant between the investigated systems.

Additionally, to provide a wider insight into the flexibility of the acyl chains, the order parameter was calculated. Presented values were averaged over the whole trajectories for clarity and collected in Table 2. We report values for the sn-2 unsaturated chain in the following manner: initial atoms in the acyl chain (Start), atoms before double bond (Midpoint), and the final (End). Standard deviations are not included, since in all cases are below 0.02.

Table 1. Comparison of structural and dynamic parameters between pure and complex membranes[a].

Membrane	Lipid type	MT_{P-P}	APL_1	APL_2	Interdigitation	Diffusion
		Å	Å^2	Å^2	Å	μm^2/s
IM	Total	39.9 ± 0.4	56.6 ± 1.8	58 ± 2	4.9 ± 0.4	–
	PYPE		58 ± 1	59.1 ± 0.6		12.0 ± 0.1
	PYPG		63 ± 2	61.7 ± 2.3		15.8 ± 0.1
	PVCL2		88 ± 4	78 ± 4		12.8 ± 0.1
OMA	Total	35.5 ± 0.2	194.2 ± 3.3	60.3 ± 1.9	5.0 ± 0.3	–
	PYPE		–	60.8 ± 0.7		6.4 ± 0.1
	PYPG		–	66.5 ± 2.0		6.1 ± 0.1
	LipidA		194.2 ± 3.3	–		0.6 ± 0.1
OM0	Total	37.2 ± 0.2	183 ± 2	55.6 ± 1.5	4.3 ± 0.2	–
	PYPE		–	56.4 ± 0.4		10.0 ± 0.1
	PYPG		–	59.7 ± 1.3		7.7 ± 0.2
	LipidA		183 ± 2	–		0.9 ± 0.1
LIPA	Outer	33.3 ± 0.6	183 ± 3	59 ± 1	4.9 ± 0.4	–
	PYPE		–	64.7 ± 0.6		2.8 ± 0.0
	PYPG		–	66.5 ± 1.1		3.2 ± 0.2
	LipidA		183 ± 3	–		0.3 ± 0.1
	Inner	35.0 ± 0.7	55 ± 2	62.8 ± 2.4	5.0 ± 0.3	–
	PYPE		59 ± 1	61 ± 1		8.4 ± 0.1
	PYPG		63 ± 3	65 ± 2		11.3 ± 0.2
	PVCL2		65 ± 2	75 ± 3		10.2 ± 0.3
LIP0	Outer	35.3 ± 0.4	182.7 ± 4.4	59.6 ± 1.1	7.1 ± 0.5	–
	PYPE		–	63.9 ± 0.5		3.4 ± 0.1
	PYPG		–	67 ± 1		4.2 ± 0.1
	LipidA		182.7 ± 4.4	–		0.2 ± 0.1
	Inner	38.8 ± 0.4	61.3 ± 2.2	65 ± 2	5.9 ± 0.2	–
	PYPE		62.4 ± 0.6	63.3 ± 0.7		7.0 ± 0.1
	PYPG		60.1 ± 1.5	70 ± 2		9.0 ± 0.1
	PVCL2		75 ± 3	82 ± 3		12.7 ± 0.1

[a]MT_{P-P} - membrane thickness measure between phosphorous atoms from opposite leaflets; APL1, APL2 - the area per lipid on the upper and lower leaflet, respectively; IM – inner membrane; OM0 – outer membrane without antigens; OMA – outer membrane with antigens; LIP0 – mimetic *E. coli* system without antigens; LIPA – mimetic *E. coli* system with O-antigens

Stress characterization was done by evaluation of the stress profile along bilayer normal and determination of the lateral pressure profile (LPP) $\pi(z)$. As a reference, we

Table 2. Acyl chain order parameter from pure and complex systems.

Membrane	Lipid type	Order parameter		
		Start	Midpoint	End
IM	PYPE	0.24	0.13	0.12
	PYPG	0.23	0.13	0.12
	PVCL	0.20	0.11	0.09
OM0	PYPE	0.24	0.13	0.12
	PYPG	0.25	0.13	0.12
	LipidA	0.20	0.15	0.10
OMA	PYPE	0.20	0.10	0.08
	PYPG	0.21	0.10	0.10
	LipidA	0.17	0.14	0.07
LIP0	PYPE	0.20	0.09	0.08
	PYPG	0.19	0.11	0.09
	PVCL	0.22	0.09	0.07
	PYPE	0.20	0.09	0.08
	PYPG	0.19	0.11	0.09
	LipidA	0.20	0.23	0.13
LIPA	PYPE	0.22	0.12	0.10
	PYPG	0.21	0.11	0.09
	PVCL	0.23	0.09	0.08
	PYPE	0.22	0.12	0.10
	PYPG	0.22	0.11	0.09
	LipidA	0.18	0.17	0.08

present Fig. 3 where the lateral pressure profile of OM's and LIPA combined with pure IM and OMA was calculated. For other systems, we collected the peak values (see Table 3).

The introduced LPPs indicate a similar tendency between basic and complex membrane. Starting from the bulk solvent, the first minor positive peak may be distinguished as a water-headgroup interface (I), indicating the repulsive forces from lipids. Further, the negative peak (II) is visible presenting glycerol region, including attractive hydrophobic forces [30], while subsequent major peak denotes the acyl chain region (III) and finish at the bilayer center nearby 6 nm (IV).

Finally, mechanical characterization of membranes is performed. Such characterization allows assessing very subtle changes induced by sugar-coating of LPS or additional membrane complexity. Both area compressibility (K_A) and bending rigidity (κ) are determined (see Table 3). All reported values are statistically significantly different.

Table 3. Mechanical and pressure properties of pure and complex systems[b].

Membrane	κ	κ_tilt	K_{A1}	K_{A2}	K_A	Lateral pressure
	kbT	kbT	mN/m	mN/m	mN/m	bar
IM	22.1 ± 0.6	10.6 ± 0.3	133 ± 20	133 ± 17	133 ± 15	647 ± 21
OMA	22.7 ± 0.4	13.3 ± 0.4	18 ± 12	84 ± 27	29 ± 10	508 ± 11
OM0	29.2 ± 0.4	17.1 ± 0.2	57 ± 6	141 ± 34	81 ± 18	670 ± 30
LIPA						
outer	25.0 ± 0.5	15.8 ± 0.6	41.7 ± 5.3	126.3 ± 5.4	62.7 ± 5.2	588 ± 16
inner	16 ± 1	4.8 ± 0.3	165 ± 24	31 ± 23	51 ± 20	695 ± 14
LIP0						
outer	26.4 ± 0.9	15.0 ± 0.7	58.8 ± 4.6	86.2 ± 7.1	70.0 ± 5.4	690 ± 26
inner	16.8 ± 0.4	7.3 ± 0.8	34 ± 9	42 ± 27	37 ± 17	533 ± 21

[b]κ – bending rigidity; κ_tilt – tilt; K_{A1}, K_{A2}, K_A – compressibility of the upper leaflet, lower leaflet, and total membrane, respectively; IM – inner membrane; OM0 – outer membrane without antigens; OMA – outer membrane with antigens; LIP0 – mimetic *E. coli* system without antigens; LIPA – mimetic *E. coli* system with antigens

4   Discussion

For a comprehensive analysis, we decided to divide the discussion section into two subsections. First, we focus on the impact of the O-antigen segment on the asymmetric outer membrane. Next, we draw attention to the discrepancies in the complete bacterial membrane systems compared to single membrane model systems.

4.1   The Effect of the O-antigen Region Presence on the Outer Membrane Parameters

In the outer membrane systems, we observe significant differences between OM0 and OMA, as the latter one is equipped with an extra O-antigen region. The presence of that structure induces membrane thickness reduction and APL extension in the upper leaflet (see Table 1). Interestingly, this directly influences the lower leaflet, where the total APL is lower than in the corresponding leaflet in OM0 system. The change of membrane thickness is proportional to the interdigitation of the acyl chains in the outer bilayer. We may conclude that the reduction in bilayer thickness is accompanied by the growth of the interdigitation [31]. The presence of O-antigens increases the interdigitation between the lipidA and PE:PG leaflets, which is followed by thickness reduction. However, analysis of the interdigitation in asymmetric membranes with the LPS layer is not straightforward. Shearer *et al.* suggested that the properties of OM systems are much more dependent on the dynamics and structure of the LPS segment [32].

The diffusion coefficient analysis showed a significant difference between OMA and OM0 systems as well. The occurrence of additional sugar coating substantially limits the mobility of the whole membrane. It remains consistent with the previous works [33–35]. The O-antigen essentially impacts the lower leaflet of the outer membrane, since PE and PG fluidity is restricted by 36% and 21%, respectively.

The calculated order parameter indicated that in both cases ordering trend is decreasing toward the bilayer center (see Table 2), our results are consistent with the ones presented by Wu *et al.* [13]. Interestingly the presence of the O-antigen segment affects membranes as values on both leaflets are lower.

Both OMA and OM0 exhibit a similar pressure trend along the bilayer normal (see Fig. 3). Noteworthy, much higher lateral stress was denoted at the lower leaflet at the OM0 system, reaching the top value of 670 ± 30 bar, while on OMA only 508 ± 11 bar was observed. Since the presented membranes are not symmetric, the lateral pressure on the upper and lower leaflets varies, however, both in a similar manner. A slightly noticeable shift at the bilayer center represents the interdigitation parameter of both membranes and remains consistent with values in Table 1. Since the interdigitation in OMA acyl chains is more intense, the plot downhill is deeper. Since the presence of the antigens in the membrane decreases the lateral pressure and lateral diffusion, this change could have a significant influence on the behavior of the system. Changes in both parameters could influence for instance membrane transport [36].

The presence of antigens in the LPS leaflet induced significant mechanical changes (see Table 3). All of the parameters - bending rigidity, tilt, and compressibility - were lower when O-antigens were present compared to the membrane without antigens. Such a difference is not surprising, as additional O-antigens are in the water part of the system, hence exposing the leaflet to additional repulsive forces, making the structure less resistant and exhibiting more fluctuations. Since lateral diffusion and lateral pressure are lower when antigens are present it can be concluded that membrane is, at least in the interphase region, more ordered. Such a conclusion cannot be made for the acyl chain region, as interdigitation increases when O-antigens are present.

4.2 The Comparison of *E. Coli* Membrane Models

The models presented in this study require a detailed analysis of their topological and mechanical characteristics. To this end, we decided to compare those properties for both inner and outer membranes from the *E. coli* models to pure ones. Taking into account the entire set of membranes, LIPA and LIP0 exhibit reduced bilayer thickness in both inner and outer membrane cases. Major differences we observe between pure IM and LIPA where the thickness reduction was supported with cardiolipin (CL) APL decrement of 12.3% (4.9 Å) and 26.1% (23 Å^2), respectively (see Table 1). In LIP0 structure modifications occurred in the inner membrane and the lower leaflet of the outer one when comparing APL and thickness parameters. Interestingly, significant reduction may be observed in the upper leaflet of the LIPA outer membrane, since lipidA reduced APL by 10 Å^2. Analysis of the interdigitation parameter between acyl chains from opposite leaflets seems to be slightly different than before. Obtained values from the LIPA system did not vary enough and were not significant compared to IM and OMA models with extra O-antigens. Thus, this parameter is not sensitive to the complexity of the membranes. However, the opposite pathway has been presented in LIP0 system. The interdigitation pitched up by 20.4% (to 5.9 ± 0.2 Å) on IM and by 65.1% (to 7.1 ± 0.5 Å) on OM, respectively. Similarly, as before this phenomenon is inversely correlated to the bilayer thickness, where the decrease is accompanied by the interdigitation increase. We confirmed that the presence of O-antigens escalates the interdigitation between acyl chains from the opposite leaflet, followed by thickness reduction.

Furthermore, investigation of the diffusion coefficient revealed that in comprehensive models LIPA and LIP0 the mobility of the lipid particles is substantially limited. In LIP0, PE and lipidA fluidity was almost three and four times reduced, compared to OM0. A similar situation appears when analyzing LIPA and OMA, PE and lipidA mobility is limited more than twice in both cases. In pure OMA long sugar chains O-antigen reduce the fluidity of the whole membrane, which is accompanied by a corresponding interdigitation parameter. Further, the difference in fluidity of OMA and OM0 outer membranes was reduced in the whole bacteria systems.

Noteworthy the ordering of acyl chains is not as clear as, on pure outer membranes, observed fluctuations in several cases are not statistically significant (see Table 2). However, we indicate that the sn-2 ordering in the inner membrane of both LIPA and LIP0 compared to pure IM significantly decreased.

Afterward, we compared the LPPs of the LIPA and LIP0 to indicate the stress tensor contrasts resulted from O-antigen presence (see Fig. 3). Pure IM exhibits higher lateral stress at the lower leaflet since extreme values are reached compared to the LIPA membrane. Marginal shifts at the bilayer center represent the interdigitation of both membranes and remain consistent with values in Table 1. Finally, LIPA reaches the top stress value of 695 ± 14 bar at the upper leaflet of the inner membrane. Pressure on the outer membrane was lower than in the case of the inner. In LIPA the total lateral pressure in the inner and outer membrane is higher compared to pure ones (IM and OMA) and it was interestingly distributed mostly on adjacent leaflets. In our opinion growth of pressure on the upper leaflet from the inner membrane and the lower leaflet from the outer membrane supports the complex system formation, while the highest lateral stress occurs there.

Moving forward to the mechanical characterization the differences in mechanical properties are also statistically significant when the additional membrane is present in the system (see Table 3). Obtained κ values from pure systems are consistent with those delivered by Hsu et al. [33]. In the case of IM bending rigidity was lower in IM of both LIPA and LIP0 when compared to the model IM system. The opposite tendency was observed in the case of area compressibility where IM of LIPA and LIP0 had, in general, lower values than in the model system. Similar to Jefferies et al. we notice that various LPS composition differ in the matter of mechanical strength or mobility [37]. Differences can be observed in the case of OM, however, due to the presence of antigens, the tendency is less straightforward. Bending rigidity values of OM in LIPA and LIP0 systems are in between the values of the model OM system with and without O-antigens. This suggests that additional bilayer in the system stabilizes the system with antigens but also increases the whole dynamics of the membrane without antigens. This is also valid for the area compressibility parameter of whole membranes, however, became more complicated to evaluate when individual leaflet compressibilities were taken into account. Nevertheless, it should be noted that the presence of the second membrane in the simulated system strongly affected the mechanical behavior of both membranes when compared to single membrane model systems, and should be considered in numerical studies for better biological context.

5 Conclusions

In this paper, we performed a detailed study of structural, mechanical and stress parameters of lipid membranes mimicking the *E. coli* dual membrane system. We showed the changes of numerically determined parameters with progressive complexity of the membrane systems. We presented that LPS-rich outer membrane properties strongly depend on the structure of LPS itself, changing dramatically each of the investigated parameters. Furthermore, we showed that the presence of the second (inner) membrane, mimicking the OM/IM relation in *E. coli*, significantly influenced primary membrane properties as well. Such changes may be crucial for interaction origins between particles and the membrane. As a result, common biological phenomena could not be observed numerically - or will behave differently from reality - if the simplified membrane model is used in the simulation. In future perspectives, the interactions of membrane-active particles and membranes in various membrane mimetic systems should be investigated.

Acknowledgements. M.R and S.K. acknowledge support from National Science Centre (grant number 2015/19/B/NZ7/02380), D.D. acknowledge support from National Science Center (grant number 2018/30/E/NZ1/00099).

References

1. Crémet, L., et al.: Comparison of three methods to study biofilm formation by clinical strains of Escherichia coli. Diagn. Microbiol. Infect. Dis. **75**, 252–255 (2013). https://doi.org/10. 1016/j.diagmicrobio.2012.11.019
2. Tenaillon, O., Skurnik, D., Picard, B., Denamur, E.: The population genetics of commensal Escherichia coli. Nat. Rev. Microbiol. **8**, 207–217 (2010). https://doi.org/10.1038/nrmicr o2298
3. Malanovic, N., Ön, A., Pabst, G., Zellner, A., Lohner, K.: Octenidine: novel insights into the detailed killing mechanism of Gram-negative bacteria at a cellular and molecular level. Int. J. Antimicrob. Agents **56**, 106146 (2020). https://doi.org/10.1016/j.ijantimicag.2020.106146
4. Alvarez-Marin, R., Aires-de-Sousa, M., Nordmann, P., Kieffer, N., Poirel, L.: Antimicrobial activity of octenidine against multidrug-resistant Gram-negative pathogens. Eur. J. Clin. Microbiol. Infect. Dis. **36**(12), 2379–2383 (2017). https://doi.org/10.1007/s10096-017-3070-0
5. Koburger, T., Hubner, N.-O., Braun, M., Siebert, J., Kramer, A.: Standardized comparison of antiseptic efficacy of triclosan, PVP-iodine, octenidine dihydrochloride, polyhexanide and chlorhexidine digluconate. J. Antimicrob. Chemother. **65**, 1712–1719 (2010). https://doi.org/ 10.1093/jac/dkq212
6. Wang, X., Quinn, P.J.: Endotoxins: lipopolysaccharides of gram-negative bacteria. Subcell. Biochem. **53**, 3–25 (2010). https://doi.org/10.1007/978-90-481-9078-2_1
7. Raetz, C.R.H., Whitfield, C.: Lipopolysaccharide endotoxins (2002). https://doi.org/10.1146/ annurev.biochem.71.110601.135414

8. Ebbensgaard, A., Mordhorst, H., Aarestrup, F.M., Hansen, E.B.: The role of outer membrane proteins and lipopolysaccharides for the sensitivity of Escherichia coli to antimicrobial peptides. Front. Microbiol. **9**, 2153 (2018). https://doi.org/10.3389/fmicb.2018.02153

9. Erridge, C., Bennett-Guerrero, E., Poxton, I.R.: Structure and function of lipopolysaccharides (2002). https://doi.org/10.1016/S1286-4579(02)01604-0

10. Sohlenkamp, C., Geiger, O.: Bacterial membrane lipids: diversity in structures and pathways (2015). https://doi.org/10.1093/femsre/fuv008

11. Epand, R.M., Epand, R.F.: Lipid domains in bacterial membranes and the action of antimicrobial agents (2009). https://doi.org/10.1016/j.bbamem.2008.08.023

12. De Vivo, M., Masetti, M., Bottegoni, G., Cavalli, A.: Role of molecular dynamics and related methods in drug discovery (2016). https://doi.org/10.1021/acs.jmedchem.5b01684

13. Wu, E.L., et al: Molecular dynamics and NMR spectroscopy studies of E. coli lipopolysaccharide structure and dynamics. Biophys. J. **105**, 1444–1455 (2013). https://doi.org/10.1016/j.bpj.2013.08.002

14. Hwang, H., Paracini, N., Parks, J.M., Lakey, J.H., Gumbart, J.C.: Distribution of mechanical stress in the Escherichia coli cell envelope. Biochim. Biophys. Acta - Biomembr. **1860**, 2566–2575 (2018). https://doi.org/10.1016/j.bbamem.2018.09.020

15. Piggot, T.J., Holdbrook, D.A., Khalid, S.: Conformational dynamics and membrane interactions of the E. coli outer membrane protein FecA: a molecular dynamics simulation study. Biochim. Biophys. Acta (BBA) – Biomem. **1828**(2), 284–293 (2013). https://doi.org/10.1016/j.bbamem.2012.08.021

16. Chen, J., Zhou, G., Chen, L., Wang, Y., Wang, X., Zeng, S.: Interaction of graphene and its oxide with lipid membrane: a molecular dynamics simulation study. J. Phys. Chem. C. **120**, 6225–6231 (2016). https://doi.org/10.1021/acs.jpcc.5b10635

17. Wu, E.L., et al.: CHARMM-GUI membrane builder toward realistic biological membrane simulations (2014). https://doi.org/10.1002/jcc.23702

18. Kanonenberg, K., et al.: Shaping the lipid composition of bacterial membranes for membrane protein production. Microb. Cell Fact. **18**, 131 (2019). https://doi.org/10.1186/s12934-019-1182-1

19. Szatmári, D., et al.: Intracellular ion concentrations and cation-dependent remodelling of bacterial MreB assemblies. Sci. Rep. **10**, 12002 (2020). https://doi.org/10.1038/s41598-020-68960-w

20. Strauss, J., Burnham, N.A., Camesano, T.A.: Atomic force microscopy study of the role of LPS O-antigen on adhesion of E. coli. J. Mol. Recognit. **22**(5), 347–355 (2009). https://doi.org/10.1002/jmr.955

21. Wu, E.L., et al.: E. coli outer membrane and interactions with OmpLA. Biophys. J. **106**, 2493–2502 (2014). https://doi.org/10.1016/j.bpj.2014.04.024

22. Klauda, J.B., et al.: Update of the CHARMM all-atom additive force field for lipids: validation on six lipid types. J. Phys. Chem. B. **114**, 7830–7843 (2010). https://doi.org/10.1021/jp101759q

23. Abraham, M.J., et al.: Gromacs: high performance molecular simulations through multi-level parallelism from laptops to supercomputers. SoftwareX. **1–2**, 19–25 (2015). https://doi.org/10.1016/j.softx.2015.06.001

24. Guixà-González, R., et al.: MEMBPLUGIN: studying membrane complexity in VMD. Bioinformatics **30**, 1478–1480 (2014). https://doi.org/10.1093/bioinformatics/btu037

25. Giorgino, T.: Computing diffusion coefficients in macromolecular simulations: the diffusion coefficient tool for VMD. J. Open Source Softw. **4**(41), 1698 (2019). https://doi.org/10.21105/joss.01698

26. Vanegas, J.M., Torres-Sánchez, A., Arroyo, M.: Importance of force decomposition for local stress calculations in biomembrane molecular simulations. J. Chem. Theory Comput. **10**, 691–702 (2014). https://doi.org/10.1021/ct4008926

27. Bacle, A., Gautier, R., Jackson, C.L., Fuchs, P.F.J., Vanni, S.: Interdigitation between triglyc-erides and lipids modulates surface properties of lipid droplets. Biophys. J. **112**, 1417–1430 (2017). https://doi.org/10.1016/j.bpj.2017.02.032
28. Drabik, D., Chodaczek, G., Kraszewski, S., Langner, M.: Mechanical properties determination of DMPC, DPPC, DSPC, and HSPC solid-ordered bilayers. Langmuir **36**, 3826–3835 (2020). https://doi.org/10.1021/acs.langmuir.0c00475
29. Doktorova, M., LeVine, M.V., Khelashvili, G., Weinstein, H.: A new computational method for membrane compressibility: bilayer mechanical thickness revisited. Biophys. J. **116**, 487–502 (2019). https://doi.org/10.1016/j.bpj.2018.12.016
30. Marsh, D.: Lateral pressure profile, spontaneous curvature frustration, and the incorporation and conformation of proteins in membranes. Biophys. J. **93**, 3884–3899 (2007). https://doi.org/10.1529/biophysj.107.107938
31. Devanand, T., Krishnaswamy, S., Vemparala, S.: Interdigitation of lipids induced by mem-brane–active proteins. J. Membr. Biol. **252**(4–5), 331–342 (2019). https://doi.org/10.1007/s00232-019-00072-7
32. Shearer, J., Khalid, S.: Communication between the leaflets of asymmetric membranes revealed from coarse-grain molecular dynamics simulations. Sci. Rep. **8**, 1805 (2018). https://doi.org/10.1038/s41598-018-20227-1
33. Hsu, P.C., Samsudin, F., Shearer, J., Khalid, S.: It is complicated: curvature, diffusion, and lipid sorting within the two membranes of escherichia coli. J. Phys. Chem. Lett. **8**, 5513–5518 (2017). https://doi.org/10.1021/acs.jpclett.7b02432
34. Domínguez-Medina, C.C., et al.: Outer membrane protein size and LPS O-antigen define protective antibody targeting to the Salmonella surface. Nat. Commun. **11**, 1–11 (2020). https://doi.org/10.1038/s41467-020-14655-9
35. Piggot, T.J., Holdbrook, D.A., Khalid, S.: Electroporation of the E. coli and S. aureus mem-branes: molecular dynamics simulations of complex bacterial membranes. J. Phys. Chem. B. **115**, 13381–13388 (2011). https://doi.org/10.1021/jp207013v
36. Camley, B.A., Lerner, M.G., Pastor, R.W., Brown, F.L.H.: Strong influence of periodic bound-ary conditions on lateral diffusion in lipid bilayer membranes. J. Chem. Phys. **143**, 243113 (2015). https://doi.org/10.1063/1.4932980
37. Jefferies, D., Shearer, J., Khalid, S.: Role of o-antigen in response to mechanical stress of the E. coli outer membrane: insights from coarse-grained MD simulations. J. Phys. Chem. B. **123**(17), 3567–3575 (2019). https://doi.org/10.1021/acs.jpcb.8b12168

PathMEx: Pathway-Based Mutual Exclusivity for Discovering Rare Cancer Driver Mutations

Yahya Bokhari[1] and Tomasz Arodz[2]([envelope]) [ID]

[1] Department of Biostatistics and Bioinformatics, King Abdullah International Medical Research Center, Riyadh, Saudi Arabia
BukhariY@ngha.med.sa
[2] Department of Computer Science, Virginia Commonwealth University, Richmond, VA 23284, USA
tarodz@vcu.edu

Abstract. The genetic material we carry today is different from that we were born with: our DNA is prone to mutations. Some of these mutations can make a cell divide without control, resulting in a growing tumor. Typically, in a cancer sample from a patient, a large number of mutations can be detected, and only a few of those are drivers - mutations that positively contribute to tumor growth. The majority are passenger mutations that either accumulated before the onset of the disease but did not cause it, or are byproducts of the genetic instability of cancer cells. One of the key questions in understanding the process of cancer development is which mutations are drivers, and should be analyzed as potential diagnostic markers or targets for therapeutics, and which are passengers. We propose PathMEx, a novel method based on simultaneous optimization of patient coverage, mutation mutual exclusivity, and pathway overlap among putative cancer driver genes. Compared to state-of-the-art method Dendrix, the proposed algorithm finds sets of putative driver genes of higher quality in three sets of cancer samples: brain, lung, and breast tumors. The genes in the solutions belong to pathways with known associations with cancer. The results show that PathMEx is a tool that should be part of a state-of-the-art toolbox in the driver gene discovery pipeline. It can help detect low-frequency driver genes that can be missed by existing methods.

Keywords: Somatic mutations · Cancer pathways · Driver mutations

1 Introduction

Human cells are prone to mutations, and some of these may transform the cell into one that divides indefinitely and has the ability to invade other tissues [3], resulting in cancer. For most human cancers to develop, a sequence of between two and eight mutations that target genes involved in specific cell functions

Supported by NSF grant IIS-1453658.

is needed [48]. Such mutations, which confer growth advantage to cells and are causally implicated in oncogenesis, are referred to as driver mutations [4]. Known somatic mutations linked to cancer, often with additional information such as known therapies that target the mutation, are being gather in databases [12, 29, 30, 42] that can be used in selecting patient treatment. Newly identified driver genes can also be screened using druggability indices [10], and considered for being targets for drug repositioning [34], leading the way to new therapeutic modalities. Thus, experimental and computational techniques for discovering driver genes are of great interest.

In recent years, the ability to discover driver mutations has advanced greatly due to availability of large datasets generated using second-generation sequencing techniques [41]. The Cancer Genome Atlas (TCGA) [50] and other similar projects perform sequencing of matched tumor and normal samples from hundreds of patients with a given tumor type, allowing for detection of somatic mutations present in tumor tissue. However, even with the increasing availability of data, the problem of identifying driver mutations and genes that harbor them (called driver genes) remains a challenge.

The main issue hampering discovery of driver mutations from sources such as TCGA is that majority of somatic mutations acquired in human cells throughout life are not causally linked to cancer - these are often referred to as passenger mutations. It has been estimated that half or more of all mutations observed in patients' cancer tissues originate prior to the onset of cancer [44]. In addition to these pre-existing mutations, cancer cells exhibit a mutator phenotype, that is, and increase mutation rate [32]. This further contributes to the dominance of passenger mutations over driver mutations in observed cancer tissue samples. Altogether, while the number of driver mutations in a tumor is typically small – a recent analysis of TCGA data shows it to be between 2 and 6 in most tumors [20] – the total number of somatic mutations present in a single patient can range between 10 to above 100, depending on tumor type and patient age. Most mutations in a cancer tissue sample are thus passenger mutations that do not contribute positively to cancer growth.

On common approach to separate driver from passenger mutations is to calculate the background mutations rate that would be exhibited by passenger mutations, and consider those mutations that are encountered more frequently as drivers. This approach typically considers mutations a result of a Poisson process, which allows for quantifying the statistical significance of any deviations from the background mutation rate. For example, MutSig [9] uses a constant mutation rate across all genes, and can also use methods for functional predictions of mutation significance, such as SIFT [37], CHASM [7], Polyphen-2 [1] and MutationAssessor [40], to account for the fact that mutations differ substantially in their effects on the mutated protein [2]. MutSigCV [24] uses factors such as chromatin state and transcriptional activity to estimate gene-specific background mutation rates. PathScan [51] utilizes a Poissonian mutation model that involves gene lengths, and for a gene set given by the user calculates the probability of observing that many mutations or more under a null hypothesis that the mutations are passengers. If the probability is low across many samples, the genes are considered driver genes. MuSiC [14] extends PathScan by adding knowledge about correlation between mutation rates and factors including clinical variables such as age, molecular variables such as Pfam family to which the genes belong, and sequence correlates such as base composition of the site and proximity among mutation sites. DrGaP tool [17] considers 11 different types of mutation types, with factors including G/C content near the mutation site and methylation status of the site, in estimating the background mutation rate. Detection of driver genes using the gene-centric methods mentioned above is complicated by the fact that rarely a single driver gene is mutated across many patients with a given tumor. Only few genes, such as TP53 or BRCA1, are mutated in large fraction of cases. Most of individual genes are mutated in less than 5% of patients suffering form the same cancer type [39]. Thus, large number of samples is required to detect statistically significant deviations from background mutation rates.

To alleviate the problems associated with relying only on mutation frequencies of individual genes, a new approach of using patterns of mutations spanning multiple genes has emerged in recent years. It has been observed that in many types of tumors, only one mutation per pathway is needed to drive oncogenesis [35,47,52]. Thus, the minimal set of mutated genes required for cancer to develop would consist of several sets of genes, each corresponding to a critical pathway such as angiogenesis. Within each gene set, exactly one gene would be mutated in each patient. That is, all patients would be covered by a mutation in a gene from the set, and there would be no excess coverage, that is, no patient will have more mutations than one in the genes from the set. This pattern has been often referred to as mutual exclusivity within a gene set. In actual patient data, additional mutations in driver genes may be present, especially in older patients or in cases of slow growing tumors. Also, some of the mutations may be missed due to observation errors. Thus, instead of detecting the presence or absence of mutual exclusivity in a set of genes covering all patients, driver detection algorithms involve a score that penalizes for deviations from a driver pattern, that is, for zero mutations in a patient, or for more than one mutation. Finding the optimal set of genes with respect to such a score has been shown to be an NP-hard problem [46], and heuristic search procedures are utilized to find a set of genes closest to the high-coverage mutual exclusivity pattern. The approach of finding a gene set through a pattern search procedure has been used by several tools, including Dendrix [46] and Multi-Dendrix [25], and RME [36]. Further methods extend this approach by helping deal with observation errors in the data [43], with cancer subtypes [26], and with computational efficiency of the search for driver genes [6,53].

Further advances in driver gene detection methods resulted from observations that show that cancer driver mutations are not confined to a specific set of loci but, instead, differ substantially in individual cases. Only when seen from the level of pathways, that is, genes related to a specific cellular process, a clearer picture emerges. This evidence has given rise to network-oriented driver detection methods, such as HotNet [45], which incorporates protein-protein networks and uses a heat diffusion process, in addition to gene mutation frequency, to detect a driver subnetwork. Another network-based technique, MEMo [11], utilizes mutation frequency in individual genes together with gene interactions to form highly mutated cliques, and then filters the cliques using mutual exclusivity principle. We have recently proposed QuaDMutNetEx [5], a method that utilizes human protein-protein interaction networks in conjunction with mutual exclusivity. These methods involve a graph with genes as nodes, and gene-gene or protein-protein interactions as edges, and do not incorporate the existing knowledge of how groups of genes and edges connect into larger functional pathways.

Here, we propose PathMEx, a novel driver gene detection technique that combines the pattern-based and pathway-based detection approaches. It is built around simultaneous optimization of patient coverage, mutual exclusivity and pathway overlap among putative cancer driver genes. We evaluated our method on three cancer mutation datasets obtained from literature and from the Cancer Genome Atlas. Compared to the state-of-the-art tool Dendrix, our method shows higher values of the Dendrix score, a metric used to judge the quality of cancer driver gene sets.

2 Methods

The proposed algorithm for detecting driver mutations in cancer operates at the gene level. That is, on input, we are given an n by p mutation matrix G, where n is the number of cancer patients with sequenced cancer tissue DNA and sequenced matched normal tissue, and p is the total number of genes explored. The matrix is binary, that is, $G_{ij} = 1$ if patient i has a non-silent somatic mutation in gene j; otherwise, $G_{ij} = 0$. More generally, G_{ij} can also be set to one if a gene is part of region with a copy-number alteration, or has a mutation in its regulatory region, although such data is less readily available compared to mutations in gene coding regions. A row vector G_i represents a row of the matrix corresponding to patient i. The solution we seek is a sparse binary vector x of length p, with $x_j = 1$ indicating that mutations of gene j are cancer driver mutations. In the proposed approach, the solution vector should capture driver genes that are functionally related, for example are all part of a pathway that needs to be mutated in oncogenesis. If we want to uncover all driver genes, we should apply the algorithm multiple times, each time removing the genes found in prior steps from consideration. We will often refer to the nonzero elements of x as the mutations present in x.

In designing the algorithm for choosing the solution vector x, we assumed that any possible vector is penalized with a penalty score based on observed patterns of driver mutations in human cancers. We expect that each patient has at least one mutation in the set of genes selected in the solution; however, in some cases, the mutation may not be detected. Also, while several distinct pathways need to be mutated to result in a growing tumor, typically one mutation in each of those pathways suffices. The chances of accumulating additional mutations in the already mutated pathway before the cancer is detected are low, and decrease with each additional mutation beyond the first one. We capture this decreasing odds through an increasing penalty associated with solution vector x given the observed mutations G_i in patient i

$$E(G_i, x) = |G_i x - 1|. \tag{1}$$

The term $G_i x$, that is, the product of row vector G_i and the solution vector x, captures the number of mutations from solution x present in patient i. We incur no penalty if the number of mutated genes from x in a given patient is one. If the patient is covered by no mutations, the penalty is one. If the patient is covered by more than one mutation, the penalty is equal to the number of mutations in excess of the one required for cancer to develop.

We also expect the genes in the solution to be functionally related, that is, we expect high pathway overlap in the solution. To capture this, we provide a reward (i.e., a negative penalty) for genes in the solution that belong to the same pathway. For a gene j, we denote by P_j the set of pathways that contain j. Further, for a gene j, we can define the set of co-pathway genes, Π_j, that is, the set of genes that share a pathway with j, as

$$\Pi_j = \{k : k \neq j, P_k \cap P_j \neq \emptyset\}. \tag{2}$$

To promote selection of genes from the same pathway, for every gene j we define a pathway overlap term added to the objective function that is being minimized

$$O(j, x) = \max(-x_j, - \sum_{k \in \Pi_j} x_k). \tag{3}$$

If gene j is selected to be part of the solution, and it shares pathways with at least one of other genes in the solution, the objective will be decreased by 1.

The final objective function being minimized is a combination of the high-coverage mutual exclusivity terms and the pathway overlap terms

$$L(G, x) = \sum_{i=1}^{n} E(G_i, x) + \sum_{j=1}^{p} O(j, x) \tag{4}$$

$$= \sum_{i=1}^{n} |G_i x - 1| + \sum_{j=1}^{p} \max(-x_j, - \sum_{k \in \Pi_j} x_k). \tag{5}$$

We also introduce a limit on the number of genes in the solution, K, by requiring $\sum_{j=1}^{p} x_j \leq K$. The solution x is a binary indicator vector, where elements $x_j = 1$ correspond to genes being selected as the set of driver genes. Below, if we need to express solution as a set instead of an indicator vector, we will use $Z_x = \{j : x_j = 1\}$.

The problem of minimizing the non-linear objective function $L(G, x)$ over possible solution vectors x can be reformulated into a constrained mixed integer linear program

$$\text{minimize} \atop x,u,v \qquad \sum_{i=1}^{n} u_i + \sum_{j=1}^{p} v_j \qquad\qquad (6)$$

$$\text{subject to} \qquad
\begin{aligned}
x_j &\in \{0,1\} & 1 &\le j \le p \\
G_i x - 1 &\le u_i & 1 &\le i \le n \\
1 - G_i x &\le u_i & 1 &\le i \le n \\
-x_j &\le v_j & 1 &\le j \le p \\
-\sum_{k \in \Pi_j} x_k &\le v_j & 1 &\le j \le p
\end{aligned}$$

$$\sum_{j=1}^{p} x_j \le K$$

with p binary variables, $n + p$ continuous variables, and $2n + 2p + 1$ inequality constraints. That is, the size of the problem grows linearly with the number of samples, n, and the number of genes, p.

Mixed-integer linear programs (MILP) are known to be NP-hard in general. However, the optimal solution can be obtained quickly for problems of small size. For cancer driver detection problems involving a large number of genes, where exact solutions are not available in any reasonable time, we designed a meta-heuristic algorithm, PathMEx, that combines network-based exploration of the solution space with optimal search for small subproblems.

Algorithm. PathMEx

1: **procedure** PathMEx(G, C, K, T, s, s_p)
2:    $\chi =$ RandomSubset($s, \{1, ..., p\}$)
3:   **for** $t \leftarrow 1, ..., T$ **do**
4:      $G^\chi =$ Select Columns χ_t from G
5:      $Z_x =$ Minimize $L(G^\chi, x)$ (eq. 5) using eq. 6 MILP
6:      $\Pi_Z = \{k : k \notin Z_x, P_k \cap P_j \ne \emptyset, j \in Z_x\}$
7:      $\chi' = Z_x \cup$ RandomSubset($s_p - |Z_x|, \Pi_Z$)
8:      $\chi = \chi' \cup$ RandomSubset($s - |\chi'|, \{1, ..., p\} \setminus \chi'$)
9:   **end for**
10:   **return** Z_x
11: **end procedure**

If the problem is small enough, PathMEx directly solves the MILP problem and returns a globally optimal solution. In other cases, the main PathMEx algorithm goes through T iterations, as shown in the pseudocode. In each iteration PathMEx considers a candidate set χ of s genes, where s is chosen to make the problem tractable for a MILP solver. In our tests, we set $s = 200$. In each iteration, a subproblem involving only genes from χ is solved by a MILP solver, and a globally optimal subset $Z_x \in \chi$ is selected as the current solution. The solution set Z_x has up to K genes. A new candidate set χ is created by keeping all genes in the solution Z_x, and choosing additional genes to make the size of the new candidate set equal to s. These include up to s_p genes that are either in Z_x or are randomly selected from all the pathways that contain genes from Z_x. It also includes other genes selected at random until the candidate set size reaches s.

3   Results and Discussion

We evaluated the proposed algorithm using cancer mutation data from the Cancer Genome Atlas (TCGA) [50] and from literature. We used two datasets that were originally used by the authors of Dendrix [46]: somatic mutations in lung cancer (LUNG), and a dataset relating to Glioblastoma Multiforme (GBM) that includes not only somatic mutations but also copy number alternations. We also used a larger dataset of somatic mutations in samples from Breast Invasive Carcinoma (BRCA) downloaded from TCGA, in which, following standard practice [24], we removed known hypermutated genes with no role in cancer, including olfactory receptors, mucins, and a few other genes such as the longest human gene, titin. The characteristics of the datasets are summarized in the Table 1.

Table 1. Summary of datasets used in testing PathMEx.

Dataset	Samples (n)	Genes (p)	Mutations
GBM	84	178	809
LUNG	163	356	979
BRCA	771	13,582	33,385

In judging the quality of a solution Z_x, that is, a set of putative driver genes, we used two metrics, coverage and excess coverage. Coverage is defined as the number of patients covered by at least one gene from Z_x divided by the total number of patients. Excess coverage is defined as the number of patients covered by more than one gene from Z_x divided by the number of patients covered by at least one gene from Z_x. These metrics together capture how well a gene set conforms to the pattern expected of driver genes. Both of the metrics range from 0 to 1. Perfect solution has coverage of 1, and excess coverage of 0, indicating that every single patient has exactly one mutation in genes from solution Z_x. We also used the Dendrix score, the objective function maximized by Dendrix [46], defined as the number of patients covered by at least one gene from Z_x set minus the coverage overlap, that is, the total count of all mutations in excess of one mutation per patient in genes from Z_x.

We ran PathMEx and Dendrix on the three datasets: GBM, LUNG, and BRCA. For two small datasets, GBM and LUNG, we explored solutions including up to 10 genes; for BRCA, a much larger dataset, we searched for solutions including up to 20 genes. PathMEx automatically picks the best solution with

Table 2. Comparison between Dendrix and PathMEx.

Method	Genes in solution	Dendrix score
GBM: Glioblastoma multiforme		
Dendrix	9	68
PathMEx	10	70
LUNG: Lung Adenocarcinoma		
Dendrix	9	106
PathMEx	10	113
BRCA: Breast Invasive Carcinoma		
Dendrix	19	392
PathMEx	20	423

Table 3. Coverage and excess coverage.

Method	Coverage	Excess coverage
GBM: Glioblastoma multiforme		
Dendrix	0.85	0.05
PathMEx	0.90	0.07
LUNG: Lung Adenocarcinoma		
Dendrix	0.74	0.12
PathMEx	0.77	0.11
BRCA: Breast Invasive Carcinoma		
Dendrix	0.56	0.10
PathMEx	0.61	0.10

size up to a given range. For Dendrix, which analyzes solutions of a fixed, user-provided size, we performed independent runs for each solution size ranging from 2 to the chosen limit (10 for GBM and LUNG, 20 for BRCA), and picked the solution with the highest value of Dendrix score. Each Dendrix run involved 10^7 iterations, as recommended by Dendrix authors. For PathMEx, which is a descent method depending on the randomized initialization, we conducted 10 runs, each consisting of 100 iterations, and picked the solution with the lowest value of the objective function among these 10 runs.

PathMEx relies on prior knowledge of biological pathways, which we obtained from the MSigDB repository of canonical pathways [27]. These include pathways from KEGG, Biocarta, Pathway Interaction Database, and Reactome. We removed 46 pathways related to disease, most notably KEGG_PATHWAYS_IN_CANCER and other cancer-specific pathways, to avoid biasing the method towards re-discovering only known cancer genes. We ended up with 1284 pathways that remained after the filtering step. Each pathway is treated as a set of genes that are members of the pathway.

Table 4. PathMEx solution gene sets and their statistical significance.

Gene set	Estimated p-value
GBM: Glioblastoma multiforme	
CDKN2B CDK4 RB1 ERBB2 TNK2 KPNA2 WEE1 CES3 INSR IQGAP1	<0.001
LUNG: Lung Adenocarcinoma	
KRAS STK11 EGFR EPHB1 MAP3K3 ABL1 PAK6 JUP CYSLTR2 FES	0.003
BRCA: Breast Invasive Carcinoma	
TP53 GATA3 MAP3K1 CDH1 MAP2K4 LOC283685 HUWE1 UBR4 ATP10A BCL6B ADCY7 TICAM1 AKT3 ELN GNAS HGF PXDN CD38 MX2 SLC13A5	0.002

The results of the tests, presented in Table 2, show that PathMEx consistently returns higher quality solutions than Dendrix. In each of the three datasets, Path-MEx reached a higher value of the Dendrix score. Table 3 shows that PathMEx also achieved higher patient coverage, while showing no consistent increase in excess coverage compared to Dendrix across the datasets.

To quantify the statistical significance of the results, we employed the randomization approach used previously [46]. For every gene, the binary column vector describing in which patient the gene is mutated is randomly reshuffled. The results of reshuffling of all genes form a new dataset, that is, a new matrix G. Each randomized dataset preserves the underlying frequencies of mutations of individual genes, but any multi-gene patterns of mutations such as mutual exclusivity may only arise by chance. We created 1024 reshuffled datasets, and ran PathMEx on each of them. As with the original non-randomized dataset, for each reshuffled dataset we performed 10 runs, and picked the solution with the lowest value of the objective function from among them. Finally, as the estimate of the p-value, we quantified the fraction of the 1024 reshuffled datasets in which the value of the objective function minimized by PathMEx (Eq. 5) is lower than or equal to the value obtained on the original non-randomized dataset. As shown in Table 4, the p-values for all three datasets are a magnitude below the 0.05 threshold.

The genes in the solutions are members of pathways known to be associated with cancer. We visualized the most enriched pathways for each dataset in Figs. 1, 2. For each dataset, among all pathways covering the genes in the solution, we first selected the pathway most enriched in solution genes, that is, the pathway with the highest ratio of pathway genes in the solution to all genes in the pathway. We then removed the genes covered by that pathway from consideration, and repeated the process, until only genes that were not members of any pathway remained not covered.

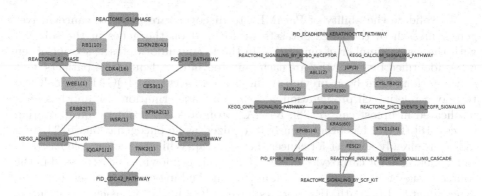

Fig. 1. Pathways covering driver genes for GBM dataset (left) and the LUNG dataset (right). Red nodes represent pathways from MSigDB, blue nodes represent genes in the PathMEx solution. For each gene, in parentheses, we provide the number of patients in the dataset that harbor a mutation in that gene. (Color figure online)

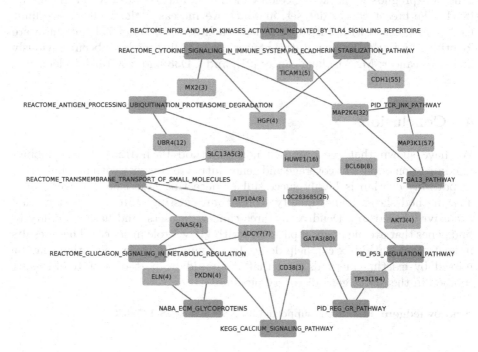

Fig. 2. Pathways covering driver genes for BRCA dataset. Red nodes represent pathways from MSigDB, blue nodes represent genes in the PathMEx solution. For each gene, in parentheses, we provide the number of patients in the dataset that harbor a mutation in that gene. (Color figure online)

To validate the ability of PathMEx to discover rare putative cancer driver genes, in each of the three datasets we focused on the genes in the solution with the fewest number of mutations. In the brain tumor dataset, six out of ten genes identified by PathMEx are each mutated in only 1 out of 84 patients. Out of these, four have been previously implicated in cancer: IQGAP1 is believed to play a role in cell proliferation and cancer transformation [19] and has been implicated in breast cancer [38], KPNA2 promotes cell proliferation in ovarian cancer [18], TNK2 has been recently recognized as an oncogenic kinase [33], and WEE1 is already a target for cancer therapy [15]. While no cancer role has been so far identified for carboxylesterase 3 (CES3), it is known to be expressed in the source tissue of our samples, the brain [16]. In the lung cancer dataset, out of 10 genes identified by PathMEx, 5 genes have only 2 mutations each in a group of 163 samples. All five genes have been previously linked to various types of cancer. Role of ABL1 in cancer is well established. FES is a known proto-oncogene [28]. PAK6 has been shown to suppress growth of prostate cancer [31]. JUP has been implicated in prostate and breast cancers [23]. Finally, expression of CYSLTR2 gene is a prognostic marker in colon cancer [49] and is causative of melanoma [8]. In the breast cancer dataset, in which we increased the solution size limit to 20, three genes with only 3 mutations each in a cohort of 771 patients were identified by PathMEx as putative cancer genes. All three have been previously linked to cancer: MX2 to lung cancer [21], and CD38 and SLC13A5 to leukemia [13,22].

4   Conclusions

We have shown that the proposed novel method PathMEx, which combines maximization of patient coverage and gene mutual exclusivity with maximization of pathway overlap is highly successful in detecting rare cancer driver genes. The method shows higher quality scores than existing state-of-the-art mutual exclusivity-based tool Dendrix on three cancer datasets, and has the ability to find genes that are members of pathways with known role in cancer. These results indicate that PathMEx can help detect low-frequency driver genes that may be missed by existing methods, and that it should be part of a state-of-the-art toolbox in the driver gene discovery pipeline.

Acknowledgements. TA is supported by NSF grant IIS-1453658.

References

1. Adzhubei, I.A., et al.: A method and server for predicting damaging missense mutations. Nat. Methods **7**(4), 248–249 (2010)
2. Arodź, T., Płonka, P.M.: Effects of point mutations on protein structure are non-exponentially distributed. Proteins Struct. Funct. Bioinf. **80**(7), 1780–1790 (2012)
3. Bertram, J.S.: The molecular biology of cancer. Mol. Aspects Med. **21**(6), 167–223 (2000)

4. Bignell, G.R., et al.: Signatures of mutation and selection in the cancer genome. Nature **463**(7283), 893–898 (2010)
5. Bokhari, Y., Alhareeri, A., Arodz, T.: QuaDMutNetEx: a method for detecting cancer driver genes with low mutation frequency. BMC Bioinf. **21**(1), 1–12 (2020)
6. Bokhari, Y., Arodz, T.: QuaDMutEx: quadratic driver mutation explorer. BMC Bioinf. **18**(1), 1–15 (2017)
7. Carter, H., et al.: Cancer-specific high-throughput annotation of somatic mutations: computational prediction of driver missense mutations. Can. Res. **69**(16), 6660–6667 (2009)
8. Ceraudo, E., et al.: Direct evidence that the GPCR CysLTR2 mutant causative of uveal melanoma is constitutively active with highly biased signaling. J. Biol. Chem. **296**, 100163 (2021)
9. Chapman, M.A., et al.: Initial genome sequencing and analysis of multiple myeloma. Nature **471**(7339), 467–472 (2011)
10. Chen, Y., et al.: Identification of druggable cancer driver genes amplified across TCGA datasets. PLoS One **9**(5), e98293 (2014)
11. Ciriello, G., Cerami, E., Sander, C., Schultz, N.: Mutual exclusivity analysis identifies oncogenic network modules. Genome Res. **22**(2), 398–406 (2012)
12. Damodaran, S., et al.: Cancer Driver Log (CanDL): catalog of potentially actionable cancer mutations. J. Mol. Diagn. **17**(5), 554–559 (2015)
13. Deaglio, S., Mehta, K., Malavasi, F.: Human CD38: a (r)evolutionary story of enzymes and receptors. Leuk. Res. **25**(1), 1–12 (2001)
14. Dees, N.D., et al.: MuSiC: identifying mutational significance in cancer genomes. Genome Res. **22**(8), 1589–1598 (2012)
15. Do, K., Doroshow, J.H., Kummar, S.: Wee1 kinase as a target for cancer therapy. Cell Cycle **12**(19), 3348–3353 (2013)
16. Holmes, R.S., Cox, L.A., VandeBerg, J.L.: Mammalian carboxylesterase 3: comparative genomics and proteomics. Genetica **138**(7), 695–708 (2010)
17. Hua, X., Xu, H., Yang, Y., Zhu, J., Liu, P., Lu, Y.: DrGaP: a powerful tool for identifying driver genes and pathways in cancer sequencing studies. Am. J. Hum. Genet. **93**(3), 439–451 (2013)
18. Huang, L., et al.: KPNA2 promotes cell proliferation and tumorigenicity in epithelial ovarian carcinoma through upregulation of c-Myc and downregulation of FOXO3a. Cell Death Dis. **4**(8), e745 (2013)
19. Johnson, M., Sharma, M., Henderson, B.R.: IQGAP1 regulation and roles in cancer. Cell. Signal. **21**(10), 1471–1478 (2009)
20. Kandoth, C., et al.: Mutational landscape and significance across 12 major cancer types. Nature **502**(7471), 333–339 (2013)
21. Kobayashi, K., et al.: Identification of genes whose expression is upregulated in lung adenocarcinoma cells in comparison with type II alveolar cells and bronchiolar epithelial cells in vivo. Oncogene **23**(17), 3089–3096 (2004)
22. Kuang, S., et al.: Genome-wide identification of aberrantly methylated promoter associated CpG islands in acute lymphocytic leukemia. Leukemia **22**(8), 1529–1538 (2008)
23. Lai, Y.H., et al.: SOX4 interacts with plakoglobin in a Wnt3a-dependent manner in prostate cancer cells. BMC Cell Biol. **12**(1), 50 (2011)
24. Lawrence, M.S., et al.: Mutational heterogeneity in cancer and the search for new cancer-associated genes. Nature **499**(7457), 214–218 (2013)
25. Leiserson, M.D., Blokh, D., Sharan, R., Raphael, B.J.: Simultaneous identification of multiple driver pathways in cancer. PLoS Comput. Biol. **9**(5), e1003054 (2013)

26. Leiserson, M.D., Wu, H.T., Vandin, F., Raphael, B.J.: CoMEt: a statistical app-roach to identify combinations of mutually exclusive alterations in cancer. Genome Biol. **16**(1), 1 (2015)
27. Liberzon, A., Subramanian, A., Pinchback, R., Thorvaldsdóttir, H., Tamayo, P., Mesirov, J.P.: Molecular signatures database (MSigDB) 3.0. Bioinformatics **27**(12), 1739–1740 (2011)
28. Lionberger, J.M., Smithgall, T.E.: The c-Fes protein-tyrosine kinase suppresses cytokine-independent outgrowth of myeloid leukemia cells induced by Bcr-Abl. Can. Res. **60**(4), 1097–1103 (2000)
29. Liu, E.M., Martinez-Fundichely, A., Bollapragada, R., Spiewack, M., Khurana, E.: CNCDatabase: a database of non-coding cancer drivers. Nucleic Acids Res. **49**(D1), D1094–D1101 (2021)
30. Liu, S.H., et al.: DriverDBv3: a multi-omics database for cancer driver gene research. Nucleic Acids Res. **48**(D1), D863–D870 (2020)
31. Liu, T., et al.: p21-Activated kinase 6 (PAK6) inhibits prostate cancer growth via phosphorylation of androgen receptor and tumorigenic E3 ligase murine double minute-2 (Mdm2). J. Biol. Chem. **288**(5), 3359–3369 (2013)
32. Loeb, L.A.: Human cancers express mutator phenotypes: origin, consequences and targeting. Nat. Rev. Cancer **11**(6), 450–457 (2011)
33. Mahajan, K., Mahajan, N.: ACK1/TNK2 tyrosine kinase: molecular signaling and evolving role in cancers. Oncogene **34**(32), 4162–4167 (2015)
34. Martinez-Ledesma, E., de Groot, J.F., Verhaak, R.G.: Seek and destroy: relating cancer drivers to therapies. Cancer Cell **27**(3), 319–321 (2015)
35. McCormick, F.: Signalling networks that cause cancer. Trends Biochem. Sci. **24**(12), M53–M56 (1999)
36. Miller, C.A., Settle, S.H., Sulman, E.P., Aldape, K.D., Milosavljevic, A.: Discover-ing functional modules by identifying recurrent and mutually exclusive mutational patterns in tumors. BMC Med. Genomics **4**(1), 1 (2011)
37. Ng, P.C., Henikoff, S.: SIFT: predicting amino acid changes that affect protein function. Nucleic Acids Res. **31**(13), 3812–3814 (2003)
38. Osman, M.A., Antonisamy, W.J., Yakirevich, E.: IQGAP1 control of centrosome function defines distinct variants of triple negative breast cancer. Oncotarget **11**(26), 2493 (2020)
39. Pon, J.R., Marra, M.A.: Driver and passenger mutations in cancer. Annu. Rev. Pathol. **10**, 25–50 (2015)
40. Reva, B., Antipin, Y., Sander, C.: Predicting the functional impact of protein mutations: application to cancer genomics. Nucleic Acids Res. **39**, gkr407 (2011)
41. Schuster, S.C.: Next-generation sequencing transforms today's biology. Nature **200**(8), 16–18 (2007)
42. Sondka, Z., Bamford, S., Cole, C.G., Ward, S.A., Dunham, I., Forbes, S.A.: The COSMIC Cancer Gene Census: describing genetic dysfunction across all human cancers. Nat. Rev. Cancer **18**(11), 696–705 (2018)
43. Szczurek, E., Beerenwinkel, N.: Modeling mutual exclusivity of cancer mutations. PLoS Comput. Biol. **10**(3), e1003503 (2014)
44. Tomasetti, C., Vogelstein, B., Parmigiani, G.: Half or more of the somatic muta-tions in cancers of self-renewing tissues originate prior to tumor initiation. Proc. Natl. Acad. Sci. **110**(6), 1999–2004 (2013)
45. Vandin, F., Upfal, E., Raphael, B.J.: Algorithms for detecting significantly mutated pathways in cancer. J. Comput. Biol. **18**(3), 507–522 (2011)
46. Vandin, F., Upfal, E., Raphael, B.J.: De novo discovery of mutated driver pathways in cancer. Genome Res. **22**(2), 375–385 (2012)

47. Vogelstein, B., Kinzler, K.W.: Cancer genes and the pathways they control. Nat. Med. **10**(8), 789–799 (2004)
48. Vogelstein, B., Papadopoulos, N., Velculescu, V.E., Zhou, S., Diaz, L.A., Kinzler, K.W.: Cancer genome landscapes. Science **339**(6127), 1546–1558 (2013)
49. Wang, D., DuBois, R.N.: Eicosanoids and cancer. Nat. Rev. Cancer **10**(3), 181–193 (2010)
50. Weinstein, J.N., et al.: The Cancer Genome Atlas pan-cancer analysis project. Nat. Genet. **45**(10), 1113–1120 (2013)
51. Wendl, M.C., et al.: PathScan: a tool for discerning mutational significance in groups of putative cancer genes. Bioinformatics **27**(12), 1595–1602 (2011)
52. Yeang, C.H., McCormick, F., Levine, A.: Combinatorial patterns of somatic gene mutations in cancer. FASEB J. **22**(8), 2605–2622 (2008)
53. Zhao, J., Zhang, S., Wu, L.Y., Zhang, X.S.: Efficient methods for identifying mutated driver pathways in cancer. Bioinformatics **28**(22), 2940–2947 (2012)

Serverless Nanopore Basecalling
with AWS Lambda

Piotr Grzesik[✉] and Dariusz Mrozek

Department of Applied Informatics, Silesian University of Technology,
ul. Akademicka 16, 44-100 Gliwice, Poland
dariusz.mrozek@polsl.pl

Abstract. The serverless computing paradigm allows simplifying operations, offers highly parallel execution and high scalability without the need for manual management of underlying infrastructure. This paper aims to evaluate if recent advancements such as container support and increased computing resource limits in AWS Lambda allow it to serve as an underlying platform for running bioinformatics workflows such as basecalling of nanopore reads. For the purposes of the paper, we developed a sample workflow, where we focused on Guppy basecaller, which was tested in multiple scenarios. The results of the experiments showed that AWS Lambda is a viable platform for basecalling, which can support basecalling nanopore reads from multiple sequencing reads at the same time while keeping low infrastructure maintenance overhead. We also believe that recent improvements to AWS Lambda make it an interesting choice for a growing number of bioinformatics applications.

Keywords: Nanopore sequencing · Bioinformatics · Serverless computing · Cloud computing · Metagenomics · AWS Lambda · Parallel computing · Basecalling

1 Introduction

In recent years, we have seen the growing popularity of the serverless computing paradigm, which was popularized by AWS Lambda offering that was made available in 2014 [17]. It allows to simplify operations, abstract away the underlying servers and reduce maintenance overhead by allowing to run and scale workflows without the need to manage the infrastructure manually, while also offering support for highly parallel execution [9]. It is also often referred to as Function-as-a-Service (FaaS) [22]. So far, it gained widespread adoption for services that require high throughput with relatively low requirements for computing power, becoming a popular choice for Web APIs and asynchronous processing [14]. However, due to the nature of most bioinformatic workflows, which often require advanced computing capabilities in terms of CPU and memory, so far, it didn't gain massive adoption, and traditional computing clusters are still the most popular architecture used to run a bioinformatics analysis. Additionally, so far, only a few computing runtimes were officially supported on major platforms

© Springer Nature Switzerland AG 2021
M. Paszynski et al. (Eds.): ICCS 2021, LNCS 12743, pp. 578–586, 2021.
https://doi.org/10.1007/978-3-030-77964-1_44

such as AWS Lambda. However, the recent introduction of support for Docker [10] containers [1], as well as expanding maximum memory that can be used by AWS Lambda function to 10,240 MB and up to 6 vCPU cores, makes it easier to run workloads with higher computing power requirements [4].

At the same time, in the past years, we observed the rapid growth of third-generation sequencing technologies that allow for cost-effective and quick genome sequencing. Nowadays, large DNA sequencing platforms offer natural transmission bridge to Cloud environments to facilitate data storage, processing, and analysis. For example, Illumina, which delivers technological solutions for genetic and genomic data analyses, promotes efficiency by streaming the sequencing data directly to the AWS cloud with their BaseSpace Sequence Hub tool [8]. This paper aims to evaluate the feasibility, performance, and cost-effectiveness of performing computations directly on AWS Lambda to see if it can be used to carry out operations such as basecalling [23] of nanopore reads. We experimentally check whether Cloud serverless computing can serve to computationally-demanding tasks related to modern DNA sequencing technologies. The paper is organized as follows. In Sect. 2, we review the related works in the area. In Sect. 3, we describe the testing environment along with bioinformatic tools considered as a part of our experiments. Section 4 contains a description of the testing methodology along with performance experiments that we carried out for selected scenarios. Finally, Sect. 5 summarizes the results and concludes our findings.

2   Related Works

In the literature, there is only a little research concerning running bioinformatics workflows with the use of serverless computing platforms. In the paper, [21], Niu et al. describe the proof of concept example of running all-against-all pairwise comparison among 20,000 human protein sequences using Striped Smith-Waterman implementation. According to their findings, it can be accomplished in about 2 min for a cost of less than one dollar. Authors conclude that the use of serverless cloud computing can be leveraged to dramatically speed up the execution time of certain tasks at a low cost. They also suggest that in a similar approach, serverless computing can be used for tasks such as sequence alignment, protein-folding, or deep learning.

Malawski et al. [20] focused on evaluating AWS Lambda, Google Cloud Functions, and HyperFlow in the context of executing scientific workflows in a serverless manner. The authors developed a prototype workflow executor functions based on the aforementioned technologies. They managed to deploy and run the Montage astronomy workflow. They suggest that AWS Lambda infrastructure offers good scalability. However, not all workflows are suitable to serverless computing architecture, and it might be worth considering a hybrid approach that combines serverless and traditional architectures.

In their research [12], Burkat et al. evaluated serverless infrastructures in the model of Container-as-a-Service as a platform for running various scientific workflows. Authors especially focused on two offerings, AWS Fargate and Google Cloud Run. For evaluation purposes, they extended the HyperFlow engine to support execution on these platforms. During experiments, the authors run four scientific workflows: Ellipsoids, Vina, KINC, and Soy-KB, which were selected due to different resource requirements. The authors conclude the paper with a claim that serverless containers can be successfully used for scientific workflows.

Joyner et al., in their article [18], propose Ripple, a dedicated programming framework that aims to allow programs that were designed for single-machine execution to take advantage of parallelism offered by serverless computing. Ripple offers an interface that allows users to express workflows of a broad spectrum of applications, such as machine learning, genomics, or proteomics. Authors port three workflows: SpaceNet building border identification, proteomics with Tide and Percolator, and DNA compression with METHCOMP, showing that using Ripple can offer significant performance benefits versus traditional cloud deployments.

John et al. [15] proposed a solution called SWEEP, which is a workflow management system that takes advantage of the serverless execution model. It is cloud-agnostic and allows users to define, run and monitor scientific workflows. The authors evaluated their system for two cases, variant calling, and satellite imagery processing. They also list the elasticity of serverless computing and lack of overhead related to cluster management in traditional computing as significant benefits of the serverless approach.

In his research [19], Lee et al. proposed a DNAVisualisation.org, a fully serverless web tool dedicated to DNA sequence visualizations. With this tool, the authors wanted to demonstrate the applicability of serverless computing in the field of molecular biology in addition to allowing the ability to visualize DNA sequences in a cost-effective manner quickly. They also suggest that while not all applications are a great fit for serverless computing, some of them might benefit from decreased costs, reduced development complexity, which can be a significant advantage over the traditional architectures.

Crespo-Cepeda et al., in their work [13], consider challenges and opportunities for AWS Lambda services in the context of bioinformatics. Their paper proposes an architecture for running CloudDmetMiner, focusing on simplifying the workflow as much as possible, based on AWS Lambda and AWS S3. Authors conclude successful experiments suggesting that serverless computing can ease the code execution by reducing the time it takes to manage and provision infrastructure manually.

In their paper [16], Jonas et al. suggest that stateless functions offered by serverless computing can allow for more straightforward parallel computation without complex management of clusters and configuration tools. Authors also introduce and evaluate a tool called PyWren, which is a framework that allows mapping selected calculations across multiple concurrently running AWS Lambda functions. They conclude the paper with a suggestion that stateless functions are a great fit for data processing for future applications.

Based on the above, we can conclude that there is a lot of interest in evaluating serverless computing architectures for various scientific workflows. However, just a few of them consider bioinformatics workflows, and not a single one of them considered performing basecalling in such an environment. This paper aims to expand knowledge in that area by evaluating basecalling workflow when using AWS Lambda and serverless computing model.

3   Testing Workflow and Environment

For the evaluation, we selected the workflow where we split Nanopore MinION FAST5 files with nanopore reads (raw signal data) into batches. These batches are later processed separately by independent Lambda functions, enabling stateless parallelism, as the functions do not have to communicate with each other during processing. During executions, Lambda calls the basecalling function, which processes the files from an AWS S3 bucket. After processing its batch, each function saves the results in a separate folder in AWS S3 bucket. The data stored in that bucket can be used for further processing. Figure 1 presents the considered workflow.

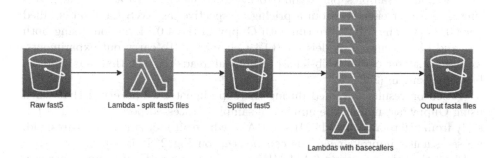

Raw fast5   Lambda - split fast5 files   Splitted fast5   Output fasta files

Lambdas with basecallers

Fig. 1. Diagram of the considered workflow.

Each Lambda function used a Docker container runtime, taking advantage of a custom Docker image based on Ubuntu 16.04 operating system with a Python 3.6 wrapper script that executes a binary with basecalling software.

For basecalling, we considered several tools. The first one was Deepnano-blitz [11], an open-source basecaller, developed by Boža, V., based on bidirectional recurrent neural networks. It was very promising, as its implementation is optimized to take advantage of Intel AVX2 instruction set [7], which is also supported by Lambda environment [6]. Unfortunately, after preliminary testing in the Lambda environment, it turned out that Lambda's lack of support for shared memory between separate processes makes it impossible to run Deepnano-blitz in its current form. We consider adjusting Deepnano-blitz in the future for the AWS Lambda environment. Next considered basecaller was Guppy [23], which is a closed-source, state-of-the-art basecaller developed by Oxford Nanopore Technologies. It has support for multiple basecalling models – fast and high accuracy. We also considered two alternative open-source basecallers, Bonito [5], and Causcalcall [24]. However, after preliminary testing, they turned out to offer much lower performance in comparison to Guppy, so we decided to focus on Guppy in our experiments and potentially considering Deepnano-blitz in the next research.

4   Performance Experiments

During experiments, we decided to measure the basecalling capabilities of the AWS Lambda computing environment with a different maximum available memory setting, which also translates to the number of virtual CPU cores that are available for the function. According to official documentation [4], available CPU power scales proportionally with memory with 1 full vCPU core at 1769 MB of memory to the maximum of 6 vCPU cores for 10240 MB of allocated memory. For each run, we recorded the number of samples processed per second, as well as the ratio of samples per second to available memory, to assess which setting is the most effective from a pricing perspective, as AWS Lambda is billed per GB/s [3]. The tests were run with Guppy in the 4.0.14 version, using both fast and high accuracy models for R9.4.1 chemistry. To carry out experiments, we used a subset of the Klebsiella pneumoniae reads dataset that was used for benchmarking in [23].

Based on results obtained during the experiments, we observed that when using Guppy fast model, the number of samples processed per second scales linearly from 256 to about 6,144 MBs of RAM, where after crossing that threshold, we see smaller improvements, as can be seen on Fig. 2. It is especially visible in Fig. 3 that after crossing 6,144 MB, we see a lower ratio of samples per second per one MB of memory. For Guppy fast model, we observed the highest ratio of samples per second per MB of memory for 2,048 MBs, with a ratio of 86.76, which means that it's the most efficient setting from the cost-effectiveness standpoint.

When considering Guppy with high accuracy models, we observed different patterns than for fast model. Except for a scenario with 6,144 MB of memory, we observe that the ratio of samples per second per MB of RAM is growing along with assigned memory, being the highest for 10,240 MBs of memory with the value of 3.74, which can be seen in Fig. 5. We also observed much lower values for samples processed per second in general compared to the fast model with a peak of 38,293 samples processed per second for maximum memory of 10,240 MBs, as presented in Fig. 4 (versus 622,154 in the fast mode for the same amount of memory). It is

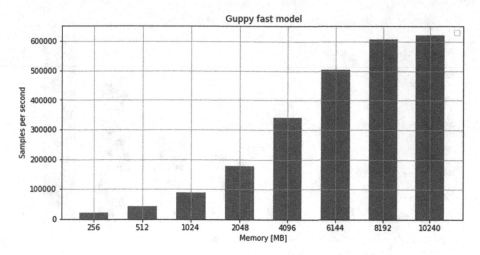

Fig. 2. Samples processed per second for Guppy fast model.

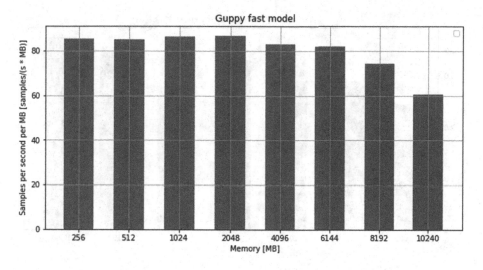

Fig. 3. Samples per second per MB of memory for Guppy fast model.

also important to note that it was impossible to run a high accuracy model with only 256 MB of memory, which was previously possible with the fast model. For all tested scenarios, we did not observe instances of failing tasks other than expected initial invocation failure related to mandatory image optimization step performed by AWS Lambda right after deployment of new version of the container image.

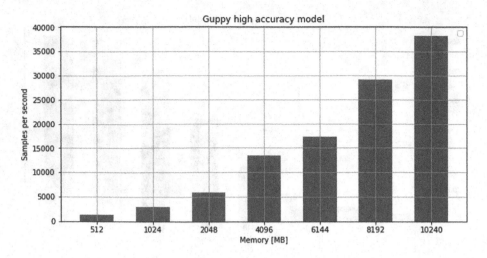

Fig. 4. Samples processed per second for Guppy high accuracy model.

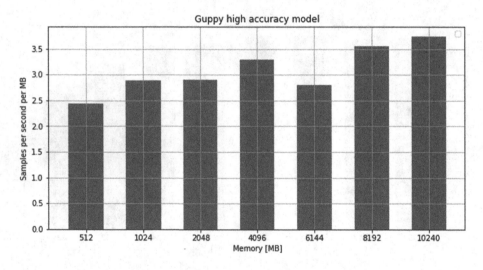

Fig. 5. Samples per second per MB of memory for Guppy high accuracy model.

5   Results Summary and Concluding Remarks

Considering successful experiments and results presented in the previous chapter, we conclude that thanks to recent advancements in AWS Lambda offering, it is now possible to run basecalling workflows in a serverless manner. For example, given the fact that the theoretical maximum of MinION Nanopore is around 2,300,000 signals per second (512 pores with around 4,500 signals read per second per pore), with real-world scenarios resulting in less than 2,000,000 signals per second, which means that 3 to 4 function instances would be able to keep up with processing that data in near real-time. Given the ability of AWS Lambda to scale up to hundreds of thousands of concurrently running functions [2], we gain the capability to quickly basecall data from multiple sequencing experiments while keeping the infrastructure maintenance overhead as low as possible. We expect even better results with CPU-optimized basecallers such as Deepnano-blitz [11]. We believe that serverless computing architectures will continue to gain more features and enable even more bioinformatic workflows in the future and that there is still a lot of room for improvement and development in that area.

Acknowledgments. The research was supported by the Polish Ministry of Science and Higher Education as a part of the CyPhiS program at the Silesian University of Technology, Gliwice, Poland (Contract No. POWR.03.02.00-00-I007/17-00) and by Statutory Research funds of Department of Applied Informatics, Silesian University of Technology, Gliwice, Poland (grant No. BK-221/RAu7/2021).

References

1. AWS Lambda container image support. Accessed 5 Feb 2021. https://aws.amazon.com/blogs/aws/new-for-aws-lambda-container-image-support/
2. AWS Lambda limits. Accessed 5 Feb 2021. https://docs.aws.amazon.com/lambda/latest/dg/gettingstarted-limits.html
3. AWS Lambda pricing. Accessed 5 Feb 2021. https://aws.amazon.com/lambda/pricing/
4. AWS Lambda support for 10240 MB and 6 vCPU cores. Accessed 5 Feb 2021. https://aws.amazon.com/about-aws/whats-new/2020/12/aws-lambda-supports-10gb-memory-6-vcpu-cores-lambda-functions/
5. Bonito basecaller repository on Github. Accessed 5 Feb 2021. https://github.com/nanoporetech/bonito
6. Creating faster AWS Lambda functions with AVX2. Accessed 5 Feb 2021. https://aws.amazon.com/blogs/compute/creating-faster-aws-lambda-functions-with-avx2/
7. How Intel® Advanced Vector Extensions 2 improves performance on server applications. Accessed 5 Feb 2021. https://software.intel.com/content/www/us/en/develop/articles/how-intel-avx2-improves-performance-on-server-applications.html
8. Augustyn, D.R., Wyciślik, Ł., Mrozek, D.: Perspectives of using cloud computing in integrative analysis of multi-omics data. Briefings Funct. Genomics 1–23 (2021, in press)
9. Baldini, I., et al.: Serverless computing: current trends and open problems. In: Chaudhary, S., Somani, G., Buyya, R. (eds.) Research Advances in Cloud Computing, pp. 1–20. Springer, Singapore (2017). https://doi.org/10.1007/978-981-10-5026-8_1
10. Bashari Rad, B., Bhatti, H., Ahmadi, M.: An introduction to Docker and analysis of its performance. IJCSNS Int. J. Comput. Sci. Netw. Secur. **17**(3), 228–235 (2017)
11. Boža, V., Perešíni, P., Brejová, B., Vinař, T.: Deepnano-blitz: a fast base caller for minion nanopore sequencers. Bioinformatics **36**, 4191–4192 (2020)
12. Burkat, K., et al.: Serverless containers - rising viable approach to scientific workflows. ArXiv abs/2010.11320 (2020)

13. Crespo-Cepeda, R., Agapito, G., Vazquez-Poletti, J.L., Cannataro, M.: Challenges and opportunities of amazon serverless lambda services in bioinformatics. In: Proceedings of the 10th ACM International Conference on Bioinformatics, Computational Biology and Health Informatics, BCB 2019, pp. 663–668. Association for Computing Machinery, New York (2019). https://doi.org/10.1145/3307339. 3343462
14. Eismann, S., et al.: A review of serverless use cases and their characteristics. arXiv 2008.11110 (2021)
15. John, A., Ausmees, K., Muenzen, K., Kuhn, C., Tan, A.: SWEEP: accelerating scientific research through scalable serverless workflows. In: Proceedings of the 12th IEEE/ACM International Conference on Utility and Cloud Computing Companion, UCC 2019, pp. 43–50. Association for Computing Machinery, New York (2019). https://doi.org/10.1145/3368235.3368839
16. Jonas, E., Pu, Q., Venkataraman, S., Stoica, I., Recht, B.: Occupy the cloud: distributed computing for the 99 Cloud Computing. In: SoCC 2017, pp. 445–451. Association for Computing Machinery, New York (2017). https://doi.org/10.1145/ 3127479.3128601
17. Jonas, E., et al.: Cloud programming simplified: a Berkeley view on serverless computing. CoRR abs/1902.03383 (2019). http://arxiv.org/abs/1902.03383
18. Joyner, S., MacCoss, M., Delimitrou, C., Weatherspoon, H.: Ripple: a practical declarative programming framework for serverless compute. CoRR abs/2001.00222 (2020). http://arxiv.org/abs/2001.00222
19. Lee, B., Timony, M., Ruiz, P.: DNAvisualization.org: a serverless web tool for DNA sequence visualization. Nucleic Acids Res. **47**, W20–W25 (2019)
20. Malawski, M., Gajek, A., Zima, A., Balis, B., Figiela, K.: Serverless execution of scientific workflows: experiments with hyperflow, AWS lambda and google cloud functions. Future Gener. Comput. Syst. **110**, 502–514 (2020). https://www. sciencedirect.com/science/article/pii/S0167739X1730047X
21. Niu, X., Kumanov, D., Hung, L.H., Lloyd, W., Yeung, K.Y.: Leveraging serverless computing to improve performance for sequence comparison. In: Proceedings of the 10th ACM International Conference on Bioinformatics, Computational Biology and Health Informatics, BCB 2019, pp. 683–687. Association for Computing Machinery, New York (2019). https://doi.org/10.1145/3307339.3343465
22. Scheuner, J., Leitner, P.: Function-as-a-service performance evaluation: a multivocal literature review. J. Syst. Softw. **170**, 110708 (2020). https://www. sciencedirect.com/science/article/pii/S0164121220301527
23. Wick, R.R., Judd, L.M., Holt, K.E.: Performance of neural network basecalling tools for oxford nanopore sequencing. Genome Biol. **20**(1), 129 (2019). https:// doi.org/10.1186/s13059-019-1727-y
24. Zeng, J., Cai, H., Peng, H., Wang, H., Zhang, Y., Akutsu, T.: Causalcall: nanopore basecalling using a temporal convolutional network. Frontiers Genet. **10**, 1332 (2020). https://www.frontiersin.org/article/10.3389/fgene.2019.01332

A Software Pipeline Based on Sentiment Analyze Narrative Medicine Texts

Ileana Scarpino, Chiara Zucco, and Mario Cannataro(✉)

Data Analytics Research Center, Department of Medical and Surgical Sciences,
University "Magna Graecia" of Catanzaro, Catanzaro, Italy
ileana.scarpino@studenti.unicz.it, {chiara.zucco,cannataro}@unicz.it

Abstract. By using social media people can exchange sentiments and emotions, allowing to understand public opinion on specific issues. Sentiment Analysis (SA) is a novel text-mining (TM) and natural language processing (NLP) methodology to extract sentiment, opinions and emotions from written texts, usually provided through social media or questionnaires. Sharing medical and clinical experiences of patients through social media, is the target of the so-called Narrative Medicine (NM). Here we report some research experiences in applying SA techniques to analyze NM texts. A problem to be faced in NM is the automatic analysis of a potentially large set of documents. Application of SA is useful for having immediate analysis and extracting information from medical literature quickly. Here we present a software pipeline based on SA and TM which allows to effectively analyze NM texts. First experimental results allow to discover topics related to diseases.

Keywords: Sentiment analysis · Text mining · Topic modeling · Narrative medicine

1 Introduction

Scientific research has shown that to make a targeted and personalized diagnostic process it is necessary to pay attention not only to the patient's illness, but also to his/her psychological and emotional state. The narration of patients and caregivers is an essential element of contemporary medicine and it is based on the active participation of the subjects involved. Through their stories, people become protagonists of their own healing process [1]. A medicine practiced with narrative competence, i.e. Narrative Medicine (NM), may be better able to recognize patients and diseases, empathize with colleagues, accompany patients and their families through the vicissitudes of the disease [2–5]. In particular, NM has many advantages: it improves clinical practice, allows a more in-depth diagnosis, promotes adherence to therapy, helps and consolidates choices, fosters relationships between patient, family and healthcare staff, improves the quality of service and

M. Paszynski et al. (Eds.): ICCS 2021, LNCS 12743, pp. 587–593, 2021.
https://doi.org/10.1007/978-3-030-77964-1_45

the therapy strategy, verifies and allows feedback on the functionality of the therapy, promotes the formation of communities that help the patient on a social and psychological level.

Although NM is a practice born in the late 1960s [6], it has been scarcely treated from a methodological point of view [7]. Only the oncological branch is projected towards the analysis of patients' narratives [8]. The methodologies that have characterized the analysis of written and oral narrative material can be grouped into three main strands: Thematic Analysis allows to count the frequency of the words and themes proposed by the patient [9]; Linguistic Analysis allows to differentiate the narratives by gender complexity [10]; Content analysis implements various procedures for quantitative survey of the narrative structure and its qualitative content [6,11].

Consequently to the social media growth, people exchange opinions, feelings and emotions and they also share personal experiences related to their diseases, by giving birth to online blogs and forums for sharing their experience, as for instance the Italian blog "Viverla Tutta" [12]. Analyzing the information present on online health forums and communities may help patients by improving the diagnostic process on the basis of similar experiences.

It is known how clinical narratives contain a moderate amount of sentiment terms that reflect the objectivity and preciseness of the clinical writing style [13]. The interest towards the extraction of opinions and emotions from textual online resourced has led to the development of Sentiment Analysis (SA), that combines Text Mining (TM) and Natural Language Processing (NLP) and whose aim is the extraction of subjective information from texts, focusing not only on the topic, but also on the opinions expressed in the texts [14]. SA tools lead to a polarity analysis that can be positive, negative or neutral [15]. Text Mining techniques are widely used to perform biomedical knowledge extraction [16] to obtain relevant information from vast online databases of health science literature or patients' electronic health records.

Previous studies have shown how sentiment analysis applied to clinical documents has the potential for assisting patients with information for self assessing treatments, providing health professionals with more insights into patients' health conditions, or even managing relations between patients and doctors [17].

In this paper we present an application of several NLP and TM methods to examine various aspects of the coexistence between patients with their illness. The main contribution is the proposal of a semi-automatic software pipeline for narrative medicine, a domain in which automatic approaches for the analysis are still poor. The rest of the paper is organized as follows. Section 2 presents the proposed software pipeline. Section 3 presents the final results showing both clinical and predominantly emotional aspects. Finally Sect. 4 presents the conclusions and future work.

2 Narrative Medicine Analysis Pipeline

The proposed analysis pipeline includes two main stages, NLP methods for the preprocessing of the input NM texts, and TM methods to analyze them (see Fig. 1). The analysis pipeline is implemented in *Python*.

The experimentation of the pipeline has been performed by using textual sources extracted from the "Viverla tutta" data source, a blog dedicated to NM. A total of 12 texts, each one written by a patient, related to three diseases (fibromyalgia, cancer and diabetes) were selected. Then 4 patient testimonials were selected for each kind of disease, as shown in Table 1.

Table 1. Data files from the Italian blog "Viverla tutta" [12] are grouped by the common disease of patients.

Disease	File name	File dimension (Bytes)
Fibromyalgia	Patient_1.txt	2,138
	Patient_2.txt	2,962
	Patient_3.txt	3,892
	Patient_4.txt	2,111
Tumor	Patient_5.txt	1,177
	Patient_6.txt	2,710
	Patient_7.txt	2,705
	Patient_8.txt	2,327
Diabetes	Patient_9.txt	1,178
	Patient_10.txt	892
	Patient_11.txt	1,513
	Patient_12.txt	6,260

The main steps of the pipeline are described below. After loading and reading text files, we proceed with data preparation, which is particularly important in text mining, because it works on textual data that must be made suitable to the subsequent steps. Pre-processing consists of several stages, i.e. lowercasing, punctuation and special characters removal, tokenization, stemming, stopwords removal and Part-of-Speech tagging. In particular, the *Nltk*[1], the *SpaCy*[2] and the *gensim*[3] *Python* library for the Italian language have been used for text preprocessing.

Term Frequency - Inverse Document Frequency (TF-IDF) statistically quantifies how important a term is within a document in relation to other similar documents. It replaces number of occurrences with a weighted frequency value. TF-IDF algorithm associates greater importance to less frequent but more relevant terms.

Topic modeling is a technique used in NLP. It is a statistical model that searches for topics from a corpus of texts. The topics are not known a priori but they are identified by the algorithm based on word frequency in the documents.

---

[1] https://www.nltk.org/.
[2] https://spacy.io.
[3] https://radimrehurek.com/gensim/index.html.

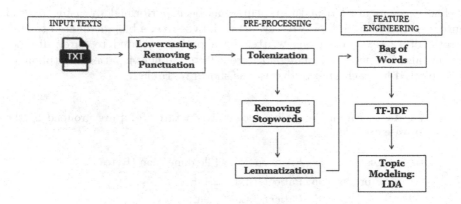

Fig. 1. The proposed pipeline for the analysis of NM texts.

Table 2. TF-IDF of terms in the analyzed dataset. Terms carrying more information are underlined. Underlined values have more weight in the document of belonging than all the others texts. Empty cells represent terms missing in some documents but present in others.

Disease	Terms	Patient 1	Patient 2	Patient 3	Patient 4
Fibromyalgia	Illness	0.1155	0.2652	0.1269	
	Syndrome			0.25377	
	Fibromyalgia	0.1155		0.29607	
	Rheumatoid arthritis		0.1326		
	Inflammation			0.0845	
	Pain	0.05774		0.25377	0.22829
	Sensibility	0.05774		0.08459	0.05707
	Life	0.11547	0.13258	0.0423	0.22829
		Patient 5	Patient 6	Patient 7	Patient 8
Tumor	Tumor	0.0811	0.0971	0.1918	0.0508
	Carcinoma	0.0811	0.0486	0.0479	0.0508
	Chemotherapy		0.0971	0.0959	0.1017
	Radiotherapy	0.0811		0.4795	
	Surgery		0.0486	0.0959	0.1525
	Relapses		0.0486		
	Defeat			0.0959	
	Life	0.0811	0.1943	0.2877	0.3558
		Patient 9	Patient 10	Patient 11	Patient 12
Diabetes	Illness			0.207	0.3105
	Diabetes		0.2572		0.3103
	Hyperglycemia		0.1715		0.0282
	Life		0.0857	0.138	0.2823
	Win	0.0814	0.0857		0.0282

In particular, Latent Dirichlet Allocation (LDA) is an unsupervised learning algorithm that allows to identify a certain number of topics, called "latent topics" in a corpus of documents. LDA generates a set of words called "keywords", to which a weight is associated. A higher weight discriminates the most relevant topics, whose number needs to be set before training. In the present analysis the number of topics has been set equal to the number of documents to avoid algorithm generating repetitions or similar topics. Following the training of LDA, perplexity and coherence score have been calculated as evaluation metrics to assess LDA model's clarity. Perplexity is a well known intrinsic evaluation metric. Topic Coherence measures, score a single topic by measuring the degree of semantic similarity between high scoring words in the topic. These measurements help distinguish between topics that are semantically interpretable topics and topics that are artifacts of statistical inference [18,19].

3 Results

Two types of studies were conducted: first patient testimonials released from the blog "Viverla Tutta" were analyzed together by relating terms specific to each disease so that it was possible to test that the focus was specific for each disease. Subsequently, texts were analyzed one-to-one to verify if some words were more present in one document than other.

Table 2 summarizes the TF-IDF results. The underlined values refer to terms that are the most relevant in comparison with all the documents analyzed, and therefore they carry more information. The empty cells instead represent the missing terms in the documents and therefore words that are not written by a patient in a given document.

Analyzing Table 2 by disease it can be seen that each patient, despite being affected by the same pathology, lives illness in a personal way by adopting different terms for its definition. Some fibromyalgia patients talk about "illness", as for instance Patient 2. Patient 3 talks about "syndrome", terms that is absent in other documents. He/she describes physical symptoms as "numbness", "tingling", "photophobia", "intolerance", "sleep", rather than focusing on the consequences that disease had in his/her life. The word "pain" predominates in documents 3 and 4 while the terms present in all documents are those related to the patient's experience such as "living", "life", "past".

Every single patient of the considered subset "Tumor" defines his/her disease as "tumor" and talks about of "carcinoma". The terms present in all the sources are "live" and "life", showing the desire of dealing with the disease and conquer it. In addition, there is the possibility that some subjects have undergone a "surgery" ; the word "relapse" occurred only in a document means that at least one patient has developed relapses.

With respect to diabetes patients, the term "diabetes" has an important frequency value in document 4 and it compares in document 2 in which "drink" and "water" are found, recalling that among the symptoms of diabetes there is "polydipsia". Other characteristic features of this disease are "blood", "sugar", "glucometer", "insulin".

LDA similarity was applied to compare how similar documents are in topics from their topic distributions. It has been allocated how much each document belongs to 1st, 2nd, 3rd or 4th topic, through numerical value indicating strength with which the text is related to a given topic. We were expecting that the main focus was the disease told by patients, but amount of verbs is also relevant. Observing carefully the LDA results, it is possible to find two types of results, in some topics a more clinical aspect and in others a more emotional one is found. Perplexity and coherence scores did not give surprising results, as shown in Table 3. By decreasing the number of topics analyzed, the probability that a document could be associated with multiple topics increases.

Table 3. Perplexity and Coherence score: intrinsic evaluation metrics of the model. The log perplexity results indicate poor human interpretable topics. Coherence scores represent a probability that of course vary in a $[0, 1]$ range. Therefore the overall results are not satisfying.

Disease	Perplexity	Coherence score
Fibromyalgia	-7.556	0.394
Tumor	-7.503	0.304
Diabetes	-7.580	0.367

4   Conclusions

A first contribution of the paper is the proposal of a software pipeline, based on different *Python* libraries, for the analysis of NM texts using NLP and TM techniques. The pipeline has been applied to analyze some NM texts provided by several patients affected by three pathologies: fibromyalgia, cancer and diabetes. Currently, the pipeline implementation is not publicly available, but it will be released to be public as a future development of the study. The presented information extraction process allowed to analyse various aspects of the coexistence of patients with their illness. The TF-IDF analysis showed that in patients with the same pathological condition, the narration of their experience was focused on discordant aspects that depend on personality traits which leads them to live their experience in a different way. On the other hand, there are also similar aspects between various patients though they are suffering from different pathologies. The identified relevant words vary from clinical to emotional topic.

From an LDA perspective, the unsatisfactory results should be improved in future works by considering a larger number of textual NM sources. Moreover, to overcome any limitations due to the use of articulated texts, the pre-processing may be improved, enhancing different machine learning techniques to improve performance.

The proposed pipeline can be tested and used in various application as for instance to help patients by improving the diagnostic treatment on the basis of similar experiences. As stated before, the main limitation of the study is the small number of analyzed text. In future works, we plan to extend the dataset by adding more textual data and by also increasing the number of considered diseases. As a second future development, we are interested in evaluating whether a correlation between extracted topics and the disease severity may be assessed, and to also consider the sentiment related to NM texts.

References

1. Hurwitz, B.: Narrative [in] medicine. In: Spinozzi, P., Hurwitz, B. (eds.) Discourses in the Biosciences, vol. 8, pp. 13–30. Vandenhoeck & Ruprecht Unipress, Gottingen (2011)
2. Evans, M.: Reflections on the humanities in medical education. Med. Educ. **36**(6), 508–13 (2002)
3. Charon, R.: Narrative Medicine: Honoring the Stories of Illness. Oxford University Press, New York (2006)
4. Zannini, L.: Medical humanities e medicina narrativa. Nuove prospettive nella formazione dei professionisti della cura. Raffaello Cortina Editore, Milano (2008)
5. Bernegger, G., Castiglioni, M., Garrino, L.: Un medico tra radure, tigri e jazz. A colloquio con Rita Charon. Rivista per le Medical Humanities **28**(8), 67–77 (2014)
6. Gottschalk, L., Gleser, G.C.: The Measurement of Psychological States Through the Content Analysis of Verbal Behavior. University of California Press, Berkeley (1969)
7. Overcash, JA.: Narrative research: a review of methodology and relevance to clinical practice. Crit. Rev. Oncol. Hematol. **48**(2), 179–184 (2003)
8. Jordens, C.F., Little, M., Paul, K., Sayers, E.J.: Life disruption and generic complexity: a social linguistic analysis of narratives of cancer illness. Soc. Sci. Med. **53**(9), 1227–1236 (2001)
9. Owen, W.F.: Interpretative themes in relational communication. Q. J. Speech **70**(3), 274–287 (1984)
10. Bakhtin, M.M.: The problem of speech genres. In: Emerson, C., Holquist, M. (eds.) Speech Genres and Other Late Essays. University of Texas Press, Austin (1986)
11. Weber, R.P.: Basic Content Analysis. SAGE, Newbury Park (1990)
12. http://www.viverlatutta.it/. Accessed 31 Mar 2021
13. Deng, Y., Stoehr, M., Denecke, K.: Retrieving attitudes: sentiment analysis from clinical narratives. In: MedIR@ SIGIR, pp. 12–15 (2014)
14. Vinodhini, G., Chandrasekaran, R.M.: Sentiment analysis and opinion mining: a survey. Int. J. **2**(6), 282–292 (2012)
15. Zucco, C., Calabrese, B., Agapito, G., Guzzi, P.H., Cannataro, M.: Sentiment analysis for mining texts and social networks data: methods and tools. Wiley Interdisc. Rev. Data Min. Knowl. Discovery **10**(1), e1333 (2020)
16. Neustein, A., Sagar Imambi, S., Rodrigues, M., Teixeira, A., Ferreira, L.: Text Mining of Web-Based Medical Content. De Gruyter (2014)
17. Liu, S., Lee, I.: Extracting features with medical sentiment lexicon and position encoding for drug reviews. Health Inf. Sci. Syst. **7**(1), 1–10 (2019)
18. Blei, D.M., Ng, A.Y., Jordan, M.I.: Latent Dirichlet allocation. J. Mach. Learn. Res. **3**, 993–1022 (2003). https://doi.org/10.1162/jmlr.2003.3.4-5.993
19. Griffiths, T.L., Steyvers, M.: Finding scientific topics. Proc. Nat. Acad. Sci. United States of America **101**(Suppl 1), 5228–5235 (2004). https://doi.org/10.1073/pnas.0307752101

Author Index

Printed in the United States
by Baker & Taylor Publisher Services